September 19–22, 2011
Mountain View, California, USA

 **Association for
Computing Machinery**

Advancing Computing as a Science & Profession

DocEng 2011

Proceedings of the 2011 ACM Symposium on
Document Engineering

Sponsored by:
ACM SIGWEB

In cooperation with:
ACM SIGDOC

Supported by:
Google, Adobe, Hewlett-Packard, and Xerox

**Association for
Computing Machinery**

Advancing Computing as a Science & Profession

The Association for Computing Machinery
2 Penn Plaza, Suite 701
New York, New York 10121-0701

Notice to Past Authors of ACM-Published Articles
ACM intends to create a complete electronic archive of all articles and/or other material previously published by ACM. If you have written a work that has been previously published by ACM in any journal or conference proceedings prior to 1978, or any SIG Newsletter at any time, and you do NOT want this work to appear in the ACM Digital Library, please inform permissions@acm.org, stating the title of the work, the author(s), and where and when published.

ISBN: 978-1-4503-0863-2

Additional copies may be ordered prepaid from:

ACM Order Department
PO Box 30777
New York, NY 10087-0777, USA

Phone: 1-800-342-6626 (USA and Canada)
+1-212-626-0500 (Global)
Fax: +1-212-944-1318
E-mail: acmhelp@acm.org
Hours of Operation: 8:30 am – 4:30 pm ET

ACM Order Number: 605114

Printed in the USA

General Chair's Welcome

It is with great pleasure that I welcome you to the 11th *ACM Symposium on Document Engineering – DocEng 2011*, which is being held during September 19-22 in Mountain View at the Google "Googleplex". DocEng continues to provide the opportunity to researchers across a broad range of institutions to share their work and knowledge in the field of Document Engineering. This year, DocEng focused on the topics of operations on documents, document representations and document-centric applications. Successful DocEng conferences have been held annually around the world: in Manchester (2010), Munich (2009), São Paulo (2008), Winnipeg (2007), Amsterdam (2006), Bristol (2005), Milwaukee (2004), Grenoble (2003), McLean (2002), and Atlanta (2001).

I would like to thank Google for generously providing the facilities for the symposium, as well as providing volunteers to help support the conference throughout the program. Special thanks go to Neil Fraser, DocEng 2011 Local Organization chair, for organizing so much for the conference.

I am very pleased that DocEng 2011 has two outstanding keynote presenters. Dr. John Warnock, co-chair of the Board of Directors and co-founder of Adobe Systems, presents "The Evolving Form of Documents", providing insights into the challenges faced by publishers today. Dr. Mark Davis, from Google and President of the Unicode Consortium, is presenting "Google's International Bloopers… and How We Fixed One", which provides insights into the complexity of internationalization.

Putting together DocEng 2011 was a team effort. I would first like to thank the authors for providing the content of the program. I am very grateful to my program chair, Frank Tompa, for putting together a broad and diverse program and to the program committee, who worked very hard in reviewing papers and providing feedback for authors. I would also like to thank the steering committee for continuing to provide guidance to the organization of the symposium. Finally, thanks go to our sponsor, ACM SIGWEB, and our generous corporate supporters, Google, Adobe Systems, HP Labs and Xerox XRCE.

I hope that you find DocEng 2011 to be as interesting and thought-provoking as our previous symposia. I invite you to take this valuable opportunity to share ideas and to network with your fellow researchers and practitioners from around the world. Most importantly, I hope you all have a fun and exciting time at the Googleplex in Mountain View.

Matthew Hardy
DocEng 2011 General Chair
Adobe Systems
San Jose, USA

Program Chair's Welcome

We're extremely pleased with the technical program again this year. The call for papers attracted 71 submissions, which were reviewed by 34 program committee members and 14 external reviewers. After careful review, we chose 38 papers, for an acceptance rate just over 53%.

This year we're excited to present a choice of six half-day workshops before the symposium begins. The workshops span the spectrum from *how-to* to *so-what* and *gee-whiz*:

- *Version Control* presented by Neil Fraser (Google)
- *Multimedia Document Processing in an HTML5 World* presented by Dick Bulterman (CWI + VU Amsterdam + W3C), Rodrigo Laiola (CWI), Pablo Cesar (CWI), Ethan Munson (Univ. Wisconsin-Milwaukee) and Maria da Graça Pimentel (Univ. São Paulo)
- *Secure Document Engineering* presented by Helen Balinsky and Steven Simske (HP Labs)
- *Making Accessible PDF Documents* presented by Heather Devine, Andres Gonzales and Matthew Hardy (Adobe Systems)
- *Documenting Social Networks* presented by Maria da Graça Pimentel (Univ. Sao Paulo)
- *Google Mystery Workshop* presented by John Day-Richter (Google)

We are happy to include the last of these as a late-breaking surprise.

DocEng 2011 includes two keynote talks. The first is given by Dr. John E. Warnock, co-chairman of the Board of Directors of Adobe Systems, Inc. and one of its co-founders. Conference attendees have been looking forward to hearing a key person behind PostScript and PDF, as well as the inventor of Warnock's algorithm for rendering surfaces in computer graphics, describe challenges behind multimedia authoring. The second keynote is presented by Google's Dr. Mark Davis, co-founder of the Unicode project and president of the Unicode Consortium since its incorporation in 1991. We are anxious to learn about some of the difficulties behind internationalization from a leader in that field.

The symposium also includes 19 full papers, eight short papers, and 11 posters and demos. In addition, seven of the 27 presentations are also supplemented by demos. The papers are organized into ten sessions, covering diverse problems in document engineering that span document manipulation and analysis, document representations, and document-centric applications. You are strongly encouraged to read about the exciting work as reported in these proceedings.

I extend my personal thanks to all the members of the program committee and to the external reviewers for their diligence and hard work. We are also indebted to all our sponsors, and especially to our host Google, for making the symposium possible. Finally, we thank all the contributing authors, from 16 countries, for making DocEng a valuable venue.

We're sure that the work presented at DocEng 2011 will stimulate you to learn more about document engineering and to build on the ideas in making your own contributions to the field. We hope we meet you in Mountain View this year and again during follow-up symposia in the years to come.

Frank Wm. Tompa
DocEng 2011 Program Chair
Waterloo, Canada

Table of Contents

Keynote Address
Session Chair: David Brailsford *(University of Nottingham)*

Session 1: Optimizing Layouts
Session Chair: Anthony Wiley *(HP Labs)*

Session 2: Multimedia Presentations
Session Chair: Helen Balinsky *(HP Labs)*

Session 3: Demos and Posters
Session Chair: Michael Gormish *(Ricoh Innovations)*

Keynote Address 2

Session 4: Editing

Session 5: Visual Analysis

Session 6: Flowing Content into Layout

Session 7: Metadata

Session 8: Summarization
Session Chair: Steven Bagley *(University of Nottingham)*

Session 9: Tailored and Adaptive Layout
Session Chair: John Lumley *(NHS)*

Session 10: Deviance Control
Session Chair: Kim Marriott *(Monash University)*

Session Workshops

DocEng2011 Symposium Organization

Conference Chair: Matthew Hardy *(Adobe Systems, Inc., USA)*

Program Chair: Frank Wm. Tompa *(University of Waterloo, Canada)*

Local Arrangements Chair: Neil Fraser *(Google, Inc., USA)*

Steering Committee Chair: Peter King *(University of Manitoba, Canada)*

Steering Committee: David Brailsford *(University of Nottingham, UK)*
Dick Bulterman *(CWI, Netherlands)*
Rolf Ingold *(University of Fribourg, Switzerland)*
Michael Gormish *(Ricoh Innovations, USA)*
Ethan Munson *(University of Wisconsin-Milwaukee, USA)*
Charles Nicholas *(University of Maryland, Baltimore County, USA)*
Maria da Graça C. Pimentel *(Universidade de São Paulo, Brazil)*
Cécile Roisin *(INRIA and Université Pierre-Mendès-France, France)*
Steven Simske *(HP Labs, USA)*
Jean-Yves Vion-Dury *(Xerox Research Centre Europe, France)*
Anthony Wiley *(HP Labs, USA)*

Program Committee: Apostolos Antonacopoulos *(University of Salford, UK)*
Steven Bagley *(University of Nottingham, UK)*
Helen Balinsky *(HP Labs, UK)*
Uwe Borghoff *(Universität der Bundeswehr München, Germany)*
David Brailsford *(University of Nottingham, UK)*
Dick Bulterman *(CWI, Netherlands)*
Pablo Cesar *(CWI, Netherlands)*
Boris Chidlovskii *(Xerox Research Centre Europe, France)*
Michael Collard *(University of Akron, USA)*
Cyril Concolato *(Telecom ParisTech, France)*
Gersende Georg *(Haute Autorité de Santé, France)*
Luiz Fernando Gomes Soares *(PUC-Rio, Brazil)*
Michael Gormish *(Ricoh Innovations, USA)*
Roger Hersch *(École Polytechnique Fédérale de Lausanne, Switzerland)*
Peter King *(University of Manitoba, Canada)*
John Lumley *(UK)*
Simone Marinai *(Università degli Studi di Firenze, Italy)*
Kim Marriott *(Monash University, Australia)*
Mirella Moro *(Universidade Federal de Minas Gerais, Brazil)*
Ethan Munson *(University of Wisconsin-Milwaukee, USA)*
Charles Nicholas *(University of Maryland, Baltimore County, USA)*
Moira Norrie *(ETH Zürich, Switzerland)*

DocEng 2011 Sponsor & Supporters

Sponsor:

In Cooperation:

Supporters:

The Evolving Form of Documents

Dr. John E. Warnock
Adobe Systems Inc.
San Jose, California, USA
warnock@adobe.com

The sophistication and visual richness of printed documents has made great strides over the last 25 years. The computerization of entry, layout and production has gone from a mostly manual process to a totally computerized process.

As documents start to contain multimedia components (sound, animation, video) and become totally "electronic", and as devices to view these new documents become more varied in size shape and capabilities, the authoring of these documents has become a difficult challenge.

This talk will address those challenges, and discuss whether the evolving web tools are moving in the right direction.

Categories and Subject Descriptors:
D.2.0 [Software Engineering]: General - *standards*

General Terms:
Design, Standardization

Keywords:
electronic document, document design, document implementation

Probabilistic Document Model for Automated Document Composition*

Niranjan
Damera-Venkata
Hewlett-Packard Laboratories
1501 Page Mill Road
Palo Alto, CA 94304
damera@hpl.hp.com

José Bento
Stanford University
Dept. of Electrical Engineering
Stanford, CA 94305
jbento@stanford.edu

Eamonn O'Brien-Strain
Hewlett-Packard Laboratories
1501 Page Mill Road
Palo Alto, CA 94304
eob@hp.com

ABSTRACT

We present a new paradigm for automated document composition based on a generative, unified probabilistic document model (PDM) that models document composition. The model formally incorporates key design variables such as content pagination, relative arrangement possibilities for page elements and possible page edits. These design choices are modeled jointly as coupled random variables (a Bayesian Network) with uncertainty modeled by their probability distributions. The overall joint probability distribution for the network assigns higher probability to good design choices. Given this model, we show that the general document layout problem can be reduced to probabilistic inference over the Bayesian network. We show that the inference task may be accomplished efficiently, scaling linearly with the content in the best case. We provide a useful specialization of the general model and use it to illustrate the advantages of soft probabilistic encodings over hard one-way constraints in specifying design aesthetics.

Categories and Subject Descriptors

I.7.4 [**Computing Methodologies**]: Document and Text Processing:Electronic Publishing

General Terms

Algorithms, Design

Keywords

automated publishing, layout synthesis, variable templates

1. INTRODUCTION

In order to compose documents, one must make aesthetic decisions on how to paginate content, how to arrange document elements (text, images, graphics, sidebars etc.) on each page, how much to crop/scale images, how to manage whitespace etc. These decision variables are not mutually exclusive, making the aesthetic

---

*All photographs used have Creative Commons licenses and are copyright their owners: Mike Baird (boat), D Sharon Pruitt (punk girl and veiled girl), and Steve Brace (owl).

graphic design of documents a hard problem often requiring an expert design professional. While professional graphic design works well for the traditional publishing industry where a single high quality document may be distributed to an audience of millions, it is not economically viable (due to its high marginal cost) for the creation of highly personalized documents formatted for a plethora of device form factors.

Automated document composition attempts to transform raw content automatically into documents with high aesthetic value. This has been a topic of much research [10] [6] and is the focus of this paper. In this paper we propose a new way to think about automated document composition based on a generative probabilistic document model (PDM) that incorporates the key design choices described above as random variables with associated probability distributions. The model is generative in the sense that documents may be produced as samples from an underlying probability distribution. We model the coupling between the design choices explicitly using a Bayesian network [12]. At the core of our model is the idea of a probabilistic encoding of aesthetic design judgment with uncertainty encoded as *prior* probability distributions. For example a designer may specify that the whitespace between two page elements have a mean (desired value), and a variance (flexibility). If the variance is large then a larger range of values is tolerated for the whitespace. A small variance implies a tighter range of values. As a consequence of the PDM model we show that automated document composition may be efficiently realized as an inference problem in the network where the inference task is to simultaneously find parameter estimates that maximize the joint probability distribution of all network variables. Probabilistic specification of design intent allows the inference procedure to make appropriate design tradeoffs while laying out content. While the resulting layout may require deviation from the designer's most desired parameter settings, parameters with tighter variances deviate less than parameters with larger variances. Such soft encodings are in contrast with state of the art template selection methods that use hard one-way constraints to encode layout design preferences [7]:

1. We explicitly model the editing process using template parameters that are automatically and *actively* adjusted. This allows our templates to *continuously* (and optimally) morph to fit content. A continuous distribution of the whitespace possibilities between even two items on a page would require an infinite number of distinct one-way constraint templates to encode. With one-way constraints, this results in a template explosion due to the need for (1) templates (and/or alternate versions of content) to cover for variability *within* a relative arrangement of elements (2) templates to cover variability *between* relative arrangements. The PDM model in

this paper shows how (1) can be addressed effectively. New template parameterizations (that are beyond the scope of this paper, but within the scope of the general PDM framework) may be needed to address the latter case.

2. We capture local aesthetic design judgments of designers explicitly using prior probability distributions of template parameters. This allows designers to capture flexibility in parameter settings allowing greater freedom in choosing the parameters to fit content. One-way constraints are a much more rigid design specification.

Our work is greatly influenced by similar work on document content modeling and analysis. Probabilistic graphical models (especially Bayesian network models) are very popular in the text mining, document retrieval and document content understanding communities. For example the probabilistic models for text content in documents proposed by Hoffman [5] and Blei et. al. [3] model words in a document by first sampling a topic from a topic distribution and then sampling words from specific topic distributions. While these models are also referred to as probabilistic document models, they model the actual text content and not document composition. Our aim was to bring the flexibility of such an approach to bear on the document composition problem.

2. RELATED WORK

The survey papers by Lok et al. [10] and Hurst et al. [6] provide a comprehensive background on automated document composition technologies. We only discuss closely related work in this section.

At the page composition level, constraint solvers are often used in variable data printing (VDP) to accommodate content block size variations via *active* templates [9]. In VDP, templates are designed for specific content by a professional graphic designer. The template containers can be *nudged* to accommodate content size variations. The adjustment of the width/height of containers is based on a set of constraints constraining relative positions of content blocks [2]. The set of constraints are solved using a constraint solver [1] to determine final block positions. Since all page layouts fitting constraints are considered equally good, such constraints are often called *hard* constraints. Layout synthesis is reduced to the problem of generating a solution consistent with all the constraints/rules. The document description framework (DDF) [11] is a format (XML) innovation that extends the notion of variable templates to allow variations in number of content blocks (flows) within fixed template regions. DDF is not an optimization framework but rather a format that allows a designer to program rules encoded within XML that may be used to determine how to flow content into local template regions.

While most of the automated layout handled by variable data printing (VDP) is often page based (i.e. content is specified a page at a time) an important component of general document quality is pagination which determines what content to place on each page. This influences how close the referenced figures, sidebars etc. are to the text blocks that referenced them. LaTeX uses a first fit sequential approach for floating objects. It places a floating object on the first page on which it fits, after it has been cited. Bruggeman-Klein et. al. [4] extended the pioneering work of Plass [13] to solve for the optimal allocation of content to pages using dynamic programming to minimize the overall number of page turns (in looking for referenced items). However, these pagination methods do not consider the impact content allocation has on document composition quality.

Jacobs et. al. [7] use dynamic programming to select a template to compose each page of the document from a template li-

brary using dynamic programming. Content is flowed into these templates using one-way constraints [15]. One-way constraints are simply mathematical expressions in terms of other constraint variables. Constraint dependencies may be expressed as a directed acyclic graph. Greedy traversal of this graph computes a layout of all elements. The highest level constraint variables are read only (ex: page.width and page.height). As an example, to flow a title and body text into a template, the constraint system may evaluate[1]:

$$title.top = page.height/10$$
$$title.bottom = title.top + < rendered\ Title\ height >$$
$$body.top = title.bottom + 10$$

Since the absolute position on the page of different blocks change when page dimensions or content changes the authors refer to their layouts as *adaptive* layouts [7]. Templates are scored based on how well content fills the containers, and how many widowed and orphaned text there are. Schrier et al. [14] develop a template description language that automatically generates several templates from a small set of flexible template descriptions. Such template generation methods are complementary to the models described in this paper.

Our goal in this paper is to formally unify under one framework the design choices explored in the prior work, including page level adjustments [9], pagination [4] and choice of relative content arrangements [7]. As discussed in the introduction, probabilistic modeling is a natural choice for encoding soft aesthetic judgments which provides added flexibility.

3. PROBABILISTIC DOCUMENT MODEL

This section introduces the basic concept of PDM and its representation as a Bayesian network. We pose automated document composition as an inference problem over the Bayesian network and derive an algorithm for optimal document composition.

3.1 Notation

We use the following general mathematical notation: random variables (uppercase, ex: T), sample realizations of a random variable (lowercase of corresponding random variable, ex: t), matrices (bold uppercase, ex: $\mathbf{X}$), vectors (bold lowercase, ex: $\mathbf{h}$), random sets (uppercase script, ex: $\mathcal{A}$), sample realizations of a random set (lowercase script of corresponding random set, ex: a). X^* denotes the *optimal* sample value of random variable X. $|\mathcal{A}|$ is a count of the number of elements in the set $\mathcal{A}$. We use $\sim$ to indicate the distribution of a random variable. Thus, $X \sim N(\bar{X}, \sigma^2)$ indicates that random variable X is normally distributed with mean $\bar{X}$ and variance σ^2). A normal probability distribution of X may also be written compactly as $\mathbb{P}(X) = N(X|\bar{X}, \sigma^2)$. We use $\approx$ to indicate normal variation around a desired equality. Thus $X \approx x$ is equivalent to $X - x \sim N(0, \sigma^2)$.

We represent the given set of all the units of content to be composed (ex: images, units of text, sidebars etc.) by a finite set c that is a particular sample of content from a random set $\mathcal{C}$ with sample space comprising sets of all possible content input sets. Text units could be words, sentences, lines of text or whole paragraphs.

We denote by c' a set comprising all sets of discrete content allocation possibilities over one or more pages starting with and including the first page. Content subsets that do not form valid allocations (e.g. allocations of non-contiguous lines of text) do not exist in c'. If there are 3 lines of text and 1 floating figure to be composed, $c = \{l_1, l_2, l_3, f_1\}$ while $c' = \{\{l_1\}, \{l_1, l_2\}, \{l_1, l_2, l_3\}, \{f_1\},$

---

[1] heights are measured from the top of the page

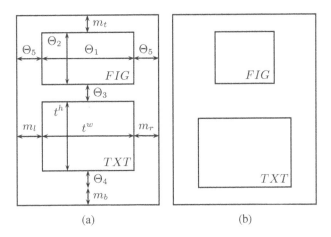

(a)   (b)

Figure 1: (a) Example Probabilistic page template containing two elements (a figure and a text stream) with random parameters $\Theta = [\Theta_1, \Theta_2, \Theta_3, \Theta_4, \Theta_5]^T$, and (b) A *static* template derived by sampling random variable Θ. In this example, values m_t, m_b, m_l, m_r are constants describing the margins, and t^w and t^h are constants describing the width and height of a block of text.

$\{l_1, f_1\}, \{l_1, l_2, f_1\}, \{l_1, l_2, l_3, f_1\}\} \cup \{\emptyset\}$. Note that the specific order of elements within an allocation set is not important since $\{l_1, l_2, f_1\}$ and $\{l_1, f_1, l_2\}$ refer to an allocation of the same content. However allocation $\{l_1, l_3, f_1\} \notin c'$ since lines 1 and 3 cannot be in the same allocation without including line 2. c' includes the empty set to allow the possibility of a null allocation. In general, if there are $|c_l|$ lines of text and $|c_f|$ figures in c, we have $|c'| = (|c_l| + 1)(|c_f| + 1)$, if figures are allowed to float freely. Thus, $|c'| \gg |c|$ even when c has a moderate number of figures.

We represent the index of a page by $i \geq 0$. C_i is a random set representing the content allocated to page i. $C_{\leq i} \in c'$ is a random set of content allocated to pages with index 0 through i. Hence we have $C_{\leq i} = \cup_{j=0}^{i} C_j$. If $C_{\leq i} = C_{\leq i-1}$ then $C_i = \emptyset$ (page i has no content allocated to it). We define $C_{\leq -1} = \emptyset$ for convenience, so that all pages $i \geq 0$ have valid content allocations to the previous $i - 1$ pages.

3.2 Representation

Traditionally, a page template is an abstract representation of page composition, consisting of *copy holes* for different page elements (such as figures, text streams, sidebars etc.). A template encodes the absolute positions of all elements to be placed on the page. A page composition using a particular template is produced by simply pasting images (either by cropping or scaling) into the image holes and flowing text into the text holes. A template thus determines the absolute positions of all page elements. We refer to such templates as *static* templates.

In this paper we expand the classical notion of a *static* template to what we call a *probabilistic* template. A *probabilistic* template is parameterized by parameters (ex: whitespace between elements, image dimensions) that are themselves random variables. A random sampling of the random variables representing the template gives rise to a particular *static* template. Thus a probabilistic template represents only relative positions of page elements. A page layout obtained from a *probabilistic* template is allowed to continuously morph (as template parameters are sampled). Fig 1 illustrates the concept of a *probabilistic* page template.

According to PDM the ith page of a *probabilistic* document may

be composed by first sampling random variable T_i representing choice from a library of page templates representing different relative arrangements of content, sampling a random vector Θ_i of template parameters representing possible edits to the chosen template, and sampling a random set C_i of content representing content allocation to that page (a.k.a. pagination).

A template t_i for page i is sampled from a probability distribution $\mathbb{P}_i(T_i)$ over a set Ω_i of template indices with $|\Omega_i|$ possible template choices. This formulation allows for example, first, last, even and odd page templates to be sampled from different sublibraries of the overall template library. Once a template is sampled we may sample its parameter vector θ_i from the conditional multi-variate probability distribution $\mathbb{P}(\Theta_i|t_i)$. This distribution may be regarded as a *prior* probability distribution that determines the *prior* uncertainty (before seeing content) of template parameters. Graphic design knowledge regarding parameter preferences may be directly encoded into this distribution. Thus sampling from this distribution makes *aesthetic* parameter settings more likely. Finally, the allocation for the current and previous pages $c_{\leq i}$ is sampled from a probability distribution $\mathbb{P}(C_{\leq i}|c_{\leq i-1}, \theta_i, t_i)$. This distribution reflects how well the content allocated to the current page fits when $T_i = t_i$ and template parameters $\Theta_i = \theta_i$. The allocation to the previous pages affects the probability of an allocation for the current page via the logical relationship content in previous pages has to content on the current page. For example, previous page allocation allows us to determine if a figure or sidebar appearing on the current page is referenced in a prior page (a dangling reference). The current page allocation can be obtained as $c_i = c_{\leq i} - c_{\leq i-1}$. In summary a random document can be generated from the probabilistic document model by using the following sampling process for page $i \geq 0$ with $c_{\leq -1} = \emptyset$:

$$\text{sample template } t_i \quad \text{from} \quad \mathbb{P}_i(T_i) \quad (1)$$
$$\text{sample parameters } \theta_i \quad \text{from} \quad \mathbb{P}(\Theta_i|t_i) \quad (2)$$
$$\text{sample content } c_{\leq i} \quad \text{from} \quad \mathbb{P}(C_{\leq i}|c_{\leq i-1}, \theta_i, t_i) \quad (3)$$
$$c_i \quad = \quad c_{\leq i} - c_{\leq i-1} \quad (4)$$

The sampling process naturally terminates when the content runs out. Since this may occur at different random page counts each time the process is initiated, the document page count I is itself a random variable defined by the minimal page number at which $C_{\leq i} = c$. Formally, $I = 1 + \operatorname*{argmin}_{\{i:C_{\leq i}=c\}} i < \infty$. A document $\mathcal{D}$ in PDM is thus defined by a triplet $\mathcal{D} = \{\{C_{\leq i}\}_{i=0}^{I-1}, \{\Theta_i\}_{i=0}^{I-1}, \{T_i\}_{i=0}^{I-1}\}$ of random variables representing the various design choices made in equations (1)-(4).

For a specific content c, the probability of producing document $\mathcal{D}$ of I pages via the sampling process described in this section is simply the product of the probabilities of all design (conditional) choices made during the sampling process. Thus,

$$\mathbb{P}(\mathcal{D}; I) = \prod_{i=0}^{I-1} \mathbb{P}(C_{\leq i}|C_{\leq i-1}, \Theta_i, T_i)\mathbb{P}(\Theta_i|T_i)\mathbb{P}_i(T_i) \quad (5)$$

The probability distribution of equation (5) may be represented as a directed graph representing the relationship between the design choices as shown in Fig. 2. The distribution can be generated from the graph by simply multiplying the conditional probability distributions of each node conditioned only on its parents. Such a model is called a Bayesian network model for the underlying probability distribution [12]. The model is generative, in the sense that a sequential process (as described above) can be used to generate documents from the model. The documents generated by this process are simply samples drawn from the probability distribution described above. Further, in a Bayesian network every node is conditionally independent of its non-descendants given its parents [12]. Thus our PDM model implicitly assumes that the allocation to page i is independent of the templates and parameter selections of all previous pages, given the content allocations to previous pages and the template and template parameters for the current page.

Although the sampling procedure described in this section generates documents with various probabilities (higher probabilities translate to higher quality) we are interested only in finding the document that has the highest probability. Our goal is to compute the optimizing sequences of templates, template parameters, and content allocations that maximize overall document probability. It is clear that a naive approach that generates and scores all possible documents and picks the one with the maximum probability is infeasible since there are an infinite number of possible documents. In fact the structure of the graph (representing conditional independence in the joint probability distribution) may be used to derive an efficient Bayesian inference procedure (essentially equivalent to dynamic programming) that computes the optimal solution. We turn to model inference next.

3.3 Model Inference

We refer to the task of computing the optimal page count and the optimizing sequences of templates, template parameters, content allocations that maximize overall document probability as the model inference task.

$$(\mathcal{D}^*, I^*) = \underset{\mathcal{D}, I \geq 1}{\operatorname{argmax}} \mathbb{P}(\mathcal{D}; I) \qquad (6)$$

The key to efficient inference in Bayesian networks is the fact that although all the variables influence each other in general, a variable is only directly influenced by a few neighbors at most. This allows us to break up the desired maximization sequentially into several sub-problems. We start by gathering terms in (5) involving Θ_i and maximizing over Θ_i.

$$\Psi(\mathcal{C}_{\leq i}, \mathcal{C}_{\leq i-1}, T_i) = \max_{\Theta_i} \mathbb{P}(\mathcal{C}_{\leq i} | \mathcal{C}_{\leq i-1}, \Theta_i, T_i) \mathbb{P}(\Theta_i | T_i) \quad (7)$$

The maximization over Θ_i effectively eliminates Θ_i from further consideration, resulting in a function Ψ of the remaining variables in the RHS of (7). Now grouping the remaining terms in (5) involving T_i with $\Psi(\mathcal{C}_{\leq i}, \mathcal{C}_{\leq i-1}, T_i)$ and maximizing over T_i we have

$$\Phi_i(\mathcal{C}_{\leq i}, \mathcal{C}_{\leq i-1}) = \max_{T_i \in \Omega_i} \Psi(\mathcal{C}_{\leq i}, \mathcal{C}_{\leq i-1}, T_i) \mathbb{P}_i(T_i) \quad (8)$$

Note that the function Φ_i depends on i since the maximization over allowed templates for each page occurs over distinct sub-libraries Ω_i that depend on i. We can now rewrite the maximization of (5)

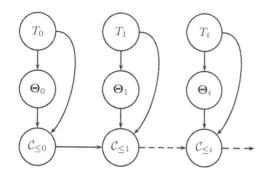

Figure 2: PDM as a graphical model.

purely in terms of functions Φ_i of content allocations as

$$\max_{\mathcal{D}, I \geq 1} \mathbb{P}(\mathcal{D}; I) = \max_{I \geq 1} \max_{\{\mathcal{C}_{\leq i}\}_{i=0}^{I-1}} \prod_{i=0}^{I-1} \Phi_i(\mathcal{C}_{\leq i}, \mathcal{C}_{\leq i-1}) \quad (9)$$

We may now sequentially maximize over content allocations $\mathcal{C}_{\leq i}$ for $i = 0, 1, \cdots, I-1$. First, grouping terms involving $\mathcal{C}_{\leq 0}$ in (9) and maximizing over $\mathcal{C}_{\leq 0}$ we have

$$\tau_1(\mathcal{C}_{\leq 1}) = \max_{\mathcal{C}_{\leq 0}} \Phi_1(\mathcal{C}_{\leq 1}, \mathcal{C}_{\leq 0}) \Phi_0(\mathcal{C}_{\leq 0}, \mathcal{C}_{\leq -1}) \quad (10)$$

$$= \max_{\mathcal{C}_{\leq 0}} \Phi_1(\mathcal{C}_{\leq 1}, \mathcal{C}_{\leq 0}) \Phi_0(\mathcal{C}_{\leq 0}, \emptyset) \quad (11)$$

Then maximizing over content allocations $\mathcal{C}_{\leq 1}$ we have

$$\tau_2(\mathcal{C}_{\leq 2}) = \max_{\mathcal{C}_{\leq 1}} \Phi_2(\mathcal{C}_{\leq 2}, \mathcal{C}_{\leq 1}) \tau_1(\mathcal{C}_{\leq 1}) \quad (12)$$

We can easily see that this process is governed by the following general recursion for $i \geq 1$

$$\tau_i(\mathcal{C}_{\leq i}) = \max_{\mathcal{C}_{\leq i-1}} \Phi_i(\mathcal{C}_{\leq i}, \mathcal{C}_{\leq i-1}) \tau_{i-1}(\mathcal{C}_{\leq i-1}) \quad (13)$$

with the added definition $\tau_0(\mathcal{C}_{\leq 0}) = \Phi_0(\mathcal{C}_{\leq 0}, \emptyset)$. Note that at the end of the recursive computation of the functions τ_i, since $\mathcal{C}_{\leq I-1} = c$, the entire content to be composed, $\tau_{I-1}(c)$ is only dependent on I. Therefore we can finally write down the maximum of (5) as

$$\max_{I \geq 1} \max_{\mathcal{D}} \mathbb{P}(\mathcal{D}; I) = \max_{I-1 \geq 0} \tau_{I-1}(c), \quad (14)$$

The optimal page count can now be determined as the corresponding maximizer over all last page possibilities (where all content $\mathcal{C}$ has been allocated). Thus,

$$I^* = 1 + \underset{i \geq 0}{\operatorname{argmax}} \tau_i(c) \quad (15)$$

We now discuss how the optimal allocations $\mathcal{C}_{\leq i}^*$ can be inferred. First note that for the final page (page number $I^* - 1$) we must have $\mathcal{C}_{\leq I^*-1}^* = c$. From $\mathcal{C}_{\leq I^*-1}^*$ we can compute the optimal allocations to the previous pages, $\mathcal{C}_{\leq I^*-2}^*$ by substituting the known $\mathcal{C}_{\leq I^*-1}^*$ in the recursion (13) for τ_{I^*-1} and finding the value of $\mathcal{C}_{\leq I^*-2}$ that maximizes τ_{I^*-1}. Specifically,

$$\mathcal{C}_{\leq I^*-2}^* = \underset{\mathcal{C}_{\leq I^*-2}}{\operatorname{argmax}} \Phi_{I^*-1}(\mathcal{C}_{\leq I^*-1}^*, \mathcal{C}_{\leq I^*-2}) \tau_{I^*-2}(\mathcal{C}_{\leq I^*-2})$$

$$(16)$$

In general we can set up a recursion that allows us to solve for optimal allocations for all I^* pages. Once the allocations for each page are determined, we may look up the optimal template for each page by finding the template T_i^* that corresponds to $\Phi(\mathcal{C}_{\leq i}^*, \mathcal{C}_{\leq i-1}^*)$. Once the template and the allocation are known, optimal template parameters Θ_i^* may be computed. This procedure is given below for $i = I^* - 1, \cdots, 0$:

$$\mathcal{C}_{\leq i-1}^* = \underset{\mathcal{C}_{\leq i-1}}{\operatorname{argmax}} \Phi_i(\mathcal{C}_{\leq i}^*, \mathcal{C}_{\leq i-1}) \tau_{i-1}(\mathcal{C}_{\leq i-1}) \quad (17)$$

$$T_i^* = \underset{T_i}{\operatorname{argmax}} \Psi(\mathcal{C}_{\leq i}^*, \mathcal{C}_{\leq i-1}^*, T_i) \mathbb{P}_i(T_i) \quad (18)$$

$$\Theta_i^* = \underset{\Theta_i}{\operatorname{argmax}} \mathbb{P}(\mathcal{C}_{\leq i^*} | \mathcal{C}_{\leq i-1}, \Theta_i, T_i^*) \mathbb{P}(\Theta_i | T_i^*) (19)$$

$$\mathcal{C}_i^* = \mathcal{C}_{\leq i}^* - \mathcal{C}_{\leq i-1}^* \quad (20)$$

Once the optimal page count, content allocations to pages, templates and template parameters are determined the inference task is complete. The solution found by this Bayesian network inference approach is a globally optimal maximizer of (5). The order of variable elimination affects efficiency but not optimality [12].

It is important to note that while a document composed using the PDM inference algorithm described here has its page count optimally selected, the user may want to constrain the page count to a specified number of pages. To force a page count of I_f we only need to compute (17)-(20) with $I^* \leftarrow I_f$. This simply corresponds to a maximization of the conditional distribution $\mathbb{P}(\mathcal{D}|I)$.

The optimal document composition algorithm derived above may be succinctly summarized as a two pass process. In the forward pass we *recursively* compute, for all valid content allocation sets $\mathcal{A}, \mathcal{B} \in c'$ with $\mathcal{A} \supseteq \mathcal{B}$ the following coefficient tables:

$$\Psi(\mathcal{A}, \mathcal{B}, T) = \max_{\Theta} \mathbb{P}(\mathcal{A}|\mathcal{B}, \Theta, T)\mathbb{P}(\Theta|T) \quad (21)$$

$$\Phi_i(\mathcal{A}, \mathcal{B}) = \max_{T \in \Omega_i} \Psi(\mathcal{A}, \mathcal{B}, T)\mathbb{P}_i(T), i \geq 0, \quad (22)$$

$$\tau_i(\mathcal{A}) = \max_{\mathcal{B}} \Phi_i(\mathcal{A}, \mathcal{B})\tau_{i-1}(\mathcal{B}), i \geq 1 \quad (23)$$

with $\tau_0(\mathcal{A}) = \Phi_0(\mathcal{A}, \emptyset)$. Computation of (23) depends on (22) which in turn depends on (21). In the backward pass we use these coefficients used to infer the optimal document using (17)-(20).

The innermost function $\Psi(\mathcal{A}, \mathcal{B}, T)$ is essentially a score of how well content in the set $\mathcal{A} - \mathcal{B}$ is suited for template T. It is the maximum of a product of two terms. The first term $\mathbb{P}(\mathcal{A}|\mathcal{B}, \Theta, T)$ represents how well content fills the page and respects figure references while the second term $\mathbb{P}(\Theta|T)$ assesses how close, the parameters of a template are to the designer's *aesthetic* preference. Thus the overall probability (score) is a tradeoff between page fill and a designer's aesthetic intent. When there are multiple parameters settings that fill the page equally well, the parameters that maximize the prior (and hence are closest to the template designer's desired values) will be favored.

Note also, that $\Psi(\mathcal{A}, \mathcal{B}, T) \propto \max_{\Theta} \mathbb{P}(\Theta|\mathcal{A}, \mathcal{B}, T)$ where the proportionality follows from Bayes rule. Thus $\Psi(\mathcal{A}, \mathcal{B}, T)$ is proportional to the maximum *a posteriori* (MAP) probability of the template parameters given content and a template. The maximizer Θ^* is the corresponding MAP estimate. $\mathbb{P}(\mathcal{A}|\mathcal{B}, \Theta, T)$ represents the likelihood function of a particular allocation that refines our *prior* belief regarding the template parameters, $\mathbb{P}(\Theta|T)$ upon seeing $\mathcal{A}$ and $\mathcal{B}$. Efficiency in computing $\Psi(\mathcal{A}, \mathcal{B}, T)$ may be obtained by a) screening allocations $\mathcal{A}$ and $\mathcal{B}$ to avoid computation altogether for invalid allocations (ex: figure or sidebar occurring before its reference, widowed or orphaned text chunks etc.) b) screening templates for compatibility (ex: content with two figures cannot be allocated to a template with only one figure) c) screening for too much and too little content. These screening approaches significantly reduce the number of cases for which the expensive optimization of (21) needs to be performed.

The function $\Phi_i(\mathcal{A}, \mathcal{B})$ scores how well content $\mathcal{A} - \mathcal{B}$ can be composed onto the ith page, considering all possible relative arrangements of content (templates) allowed for that page. $\mathbb{P}_i(T)$ allows us to boost the score of certain templates, increasing the chance that they will be used in the final document composition.

Finally functions $\tau_i(\mathcal{A})$ is a pure pagination score of the allocation $\mathcal{A}$ to the first i pages. The recursion (23) basically says that the pagination score for an allocation $\mathcal{A}$ to the first i pages, $\tau_i(\mathcal{A})$ is equal to the product of the best pagination score over all possible previous allocations $\mathcal{B}$ to the previous $(i-1)$ pages with the score of the current allocation $\mathcal{A} - \mathcal{B}$ to the ith page, $\Phi_i(\mathcal{A}, \mathcal{B})$. Note that the pure pagination score $\tau_i(\mathcal{A})$ encapsulates dependency on relative arrangements of content (represented by templates T) and possible template edits (Θ) via recursive dependency on $\Phi_i(\mathcal{A}, \mathcal{B})$ and $\Psi(\mathcal{A}, \mathcal{B}, T)$.

The PDM inference framework may easily be extended to handle alternate and optional content. Alternate images may be handled

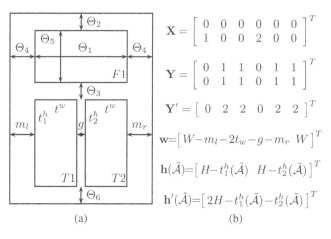

Figure 3: **(a) Example Probabilistic page template containing one figure and two linked text stream blocks with random parameters $\Theta = [\Theta_1, \Theta_2, \Theta_3, \Theta_4, \Theta_5, \Theta_6]^T$. (b) Matrix representation**

by simply evaluating the RHS of (21) for all possible alternatives and choosing the maximum value for $\Psi(\mathcal{A}, \mathcal{B}, T)$. Optional images may be handled by allowing content allocations with optional images to match templates with or without available template slots to render them. Templates without available slots for any images may simply ignore optional images. However the optional images are still deemed to have been allocated.

The entire inference process is essentially dynamic programming with the coefficients $\Psi(\mathcal{A}, \mathcal{B}, T)$, $\Phi_i(\mathcal{A}, \mathcal{B})$, and $\tau_i(\mathcal{A})$ playing the role of dynamic programming tables. The optimization itself may also be considered analogous to the minimization of an energy function proportional to the negative logarithm of (5). In this case we simply replace products with sums, throughout but the essential math remains the same. Inference complexity is analyzed in the Appendix.

4. MODEL PARAMETERIZATION

The results of Section 3 are valid for arbitrary template parameterizations and general probability distributions. This section focuses one specific efficient parameterization of the general PDM model to illustrate the value of probabilistic modeling with aesthetic priors. This specialization can express arbitrary *soft* linear parameter relationships probabilistically. It can support multicolumn layouts (with sidebars) but without element overlaps. This is still an important class of documents. This is not a limitation of the general model. This formulation is convenient since it makes the the innermost sub-problem of model inference, given by equation (21) particularly efficient.

4.1 Template Representation

Content fit to a template is assessed along all *paths* that go from top to bottom and left to right on a page. X-paths and Y-paths are computed for each template. For the example in Figure 3(a) we define two Y-paths ($\mathcal{Y}_1 = \Theta_2 \rightarrow \Theta_5 \rightarrow \Theta_3 \rightarrow t_1^h \rightarrow \Theta_6$, $\mathcal{Y}_2 = \Theta_2 \rightarrow \Theta_5 \rightarrow \Theta_3 \rightarrow t_2^h \rightarrow \Theta_6$) and two X-paths ($\mathcal{X}_1 = m_l \rightarrow t_w \rightarrow g \rightarrow t_w \rightarrow m_r$, $\mathcal{X}_2 = \Theta_4 \rightarrow \Theta_1 \rightarrow \Theta_4$).

Paths have constant and variable parameter components. In the PDM model described in Section 3 each template had fixed text elements and potentially variable image and whitespace parameters. We sampled the content for a page after the template parameters were sampled. In our model inference procedure however the situation is reversed. When we compute $\Psi(\mathcal{A}, \mathcal{B}, T)$ we are given the allocation to the current page $\tilde{\mathcal{A}} = \mathcal{A} - \mathcal{B}$ and then solve for the maximizing template parameters. Thus, text heights on a template can vary with the content allocation $\tilde{\mathcal{A}}$ to the template. However, we still regard text heights as constants since, while they vary with $\tilde{\mathcal{A}}$ they do not vary as the variable parameters change.

Variable path components of the N X-paths and M Y-paths comprising V random parameters may be described by matrices $\mathbf{X} = [\mathbf{x}_1 \; \mathbf{x}_2 \; \cdots \; \mathbf{x}_N]$ and $\mathbf{Y} = [\mathbf{y}_1 \; \mathbf{y}_2 \; \cdots \; \mathbf{y}_M]$ consisting of column vectors of dimension $V \times 1$ for each path. The element (v, n) in matrix $\mathbf{X}$ is the number of times random parameter Θ_v appears in the nth X-path. Similarly, the element (v, m) in matrix $\mathbf{Y}$ is the number of times random parameter Θ_v appears in the mth Y-path. $\mathbf{X}$ and $\mathbf{Y}$ matrices corresponding to the example template in Figure 3(a) are shown in Figure 3(b).

Residual path heights and widths represent the *adjustable* portion of each of the paths. These are the path heights and widths after subtracting constant path elements. Residual path sizes are represented by the vectors $\mathbf{h}(\tilde{\mathcal{A}})$ and $\mathbf{w}$ in Figure 3(b) where H and W represent the height and width of the page respectively. Note the dependence of the residual path heights on $\tilde{\mathcal{A}}$ due to the dependency of the text heights on $\tilde{\mathcal{A}}$ as discussed earlier.

In many cases we have *multi-column flows* where text flows from one column to the next. In this case, the Y-paths with linked text flows are be grouped into a single path by simply summing the corresponding columns of $\mathbf{Y}$ and $\mathbf{h}$ giving rise to new matrices $\mathbf{Y}'$ and $\mathbf{h}'$ respectively.

Thus, we represent a probabilistic template t by matrices $\mathbf{X}_t$, $\mathbf{Y}_t$, $\mathbf{Y}'_t$ and vectors $\mathbf{h}_t(\tilde{\mathcal{A}})$, $\mathbf{w}_t$ and $\mathbf{h}'_t(\tilde{\mathcal{A}})$. Figure 3(b) gives the complete matrix representation of the example template in Figure 3(a). This representation will prove useful when we discuss the parameterization of the probability distributions in PDM, next.

4.2 Parameterized probability distributions

In this section we parameterize probability distributions used in the PDM model of (1)-(3) to make various conditional design choices more or less probable.

4.2.1 Aesthetic priors

We use a multinomial distribution for the template probabilities. This gives us the flexibility of making certain templates more or less likely to be used in creating a document. Thus $\mathbb{P}_i(T_i = t_i) = p_{t_i} = 1/|\Omega_i|, \; \forall t_i$, if there is no *a priori* preference for a template.

We model the *prior* distribution of template parameters $\mathbb{P}(\boldsymbol{\Theta}|T)$ in equation (2) to be a multi-variate Normal distribution:

$$\mathbb{P}(\boldsymbol{\Theta}|T = t) = \mathrm{N}(\boldsymbol{\Theta}|\bar{\boldsymbol{\Theta}}_t, \boldsymbol{\Lambda}_t^{-1}) \qquad (24)$$

where $\bar{\boldsymbol{\Theta}}_t$ is the mean of $\boldsymbol{\Theta}|T = t$. $\boldsymbol{\Lambda}_t$ represents the precision (inverse of covariance) matrix. For the rest of the paper we will use the shorthand $\boldsymbol{\Theta}_t$ for $\boldsymbol{\Theta}|T = t$ (the parameters of a particular template t [2]).

While the full multivariate distribution of equation (24) is quite general, it is hard for a designer to specify a full covariance matrix. It is easier to specify the means, variances (hence precisions), minimum and maximum values of each random parameter. This leads to a diagonal precision matrix $\boldsymbol{\Lambda}_{t_d} = diag[\lambda_{t1}, \lambda_{t2}, \cdots, \lambda_{tV}]$.

It is also desirable to consider linear relationships among parameters (ex: ratios) whose aesthetics should be encoded. Thus we consider general *soft* linear relationships expressed by

$$\mathbf{C}_t \boldsymbol{\Theta}_t - \mathbf{d}_t \sim \mathrm{N}(\mathbf{0}, \boldsymbol{\Delta}_{t_d}^{-1}) \qquad (25)$$

---

[2] not to be confused with $\boldsymbol{\Theta}_i$ used in Section 3.2 which refers to the parameters sampled for the ith page

for general $\mathbf{C}_t$ and $\mathbf{d}_t$ or equivalently, $\boldsymbol{\Theta}_t \sim \mathrm{N}(\bar{\boldsymbol{\Theta}}_t, \boldsymbol{\Lambda}_t^{-1})$, with

$$\boldsymbol{\Lambda}_t = \mathbf{C}_t^T \boldsymbol{\Delta}_{t_d} \mathbf{C}_t \qquad (26)$$

$$\bar{\boldsymbol{\Theta}}_t = \boldsymbol{\Lambda}_t^{-1} \mathbf{C}_t^T \boldsymbol{\Delta}_{t_d} \mathbf{d}_t \qquad (27)$$

Note that in the case $\mathbf{C}_t = \mathbf{I}$, we have $\bar{\boldsymbol{\Theta}}_t = \mathbf{d}_t$ and $\boldsymbol{\Lambda}_t = \boldsymbol{\Delta}_{t_d} = \boldsymbol{\Lambda}_{t_d}$, so this formulation is a more general than simple mean and diagonal precision specification. This representation is particularly useful in developing priors for image scaling and re-targeting as discussed in Section 4.2.2.

4.2.2 Incorporating image scaling and re-targeting

A template may have parameters Θ_w and Θ_h that describe the width and height of an image. However, these cannot vary independently. In this case we have

$$\Theta_w \approx a \, \Theta_h \quad , \quad \text{with precision } \rho \qquad (28)$$

This is simply a linear relationship that may be handled by encoding it into the prior as described in Section 4.2.1. If we do not want to allow the aspect ratio of an image to change (in the case of pure image scaling) we let $\rho \to \infty$. On the other hand, we may use image re-targeting algorithms to allow the image aspect ratio to change. In this case the value of ρ will determine if we allow small (by setting ρ to a large value) or large (by setting ρ to a small value) changes in aspect ratio. We use the re-targeting algorithm described in [8] to generate the results presented in this paper.

4.2.3 Content allocation likelihood

The probability distribution that determines the likelihood of an allocation as described by equation (3) is represented by the following distribution

$$
\begin{aligned}
\mathbb{P}(\mathcal{A}|\mathcal{B}, \boldsymbol{\Theta}, T) \quad \propto \quad & \exp(-\gamma |R(\mathcal{A}, \mathcal{B})|) \times \\
& \mathrm{N}(\mathbf{w}_t | \mathbf{X}_t^T \boldsymbol{\Theta}_t, \alpha^{-1}\mathbf{I}) \times \\
& \times \mathrm{N}(\mathbf{h}'_t(\mathcal{A} - \mathcal{B}) | \mathbf{Y}'^T_t \boldsymbol{\Theta}_t, \beta^{-1}\mathbf{I}) \quad (29)
\end{aligned}
$$

In the above equation, $|R(\mathcal{A}, \mathcal{B})|$ represents the number of dangling references due to the allocation $\mathcal{A} - \mathcal{B}$ to the current page and $\mathcal{B}$ to the previous pages. The constant γ represents an exponential weighting factor that represents how much to penalize mismatched references in the probability distribution. For good page fill in the Y and X directions the heights and widths of all Y-path groups must satisfy $\mathbf{h}'_t(\mathcal{A} - \mathcal{B}) - \mathbf{Y}'^T_t \boldsymbol{\Theta}_t \approx \mathbf{0}$ and X-paths must satisfy $\mathbf{w}_t - \mathbf{X}_t^T \boldsymbol{\Theta}_t \approx \mathbf{0}$. The normal distributions above simply assign high probability to these events and lower probability for deviations from ideal. The constants α and β are precision parameters of the normal distribution with diagonal precision matrices ($\mathbf{I}$ is the identity matrix) that control the degree to which we want to produce full pages.

4.3 MAP estimation of template parameters

The particular parameterization of the prior and likelihood given above makes the computation of $\Psi(\mathcal{A}, \mathcal{B}, T)$ particularly efficient since the posterior distribution formed from the product of $\mathbb{P}(\boldsymbol{\Theta}|T)$ (24) and $\mathbb{P}(\mathcal{A}|\mathcal{B}, \boldsymbol{\Theta}, T)$ (29) is a multi-variate Normal distribution in $\boldsymbol{\Theta}_t$. With some algebraic manipulation (not shown here due to space constraints) we can compute the optimal MAP estimate in closed form by simply calculating the mean of this product posterior distribution. This results in the following closed form solution for $\boldsymbol{\Theta}_t^* = \mathbf{A}_t^{-1}\mathbf{b}_t$, where:

$$\mathbf{A}_t = \boldsymbol{\Lambda}_t + \alpha \mathbf{X}_t \mathbf{X}_t^T + \beta \mathbf{Y}'_t \mathbf{Y}'^T_t \qquad (30)$$

$$\mathbf{b}_t = \boldsymbol{\Lambda}_t \bar{\boldsymbol{\Theta}}_{\mathbf{t}} + \alpha \mathbf{X}_t \mathbf{w}_t + \beta \mathbf{Y}'_t \mathbf{h}'_t(\mathcal{A} - \mathcal{B}) \qquad (31)$$

In general, however, there may be bound constraints on some of the components of Θ_t (ex: figure dimensions cannot be negative). To incorporate these constraints we solve the following bound-constrained least-squares quadratic program.

$$\Theta_t^* = \underset{\{\Theta_t : \mathbf{l} \leq \Theta_t \leq \mathbf{u}\}}{\operatorname{argmax}} (\mathbf{A}_t \Theta_t - \mathbf{b}_t)^T (\mathbf{A}_t \Theta_t - \mathbf{b}_t) \quad (32)$$

where $\mathbf{l}$ and $\mathbf{u}$ are lower and upper bound vectors constraining Θ_t.

Note that our likelihood formulation of equation (29) attempts to ensure that the content fits the X-paths and the aggregate Y-path flows represented by $\mathbf{Y}'_t$. In an ideal world if text flow is continuous this would be all that is required. Y-paths $\mathbf{Y}_t$ were left out of this optimization because text flow across columns is discrete and it is unclear how text allocated to a page is to be distributed across columns. Using $\mathbf{Y}'_t$ instead of $\mathbf{Y}_t$ allows us to get around this issue by considering how the whole flow (aggregating across columns) fits the page. This approach of course implicitly assumes that the flow is continuous. This is not true for text, but is usually a good approximation.

Fortunately, in practice, we may refine this approximation since Θ_t^* computed using equation (32) effectively converts the probabilistic template into a static template. This means that the text block sizes are now known, so we can distribute the text across the blocks. This may be done without rendering the text if we knew the number of lines to be allocated and the font size. This procedure will produce actual text block height estimates (taking into account discrete line breaks across columns). We can therefore re-compute $\mathbf{h}_t$ using the correct text heights for text blocks in every path and re-solve for Θ_t using equation (32) with $\mathbf{Y}_t$ substituted for $\mathbf{Y}'_t$ and $\mathbf{h}_t(\mathcal{A} - \mathcal{B})$ substituted for $\mathbf{h}'_t(\mathcal{A} - \mathcal{B})$.

5. PRACTICAL CONSIDERATIONS

The input *raw* content to the document composition engine is an XML file. The XML elements match the element types that are allowed on a page template. In our examples we allow three element types including text blocks, figures and sidebars (a grouping of figures and/or text blocks that must appear together on a page). The XML file also encodes figure/sidebar references via an XML attribute indicating the id of the text block that referenced it. Each content XML file is coupled with a style sheet. Content blocks within the content XML have attributes that denote their type. For example, text blocks may be tagged as head, subhead, list, para, caption etc. The document style sheet defines the type definitions and the formatting for these types. Thus the style sheet may require a heading to use Arial bold font with a specified font size, linespacing etc. The style sheet also defines overall document characteristics such as, margins, page dimensions etc.

We use a GUI based authoring tool to author templates and style sheets. Note that the template library design task is fixed overhead cost. The design process a graphic designer must go through to create a template includes a) content block layout, b) specification of linked text streams and c) specification of prior probability distributions (mean-precision and min-max). For images the designers set min-max of height/width and the precision of the aspect ratio (we use the native aspect ratio as the mean). Note that it is the relative values of variances (precisions) that matter. So a designer could use order of magnitude precision changes to indicate preference. For example if a whitespace precision is set to 100, setting figure aspect precision to 10 would allow figure aspect ratio to change much more freely than whitespace. We automatically compute overall prior of (24) using equations (26) and (27). All other distributions and probabilities are calculated computationally without designer involvement.

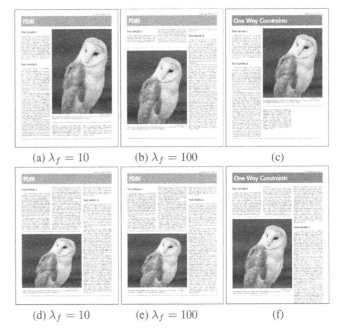

(a) $\lambda_f = 10$ (b) $\lambda_f = 100$ (c)

(d) $\lambda_f = 10$ (e) $\lambda_f = 100$ (f)

Figure 4: Soft probabilistic encodings vs. one way constraints. *Owl: ©Steve Brace (http://www.flickr.com/photos/steve_brace/)*

To run PDM inference we need to generate several discrete content sets $\mathcal{A}$ and $\mathcal{B}$ from the input content. Since figures and sidebars cannot break across pages, it is straightforward to allocate them to sets. A single text block in the content stream may be chunked as a whole if it cannot flow across columns or pages (ex: headings, text within sidebars etc.). However if the text block attribute indicates that it is allowed to flow (paragraphs, lists etc.), it must be decomposed into smaller atomic chunks for set allocation. We use a line based text chunk decomposition in our experiments. We assume that text width is selected from a discrete set of widths allowed for a template library. The height of a chunk is determined by rendering the text chunk at all possible text widths using the specified style sheet style in a preprocess rendering pass. We use the open source desktop publishing software Scribus as the rendering engine and are able to query the number of rendered lines via an API. We use the number of lines and information regarding the font style, line spacing etc. to calculate the rendered height of a chunk. Thus when allocating a text chunk to a text element of a template, we may simply look up its height using the chunk index and template text element width. We can do this if text column widths are known and text is not allowed to wrap around figures. The choice of lines as the unit of text allocation restricts the class of layouts to multi-column layouts with no text wrapping around figures, since when wrap-around is allowed a paragraph may break into different number of lines depending on the extent of wrap-around and so would not be a reliable unit of allocation. Since our specific template parameterization also does not allow text wrap-around, line based allocation is an efficient choice. This is however not a restriction on the generality of the PDM model.

6. EXPERIMENTAL RESULTS

In this section we illustrate the performance PDM based document composition with example 1 and 2-page document compositions (due to space restrictions) shown in Figures 4 and 5. Our template library is designed for 2-page News content with multiple articles in a 3-column format. Each article has at most one figure associated with it and must appear naturally adjacent to article text.

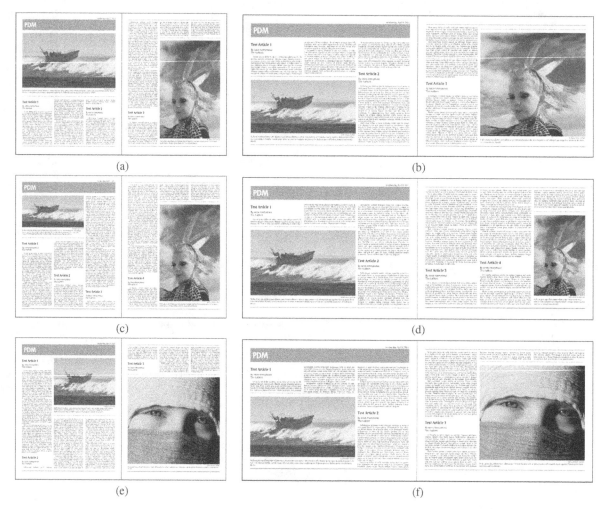

Figure 5: PDM compositions with variations in content and form factor. Each row has exactly the same content with variation only in form factor. (a) and (b) show three articles composed onto two pages. (c) and (d) have a new article inserted between the original articles 2 and 3. (e) and (f) have a new figure with a different aspect ratio substituted for the original figure in article 3. Our algorithm actively paginates content, selects content arrangements (templates), re-targets images, re-flows text and manages whitespace to fit the content in each case. The same template library is used for both form factors. Since we only encode relative position in templates we only need to set H and W to the appropriate values for the new form factor. *Punk girl and Veiled girl: ⓒD Sharon Pruitt (http://www.flickr.com/photos/pinksherbet/), Boat: ⓒMike Baird (http://www.flickr.com/photos/mikebaird/)*

Figure 6: First six pages of the HP Labs 2010 annual report generated using PDM inference. The entire report can be downloaded from http://www.hpl.hp.com/news/2011/jan-mar/annual_report.html

Our template library consists of a total of 133 probabilistic templates including 83 first page and 50 back page templates encoding various relative arrangements of figures, article titles, and text streams. Since the design intent was to have images span columns, only the height of the images was parameterized with a mean equal to the height that maintained aspect ratio. Most whitespace was parameterized with a desired mean of 8 points between text and margins and 16 points between text and figures. Whitespace precision was set to 100 points^{-2}. The precision parameters for the likelihood function in equation (29) are set as follows, $\alpha = \beta = 1000$ and $\gamma = 10000$. These settings reflect our desire to produce full pages and force figures to appear on the same page as their corresponding articles (since the corresponding probability distributions have a tight spread).

Fig. 4 illustrates the value of soft probabilistic encodings vs. hard one way constraint based specification (ex: [7]). To simulate a constraint-based template evaluation system, we set the parameter values of all the templates in our library to their mean values (most aesthetically desired values). This gives us a set of *static* templates that can be interpreted using one-way constraints. When a figure is flowed into an image copy hole, the height of the hole is adjusted to maintain the aspect ratio, and the text below it is constrained as a result. We evaluate the posterior probability of the mean parameter settings and use this to score templates.

Each row of Fig. 4 shows compositions of the same content. The second row has more content than the first row. Each PDM template has precision matrix of (24) set to $\Lambda_t = diag[\lambda_f, \lambda_w, \cdots, \lambda_w]$. The precision of whitespace, $\lambda_w = 100$ for all the PDM compositions. Along each row for PDM, the precision of the figure height λ_f is varied from a small value $\lambda_f = 10$ to a value equal to the whitespace precision $\lambda_f = 100$. Thus the PDM compositions (a) and (d) use mainly figure height (hence, aspect ratio) changes to fit content since it has a low precision relative to whitespace variation. The compositions (b) and (e) allow both figure aspect ratio and whitespace to vary from nominal desired values to fit content since their precisions are the same [3]. While changes in image aspect ratio and whitespace lower the contribution of prior parameter probability to the overall document probability (since they deviate from desired mean values), this impact is offset by relatively higher contributions for pagination or content fit. Of course, if possible, the algorithm will attempt to preserve the mean values of all quantities. In contrast, One one way constraints are much more rigid, not allowing figure aspect ratio and whitespace to change from desired value to fit content. This often results in voids (c) or overflows (d). One approach to add more flexibility to one way constraint methods is to allow several alternate versions of images and/or filler images [7]. Even if alternates were allowed, since each image has a discrete aspect ratio, it is very easy to make content by adding/deleting lines where voids and overflows occur. In general, one way constraint based composition requires huge template libraries so that a good sequence of templates can be found in the library [14]. For PDM compositions, since the templates are actively adjusted to fit content, consistently good results can be achieved even with modest template libraries.

Fig. 5 illustrates 2 page document compositions with variations in content along a column and form factor along a row. Note how we are able to actively paginate content, choose appropriate arrangements and actively re-target images to fit the content. In general our parameter prior favors aspect ratio changes to changes in whitespace since $\lambda_f = 10$ in this case.

As a real world example, we also used the PDM inference al-

---

[3](d) and (e) produced the same composition since the whitespace in (e) could not be further reduced since its min value was reached

gorithm to create the 64 page HP Labs 2010 annual report using a template library of 37 templates. Figure 6 shows the first six pages of the report. The report includes richer formatting including sidebars and profile pictures with text overlay. Changes in image aspect ratio were not allowed in this case, so we had to parameterize whitespace precision around images to allow more flexible image resizing. A high emphasis was placed on referenced ($\gamma = 10000$) figures and sidebars occurring on the same spread as their text reference. Note how the algorithm automatically resizes the images to allow the figures to appear close to their references. While the same 1-figure template was chosen for pages 4 and 6, the algorithm automatically scaled down the image (allowing modest whitespace on its left and right) on page 4 to allow it to be close to the *HP Labs Singapore* heading.

Our code for these experiments was written in MATLAB (unoptimized for efficiency and for-loops). The longest 2-page content consisting of 171 lines of text and two figures took around 45 seconds while the report (1719 lines of text, 14 floating figures and 38 floating sidebars) took around 1 hour. Experiments were performed on a Linux machine with single core 3GHz Intel Xeon processor with 3GB of RAM. We expect that an order of magnitude performance improvement can be achieved with optimized C code.

7. CONCLUSIONS

This paper presented a probabilistic framework for adaptive document layout that supports automated generation of paginated documents for variable content. We attempted to address one of the main weaknesses of template based automated document composition, the fact that one-way constraint-based templates can be overly rigid. Our approach uniquely encodes soft constraints (aesthetic priors) on properties like whitespace, exact image dimensions and image rescaling preferences. Our main contribution is a probabilistic document model (PDM) that combines all of these preferences (along with probabilistic formulations of content allocation and template choice) into a unified model. In addition, we describe a model inference algorithm for computing the optimal document layout for a given set of inputs. PDM opens the door to leveraging probabilistic machinery and machine learning algorithms for both inference and learning of parameterizations directly from examples. This may reduce if not eliminate the need for designer involvement in creating templates and specifying prior distributions.

8. REFERENCES

[1] G. J. Badros, A. Borning, and P. J. Stuckey. The cassowary linear arithmetic constraint solving algorithm. *ACM Trans. Comput.-Hum. Interact.*, 8(4):267–306, 2001.

[2] G. J. Badros, J. J. Tirtowidjojo, K. Marriott, B. Meyer, W. Portnoy, and A. Borning. A constraint extension to scalable vector graphics. In *WWW '01: Proceedings of the 10th international conference on World Wide Web*, pages 489–498, New York, NY, USA, 2001. ACM.

[3] D. M. Blei, A. Y. Ng, and M. I. Jordan. Latent dirichlet allocation. *J. Mach. Learn. Res.*, 3:993–1022, 2003.

[4] A. Brüggemann-Klein, R. Klein, and S. Wohlfeil. On the pagination of complex documents. In *Computer Science in Perspective: Essays Dedicated to Thomas Ottmann*, pages 49–68, New York, NY, USA, 2003. Springer-Verlag New York, Inc.

[5] T. Hofmann. Probabilistic latent semantic indexing. In *SIGIR '99: Proceedings of the 22nd annual international ACM SIGIR conference on Research and development in*

information retrieval, pages 50–57, New York, NY, USA, 1999. ACM.

[6] N. Hurst, W. Li, and K. Marriott. Review of automatic document formatting. In *DocEng '09: Proceedings of the 9th ACM symposium on Document engineering*, pages 99–108, New York, NY, USA, 2009. ACM.

[7] C. Jacobs, W. Li, E. Schrier, D. Bargeron, and D. Salesin. Adaptive grid-based document layout. *ACM Transactions on Graphics*, 22(3):838–847, Jul. 2003.

[8] Z. Karni, D. Freedman, and C. Gotsman. Energy-based image deformation. In *Proceedings of the Symposium on Geometry Processing*, SGP '09, pages 1257–1268, Aire-la-Ville, Switzerland, Switzerland, 2009. Eurographics Association.

[9] X. Lin. Active layout engine: Algorithms and applications in variable data printing. *Comput. Aided Des.*, 38(5):444–456, 2006.

[10] S. Lok and S. Feiner. A survey of automated layout techniques for information presentations. In *SmartGraphics '01: Proceedings of SmartGraphics Symposium '01*, pages 61–68, New York, NY, USA, 2001. ACM.

[11] J. Lumley, R. Gimson, and O. Rees. A framework for structure, layout & function in documents. In *DocEng '05: Proceedings of the 2005 ACM symposium on Document engineering*, pages 32–41, New York, NY, USA, 2005. ACM.

[12] J. Pearl. *Probabilistic reasoning in intelligent systems: networks of plausible inference*. Morgan Kaufmann Publishers Inc., San Francisco, CA, USA, 1988.

[13] M. F. Plass. *Optimal pagination techniques for automatic typesetting systems*. PhD thesis, Stanford University, Stanford, CA, USA, 1981.

[14] E. Schrier, M. Dontcheva, C. Jacobs, G. Wade, and D. Salesin. Adaptive layout for dynamically aggregated documents. In *IUI '08: Proceedings of the 13th international conference on Intelligent user interfaces*, pages 99–108, New York, NY, USA, 2008. ACM.

[15] B. T. Vander Zanden, R. Halterman, B. A. Myers, R. McDaniel, R. Miller, P. Szekely, D. A. Giuse, and D. Kosbie. Lessons learned about one-way, dataflow constraints in the garnet and amulet graphical toolkits. *ACM Trans. Program. Lang. Syst.*, 23(6):776–796, 2001.

APPENDIX

A. INFERENCE COMPLEXITY

The forward pass given by equations (21)-(23) dominates the asymptotic complexity of PDM inference. A naive computation of the table $\Psi(\mathcal{A}, \mathcal{B}, T)$ has asymptotic complexity $O(|\Omega||c'|^2)$ (where $|\Omega| = |\bigcup_i \Omega_i|$) since we must loop over all sets $\mathcal{A}, \mathcal{B} \in c'$ with $\mathcal{A} \supseteq \mathcal{B}$ and over all $|\Omega|$ distinct templates in the library. Recall that c' is the set of all legal content allocations (Section 3.1).

We observe that we can effectively bound the set $\mathcal{A} - \mathcal{B}$ that represents the content allocated to a page. This assumption implies that we do not need to loop over all legal subsets $\mathcal{A}$ and $\mathcal{B}$ in building $\Psi(\mathcal{A}, \mathcal{B}, T)$, but only those that are *close enough* so that the content $\mathcal{A} - \mathcal{B}$ can reasonably be expected to fit on a page. In general, for each $\mathcal{A}$ the allowed $\mathcal{B}$'s are in a neighborhood $\mathcal{N}_{\mathbf{f}}(\mathcal{A}) = \{\mathcal{B} : \delta(\mathcal{A} - \mathcal{B}) \leq \mathbf{f}\}$. The function $\delta(\mathcal{A} - \mathcal{B})$ returns a vector of the counts of various page elements in the set $\mathcal{A} - \mathcal{B}$. $\mathbf{f}$ is a vector that expresses what it means to be *close* by bounding the numbers of various page elements allowed on a page. For example

we may set $\mathbf{f} = [100 \text{ (lines)}, 2 \text{ (figures)}, 1 \text{ (sidebar)}]^\mathsf{T}$. This will eliminate an allocation comprising 110 lines of text and 1 sidebar. This page capacity bound improves the complexity of computing $\Psi(\mathcal{A}, \mathcal{B}, T)$ to $O(|\Omega||c'|)$.

Once the table $\Psi(\mathcal{A}, \mathcal{B}, T)$ has been computed, the computation of $\Phi_i(\mathcal{A}, \mathcal{B})$ for each i is $O(|\Omega_i||c'|)$ since the maximization over templates only occurs over a sub-library containing $|\Omega_i|$ templates. The computation of $\Phi_i(\mathcal{A}, \mathcal{B})$ for all i is thus $O(\max_i(|\Omega_i|)|c'|\hat{I})$, where $\hat{I}$ is the estimated page count for the forward pass. Since $\hat{I} \propto |c|$ this complexity may also be expressed as $O(\max_i(|\Omega_i|)|c'||c|)$.

Once all the tables $\Phi_i(\mathcal{A}, \mathcal{B})$ have been computed, we recursively compute $\tau_i(\mathcal{A})$ for *all* $\mathcal{A}, i$ with $\tau_0(\mathcal{A}) = \Phi_0(\mathcal{A}, \emptyset)$. This computation has complexity $O(|c'||c|)$ since we loop over every $\mathcal{A}$ and all pages. Thus, overall asymptotic algorithm complexity of PDM inference (under the mild assumption that page capacity is bounded) is $O(|\Omega||c'||c|)$ or $O(|\Omega||c|^3)$, since[4] $|c|^2 > |c'|$.

If we make a further assumption that $\Phi_i(\mathcal{A}, \mathcal{B})$ is independent of i (i.e. templates for all pages are drawn from the same template library) then $|\Omega_i| = |\Omega|$ and $\Phi_i(\mathcal{A}, \mathcal{B}) = \Phi(\mathcal{A}, \mathcal{B})$. Thus the computation of $\Phi(\mathcal{A}, \mathcal{B})$ now has complexity $O(|\Omega||c'|)$. This does not change overall asymptotic complexity since the computation of $\tau_i(\mathcal{A})$ for *all* $\mathcal{A}, i$ is still $O(|c'||c|)$ or $O(|c|^3)$.

However, if we are seeking to automatically determine optimal page count we only need to compute and store $\tau(\mathcal{A}) = \max_{i \geq 0} \tau_i(\mathcal{A})$ since, from (15), we have

$$\max_{i \geq 0} \tau_i(c) = \max_{i \geq 1} \left\{ \tau_0(c), \max_\mathcal{B} \Phi(c, \mathcal{B}) \tau_{i-1}(\mathcal{B}) \right\}$$

$$= \max \left\{ \tau_0(c), \max_\mathcal{B} \Phi(c, \mathcal{B}) \underbrace{\max_{i \geq 0} \tau_i(\mathcal{B})}_{\tau(\mathcal{B})} \right\}$$

If we allow $\mathcal{B} = \emptyset$ above (so that $\tau_0(c) = \Phi(c, \emptyset)$), define $\tau(\emptyset) = 1$ and substitute $\mathcal{A}$ for c we have the much more succinct general recursion for the computation of $\tau(\mathcal{A})$ and page count $Pg(\mathcal{A})$

$$\tau(\mathcal{A}) = \max_\mathcal{B} \Phi(\mathcal{A}, \mathcal{B}) \tau(\mathcal{B}) \tag{33}$$

$$\mathcal{B}^* = \operatorname*{argmax}_\mathcal{B} \Phi(\mathcal{A}, \mathcal{B}) \tau(\mathcal{B}) \tag{34}$$

$$Pg(\mathcal{A}) = 1 + Pg(\mathcal{B}^*) \tag{35}$$

with $Pg(\emptyset) = 0$. Thus the complexity of computing $\tau(\mathcal{A})$ for *all* $\mathcal{A}$ is $O(|c'|)$. Overall algorithm complexity now becomes $O(|\Omega||c'|)$ or $O(|\Omega||c|^2)$. In fact, we can generalize these asymptotic results even more, to support the common case that templates for even and odd pages are drawn from distinct libraries. In this case we need to compute and store $\Phi_{odd}(\mathcal{A}, \mathcal{B})$, $\Phi_{even}(\mathcal{A}, \mathcal{B})$, $\tau_{odd}(\mathcal{A})$ and $\tau_{even}(\mathcal{A})$. The additional computational complexity does not grow with content and so has no impact on asymptotic inference complexity.

Finally, if we make the assumption that content follows a linear ordering (i.e. text and figures are organized and allocated within a single flow order), then $|c'| = |c| + 1$ (the +1 is to include the empty set). This assumption implicitly means that a figure that appears after a text block must be allocated immediately after it. In practice this forces figures to appear on the same page or on the next page instead of allowing them to float freely. This extra assumption means that the best case complexity of PDM inference is $O(|\Omega||c|)$. Thus in the case of linear content ordering, finite sub-libraries and bounded page capacity, the task of infering the optimal document is linear in content and the size of the template library.

---

[4] $|c|^2 = (|c_l| + |c_f|)^2 > |c'| = (|c_l| + 1)(|c_f| + 1)$ as $|c| \to \infty$

Building Table Formatting Tools

Mihai Bilauca
CSIS Department
University of Limerick
Limerick, Ireland
mihai@bilauca.net

Patrick Healy
CSIS Department
University of Limerick
Limerick, Ireland
patrick.healy@ul.ie

ABSTRACT

In this paper we present an overview of the challenges to overcome when developing table authoring tools, including a review of logical table models, typographical issues and automated table layout optimization. We present a Table Drawing Tool prototype which implements an automated solution for the table layout optimization problem for tables with spanning cells using a mathematical modelling method. We report on the performance improvements of this new optimization method compared to previous solutions.

Categories and Subject Descriptors

I.7.2 [**Document and Text Processing**]: Document Preparation—*Format and notation, Photocomposition/typesetting*; D.3.3 [**Programming Languages**]: Language Constructs and Features—*Constraints*

General Terms

Algorithms,Experimentation,Languages

Keywords

table layout, constrained optimization

1. INTRODUCTION

Tables are a simple yet efficient way of presenting information while preserving the logical relationship between the data items. Thus, it is no surprise that they are widely used and well supported in all modern document authoring tools.

Authoring tables is hard and the need to improve the table layout tools has been highlighted for many years. As WYSIWYG tools have become increasingly popular they provide powerful features for controlling the presentation parameters. However, WYSIWYG tools provide limited or no features for controlling the logical table structure. The advantages of separating the logical structure of a table from its physical layout have already been identified[6, 17, 18]. When

editing a document, the author is primarily concerned with the document's logical structure and only at a later stage with its presentation. Similarly, the author needs to define the logical table structure independent of its presentation. Using WYSIWYG tools the authors face difficult tasks when the logical structure of a table must be changed. For example, when a short table is too wide to fit in the available space, the recommended solution is to swap the table headers and present the information in a vertical manner. With WSYWIG tools the author has to manually perform this task by copying, moving or re-entering content. This can be time consuming and at a high risk of introducing errors. Using a tool that allows the author to alter the logical structure of the table this task is simply a matter of selecting the horizontal and vertical headers.

The common purpose of tables is to represent logical relationships between data items. Usually, a table is represented as a two dimensional rectangular grid with rectangular cells arranged in rows and columns. There are cases when using other shapes will show these relationships in a more meaningful way. Figure 2 shows an example of a *circular table* where the cells are not of a rectangular shape and also there are no rows or columns. It is obvious that in this case a new range of typographical problems need to be addressed.

In this paper we present an overview of the challenges to overcome when developing table authoring tools. We review the logical table models proposed in the recent years and the typographical issues which need to be addressed when developing a Table Drawing Tool prototype: parameters propagation in a hierarchical document structure, line styles and line intersection for drawing cells or inner tables, table layout elements.

In this context we discuss automated table layout optimization. Often, tables can be larger than the space available and the user is required to find a layout that respects various aesthetic criteria, i.e. minimize table height, equal column widths, fixed size for some columns widths or row heights. To find this layout the user has to repetitively change cell configurations but each change can trigger a chain of changes in other cells. This is especially relevant for tables that contain *spanning cells*. When a cell spans a number of columns, any change in its configuration will affect all the cells in the columns that it spans. Even more relevant is the case when cells span over a common set of columns. Any change in a spanning cell will affect not only the columns it spans but also, in a cascading effect, all the cells in the common columns. We present a mathematical programming method for solving this problem. As this model outperforms previously published models[4] where tables with maximum 1,600 cells (40x40) have been tested, the increased performance allows us to report on results for even larger cases - tables with 3,600 cells(60x60).

Outline of the paper. In Section 2 we review the logical table models and describe the typographical issues in developing a table drawing prototype. In Section 3 we describe a solution for the table layout optimization problem giving examples with OPL. In Section 4 we show our experimental results and conclusions in Section 5.

2. DRAWING TABLES

Drawing tables can easily be underestimated but the problems that need to be addressed in this process are the source of the large number of inaccuracies and errors that affect all the current tables tools[5]. In this section we present a prototype called Table Drawing Tool (TDT) and outline the typographical problems that need to be resolved for accurate and aesthetic representations of tables. Research has shown that table tools should clearly separate the logical structure from the presentation, that is table topology rules (arrangement and ordering of data items) and table typography (visual parameters).

Therefore, in this section we study the table authoring problem by first reviewing in Section 2.1 the proposed logical table models and in Section 2.2 we study table presentation problems part of the development of a Table Drawing Tool prototype.

2.1 Logical Table Models

Beach[3] defines the logical table model as a document object as defined by the Tioga document manipulation system. Each document object has a set of dimensions and it must know how to format itself. Thus, it must provide a function for determining its layout and a function for rendering the object. The table model proposed by Beach separates the *table topology* from the *table geometry*. Beach points out that to represent tables a hierarchical model is not ideal as tables are not always hierarchical. Therefore he defines the table topology - the arrangement or ordering of data items - based on a grid. The table geometry is computed based on the dimensions of the table entries and the table topology.

Cameron[6] studied the mental processes involved in authoring tables. He identifies three processes: *structure editing*, *content editing* and *visual editing*. During structure editing, the authors define the logical dimensions of the data, its classifications and other complex operations such as data transposition or re-arrangement of classification data entries. During content editing authors simply alter the individual data entries which can be numeric, textual, graphics or mathematical expressions. Formatting parameters such as text font, colours, rules and alignment are set during the visual formatting process.

In her 1989 study, Christine Vanoirbeek[17] shows through examples the importance of the separation between the logical and the presentation layers. She suggests a logical model where tables are a collection of data items logically connected in a multidimensional space. Each dimension is divided in *rubrics* which classify data items based on some established criteria. The physical layout of a table is a projection from the multidimensional logical space into a two-dimensional visual space. This model is then further expanded by the introduction of *sub-rubrics* and *merged items*.

A sub-rubric is utilized as a way to sub-divide the logical dimension. Merged items are used for data items which are connected to multiple rubrics. The table model is then demonstrated in a prototype for formatting an expandable document model. The document's structure is defined using a language based on attribute grammars. This allows formatting parameters to be associated to the logical components of the document.

Wang[18] proposed a table model that separates the layout structure from the logical structure. The layout structure defines the tabular topology and the typographic style. By using well understood mathematical expressions the logical model proposed by Wang[18] separates its representation and implementation from its mathematical formulation. An abstract table is defined as a tuple $< C, \delta >$ where C is a set of categories (a set of labelled domains) and δ is a mapping of data entries in the table. Wang also divides the logical operations on abstract tables in *table operations*, *category operations* and *label and entry* operations. Table operations modify the number of categories in a table thus affecting the number of dimensions and possibly its size. Category operations alter sub-categories (sub-headings) of a table and label and entry operations deal with the content of the table. As the logical model is defined using unordered sets the ordering of categories and entries is seen as a topology concern and it is not included in the abstract model.

Parnas expands the use of tables from simply organizing data to providing meaningful expressions which can be evaluated and validated. He shows how *tabular expressions* - mathematical expressions in tabular format - can be used to precisely document software systems. In the logical model proposed by Jin and Parnas[10] indexation is used to separate the table semantics from the table appearance. Tables are defined using mathematical expressions, as an indexed set of grids and each grid is an indexed set of expressions. Thus a table is a triple $< GS, I, x >$, with GS an indexed set of grids, an index set I and a mapping x between GS and I. A grid is a triple $< SetExp, J, y >$ that is a set of expressions $SetExp$, an indexed set J and a mapping between $SetExp$ and J. To precisely define the meaning of the tabular expression a restriction schema and one or more evaluation schemas are provided.

It has been accepted by all authors that it is hard or almost impossible to define a general logical model that covers all types of tables without clearly defining what a table is. This is not easy, as Vanoirbeek explains, because it is not easy to distinguish between tables and other components of a document that follow a tabular pattern.

It can be observed that while the initial table programs were only dealing with basic presentation of tables research has highlighted the need for an abstract logical table model in order to deal with more complex structural operations. The logical table model has evolved during the years from a simple grid to multidimensional logical structures such as *trees* suggested by Vanoirbeek, *sets* proposed by Wang and indexed sets of mathematical expressions which can be evaluated as proposed by Parnas.

2.2 Prototype of a Table Drawing Tool. Typographical Problems

In this section we present a Table Drawing Tool prototype - TDT - which has been developed as part of a larger project, a tabular expressions tool developed at University

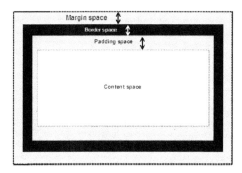

Figure 1: The *box* element defines margins, borders, padding and content spaces

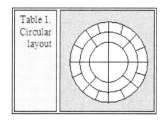

Figure 2: A circular table with caption placed to the left

of Limerick. *Tabular expressions* are represented by tables where data items are well understood mathematical expressions. Tabular expressions add another level of complexity to tables by defining the meaning of the data items. Thus, tables can be evaluated and validated to ensure that, for example, the data values in the header are unique or data ranges are completely and correctly defined. They have been used in practice to document software systems with a high level of accuracy in order to improve quality and reduce development costs[14]. The tabular expressions tool uses the logical table model defined by Parnas where tables are defined as an indexed set of grids and a grid is an indexed set of expressions as explained in section 2.1. This tool has four independent modules: the *table input tool* that deals with problems related to data input, a *kernel module* which handles the logical model, an *evaluation module* which is responsible for evaluating expressions for completeness and correctness and a *table output (drawing) tool - TDT*. TDT is concerned only with drawing tables. It is not aware of the data input methods or the logical table model nor it should be. It takes input data in XML format and it draws tables to a *rendering device* depending on the *table layout* using a selected *table layout optimizer*. This clear separation of concerns between various modules allows us to modify any module independently without affecting or relying on the functionality of the other modules. Thus, we can define any table layout, that can be drawn on any rendering device and apply any table layout optimization technique as long as a predefined set of input/output functions are provided. TDT currently draws *rectangular layout* tables on a *PDF device*. The layout is optimized by an *OPL optimizer* which finds the minimum height layout for a given page width.

A *device* is an abstract object that must provide functions to assist the rendering process: open and close the device, set colours, fill colours, line styles with all the properties described above, font settings (font family, style, size), vertical and horizontal alignment, text spacing (word, character and line spacing), drawing bezier curves when iregular shapes are required and insert images. It is important for any device object to indicate if it supports paging. For example, the table layout algorithm should be able to split tables across pages and it needs information regarding the page size and its default settings such as margins, borders, paddings and any header and footer areas. The device must also be able to load its default settings from the XML file provided as input. TDT uses the iText[12] open source PDF library to generate PDF documents. This devices loads specific settings for

PDF documents such as the title, author, keywords, initial page index and initial zoom.

Why a hierarchical model

Beach acknowledged that "Table entries can be arbitrary document objects such as text, illustrations, mathematical notations or other tables" [3](page 5-2). Therefore a table layout tool needs to be as powerful as a page layout tool. To deal with different types of objects TDT implements a hierarchical structure, similar to the CSS model[15], where each *container* has a parent and an optional list of inner containers. The root container holds document wide settings such as custom lines styles, numbering styles, table caption styles and other global settings. This model allows:

hierarchical propagation of typographical parameters. Line style, background colour and pattern, font size, etc. can be set at the table level and then all the cells inherit these settings unless they are set specifically for that cell.

layered control of typographical parameters. When objects overlap the user needs to control what part of an object is visible. In the case of a table with a rectangular layout, where settings are applied at table, row, column or cell level the user needs to prioritize which settings will take priority. This can be achieved by organizing settings in layers and simply changing their order. In CSS specifications rows are prioritized over columns[15] but we believe that this option should be provided to the user and not imposed by the system.

It is each container's responsibility to format itself; each container must implement *computeLayout* and *render* on *device* functions. For example, rectangular tables implement a layout algorithm that organizes cells in rows and columns, uses the global line settings, caption settings and numbering styles. Their *render* function uses formatting parameters such as colour, font family and size, etc. to render the content on a *device*.

The Box

For drawing tables, TDT implements a *box* model similar to the CSS model where *margin*, *border*, *padding* settings can be set for each top, bottom, left and right sides displayed in Figure 1. There are many published table tools prototypes starting with a special program running on an IBM 7090 computer in 1962, *TABPRINT* mentioned by Bartnett, *Tbl* in 1979, Beach in 1985, Vanoirbeek in 1989 but only recently WYSIWYG tools, browsers and specialized Desktop Publishing software deal this level of typographical detail.

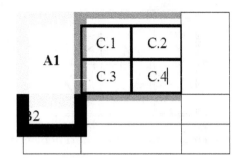

Figure 3: Borders in Open Office.org 3

This flexible model creates a number of problems that need to be addressed:

table layout In TDT a table has two components: a *caption* and a *layout*. The caption has a *label, number* and *caption text*. The number is formatted using any of the global numbering styles. The label and number are especially relevant when, due to space constraints, inner tables are displayed outside the parent table. In this case the inner table's label is used as reference in the parent table. The table's layout can be of any type: *rectangular, circular* or any other shape as long as it implements the *computelayout* and *render* functions. The caption can be placed on any side of the layout as shown in Figure 2;

laying containers, fixed and floating positioning As table entries can contain any type and number of objects a drawing algorithm needs to determine how to layout these objects. Knuth[11] documented the problem in detail and his research unearthed subsequent difficult problems such as paragraphing. The main idea (which is also implemented in TDT) is to lay floating (non-fixed) containers on lines with a maximum width given by the available space then vertically align these containers on top, centre, bottom or text base lines. The complexity of this problem is increased when some containers have fixed positions which makes harder the aesthetical arrangement of floating containers. Marriott *et al.*[13] discusses in detail the problem of laying out floating containers in multi-column documents.

lines: styles and joins Lines are defined by their *pattern, line cap, line join* and *miter limit*[1]. The pattern is defined as a *dash array* and a *phase*. The dash array specifies the lengths of alternating dashes and gaps. The phase specifies the starting point of the pattern. When two lines meet at an angle the join point needs to be aesthetically drawn. Also, there is a need to define how a continuous line will start and end be defining the *line cap*. The *miter limit* is a maximum ratio between the miter length and the line width.

borders Without a clear definition of the lines styles, dealing with borders can be a major cause of frustration when authoring tables as borders. A number of problems need careful consideration:

border space should the space required by the border width be added to the box dimensions, taken

from the content space or it should be divided equally between the two? If the border width is divided equally, then the case with an odd border width needs to be clarified.

border alignment Because cells in a rectangular table are aligned based on a grid layout the join between borders of different widths, colours and styles must be established;

border priority if borders of different colours overlap, which border will take priority: top border over the bottom border, the border of the last modified cell or based on a user setting.

In Figure 3, a table generated with Open Office shows a number of problems: the text in cell *B2* is drawn over by the border because half the border space is taken from the content space. However this is not consistent with the way the horizontal border where the border is placed is taken inside the box for the top border and outside the box for the bottom border. Also, the priority is given to cell *B2* over the cell *A1* but only for the left and right borders.

margins Dealing with margins needs careful consideration. When two adjacent boxes have margins the drawing algorithm needs to decide if the margins should be added or they should merge by selecting the largest margin value. Also, the algorithm must decide how to deal with margins for the case of inner boxes;

coordinates space In any drawing system it is essential to define the co-ordinate space, the origin, axes and the unit in use. For convenience we selected for TDT a coordinate system where the origin is in the bottom-left corner and the using units are defined as points (1/72 of an inch) in both horizontal vertical direction.

text direction Text is normally displayed from left-to-right but other text directions are required depending on the culture: right-to-left or top-to-bottom. This is particulary relevant when the text needs to be rotated to reduce horizontal space or in determining the number of text configurations.

equations TDT allows users to insert equations provided in MathML format. *WebEQ*, a specialized library for rendering mathematical expressions developed by Design Science is used to render each equation as an image which is then inserted in the table layout.

images When inserting images TDT must translate their dimensions, which are usually reported in pixels in its own co-ordinate system and units (point)

3. MODELING THE TABLE LAYOUT OPTIMIZATION PROBLEM FOR TABLES WITH SPANNING CELLS

Often, tables are wider than the available space and the user is required to adjust the configuration of each cell until it finds a layout that satisfies all the visual and space constraints. This process is time consuming and prone to errors as the designer may unintentionally and unconsciously change the meaning of the data. A problem that has been

Table A.1 Optimized layout for a table with spanning cells containing text for a fixed page width of 190pt

This document is the	second published release	of the Software Requirements	of the A-7E Aircraft. The	first release
published	new approach to specifying requirements for			form
document has	been	perhaps the	most	successful of the publications
of NRL's Software Cost	Reduction project in terms of the	interest generated and the	number of copies	requested since
its introduction. In spite	of its success (in	a sense, because of it)	the specification	in many details over the years.
This is not the	result of flaws in its design,	but the fulfillment of	creators'	vision that
the requirements	be a "living document;"	i.e., that it	would serve as	the primary reference document for system
designers, as well as the	authoritative	"test to" document for program validation,		
document has served	these purposes as	it has changed over	the years as	requirements became better
Further, since the document	is intended to serve as a	model	document, we have felt	free

|◄——————— 190 pt ———————►|

Figure 4: Table layout with minimum height for a given page width and spanning cells

Table B. A layout with a nested table

This document is	the second published	release				the
A-7E Aircraft. The first release published	in November 1978, introduced a	new approach				for real-time embedded systems in
the form of an engineering	model. That document	has	been perhaps the most	successful of the publications of NRL's Software		Cost Reduction
		I requested since its introduction. In	spite of its	success (in a		
		in many	details over the years. This is	not the result of flaws in		
project of its creators' vision	that the requirements	should be a "living document;" i.e.,				for system designers, as well as

|◄——————— 250 pt ———————►|

Figure 5: Table layout with nested tables

various rectangles or *configurations* with height h and width w. Each cell i,j with $1 \leq i \leq m$ and $1 \leq j \leq n$ has a configuration set C_{ij} of the form $\{(h_{ij}^k, w_{ij}^k) \mid 1 \leq k \leq |C_{ij}|\}$. Finding a layout $\mathcal{L}$ for table $\mathcal{T}$ is finding for each cell i,j a configuration from $C_{i,j}$ with the index $l_{i,j}, 1 \leq l_{i,j} \leq |C_{ij}|$ to be used when displaying the table. The cell height becomes $h_{i,j} = h_{i,j}^{l_{i,j}}$ and the cell width is $w_{i,j} = w_{i,j}^{l_{i,j}}$. The row height h_i is the maximum height for that row's cells and the column width w_j is the maximum width.

$$h_i = \max_j h_{ij}$$

$$w_j = \max_i w_{ij}$$

Therefore, the layout's height is $height(\mathcal{L}) = \sum_{i=1}^m h_i$ and the layout's width is $width(\mathcal{L}) = \sum_{j=1}^n w_j$.

From a layout perspective, when a cell i, j spans a number of columns there is an interdependency between the cell's width $w_{i,j}$ and the width of the columns it spans, i.e. the cell's width must be equal to the sum of the spanned columns widths. Similarly, when a cell spans a number of rows the cell's height $h_{i,j}$ must be equal to the sum of the spanned rows heights.

$$h_{i,j} = \sum_{r=i}^{i+S_{i,j}^h} h_r \qquad (1)$$

$$w_{i,j} = \sum_{c=j}^{j+S_{i,j}^w} w_c \qquad (2)$$

where $S_{i,j}^h$ is the number of additional spanned rows and $S_{i,j}^w$ the number for additional spanned columns by each cell i,j.

When tables contain multiple spanning cells, these cells can span over a disjoint sets of columns or, when the spanning cells are on different rows, the spanned columns can overlap. This adds another level of complexity to the problem. In Figure 4 cell 2,2 spans over three columns while cell 8,3 also spans three columns with two of the spanned columns in common. It is easy to observe the interdependency between these 2 spanning cells. When solving the table layout problem, any change in the configuration of cell 2,2 or 8,3 will directly affect not only the configuration of the columns each cell spans but indirectly the configuration of the columns spanned by both cells.

The complexity of the problem is further increased in the case of nested tables. To minimize the layout of the 4x4

studied as part of automating this process is the *table layout optimization problem*, finding the table with the minimal height for a given page width. For tables that contain text, it has been demonstrated that this problem is NP-hard [18, 2], since the table shape can change when any cell configuration is changed; an exponential number of possible table shapes may result. This type of problem requires significant computational time to find the exact solution. A number of heuristic methods have been recently developed by Hurst and Marriott[9, 8] for cases where a satisfactory layout is required in a short period of time, i.e. in dynamic document presentation such as Internet browsers and mobile devices. A comprehensive review of table formatting methods has been published by Hurst, Li and Marriott [7].

There are many cases where knowing the precise solution to the table layout problem is desirable. Finding the exact layout with the minimum table height can reduce printing costs. When printing a book, the computational time required to find a table layout that can be printed in one page can be offset against the costs with additional design, printing and lost of clarity if the table is printed on multiple pages.

In this section we present an optimization model for the table layout problem for tables that contain *spanning cells*. Figure 4 shows a table with 5 columns and 10 rows, 2 spanning cells and a variable number of text configurations. Using our optimization model we determined the layout with the minimum height when the page width was constrained to 190 points. It can be observed that any change in the configuration of any cell will cause a chain of changes which will result in a layout with at least the same height or larger.

For simple tables (without spanning cells), the *table layout optimization problem* is to find the layout $\mathcal{L}$ of a table $\mathcal{T}$ with minimum $height(\mathcal{L})$ such that $width(\mathcal{L}) \leq W$ where W is a given page width. A similar definition is given by Anderson and Sobti [2]. Table $\mathcal{T}$ is a matrix with m rows and n columns that contains $m \times n$ cells. The content of a cell can have many configurations, i.e. if the cell contains text then it can be split on consecutive lines and therefore it generates

Ex pe ndi tur e	Income	
-	-	-
-	-	-

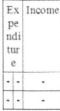

(a) Glue is required for narrow columns.

Expenditure	Income
-	-
-	-

(b) The *glue* is displayed in a gray background

Figure 6: A spanning cell over narrow columns

Expenditure		Income
€999,999	€888,888	-
	-	-

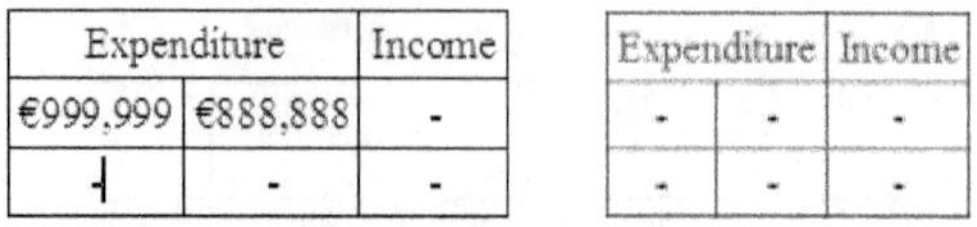

(a) Cell 1,1 spans wide columns

Expenditure	Income	
-	-	-
-	-	-

(b) Cell 1,1 spans narrow columns

Figure 7: Table with spanning cell

table displayed in Figure 5 as a solution of the table layout problem, all the cell configurations are required. Computing the configurations for cell **3,3** cannot be done in isolation as the inner table's configurations are directly related to the configuration of the parent table. Minimizing the inner table's height and using that configuration when minimizing the parent table's height is not sufficient as any change will affect both tables. Therefore, we suggest as a possible solution merging the inner tables recursively in the parent table and finding a solution for this larger problem. For the table presented in Figure 5 we now have to minimize the layout of a 6x6 table. If the inner table has a caption and or footer notes the merging process becomes even more difficult.

An interesting case is presented In Figure 6(a) where cell **1,1** has only one configuration which is wider than its spanning columns. When computing the width of a spanning cell two cases must be considered:

case 1 when the total width of the spanned columns is larger than the width of the spanning cell as presented in Figure 7(a);

case 2 when the total width of the spanned columns is less than the width of the spanning cell as presented in Figure 7(b).

In the first case, as the widest cell configuration is selected for the spanning cell, the responsibility of drawing it using the dimensions of the spanned columns is left with a post-processing drawing program.

For the second case we introduce the *glue* variable, similar to the model introduced by Knuth when determining the required character and word space for aligning text in paragraphs[11]. The glue represents the amount that the spanned columns need to stretch in order to have a total width equal with the width of the spanning cell. In Figure 6(b) the glue is presented in a gray background.

3.1 Integer Programming Definition

We first define the table layout problem for simple tables as a basic mixed-integer optimization problem. We determine the value of binary variables associated with each cell configuration such that the table height is minimized, the table width is less than the given page width W and for each cell i,j only one configuration is selected.

$$\text{minimize} \sum_{i=1}^{m} \max_{j=1}^{n} h_{i,j,k} \cdot x_{i,j,k} \qquad (3a)$$

$$\text{subject to}$$

$$\sum_{j=1}^{n} \max_{i=1}^{m} w_{i,j,k} \cdot x_{i,j,k} \leq W \qquad (3b)$$

$$\sum_{k} x_{i,j,k} = 1, \qquad \forall i,j \qquad (3c)$$

$$\text{where } x \in \{0,1\}, \quad 1 \leq k \leq |C_{i,j}|$$

In our previous work[4] we showed that the running time of this basic model is unsatisfactory for larger tables. We now present an improved strategy where we use additional binary variables for selecting row heights and column widths from a predetermined set. Then, we select only those cell configurations that can fit in the selected column widths and row heights. We show in Section 4 that the significant performance improvement of this strategy allows us to experiment with even larger tables. As the objective function of the model is to minimize table height this will ensure that the shortest and therefore the widest cell configurations are selected. Starting from the set of cell configurations $C_{i,j}$ we determine for each column j the set of possible column widths $\mathcal{W}_j$ as the union of non-spanning cell width values. If the set of column widths $\mathcal{W}_j$ is:

$$\mathcal{W}_j = \{w_{i,j}^k \mid 1 \leq i \leq m, 1 \leq k \leq |C_{i,j}|, S_{i,j}^w = 0\}$$

then the set of row heights $\mathcal{H}_i$ is defined as

$$\mathcal{H}_i = \{h_{i,j}^k \mid 1 \leq j \leq n, 1 \leq k \leq |C_{i,j}|, S_{i,j}^h = 0\}$$

The width of a column w_j is now determined by finding an index c_j with $1 \leq c_j \leq |\mathcal{W}_j|$ while the height of each row h_i is determined by selecting an index r_i with $1 \leq r_i \leq |\mathcal{H}_i|$:

$$w_j = \mathcal{W}_j^{c_j}, \quad h_i = \mathcal{H}_i^{r_i}$$

The table layout problem is to minimize the table layout $height(\mathcal{L})$ by finding for each row the height with index r_i, for each column the width with index c_j and for each cell i,j the configuration with the index $l_{i,j}$ with $1 \leq l_{i,j} \leq |C_{ij}|$, such that

$$width(\mathcal{L}) \leq W$$

$$h_{i,j}^{l_{i,j}} \leq \mathcal{H}_i^{r_i}$$

$$w_{i,j}^{l_{i,j}} \leq \mathcal{W}_j^{c_j}$$

This integer programming model requires the width of a cell to be less or equal the selected width for the column. In tables with spanning cells, there are further constraints: the width of a spanning cell must be equal with the sum of widths for the columns it spans. If the cell's width is smaller the width adjustment is performed by the drawing algorithm. When the selected cell width is larger than the sum of column widths then the *glue* value is computed for each spanned column. Thus, for a spanning cell `i,j` equations 1 and 2 above are changed to:

$$h_{i,j} = \sum_{r=i}^{i+S_{i,j}^h} (h_r + g_r^h) \qquad (4)$$

$$w_{i,j} = \sum_{c=j}^{j+S_{i,j}^w} (w_c + g_c^w) \qquad (5)$$

where w_c is the width of column c, h_r the height of row r, g_c^w is the horizontal glue of column c and g_r^h is the vertical glue for each row r.

The layout width $width(\mathcal{L})$ and the layout height $height(\mathcal{L})$ are now defined as:

$$width(\mathcal{L}) = \sum_{j=1}^{n} (w_j + g_j^w) \qquad (6)$$

$$height(\mathcal{L}) = \sum_{i=1}^{m} (h_i + g_i^h) \qquad (7)$$

Constraint 3c in the integer programming model requires that for each cell `i,j` one cell configuration must be selected. To satisfy this constraint for spanned cells which have no configurations we introduce *dummy cells* having one configuration with the width and height set to 0.

The definition of `TSC` - the integer programming model for the table layout optimization problem for tables with spanning cells is:

$$\text{minimize} \sum_{i=1}^{m} h_i^r \cdot y_i^r + g_i^h \qquad (8a)$$

subject to

$$\sum_{j=1}^{n} (w_j^c \cdot z_j^c) + g_j^w \le W \qquad (8b)$$

$$h_{i,j} \le \sum_{r=i}^{i+S_{i,j}^h} h_r + g_r^h \qquad (8c)$$

$$w_{i,j} \le \sum_{c=j}^{j+S_{i,j}^w} w_c + g_c^w \qquad (8d)$$

$$\sum_k x_{i,j,k} = 1, \forall i,j \qquad (8e)$$

$$\sum_{i=1}^{m} y_i^r = 1 \qquad (8f)$$

$$\sum_{j=1}^{n} z_j^c = 1 \qquad (8g)$$

$$\qquad (8h)$$

where

$$x, y, z \in \{0,1\}, \ 1 \le k \le |C_{i,j}|, \ 1 \le c \le |\mathcal{W}_j|, \ 1 \le r \le |\mathcal{H}_i|$$

For non-spanning cells expressions 8c and 8d become $h_{i,j} \le h_i$ and $w_{i,j} \le w_j$. To simplify the constraint expressions we

Listing 1: TSC - Initial data

```
1  // sets of row/column spanning cells
2  {CellSpan} rowSpan = ...;
3  {CellSpan} colSpan = ...;
4
5  // vector of rows/columns spanned by each cell
6  int Sh[cells] = [<i,j> : s | <i,j,s> in rowSpan];
7  int Sw[cells] = [<i,j> : s | <i,j,s> in colSpan];
8  }
```

define cell dimensions as in expression 10a and 10b. Also, the height for each row and the width of each column are defined using decision variables y and z as in equations 10c and 10d.

$$h_{i,j} = \sum_k h_{i,j,k} \cdot x_{i,j,k} \qquad (10a)$$

$$w_{i,j} = \sum_k w_{i,j,k} \cdot x_{i,j,k} \qquad (10b)$$

$$h_i = \sum_{i=1}^{m} h_i^r \cdot y_i^r \qquad (10c)$$

$$w_j = \sum_{j=1}^{n} w_j^c \cdot z_j^c \qquad (10d)$$

where $h_i^r \in \mathcal{H}_i$, $w_j^c \in \mathcal{W}_j$

3.2 TSC Model with OPL

In this section we give examples for OPL, the Optimization Programming Language [16]. OPL is a modelling language designed for solving combinatorial optimization problems. Compared with other programming languages that support mathematical programming and constraint programming OPL has a major strength given by its syntax. As OPL shares many syntax features of traditional mathematical programming languages such as AMPL or GAMS, problems can be formulated in a language similar to their algebraic notation. For our solution we use the industrial implementation of OPL from IBM ILOG. The listings below show only the relevant parts of the model. The full code is available upon request.

We start our TSC model with two variables `rowSpan` and `colSpan` which represent the sets of row and column spanning cells as tuples `<i,j,s>` where s is the number of rows/columns spanned by cell `i,j` as presented in Listing 1. We build two vectors `Sh[<i,j>]` and `Sw[<i,j>]` where we store for each cell the number of additional rows and respectively columns it spans, i.e. for a non spanning cell `Sw[i,j]` contains 0 and if the cell spans 2 columns its value is 1.

It is important to note that the width of column spanning cells is ignored when determining the minimum width for each column. In a similar way, the height of row spanning cells is ignored when computing the minimum row heights as displayed in Listing 2.

We define binary decision variables as presented in Listing 3. The selected cell configuration is stored in `x[]` while `y[]` and `z[]` indicate the selected height and width for each row and column. Two vectors `Gh[]` and `Gw[]` store the vertical and horizontal glue.

Table layout dimensions are computed as presented in Listing 4. The row height is computed as defined by equa-

Listing 2: TSC - computing the set of column widths and row heights

```
1  {Pair} colWset={<j,k.w> | <i,j,k> in configs :
2                     Sw[<i,j>]==0};
3
4  {Pair} rowHset={<i, k.h> | <i,j,k> in configs :
5                     Sh[<i,j>]==0};
6  }
```

Listing 3: TSC - Decision variables

```
1  // cell configuration selector
2  dvar int x[configs] in 0..1;
3  // row height selector
4  dvar int y[rowHset] in 0..1;
5  // column width selector
6  dvar int z[colWset] in 0..1;
7  // row glue
8  dvar int Gh[rows] in 0..maxint;
9  // column glue
10 dvar int Gw[columns] in 0..maxint;
11 }
```

tion 10c by selecting a height value from the set of row heights. The table width and height are computed, as defined by equation 6 and equation 7, by adding the horizontal and vertical glue.

In Listing 5 constraint `ct2_1` ensures that only one row height is selected for each row while constraint `ct3` deals with the interdependency between the width of the spanning cell and the width of the spanned columns. For spanning cells ($Sw[<i,j>] > 0$), the width must be less or equal than the sum of all spanned column widths and glue; for non spanning cells, the width must be less or equal with the selected column width.

Constraint `ct3_1` enforces the height for both spanned and spanning cells. In a similar manner `ct3_2` enforces the width.

Constraint `ct4` ensures that at least one configuration is selected for each cell. This constraint is also valid for spanned cells because we defined dummy configurations. Constraint `ct4_1` and `ct4_2` make sure that only one height and width is selected from the sets of possible values.

The TSC model determines the amount of *glue* for each column that needs to stretch in order to match the width of its spanning cell. In the current form the TSC model determines the glue for any of the spanned columns but there is no

Listing 4: TSC - Layout dimensions

```
1  // compute row height
2  dexpr int rowH[i in rows] =
3        sum(<i,h> in rowHset) y[<i,h>] * h;
4  // compute table height
5  dexpr int tableH =
6        sum(i in rows) (rowH[i] + Gh[i]);
7  // compute column width
8  dexpr int colW[j in columns] =
9        sum(<j,w> in colWset) z[<j,w>] * w;
10 // compute table width
11 dexpr int tableW =
12       sum(j in columns) (colW[j] + Gw[j]);
13 }
```

Listing 5: TSC Model constraints

```
1  minimize tableH;
2
3  constraints
4  {
5    ct3_1: //height constraint
6    forall(i in rows, j in columns)
7     if (Sh[<i,j>] > 0)
8         cellH[<i,j>] <= sum(row in i..i+Sh[<i,j>])
9                          (rowH[row] + Gh[row]);
10    else
11        cellH[<i,j>] <= rowH[i];
12   ct3_2: // width constraint
13    forall(j in columns, i in rows)
14     if (Sw[<i,j>] > 0)
15         cellW[<i,j>] <= sum(col in j..j+Sw[<i,j>])
16                          (colW[col] + Gw[col]);
17    else
18         cellW[<i,j>] <= colW[j];
19   ct4:
20    forall(i in rows, j in columns)
21     sum(<i,j,k> in configs) x[<i,j,k>] == 1;
22   ct4_1: // only 1 selected height
23    forall(i in rows)
24     sum(<i,k> in rowHset) y[<i,k>] == 1;
25   ct4_2: // only 1 selected width
26    forall(j in cols)
27     sum(<i,k> in colWset) z[<j,k>] == 1;
28 }
```

Listing 6: TSC - Glue distribution

```
1  ...
2  dexpr int glueW[<i,j,s> in colSpan] =
3        cellW[<i,j>] - sum(col in j..j+s) colW[col];
4
5  constraints
6  {
7   ...
8   ct5: // equal glue
9    forall(<i,j,s> in colSpan)
10    forall(col in j..j+s)
11    Gw[col] <= glueW[<i,j,s>] / (s+1);
12 }
```

control on how much each column is allowed to stretch. One of the aesthetics criteria is to distribute the glue equally for each column. To implement this criterion a new constraint `ct5` can be added to the TSC model as presented in Listing 6 where the total glue `glueW` is equally distributed to all s columns spanned by cell i, j. The glue is defined as the difference between the cell's width and the sum of column widths.

Such a constraint may cause a conflict when determining the minimum height layout especially when the table contains multiple spanning cells on overlapping column sets. If a column is spanned by two cells as in Figure 4 an equal distribution of glue is not possible as the glue is constraint by the two spanning cells. The second spanning cell can have a cell configuration with a smaller height that will not be selected and thus the minimum height layout is affected by the equal distribution of glue constraint.

4. EXPERIMENTAL RESULTS

In this section we present experimental results using the TSC model and we compare its run time with a previously published MIP (Mixed Integer Programming) model [4]. In the MIP model we use a simplified strategy where we first

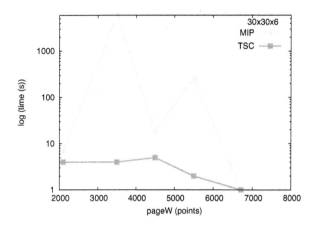

Figure 8: Running time for a 30x30 table and up to 6 words per cell.

find column width configurations from a predefined set, then we select only cell configurations that can fit in the selected column width. The MIP model only works on simple tables.

For our experiments we used IBM ILOG OPL IDE 6.3 with CPLEX engine 12.1 and CP Optimizer 2.3 running on a Windows Vista PC with a 2GHz dual core processor and 2Gb of RAM. By using IBM's ILOG Concert Technology we integrated this OPL model in our Table Drawing Tool prototype which allowed us to perform visual inspection on the computed table layouts. In our experiments we report only on the time required to find the optimal solution excluding the time required to compute cell configurations. *Paragraphing*, determining cell configurations or how to split the text on consecutive lines, is an essential aspect as it impacts on the quality and computational time of these solutions.

Our experiments showed a substantial improvement on the computational time using the TSC model.

Table 1 shows the running time for tables with 400 cells, in increasing size, up to tables with 3,600 cells (20x20 to 60x60) with maximum 6 words in each cell, for both MIP and TSC models. The runtimes are in the format h:mm:ss. $minW$ and $maxW$ represent the minimum and maximum table width, computed from the cell-width configurations. We report the running time for page width constraint to be $minW$ plus 25%, 50% and 75% of the difference between $maxW$ and $minW$. The *summary* line provides information regarding the minimum and maximum table configurations, the number of cell configurations cnf, distinct column widths $\bigcup_j \mathcal{W}_j$ and row heights $\bigcup_i \mathcal{H}_i$.

For a 20x20 table the run time was halved or reduced to under 1s when the layout width was reduced to 75% of its maximum width. The improvement in the run time can especially be observed in the case of larger tables. The TSC model finds the same solutions as the MIP model but only in a fraction of the time. In Figure 8 we show that for a 30x30 table the TSC running time for a page width of 3,500 points was only 9s compared with the running time of the MIP model of 1 hour 39 minutes and 49 seconds. A similar reduction of run time occured on a 40x40 table for a page width of 7,000 points where the running time was reduced from 1 hour 26m 10s for the MIP model to only 2 minutes and 13s for the TSC model.

As the TSC model proved to be much more efficient we

Table 1: Run time for MIP and TSC models.

Table	minW	+25%	+50%	+75%	maxW
	20x20x6				
width	1,320	2,000	2,500	3,000	4,284
height	1,040	600	440	380	200
time MIP	00:01	00:21	00:06	00:27	00:01
time TSC	00:01	00:02	00:02	00:02	00:01
summary	m=1320x1170; M=4284x200; cnf=1382; $\|\bigcup_j \mathcal{W}_j\|$=592; $\|\bigcup_i \mathcal{H}_i\|$=79				
	30x30x6				
width	2,094	3,500	4,500	5,500	6,709
height	1560	770	570	440	300
time MIP	00:07	**1:39:49**	00:19	04:15	00:01
time TSC	00:03	**0:00:09**	00:05	00:03	00:01
summary	m=2094x1780; M=6709x300; cnf=3082; $\|\bigcup_j \mathcal{W}_j\|$=1145; $\|\bigcup_i \mathcal{H}_i\|$=110				
	40x40x6				
width	2,879	5,000	6,000	7,000	9,189
height	2140	1280	780	680	400
time MIP	00:17	01:48	01:57	**1:26:10**	00:01
time TSC	00:04	00:13	00:27	**0:02:13**	00:01
summary	m=2879x2400; M=9189x400; cnf=5546; $\|\bigcup_j \mathcal{W}_j\|$=1908; $\|\bigcup_i \mathcal{H}_i\|$=149				
	50x50x6				
width	3,603	5,200	7,000	9,000	11,670
height	2,790	1,700	1,110	870	500
time MIP	n/a	n/a	n/a	n/a	n/a
time TSC	00:07	01:05	**00:54**	05:10	00:02
summary	m=3603x3000; M=11670x500; cnf=8850; $\|\bigcup_j \mathcal{W}_j\|$=2877; $\|\bigcup_i \mathcal{H}_i\|$=198				
	60x60x6				
width	4,368	7,000	9,000	12,000	13,918
height	3,3660	1,720	1,210	880	600
time MIP	n/a	n/a	n/a	n/a	n/a
time TSC	00:12	01:58	**00:58**	22:36	00:03
summary	m=4368x3600; M=13918x600; cnf=12580; $\|\bigcup_j \mathcal{W}_j\|$=3775; $\|\bigcup_i \mathcal{H}_i\|$=246				

are now able to report the results for larger tables with 2,500 cells (50x50) and even 3,600 cells (60x60). The running time of the MIP model could not be realistically reported for these large tables but using the TSC model we found solutions in acceptable running time as presented in Table 1. For a 50x50 table with 8,850 cell configurations it took only 54s to determine a layout that was constraint to 50% of its maximum width and only 58s for a 60x60 table with 12,580 configurations.

5. CONCLUSIONS AND FUTURE DIRECTIONS

In this paper we reviewed the challenges of building table tools. We highlighted that while WYSIWYG tools have been widely used, supported and developed they lack support for operations at the table structure level. We reviewed the various logical models that have been proposed to date and presentation problems in the context of documenting a table drawing prototype that implements automatic table layout. We describe a mathematical programming solution for the table layout problem with examples in OPL. We conclude that the wide spread of WYSIWYG tools has placed a large emphasis on the presentation of tables and therefore operations at the table structure level are still hard and prone to errors.

Building table tools is hard because tables can contain any type of objects: text, images, mathematical expressions and even other tables. A table layout algorithm has to be at least as powerful as a document layout but, as the space is limited and the presentation rules require stricter placement of objects, the challenges to overcome become even harder. Using a flexible presentation model such as the hierarchical model with *box* elements raises a number of problems which need to be addressed.

The TSC model presented in this paper provides fast results for the table layout problem even for tables with spanning cells. This allows us to report on larger tables with up to 3,600 cells (60x60). As far as we are aware this is the largest table that has been reported so far.

Beyond spanning cells the next most pressing problem to address is the nesting of tables. We have suggested a possible (simplified) approach to this case. This is the subject of ongoing research for us.

6. ACKNOWLEDGEMENTS

We would like to express our gratitude to Prof. David Parnas for motivating much of the content of this paper. This work was supported in part by Science Foundation Ireland under the research programme 01/P1.2/C009, Mathematical Foundations, Practical Notations, and Tools for Reliable Flexible Software.

7. REFERENCES

[1] Adobe Systems Incorporated. PDF reference version 1.7 (6th edition). http://www.adobe.com/, Nov 2006.

[2] R. J. Anderson and S. Sobti. The table layout problem. In *SCG '99: Proceedings of the fifteenth annual symposium on Computational geometry*, pages 115–123, New York, NY, USA, 1999. ACM.

[3] R. J. Beach. *Setting tables and illustrations with style.* PhD thesis, University of Waterloo, 1985.

[4] M. Bilauca and P. Healy. A new model for automated table layout. In *Proceedings of the 10th ACM symposium on Document engineering*, DocEng '10, pages 169–176, New York, NY, USA, 2010. ACM.

[5] M. Bilauca and P. Healy. Table layout performance of document authoring tools. In *Proceedings of the 10th ACM symposium on Document engineering*, DocEng '10, pages 199–202, New York, NY, USA, 2010. ACM.

[6] J. P. Cameron. A cognitive model for tabular editing. Technical report, Ohio State University, 1989.

[7] N. Hurst, W. Li, and K. Marriott. Review of automatic document formatting. In *DocEng '09: Proceedings of the 9th ACM symposium on Document engineering*, pages 99–108, New York, NY, USA, 2009. ACM.

[8] N. Hurst, K. Marriott, and D. Albrecht. Solving the simple continuous table layout problem. In *DocEng '06: Proceedings of the 2006 ACM symposium on Document engineering*, pages 28–30, New York, NY, USA, 2006. ACM.

[9] N. Hurst, K. Marriott, and P. Moulder. Toward tighter tables. In *DocEng '05: Proceedings of the 2005 ACM symposium on Document engineering*, pages 74–83, New York, NY, USA, 2005. ACM.

[10] Y. Jin and D. L. Parnas. Defining the meaning of tabular mathematical expressions. *Sci. Comput. Program.*, 75:980–1000, November 2010.

[11] D. E. Knuth. *The TeXbook.* Addison-Wesley Professional, 1986.

[12] B. Lowagie. *iText in Action.* Manning Publications Co., Greenwich, CT, USA, 2010.

[13] K. Marriott, P. Moulder, and N. Hurst. Automatic float placement in multi-column documents. In *Proceedings of the 2007 ACM symposium on Document engineering*, DocEng '07, pages 125–134, New York, NY, USA, 2007. ACM.

[14] T.Alspaugh, S.Faulk, K. Britton, R.A.Parker, D.L.Parnas, and J. Shore. Software requirements for the A7E aircraft. BS NRL/FR/5530-92-9194, Naval Research Laboratory, 1992.

[15] The World Wide Web Consortium. Cascading style sheets level 2 revision 1 (CSS 2.1) specification. http://www.w3.org/TR/CSS2/, Sept 2009. W3C Candidate Recommendation.

[16] P. Van Hentenryck. *The OPL optimization programming language.* MIT Press, Cambridge, MA, USA, 1999.

[17] C. Vanoirbeek. Formatting structured tables. In *EP92 (Proceedings of Electronic Publishing)*, pages 291–309. Cambridge University Press, 1992.

[18] X. Wang. *Tabular abstraction, editing, and formatting.* PhD thesis, University of Waterloo, Waterloo, Ont., Canada, Canada, 1996. AAINN09397.

Optimal Automatic Table Layout

Graeme Gange
Dept of CSSE
University of Melbourne
Vic. 3010, Australia
ggange@csse.unimelb.edu.au

Kim Marriott and
Peter Moulder
Clayton School of IT
Monash University
Vic. 3800, Australia
{kim.marriott, peter.moulder}
@monash.edu

Peter Stuckey
Dept of CSSE
University of Melbourne
Vic. 3010, Australia
pjs@csse.unimelb.edu.au

ABSTRACT

Automatic layout of tables is useful in word processing applications and is required in on-line applications because of the need to tailor the layout to the viewport width, choice of font and dynamic content. However, if the table contains text, minimizing the height of the table for a fixed maximum width is a difficult combinatorial optimization problem. We present three different approaches to finding the minimum height layout based on standard approaches for combinatorial optimization. All are guaranteed to find the optimal solution. The first is an A*-based approach that uses an admissible heuristic based on the area of the cell content. The second and third are constraint programming (CP) approaches using the same CP model. The second approach uses traditional CP search, while the third approach uses a hybrid CP/SAT approach, lazy clause generation, that uses learning to reduce the search required. We provide a detailed empirical evaluation of the three approaches and also compare them with two mixed integer programming (MIP) encodings due to Bilauca and Healy.

Categories and Subject Descriptors

I.7.2 [**Document and Text Processing**]: Document Preparation—*Format and notation, Photocomposition/typesetting*

General Terms

Algorithms

Keywords

automatic table layout, constrained optimization, typography

1. INTRODUCTION

Tables are provided in virtually all document formatting systems and are one of the most powerful and useful design elements in current web document standards such as (X)HTML, CSS and XSL. For on-line presentation it is not

practical to require the author to specify table column widths at document authoring time since the layout needs to adjust to different width viewing environments and to different sized text since, for instance, the viewer may choose a larger font. Dynamic content is another reason that it can be impossible for the author to fully specify table column widths. This is an issue for web pages and also for VDP in which improvements in printer technology now allow companies to cheaply print material which is customized to a particular recipient. Good automatic layout of tables is therefore needed for both on-line and VDP applications and is useful in many other document processing applications since it reduces the burden on the author of formatting tables.

However, automatic layout of tables that contain text is computationally expensive. Anderson and Sobti [1] have shown that table layout with text is NP-hard. The reason is that if a cell contains text then this implicitly constrains the cell to take one of a discrete number of possible configurations corresponding to different numbers of lines of text. It is a difficult combinatorial optimization problem to find which combination of these discrete configurations best satisfies reasonable layout requirements such as minimizing table height for a given width.

Table layout research is reviewed in Hurst, Li & Marriott [7]. Starting with Beach [3], a number of authors have investigated automatic table layout from a constrained optimization viewpoint and a variety of approaches for table layout have been developed. Almost all approaches use heuristics and are not guaranteed to find the optimal solution. They include methods that use a desired width for each column and scale this to the actual table width [17, 6, 2], methods that use a continuous linear or non-linear approximation to the constraint that a cell is large enough to contain its contents [1, 4, 9, 8, 11], a greedy approach [9] and an approach based on finding a minimum cut in a flow graph [1].

In this paper we are concerned with complete techniques that are guaranteed to find the optimal solution. While these are necessarily non-polynomial in the worst case (unless P=NP) we are interested in finding out if they are practical for small and medium sized table layout. Even if the complete techniques are too slow for normal use, it is still worthwhile to develop complete methods because these provide a benchmark with which to compare the quality of layout of heuristic techniques. For example, while Gecko (the layout engine used by the Firefox web browser) is the most sophisticated of the HTML/CSS user agents whose source code we've seen, Figure 1 shows that its layouts can be far

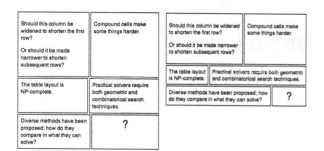

Figure 1: Example table comparing layout using Gecko (on the left) with the minimal height layout (on the right).

from the most compact as computed by one of the algorithms we present.

We know of only two other papers that have looked at complete methods for table layout with breakable text. The first is a branch-and-bound algorithm described in [19], which finds a layout satisfying linear designer constraints on the column widths and row heights. However it is only complete in the sense that it will find a feasible layout if one exists and is not guaranteed to find an optimal layout that, say, minimizes table height.[1] The second is detailed in a recent paper by Bilauca and Healy [5]. They give two MIP based branch-and-bound based complete search for simple tables.

The first contribution of this paper is to present three new techniques for finding a minimal height table layout for a fixed width. All three are based on generic approaches for solving combinatorial optimization problems that have proven to be useful in a wide variety of practical applications.

The first approach uses an $A^\star$ based approach [18] that chooses a width for each column in turn. Efficiency of the $A^\star$ algorithm crucially depends on having a good conservative heuristic for estimating the minimum height for any full table layout that extends the current layout. We use a heuristic that treats the remaining unfixed columns in the layout as if they are a single merged column each of whose cells must be large enough to contain the contents of the unfixed cells on that row. The other key to efficiency is to prune layouts that are not *column-minimal* in a sense that it is possible to reduce one of the fixed column widths without violating a cell containment constraint while keeping the same row heights.

The second and third approach are constraint programming (CP) [13] approaches. We model the problem using constraint programming and apply two different solving technologies to the same model. The second approach uses traditional CP search, while the third approach uses a hybrid solving approach, lazy clause generation [16], which combines CP and SAT technology. The advantages of the hybrid approach is that during search it learns *nogoods* that prevent it from repeating similar search later on, and it tracks *activity* of decisions, and uses an automatic search approach that concentrates on decisions likely to lead to early failure.

---

[1]Presumably one could minimize table height by repeatedly searching for a feasible solution with a table height less than the best solution so far.

Both these advantages can drastically reduce the amount of search required.

The second contribution of this paper is to provide an extensive empirical evaluation of these three approaches as well as the two MIP-based approaches of Bilauca and Healy [5]. We first compare the approaches on a large body of tables collected from the web. This comprised more than 2000 tables that were hard to solve in the sense that the standard HTML table layout algorithm did not find the minimal height layout. Most methods performed well on this set of examples and solved almost all problems in less than 1 second. We then stress-tested the algorithms on some large artificial table layout examples. In this case we found that the hybrid CP/SAT approach was the most robust approach.

2. BACKGROUND

2.1 The Table layout Problem

We assume throughout this paper that the table of interest has n columns and m rows. A *layout* (w, h) for a table is an assignment of widths, w, to the columns and heights, h, to the rows where w_c is the width of column c and h_r the height of row r. We make use of the width and height functions:

$$wd_{c_1,c_2}(w) = \sum_{c=c_1}^{c_2} w_c, \qquad wd(w) = wd_{1,n}(w),$$
$$ht_{r_1,r_2}(h) = \sum_{r=r_1}^{r_2} h_r, \qquad ht(h) = ht_{1,m}(h)$$

where ht and wd give the overall table height and width respectively.

The designer specifies how the grid elements of the table are partitioned into logical elements or *cells*. We call this the *table structure*. A *simple* cell spans a single row and column of the table while a *compound cell* consists of multiple grid elements forming a rectangle, i.e. the grid elements span contiguous rows and columns.

If d is a cell we define $rows(d)$ to be the rows in which d occurs and $cols(d)$ to be the set of columns spanned by d. We let

$$bot(d) = \max rows(d), \quad top(d) = \min rows(d),$$
$$left(d) = \min cols(d), \quad right(d) = \max cols(d).$$

and, letting $Cells$ be the set of cells in the table, for each row r and column c we define

$$rcells_c = \{d \in Cells \mid right(d) = c\},$$
$$cells_c = \{d \in Cells \mid c \in cols(d)\},$$
$$bcells_r = \{d \in Cells \mid bottom(d) = r\}$$

Each cell d has a minimum width, $minw(d)$, which is typically the length of the longest word in the cell, and a minimum height $minh(d)$, which is typically the height of the text in the cell.

The table's *structural constraints* are that each cell is big enough to contain its content and at least as wide as its minimum width and as high as its minimum height.

The table *layout style* captures what is required in a good layout. We shall focus on the *minimum height* layout style. This finds a layout for the table that, in decreasing order of importance, is no wider than the fixed maximum width, minimizes table height, and minimizes table width.

The problem we are addressing is, given a table structure and content for the table cells, to find an assignment to the column widths and row heights such that the structural constraints are satisfied and the minimum height layout style is satisfied.

For simplicity, we assume that the minimum table width is wide enough to allow the structural constraints to be satisfied. Furthermore, we do not consider nested tables nor do we consider designer constraints such as columns having fixed ratio constraints between them.

2.2 Minimum configurations

The main decision in table layout is how to break the lines of text in each cell. Different choices give rise to different width/height cell configurations. Cells have a number of *minimal configurations* where a minimal configuration is a pair (w, h) s.t. the text in the cell can be laid out in a rectangle with width w and height h but there is no smaller rectangle for which this is true. That is, for all $w' \leq w$ and $h' \leq h$ either $h = h'$ and $w = w'$, or the text does not fit in a rectangle with width w' and height h'. These minimum configurations are *anti-monotonic* in the sense that if the width increases then the height will never increase. For text with uniform height with W words (or more exactly, W possible line breaks) there are up to W minimal configurations, each of which has a different number of lines. In the case of non-uniform height text there can be no more than $O(W^2)$ minimal configurations.

A number of algorithms have been developed for computing the minimum configurations of the text in a cell [7]. Here we assume that these are pre-computed and that

$$configs_d = [(w_1, h_1), ..., (w_{N_d}, h_{N_d})]$$

gives the width/height pairs for the minimal configurations of cell d in increasing order of width. We will make use of the function $minheight(d, w)$ which gives the minimum height $h \geq minh(d)$ that allows the cell contents to fit in a rectangle of width $w \geq minw(d)$. This can be readily computed from the list of configurations.

The mathematical model of the table layout problem can be formalized as:

find w and h that minimize $ht(h)$ subject to
$$\forall d \in Cells. \quad (cw_d, ch_d) \in configs_d \ \wedge \quad (1)$$
$$\forall d \in Cells. \quad wd_{left(d), right(d)}(w) \geq cw_d \ \wedge \quad (2)$$
$$\forall d \in Cells. \quad ht_{top(d), bot(d)}(h) \geq ch_d \ \wedge \quad (3)$$
$$wd(w) \leq W \quad (4)$$

In essence, automatic table layout is the problem of finding *minimal configurations for a table*: i.e. minimal width / height combinations in which the table can be laid out. One obvious necessary condition for a table layout (w, h) to be a minimal configuration is that it is impossible to reduce the width of any column c while leaving the other row and column dimensions unchanged and still satisfy the structural constraints. We call a layout satisfying this condition *column-minimal*.

We now detail three algorithms for solving the table layout problem. All are guaranteed to find an optimal solution but in the worst case may take exponential time.

3. A* ALGORITHM

The first approach uses an $A^\star$ based approach [18] that chooses a width for each column in turn. A partial layout (w, c) for a table is a width w for the first $c - 1$ columns. The algorithm starts from the empty partial layout ($c = 1$) and repeatedly chooses a partial layout to extend by choosing possible widths for the next column.

Partial layouts also have a penalty p, which is a lower bound on the height for any full table layout that extends the current partial layout. The partial layouts are kept in a priority queue and at each stage a partial layout with the smallest penalty p is chosen for expansion. The algorithm has found a minimum height layout when the chosen minimal-penalty partial layout has $c = n + 1$ have at least as great a penalty then all other partial layouts must lead to at least as tall a layout. The code is given in function *complete-A*-search(W)* where W is the maximum allowed table width. For simplicity we assume W is greater than the minimum table width. (The minimum table width can be determined by assigning each column its min_c width from *possible-col-widths*, or can equivalently be derived from the corresponding maximum positions also used in that function.)

Given widths w for columns $1, \ldots, c - 1$ and maximum table width of W, the function *possible-col-widths(c,w,W)* returns the possible widths for column c that correspond to the width of a minimal configuration for a cell in c and which satisfy the minimum width requirements for all the cells in d and still satisfy the minimum width requirements for columns $c + 1, \ldots, n$ and allow the table to have width W.

Efficiency of an $A^\star$ algorithm usually depends strongly on how tight the lower bound on penalty is, i.e. how often (and how early) the heuristic informs us that we can discard a partial solution because all full table layouts that extend that partial layout will either have a height greater than the optimal height, or have height greater or equal to some other layout that isn't discarded.

We use a heuristic that treats the remaining unfixed columns in the layout as if they are a single merged column each of whose cells must be large enough to contain the contents of the unfixed cells on that row. We approximate the contents by a lower bound of their area. The function *compute-approx-row-heights(w,h,c,W)* does this, returning the estimated (lower bound) row heights after laying out the area of the contents of columns $c + 1, \ldots, n$ in a single column whose width brings the table width to W. Compound cells that span multiple rows, and positions in the table grid that have no cell, use a very simple lower bound of zero. (A simple refinement would be to use the product of the width requirement for the cell's column span and the height requirement for each row.)

A standard method of discarding partial solutions that must lead to solutions no better than some other partial solution is to use a *closed set*. However, for a standard closed set, the key would include the height requirements of each row span (including non-compound row spans) encountered, as well as certain column start positions. Implementers might consider how useful such a closed set would be for the inputs of interest to them. Our implementation doesn't use one.

We instead present the following more interesting method for discarding partial solutions. Partial layouts which must lead to a full layout which is not column minimal are not considered. If the table has no compound cells spanning multiple rows then any partial layout that is not column minimal for the columns that have been fixed can be discarded because row heights can only increase in the future and so the layout can never lead to a column-minimal layout. This no longer holds if the table contains cells spanning multiple rows as row heights can decrease and so a partial layout that is not column minimal can be extended to one that is column minimal. However, it is true that if the cells spanning multiple rows are ignored, i.e. assumed to have zero content, when determining if the partial layout is column minimal then partial layouts that are not column minimal can be safely discarded. The function *weakly-column-minimal(w,c)* does this by checking that none of the columns $1, \ldots, c$ can be narrowed without increasing the height of a row, ignoring compound cells spanning multiple rows.

In our implementation of *complete-A*-search*, the iteration over possible widths works from maximum v downwards, stopping once the new partial solution is either known not to be column minimal or (optionally) once the penalty exceeds a certain maximum penalty which should be an upper bound on the minimum height. Our implementation computes a maximum penalty at the start, by using a heuristic search based on [9].

Creating a new partial layout is relatively expensive (see below), so this early termination is more valuable than one might otherwise expect. However, the cost of this choice is that this test must be done before considering the height lower bounds for future cells (the remaining-area penalty), since the future penalty is at its highest for maximum v.

For the implementation of *compute-approx-row-heights*, note that $D2_r$, $area_r$ and the c' bound in the sum don't depend on w or $h0$, and hence may be calculated once off in advance; while w may be stored in cumulative form; so that the loop body can run in constant time. (Our implementation uses this approach.)

Our implementation literally evaluates *minheight* once per cell ending at c for each partial solution being added to the queue, as Figure 2 depicts. One could instead combine some of this work with finding the union of widths of interest (the second half of *possible-col-widths*), by changing the sorted list of configurations of interest for a cell to store the increase in height for its row span, so that *complete-A*-search* can update the height requirements for the current column (and in turn update the height requirements for the new partial solution) as it iterates over the combined list of widths. How significant a saving this would be would depend on how expensive the *minheight* implementation is compared to preparing and storing the new partial solution's height requirements. For it to have a significant payoff might require changing the representation of height requirements to avoid copying the full set for each new partial solution.

For *possible-col-widths*, our current implementation of the min_c calculation literally iterates over all $rcells_c$, though one could instead calculate in advance a single $minw$-like value for each column span occurring in the table, and have the min_c calculation iterate over the set of column spans ending at c (which would often be a singleton set) instead of all the cells ending at c. Less valuably, one could similarly calculate in advance a set of possible widths for each column span, and have the $widths_d$ calculation iterate over the set of column spans ending at c instead of all cells ending at c. (Again, our current implementation does not do this.)

Whereas for the max_c calculation, we do calculate in advance a single array of maximum positions (one per column), so max_c is a simple lookup.

For *weakly-column-minimal*, our implementation is comparable to what's written in the figure: we do cache h, but we still iterate over all the cells placed so far to determine column minimality. In principle, one can do better than this,

by keeping track of what height a row span must increase to allow narrowing a cell having that row span and being one of the cells forcing a column end position to be considered minimal. We haven't looked into this, though looking more might be worthwhile for uses where difficult, wide tables are common.

4. CONSTRAINT PROGRAMMING

Constraint programming (see e.g. [13]) is an approach to combinatorial satisfaction problems which combines search and inference. The constraint model is defined in terms of a *domain* of possible values for each variables, and *propagators* for each constraint. The role of a propagator is to remove values from the domains of the variables for that constraint which cannot be part of a solution. Constraint programming can implement combinatorial optimization search, by solving a series of satisfaction problems, each time looking for a better solution, until no better solution can be found and the optimal is proved.

We consider constraint satisfaction problems, consisting of a set of constraints C over n variables x_i taking integer values, each with a given finite domain $D_{orig}(x_i)$. A feasible solution is a valuation to the variables such that each x_i is within its allowable domain and all constraints are satisfied simultaneously.

A propagation solver maintains a domain restriction $D(x_i) \subseteq D_{orig}(x_i)$ for each variable, and considers only solutions that lie within $D(x_1) \times D(x_2) \times \ldots \times D(x_n)$. Propagators for the constraints C determine given the current domain whether we can remove values that cannot take part in any solution, e.g. if $x_1 \in \{1, 2, 3\}$ and $x_2 \in \{2, 3\}$ and $C = \{x_1 \geq x_2\}$ then the value $x_1 = 1$ cannot be part of any solution, so it can be eliminated. Propagation solving interleaves propagation, which repeatedly applies propagators to remove unsupported values until no further domain reduction is detected, and search which (typically) splits the domain of some variable in two and considers both the resulting sub-problems. This continues until all variables are fixed and a solution found, or propagation detects *failure* (a variable with empty domain) in which case execution backtracks and tries another subproblem.

Lazy clause generation [16] is a hybrid approach to combinatorial optimization combining finite domain propagation and Boolean satisfiability methods. A lazy clause generation solver performs finite domain propagation just as in a standard CP solver, but records the reasons for all propagations. When a failure is determined it determines a minimal set of reasons that have caused this failure and records this as a *nogood* in the solver. This nogood prevents the search from examining similar sets of choices which lead to the same inability to solve the problem.

Lazy clause generation is implemented by defining an alternative model for the domains $D(x_i)$, which is maintained simultaneously. Specifically, Boolean variables are introduced for each potential value of a variable, named $[\![x_i = j]\!]$ and similarly $[\![x_i \geq j]\!]$. Negating them gives the opposite, $[\![x_i \neq j]\!]$ and $[\![x_i \leq j - 1]\!]$. Fixing such a *literal* modifies D to makes the corresponding fact true in $D(x_i)$ and vice versa. Hence these literals give an alternate Boolean representation of the domain.

In a lazy clause generation solver, the actions of propagators (and search) to change domains are recorded in an *implication graph* over the literals. Whenever a propagator

function *possible-col-widths(c,w,W)*
$$min_c := \max_{d \in rcells_c} \{minw(d) - wd_{left(d),c-1}(w)\}$$
for $c' := n$ down to $c+1$ **do**
$$w_{c'} := \max_{d \in lcells_{c'}} \{minw(d) - wd_{c'+1,right(d)}(w)\}$$
endfor
$$max_c := W - wd_{1,c-1}(w) - wd_{c+1,n}(w)$$
for $d \in rcells_d$ **do**
$$widths_d := \{w_k - wd_{left(d),c-1}(w)|(w_k,h_k) \in configs_d\}$$
$$widths_d := \{v \in widths_d \mid min_c \leq v \leq max_c\}$$
return $(\bigcup_{d \in rcells_d} widths_d)$

function *weakly-column-minimal(w,c)*
for $r := 1$ to m **do**
$$D_r := \{d \in Cells \mid right(d) \leq c \text{ and } rows(d) = \{r\}\}$$
$$h_r := \max_{d \in D_r} \{minheight(d, wd_{left(d),right(d)}(w))\}$$
endfor
for $c' := 1$ to c **do**
$\quad cm := false$
$\quad$**for** $d \in rcells_{c'}$ s.t. $|rows(d)| = 1$ **do**
$\quad\quad$**if** $minheight(d, wd_{left(d),c'}(w) - \epsilon) > h_{bot(d)}$
$\quad\quad$**then** $cm := true$; **break**
$\quad$**endfor**
$\quad$**if not** cm **then return** false
endfor
return true

function *compute-approx-row-heights(w,h0,c,W)*
for $r := 1$ to m **do**
$\quad D1_r := \{d \in Cells \mid right(d) = c \text{ and } bot(d) = r\}$
$\quad$**if** $D1_r = \emptyset$ **then** $h1 := 0$
$\quad$**else** $h1 := \max_{d \in D1_r} \{ minheight(d, wd_{left(d),right(d)}(w))$
$$- ht_{top(d),r-1}(h) \}$$
$\quad$**endif**
$\quad D2_r := \{d \in Cells \mid c < right(d) \text{ and } rows(d) = \{r\}\}$
$$area_r := \sum_{d \in D2_r} area(d)$$
$\quad$**if** $area_r = 0$ **then** $h2 := 0$
$\quad$**else** $h2 := area_r / \left(W - \sum_{c'=1}^{\min(\{c\} \cup \{left(d)|d \in D2_r\})} w_{c'} \right)$
$\quad$**endif**
$\quad h_r := \max\{h0_r, h1, h2\}$
endfor
return h

function *complete-A*-search(W)*
create a new priority-queue q
add $(0, -1, [c \mapsto 0|c = 1..n], [r \mapsto 0|r = 1..m])$ to q
repeat
$\quad$remove lowest priority state $(p, -c, w, h)$ from q
$\quad$**if** $c = n+1$ **then return** (w, h)
$\quad widths_c := possible\text{-}col\text{-}widths(c, w, W)$
$\quad$**foreach** $v \in widths_c$ s.t.
$\quad\quad weakly\text{-}column\text{-}minimal(w[c \mapsto v], c)$ **do**
$\quad\quad w' := w[c \mapsto v]$
$\quad\quad h' := compute\text{-}approx\text{-}row\text{-}heights(w, h, c, W)$
$\quad\quad$add $(ht(h'), -(c+1), w', h')$ to q
$\quad$**endfor**
forever

Figure 2: An A* algorithm for table layout.

f changes a domain it must *explain* how the change occurred in terms of literals, that is, each literal l that is made true must be explained by a clause $L \to l$ where L is a (set or) conjunction of literals. When the propagator causes failure it must explain the failure as a *nogood*, $L \to false$, where L is a conjunction of literals which cannot hold simultaneously. Note that each explanation and nogood is a clause. The explanations of each literal and failure are recorded in the implication graph with edges from each $l' \in L$ to l.

The implication graph is used to build a nogood that records the reason for search failure. We explain the First Unique Implication Point (1UIP) nogood [15], which is standard. Starting from the initial failure nogood, a literal l (explained by $L \to l$) is replaced in the nogood by L by resolution. This continues until there is at most one literal in the nogood made true after the last decision. The resulting nogood is *learnt*, i.e. added as a clause to the constraints of the problem. It will propagate to prevent search trying the same subsearch in the future.

Lazy clause generation effectively imports Boolean satisfiability (SAT) methods for search reduction into a propagation solver. The learnt nogoods can drastically reduce the search space, depending on how often they are reused (i.e. propagate). Lazy clause generation can also make use of SAT search heuristics such as *activity-based search* [15]. In activity-based search each literal seen in the conflict generation process has its activity bumped, and periodically all activities are decayed. Search decides a literal with maximum activity, which tends to focus on literals that have recently caused failure.

4.1 A CP model for table layout

A Zinc [12] model is given below. Each cell d has a configuration variable $f[d]$ which chooses the configuration (cw, ch) from an array of tuples $cf[d]$ of (width, height) configurations defining $configs_d$. Note that $t.1$ and $t.2$ return the first and second element of a tuple respectively. The important variables are: w, the width of each column, and h, the height of each row. These are constrained to fit each cell, and so that the maximum width is not violated.

```
int: n; % number of columns
int: m; % number of rows
int: W; % maximal width
set of int: Cells; % numbered cells
array[Cells] of 1..m: top;
array[Cells] of 1..m: bot;
array[Cells] of 1..n: left;
array[Cells] of 1..n: right;
array[Cells] of array[int] of tuple(int,int): cf;

array[Cells] of var int: f; % cell configurations
array[1..n] of var int: w;  % column widths
array[1..m] of var int: h;  % row heights

constraint forall(d in Cells)(
      % constraint (2)
  sum(c in left[d]..right[d])(w[c])>=cf[d][f[d]].1
  /\ % constraint (3)
  sum(r in top[d]..bot[d])(h[r])>=cf[d][f[d]].2
);
      % constraint (4)
constraint sum(c in 1..n)(w[c]) <= W;
solve minimize sum(r in 1..m)(h[r]);
```

The Zinc model does not enforce column minimality of the solutions, but solutions will be column minimal because of the optimality condition.

The reason that we thought that lazy clause generation might be so effective for the table layout problem is the small number of key decisions that need to be made. While there may be $O(nm)$ cells each of which needs to have an appropriate configuration determined for it, there are only n widths and m heights to decide. These variables define all communication between the cells. Hence if we learn nogoods about combinations of column widths and row heights there are only a few variables involved, and these nogoods are likely to be highly reusable. We can see the benefit of nogood learning by comparing the constraint programming model, with and without learning.

EXAMPLE 4.1. *Consider laying out a table of the form*

```
aa        aa
     aa         aa   aa
     aa   aa
          aa   aa
               aa   aa
          aa         aa
```

where each aa entry can have two configurations: wide two characters wide and one line high, or high one character wide and two lines high (so cf[d] = [(2,1),(1,2)]). Assume the remaining cells have unique configuration (1,1), and there is a maximal table width of 9, and a maximal table height of 9. Choosing the configuration of cell (1,1) as wide ($f[(1,1)]$ = 1) makes $w_1 \geq 2$, similarly if cell (1,3) is wide then $w_3 \geq 2$. The effect of each decision in terms of propagation is illustrated in the implication graph in Figure 3 Now choosing the configuration of cell (2,2) as wide makes $w_2 \geq 2$ and then propagation on the sum of column widths forces each of the remaining columns to be at most width 1: $w_4 \leq 1, w_5 \leq 1, w_6 \leq 1$. Then $w_4 \leq 1$ means $h_2 \geq 2$, $h_3 \geq 2$, $h_4 \geq 2$ and $h_6 \geq 2$ since we must pick the second configuration for each of the cells in column 4. These height constraints together violate the maximal height constraint. Finite domain propagation backtracks undoing the last decision and sets the configuration of (2,2) as high ($f[(2,2)]$ = 2), forcing $h_2 \geq 2$. Choosing the configuration of (2,5) as wide makes $w_5 \geq 2$ and then propagation on the sum of column widths forces each of the remaining columns to be at most width 1: $w_2 \leq 1, w_4 \leq 1, w_5 \leq 1$. Again $w_4 \leq 1$ means the maximal height constraint is violated. So search undoes the last decision and sets the configuration of (2,5) as high.

Lets contrast this with lazy clause generation. After making the first three decisions the implication graph is shown in Figure 3. The double boxed decisions reflect making the cells (1,1), (1,3) and (2,2) wide. The consequences of the last decision are shown in dashed boxes. Lazy clause generation starts from the nogood $h_2 \geq 2 \wedge h_3 \geq 2 \wedge h_4 \geq 2 \wedge h_6 \geq 2 \rightarrow false$ and replaces $h_6 \geq 2$ using its explanation $f[(6,4)] = 2 \rightarrow h_6 \geq 2$ to obtain $h_2 \geq 2 \wedge h_3 \geq 2 \wedge h_4 \geq 2 \wedge f[(6,4)] = 2 \rightarrow false$. This process continues until it arrives at the nogood $w_4 \leq 1 \rightarrow false$ which only has one literal from the last decision level. This is the 1UIP nogood. It will immediately backjump to the start of the search (since the nogood does not depend on any other literals at higher decision levels) and enforce that $w_4 \geq 2$. Search will again make cell (1,1) wide, and on making cell (1,3) wide it will determine $w_3 \geq 2$ and consequently that $w_2 \leq 1, w_5 \leq 1$ and $w_6 \leq 1$ which again causes violation of the maximal height

constraint. The 1UIP nogood is $w_1 \geq 2 \wedge w_3 \geq 2 \rightarrow fail$, so backjumping removes the last choice and infers that $w_3 \leq 1$ which makes $h_1 \geq 2$ and $h_3 \geq 2$.

Note that the lazy clause generation completely avoids considering the set of choices (1,1), (1,3) and (2,5) wide since it already fails on setting (1,1) and (1,3) wide. This illustrates how lazy clause generation can reduce search. Also notice that in the implication graph the consequences of a configuration choice only propagate through width and height variables, and hence configuration choices never appear in nogoods.

5. MIXED INTEGER PROGRAMMING

Bilauca and Healy [5] consider using mixed integer programming (MIP) to model the table layout problem. They consider two models for the simple table layout problem and do not consider compound cells, i.e. row and column spans. Their basic model BMIP uses 01 variables (`cellSel`) for each possible configuration, to model the integer configuration choice f used in the CP model. This basic model can be straightforwardly extended to handle compound cells.

Their improved model adds redundant constraints on the column widths to substantially improve MIP solving times for harder examples. They compute the minimum width (`minW`) for each column as the maximum of the minimum widths of the cells in the column, and the minimum height (`minH`) for each row analogously. They then compute the set of possible column widths (`colWSet`) for each column from those configurations in the column which have at least width `minW` and height `minH`. Note this improvement relies on the fact that there are no column spans. While in practice this usually does provide the set of possible widths in any column-minimal layout, in general we believe the "improved model" is incorrect.

EXAMPLE 5.1. *Consider laying out a 2×2 table with cell configurations $\{(1,3),(3,1)\}$ for the top left cell, and $\{(2,2)\}$ for the remaining cells, with a width limit of 5. The minimal height of the first row is 2. The minimal width of the first column is 2. Hence none of the configurations of the top left cell are both greater than the minimal height and minimal width. The possible column widths for column 1 are then computed as $\{2\}$. The only layout is then choosing configuration (1,3) for the top left cell, giving a total height of 5. Choosing the other configuration leads to a total table height of 4.*

We can fix this model by including in the possible column widths the smallest width configuration for each cell in that column which is less than the minimum row height. For the example above this means the possible columns widths become $\{2,3\}$. Bilauca and Healy [5] give an OPL model of their "improved model", to contrast it with the CP model above we give a corresponding corrected Zinc model MIP.

```
int: m; % number of rows
int: n; % number of columns
int: W; % maximal width
array[1..m,1..n] of set of tuple(int,int): cf;

array[1..n] of int: minW = [ max(r in 1..m)
  (min(t in cf[r,c])(t.1)) | c in 1..n ];
array[1..m] of int: minH = [ max(c in 1..n)
  (min(t in cf[r,c])(t.2)) | r in 1..m ];
```

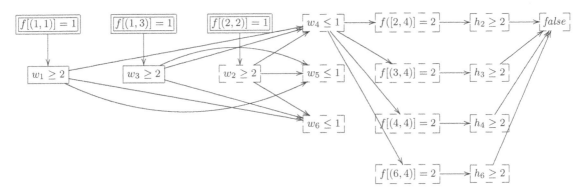

Figure 3: An implication graph for searching the layout problem of Example 4.1

```
array[1..n] of set of int: colWset =
  [ { t.1 | r in 1..m, t in cf[r,c] where
    t.1 >= minW[c] /\ (t.2 >= minH[r] \/
    t.1 == min({ u.1 | u in cf[r,c] %FIX
              where u.2 < minH[r] })) }
  | c in 1..n];

array[1..n] of array[int] of var 0..1: colSel =
  [ [ d:_ | d in colWset[c] ] | c in 1..n ];
array[1..n,1..m] of array[int,int] of var 0..1:
  cellSel =
  array2d(1..n,1..m,[ [ t:_ | t in cf[r,c] ]
                  | r in 1..m, c in 1..n ]);
array[1..n] of var int: w;
array[1..m] of var int: h;

constraint forall(r in 1..m, c in 1..n)(
  sum(t in cf[r,c])(cellSel[r,c][t]) = 1 /\
  sum(t in cf[r,c])(cellSel[r,c][t] * t.1) <= w[c] /\
  sum(t in cf[r,c])(cellSel[r,c][t] * t.2) <= h[r]);
constraint forall(c in 1..n)(
  sum(d in colWset[c])(colSel[c][d]) = 1 /\
  w[c] = sum(d in colWset[c])(colSel[c][d] * d));

constraint sum(c in 1..n)(w[c]) <= W;
solve minimize sum(r in 1..m)(h[r]);
```

Note that Bilauca and Healy also consider a CP model of the problem, but this is effectively equivalent to the MIP model because they do not make use of variable array indices (`element` constraints) that FD solvers support and which allow stronger propagation than that of the 01 encoding.

6. EVALUATION

We compare different approaches to optimal table layout: the A* algorithm of Section 3; the two constraint programming models of Section 4, namely the basic CP implementation without learning (CP-W) and the model using lazy clause generation to provide learning (CP); and BMIP the basic MIP model of [5], and MIP the (corrected) improved model of [5] described in Section 5. For the CP approaches, both CP-W and CP$_{seq}$ use a sequential search that chooses a cell that has the smallest height configuration remaining of all unfixed cells and tries to set it to that minimal height. For the lazy clause solver CP$_{vsids}$ uses the default activity based search.

The A* algorithm is written in the high-level declarative programming language Mercury. Notes in Section 3 give some idea of what optimizations have or haven't been applied to the source code of the implementation shown in these timings.

For the constraint programming approaches we used the Chuffed lazy clause generation solver (which can also be run without nogood generation). Chuffed is a state-of-the-art CP solver, which scored the most points in all categories of the 2010 MiniZinc Challenge [14] which compares CP solvers. Since Chuffed does not currently support Zinc, we created the model using the C++ modelling capabilities of Chuffed. The resulting constraints are identical to that shown in the model.

For the MIP approach, we used a script to construct a mixed integer programming model for each table, identical to that created by the Zinc model (and the (corrected) original OPL model of Bilauca and Healy), which was solved using CPLEX 12.1.

We first evaluated the various approaches using a large corpus of real-world tables. This was obtained by collecting more than 10,000 web pages using a web crawler, extracting non-nested tables (since we have not considered how to handle nested tables efficiently), resulting in over 50,000 tables. To choose the goal width for each table, we laid out each web page for three viewport widths (760px, 1000px and 1250px) intended to correspond to common window widths. Some of the resulting table layout problems are trivial to find solutions for; we retained only those problems that our A* implementation took at least a certain amount of time to solve. (Thus, the choice may be slightly biased against our A* solver, insofar as some other solver might find different examples to be hard.) This left 2063 table layout problems in the corpus. We split the corpus into **web-compound** and **web-simple** based on whether the table contained compound cells or not.

Table 1 shows the results of the different methods on the **web-simple** examples. The table shows the number of tables laid out optimally for various time limits up to 10 seconds. They show that in practice for simple tables all of the methods are very good, and able to optimally layout almost all tables very quickly. The worst method is CP-W and the evaluation clearly shows the advantage of learning for constraint programming. We find, like [5], that the improved MIP model MIP while initially slower is more robust than basic model BMIP. Overall CP$_{seq}$ is the most robust approach never requiring more than 1 second on any example. However, the performance of the A* model is surprisingly good

29

time (s)	CP-W	CP_{seq}	CP_{vsids}	BMIP	MIP	$A^\star$
≤ 0.01	1018	**1097**	1043	1009	774	972
≤ 0.10	1064	**1252**	1186	1160	1076	1168
≤ 1.00	1103	**1271**	1257	1221	1223	1262
≤ 10.00	1120	**1271**	1269	1261	1270	1269
> 10.00	151	**0**	2	10	1	2

Table 1: Number of instances from the web-simple data-set solved within each time limit.

time (s)	CP-W	CP_{seq}	CP_{vsids}	BMIP	$A^\star$
≤ 0.01	708	**713**	665	702	636
≤ 0.10	721	**774**	737	760	751
≤ 1.00	734	**788**	771	787	787
≤ 10.00	742	790	778	790	**792**
> 10.00	50	2	14	2	0

Table 2: Number of instances from the web-compound data-set solved within each time limit.

	s	CP_{seq}	CP_{vsids}	BMIP	MIP	$A^\star$
	0.00	**0.00**	**0.00**	**0.00**	**0.00**	0.02
	0.25	3.29	**0.02**	0.16	0.97	0.06
10×10	0.50	0.3	**0.01**	0.22	0.56	0.03
	0.75	0.07	**0.02**	0.55	1.18	0.09
	1.00	**0.00**	**0.00**	0.01	0.01	**0.00**
	0.00	0.02	**0.01**	**0.01**	0.04	0.19
	0.25	—	**0.86**	27.18	28.65	37.04
20×20	0.50	—	**0.07**	188.27	163.86	11.44
	0.75	—	**0.28**	43.83	40.07	62.03
	1.00	0.02	0.01	0.04	0.08	**0.00**
	0.00	0.04	**0.03**	0.04	0.07	1.53
	0.25	—	254.47	—	**253.08**	—
30×30	0.50	—	**0.38**	—	—	—
	0.75	—	**9.4**	—	—	—
	1.00	0.04	0.04	0.10	0.18	**0.00**
	0.00	0.09	**0.06**	0.07	0.20	3.55
	0.25	—	—	—	—	—
40×40	0.50	—	**1.11**	—	—	—
	0.75	—	216.67	—	—	—
	1.00	0.09	0.05	0.19	0.34	**0.02**

Table 3: Results for artificially constructed tables. Times are in seconds.

given the relative simplicity of the approach in comparison to the sophisticated CPlex and Chuffed implementations and the use of Mercury rather than C or C++.

Table 2 shows the results of the different methods on the **web-compound** examples. We compare all the previous algorithms except for MIP since it is not applicable when there are compound cells. The results are similar to those for simple tables. For these more complicated tables, the $A^\star$ approach is slightly more robust, while CP_{seq} is the fastest for the easier tables.

Given the relatively similar performance of the approaches on the real-world tables we decided to "stress-test" the approaches on some harder artificially constructed examples. We only used simple tables so that we could compare with MIP. Table 3 shows the results. We created tables of size $m \times n$ each with k configurations by taking text from the Gutenberg project edition of The Trial [10] k words at a time, and assigning to a cell all the layouts for that k words using fixed width fonts. For the experiments we used $k = 6$. We compare different versions of the layout problem by computing the minimum width $minw$ of the table as the sum of the minimal column widths, and the maximal width $maxw$ of the table as the sum of the column widths resulting when we choose the minimal height for each row. The squeeze s for table is defines as $(W - minw)/maxW$. We compare the table layout for 5 different values of squeeze. Obviously with a squeeze of 0.0 or 1.0 the problem is easy, the interesting cases are in the middle.

The harder artificial tables illustrate the advantages of the conflict directed search of CP_{vsids}. On the **web-simple** and **web-compound** corpora, the approaches with more naive search strategies, CP_{seq} and $A^\star$, performed best. However, they were unable to solve many of the harder tables from the **artificial** corpus, where CP_{vsids} solved all but one instance, sometimes 2 orders of magnitude faster than any other solver. Interestingly the MIP approach wins for one hard example, illustrating that the automatic MIP search in CPLEX can also be competitive.

The difference in behavior between the real-world and artificial tables may be due to differences in the table structure. The tables in the **web-simple** and **web-compound** corpora tend to be narrow and tall, with very few configurations per cell – the widest table has 27 columns, compared

with 589 rows, and many cells have only one configuration. On these tables, the greedy approach of picking the widest (and shortest) configuration tends to quickly eliminate tall layouts. The **artificial** corpus, having more columns and more configurations, requires significantly more search to prove optimality; in these cases, the learning and conflict-directed search of CP_{vsids} provides a significant advantage.

7. CONCLUSION

Treating table layout as a constrained optimization problem allows us to use powerful generic approaches to combinatorial optimization to tackle these problems. We have given three new three new techniques for finding a minimal height table layout for a fixed width: the first uses an $A^\star$ based approach while the second approach uses pure constraint programming (CP) and the third uses lazy clause generation, a hybrid CP/SAT approach. We have compared these with two MIP models previously proposed by Bilauca and Healy.

An empirical evaluation against the most challenging of over 50,000 HTML tables collected from the Web showed that all methods can produce optimal layout quickly.

The $A^\star$-based algorithm is more targeted than the constraint-programming approaches: while the $A^\star$ algorithm did well on the web-page-like tables for which it was designed, we would expect that more generic constraint-programming approaches would be a safer choice for other types of large tables. This turned out to be the case for the large artificially constructed tables we tested, where the approach using lazy clause generation was significantly more effective than the other approaches.

All approaches can be easily extended to handle constraints on table widths such as enforcing a fixed size or that two columns must have the same width. Handling nested tables, especially in the case cell size depends in a non-trivial way on the size of tables inside it (for example when floats are involved) is more difficult, and is something we plan to pursue.

8. ACKNOWLEDGEMENTS

The authors acknowledge the support of the ARC through Discovery Project Grant DP0987168

9. REFERENCES

[1] R. J. Anderson and S. Sobti. The table layout problem. In *SCG '99: Proceedings of the Fifteenth Annual Symposium on Computational Geometry*, pages 115–123, New York, NY, USA, 1999. ACM Press.

[2] G. J. Badros, A. Borning, K. Marriott, and P. Stuckey. Constraint cascading style sheets for the web. In *Proceedings of the 1999 ACM Conference on User Interface Software and Technology*, pages 73–82, New York, Nov. 1999. ACM.

[3] R. J. Beach. *Setting tables and illustrations with style*. PhD thesis, University of Waterloo, 1985.

[4] N. Beaumont. Fitting a table to a page using non-linear optimization. *Asia-Pacific Journal of Operational Research*, 21(2):259–270, 2004.

[5] M. Bilauca and P. Healy. A new model for automated table layout. In *Proceedings of the 10th ACM symposium on Document engineering*, DocEng '10, pages 169–176. ACM, 2010.

[6] A. Borning, R. Lin, and K. Marriott. Constraint-based document layout for the web. *Multimedia Systems*, 8(3):177–189, 2000.

[7] N. Hurst, W. Li, and K. Marriott. Review of automatic document formatting. In *Proceedings of the 9th ACM symposium on Document engineering*, pages 99–108. ACM, 2009.

[8] N. Hurst, K. Marriott, and D. Albrecht. Solving the simple continuous table layout problem. In *DocEng '06: Proceedings of the 2006 ACM symposium on Document engineering*, pages 28–30, New York, NY, USA, 2006. ACM.

[9] N. Hurst, K. Marriott, and P. Moulder. Towards tighter tables. In *Proceedings of Document Engineering, 2005*, pages 74–83, New York, 2005. ACM.

[10] F. Kafka. *The Trial*. Project Gutenberg, 1925, 2005.

[11] X. Lin. Active layout engine: Algorithms and applications in variable data printing. *Computer-Aided Design*, 38(5):444–456, 2006.

[12] K. Marriott, N. Nethercote, R. Rafeh, P. Stuckey, M. Garcia de la Banda, and M. Wallace. The design of the Zinc modelling language. *Constraints*, 13(3):229–267, 2008.

[13] K. Marriott and P. Stuckey. *Programming with Constraints: An Introduction*. MIT Press, Cambridge, Massachusetts, 1998.

[14] 2010 MiniZinc challenge. http://www.g12.csse.unimelb.edu.au/minizinc/challenge2010/results2010.html, 2010.

[15] M. W. Moskewicz, C. F. Madigan, Y. Zhao, L. Zhang, and S. Malik. Chaff: Engineering an efficient SAT solver. In *DAC '01: Proceedings of the 38th conference on Design automation*, pages 530–535, 2001.

[16] O. Ohrimenko, P. Stuckey, and M. Codish. Propagation via lazy clause generation. *Constraints*, 14(3):357–391, 2009.

[17] D. Raggett, A. L. Hors, and I. Jacobs. HTML 4.01 Specification, section 'Autolayout Algorithm'. http://www.w3.org/TR/html4/appendix/notes.html#h-B.5.2, 1999.

[18] S. Russell and P. Norvig. *Artificial Intelligence: a Modern Approach*. Prentice Hall, 2nd edition, 2002.

[19] X. Wang and D. Wood. Tabular formatting problems. In *PODP '96: Proceedings of the Third International Workshop on Principles of Document Processing*, pages 171–181, London, UK, 1997. Springer-Verlag.

A Framework with Tools for Designing Web-based Geographic Applications

The Nhan Luong, Sébastien Laborie and Thierry Nodenot
T2i – LIUPPA – Université de Pau et des Pays de l'Adour
2 Allée du Parc Montaury, 64600 Anglet, France
{thenhan.luong, sebastien.laborie, thierry.nodenot}@iutbayonne.univ-pau.fr

ABSTRACT

Many Web-based geographic applications have been developed in various domains, such as tourism, education, surveillance and military. However, developing such applications is a cumbersome task because it requires several types of components (e.g., maps, contents, indexing services, databases) that have to be assembled together. Hence, developers have to deal with different technologies and application behavior models. In order to create Web-based geographic applications and overcome these design problems, we propose a framework composed of three complementary tasks: identifying some desired data, building the graphical layout organization and defining potential user interactions. According to this framework, we have specified a unified model and we have encoded it using Semantic Web technologies, such as RDF. Through a prototype named WINDMash, we have implemented some tools that instantiate our model and automatically generate concrete Internet geographic applications that can be executed on Web browsers.

Categories and Subject Descriptors

D.2 [**Software Engineering**]: Design Tools and Techniques;
I.7.2 [**Document and Text Processing**]: Document Preparation.

General Terms

Design, Languages.

Keywords

Document generation, Authoring tool, Mashups.

1. INTRODUCTION

Currently, a fair amount of research and development has been conducted on Web-based application generation thanks to Web 2.0 technologies. Particularly in the domain of geographic information system (GIS), specific terms appeared in order to designate Internet Geographic Applications [10]: "GeoWeb", "Geospatial Web" and "Web Mapping 2.0".

Indeed, many Web-based geographic applications have been developed in different application domains (e.g., tourism, education, surveillance, military) and are using online mapping services (e.g., Google Maps[1], MapQuest[2], MultiMap[3], OpenLayers[4], Yahoo! Maps[5]).

However, developing such applications is a cumbersome task. The reasons are twofold: (1) they mix several components (e.g., maps, multimedia contents, indexing services, databases) which have to interact together and (2) developers have to deal with several technologies and different data structures as well as application models.

In this paper, we propose a framework for designing Web-based geographic applications. This framework is composed of three complementary tasks: (1) identifying the desired data that have to be manipulated by the system, (2) specifying the graphical layout of the application, and (3) defining potential end-user interactions between components. According to this framework, we have specified a unified model with three facets and proposed to encode this model using RDF [16]. We have chosen such a W3C standard because it enables to describe, to aggregate, to share and to reuse information embedded into each facet.

In order to experiment our proposal, we have developed some tools allowing designers to easily instantiate our unified model, i.e., create Web-based geographic applications. Moreover, from our proposed model, the designed descriptions can be computed into concrete Internet geographic applications and finally executed on a Web browser.

The remainder of the paper is structured as follows. In Section 2, we define the characteristics of a Web-based geographic application and illustrate an example. Moreover, we explain why it is currently difficult to design this kind of application. Thereafter, in Section 3, we describe some related existing systems. In Section 4, we specify our framework for designing Web-based geographic applications and detail in Section 5 its corresponding unified model. Section 6 demonstrates our framework through a prototype named WINDMash, especially some specific tools are presented in Section 7. Finally, we conclude in Section 8.

---

[1] http://maps.google.com

[2] http://www.mapquest.com

[3] http://classic.multimap.com

[4] http://openlayers.org

[5] http://maps.yahoo.com

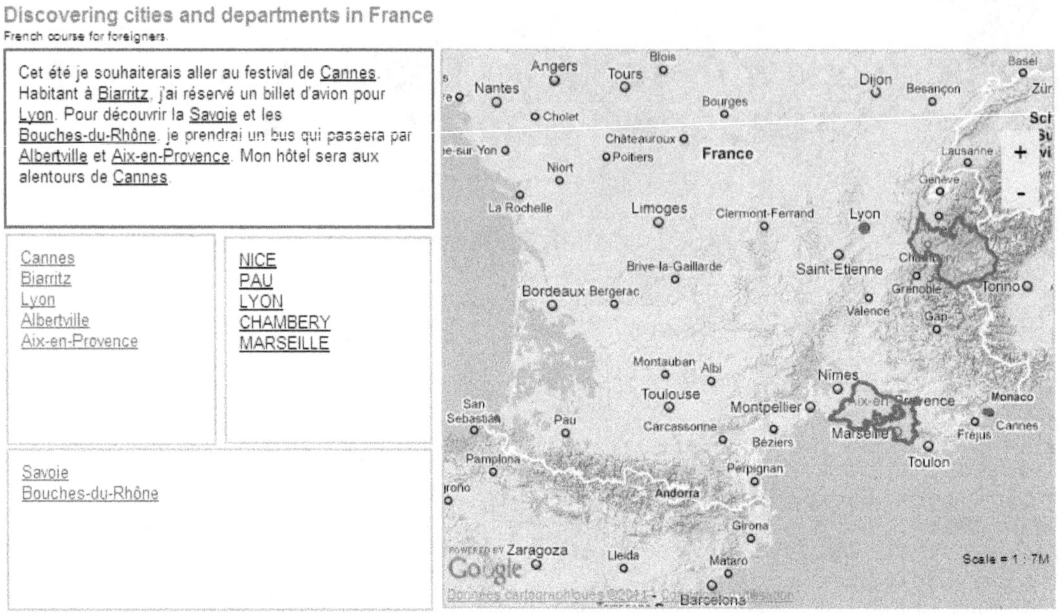

Figure 1: A Web-based geographic application for discovering French places.

2. DESIGNING WEB-BASED GEOGRAPHIC APPLICATIONS

Designing interactive Web-based geographic applications that allow, for instance, completing and/or increasing the discovery of a territory, such as in tourism or in education, is currently a cumbersome task. Indeed, such applications may generally include several components that should interact with each other, such components are:

- Mapping components with, for instance, spatial annotations and various types of geovisualization (e.g., street map, satellite view);

- Content components, such as texts, images, audios and videos. In this paper, we will exclusively focus on textual components;

- Indexing components which automatically extract spatial entities from some given contents and identify spatial metadata, like geolocation information;

- Communication components in order to exchange data with database servers, e.g., GeoNames[6].

In order to illustrate a typical Web-based geographic application, we present a use-case scenario in the following section (§2.1). This example will be used all during the paper. Furthermore, we also explain why developing such applications is not straightforward (§2.2).

2.1 A use-case scenario

Suppose a French course for foreigners where a teacher wants to explain the following concepts: Town, "Département" (a sort of province inside the country) and "Préfecture" (an administrative town inside a "Département"). More precisely, the teacher wants students to study a text written

in French[7] which contains several places names, e.g., Cannes (Town), Lyon ("Préfecture") and Savoie ("Département").

For pedagogical purposes, the teacher would like a Web-based geographic application, like the one illustrated in Figure 1. Indeed, it contains the text in French, a map containing spatial annotations and three lists of towns, "Préfectures" and "Départements". These lists should be automatically computed from the text. Furthermore, students are allowed to select a specific place in the lists. When clicking on a place, the corresponding term in the text is highlighted and the map is focused on this place.

2.2 Some requirements

Developing the interactive Web-based geographic application illustrated in Figure 1 with the behaviors described above is not a trivial task. It requires:

- Some programming skills, e.g., using JavaScript or AJAX, in order to create an interactive Web application;

- Knowledge about several databases schemas, and especially geographical databases in order to query and get spatial data, such as geolocations on a map;

- Manipulation of Web services (e.g., indexing services), particularly their inputs/outputs. Naturally, data structures have to be homogenized in order to confront and/or aggregate different services outputs.

Consequently, there is a need of a general framework which contains models and supports some tools for designing Web-based geographic applications. This framework is even more motivated in the next section.

---

[6]http://www.geonames.org

[7]The English version of the text in Figure 1 is available at the following address: http://www.iutbayonne.univ-pau.fr/~slaborie/DocEng11/EnglishTranslation.txt

3. RELATED WORK

End-user mashup programming environments are a new generation of online visual tools enabling users to quickly create, for example, Web-based applications [3]. They rely on metaphors that are easy to grasp by non-professional coders. They may bind together spreadsheets, the flow of linked processing blocks and the visual selection of GUI actions. In [5] and [21] several types of mashup environments have been summarized, some examples are: Yahoo! Pipes[8], Microsoft Popfly[9], Google Mashup Editor[10], IBM Mashup Center[11], MashMaker [8], Marmite [23], Vegemite [14], Exhibit [13] and Bill Organiser Portal [18].

However, developing the geographic application example presented in Section 2 with these mashup environments is still difficult, while impossible with some of these systems. In fact, many of them do not take into account the specification of end-user interactions and especially the system reactions on the different application components. Furthermore, these systems are not designed to exclusively build geographic applications, hence they do not propose a framework for building such kind of applications.

Some recent works on Web Engineering, especially focusing on geographic applications, proposed architectures based on mapping services. For instance, Mashlight [2] is a Web framework for creating and executing mashup applications by building data processing chains. The generated applications contain by default one mapping component with geolocation information. Besides, DashMash [6] offers to end-users more flexibility for creating geographic applications. Through a set of viewers, they can specify a graphical layout and visualize specific data inside them. Nevertheless, these systems do not allow to design end-user interactions.

SPARQLMotion[12] is a visual scripting language and an engine for semantic data processing. Scripts implementing sophisticated data services and processing, such as queries, data transformations and mashups, can be quickly assembled with user-friendly graphical tools. Such a system is using Semantic Web technologies, like RDF [16] and SPARQL [17]. However, as the previous systems, it does not provide solutions to specify end-user interactions and a general framework for constructing geographic application.

PhotoMap [22] offers graphical interfaces for navigation and query over some photo collections of users. Especially, with the map-based interface, itineraries followed by users are illustrated and photos with information are attached to specific places. Hence, end-users are able to retrieve where some photos have been taken either by clicking on specific places or by selecting a group of photos. This environment is another use-case for illustrating Web-based geographic applications. Nevertheless, this environment has been hard-coded by an expert and no design framework has been proposed by the authors.

In the next section, we propose a new framework for designing Web-based geographic applications. It should be considered to overcome the limits of the tools presented above.

8http://pipes.yahoo.com/pipes/

9http://www.deitel.com/Popfly/

10http://code.google.com/intl/en/gme/

11http://www-01.ibm.com/software/info/mashup-center/

12http://www.topquadrant.com/products/SPARQLMotion.html

4. A FRAMEWORK FOR DESIGNING WEB-BASED GEOGRAPHIC APPLICATIONS

To design a Web-based geographic application, we propose the framework depicted in Figure 2. This framework is composed of three complementary tasks:

1. Identifying the desired data that have to be manipulated by the geographic application. The data could refer to multimedia contents[13] (e.g., the text written in French in Figure 1), some extracted information (e.g., in Figure 1, the list of towns automatically identified from the text) and some computed information (e.g., in Figure 1, the lists of "Départements" and "Préfectures" which correspond to the list of towns).

2. Specifying the graphical layout of the geographic application. This layout may be composed of several viewers, such as textual viewers, list viewers or map viewers. Thanks to the data that have been defined during the previous step, viewers might display some data sets. For instance, if some data about towns are determined, these towns will be displayed on a specific map viewer.

3. Defining potential end-user interactions on the data that are contained inside viewers. For instance, if a user clicks on a specific town listed in Figure 1, this town will be highlighted in the text written in French and the map will be focused on this town.

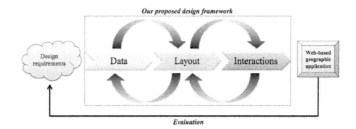

Figure 2: Our framework for designing Web-based geographic applications.

As illustrated in Figure 2, even if we have ordered the three tasks, it is possible to go backward and forward during the design process. Actually, at any time the application designer may add, modify and/or remove some data, some viewers or some interactions. Furthermore, it is also possible to design a geographic application without interactions.

When all tasks have been completed, the Web-based geographic application can be generated and executed. Obviously, it is possible to update the generated application by repeating our proposed framework.

According to the framework presented in this section, for each task illustrated in Figure 2, we have specified a model and we have encoded it using RDF [16]. Details and examples related to the use-case described in Section 2.1 are presented in the next section.

13In this paper, we only consider textual contents.

5. A UNIFIED MODEL FOR DESCRIBING AND COMPUTING WEB-BASED GEOGRAPHIC APPLICATIONS

According to our framework proposed in Figure 2, we have defined a unified model for describing Web-based geographic applications. This model is structured into three facets:

- **The data facet** (§5.1) describes the data that are manipulated by the application.

- **The graphical interface facet** (§5.2) defines the organization of the application layout.

- **The user interaction facet** (§5.3) specifies potential human-computer interactions.

Our unified model corresponds to the merge of all facet descriptions. We particularly show that these facets are complementary through the concept of *Annotation*. Furthermore, in order to encode, to aggregate, to share and to reuse each facet description, we have proposed an RDF/XML serialization [4]. Thanks to this standard, each facet can be described independently and may refer to other facet descriptions. Moreover, semantic queries can be executed on these descriptions.

Each facet will be described and exemplified in the following subsections.

5.1 The data facet

As presented in Section 2, a Web-based geographic application manipulates data. In this paper, we consider two categories of data: *Content* and *Annotation*. The former refers to multimedia contents that might be composed of several segments, while the latter concerns the description of named entities (e.g., a person, a location).

As illustrated in Figure 3 (upper part), a content may hold several annotations referring to specific segments. Furthermore, an annotation may be connected with other annotations through different properties.

In this paper, we are considering exclusively textual contents inside a geographic application. Consequently, as presented in Figure 3 (lower part), a text may be segmented into paragraphs and/or tokens. Moreover, annotations will correspond to geographic information, e.g., places. [9] stated that geographic information is composed of three complementary features: temporal, spatial and thematic. We followed the same scheme, however the examples in the paper will focus only on the spatial feature.

Figure 4 presents an RDF/XML sample that corresponds to the data facet described in Figure 3. This description indicates that the file `text.txt` (i.e., the text written in French in Figure 1) contains an annotation (`Annotation1`) about an entity named `Cannes`. This geographic entity is located in the first paragraph of the text at the 9th token. Moreover, a `geolocation` and a `geotype` are associated to this entity.

Naturally, the description of Figure 4 could be enhanced with additional information. For instance, it may express that Nice, which is another spatial entity, is the "Préfecture" of Cannes. Since Nice does not appear in the file `text.txt`, this entity will not be related to a specific segment of the text. Furthermore, the description of Figure 4 may describe other spatial entities contained in the text, such as Biarritz or Lyon.

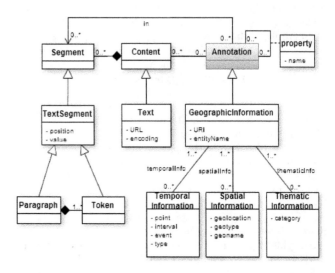

Figure 3: The data facet.

```
<rdf:RDF>
 <wm:Content rdf:about="&ex;text.txt">
  <wm:annotation>
   <wm:GeographicInformation rdf:about="&ex;
       data.rdf#Annotation1">
    <wm:entityName>Cannes</wm:entityName>
    <wm:in>
     <rdf:Description>
      <wm:start rdf:resource="&ex;text.txt#
          Par1-Token9"/>
      <wm:end rdf:resource="&ex;text.txt#Par1-
          Token9"/>
     </rdf:Description>
    </wm:in>
    <wm:spatialInfo>
     <wm:SpatialInformation rdf:about="&
         geotopia;#Cannes">
      <wm:geolocation>MULTIPOLYGON(...)</wm:
          geolocation>
      <wm:geoname>Cannes</wm:geoname>
      <wm:geotype rdf:resource="&ex;data.rdf#
          Town"/>
     </wm:SpatialInformation>
    </wm:spatialInfo>
   </wm:GeographicInformation>
   ...
  </wm:annotation>
 </wm:Content>
</rdf:RDF>
```

Figure 4: An RDF/XML sample of the data facet.

5.2 The graphical interface facet

As illustrated in Figure 1, a Web-based geographic application contains a graphical user interface (GUI). This one may be composed of several viewers organized in the presentation layout. In this paper, we consider three categories of viewers: `TextViewer`, `MapViewer` and `ListViewer`. For instance, in Figure 1 there are a `TextViewer` which contains the text written in French, a `MapViewer` which includes Google Maps and three `ListViewers` enumerating some places.

Figure 5 presents our GUI facet. In this figure, each viewer may embed several annotations. More precisely, spatial entities may be highlighted in a text, displayed on a map or enumerated in a list. In fact, segments associated to annotations are used to locate entities, geolocations are useful for maps and geonames may be employed in lists.

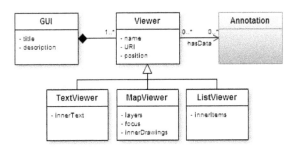

Figure 5: The GUI facet.

Figure 6 presents an RDF/XML sample that conforms to the GUI facet that we proposed in Figure 5. This description shows that the GUI contains a map viewer with a position specified by its top, left, height and width. This viewer embeds an annotation, i.e., `Annotation1`. Since the RDF merge operation has been clearly defined[14], from the RDF/XML samples of Figure 4 and 6, it is possible to display on the map the geolocation of Cannes.

5.3 The user interaction facet

A Web-based geographic application may be composed of several user interactions. For instance, in Figure 1 when a user selects a specific `Town` with a mouse click, this entity is highlighted in the text and the map zoom in on it. In [7] and [20], authors have formalized a vocabulary for defining user interactions, especially they are using the following concepts: *Event* and *Action*. An *Event* is generally characterized by two elements [20]: the type of event (e.g., click, double-click, mouseover) and where the event occurred (e.g., on one entity in a viewer). An event may trigger several *Actions*, e.g., it might be a set of visual effects in the graphical user interface, such as highlight and zoom.

Figure 7 presents the user interaction facet which is inspired from [7]. In this figure, events and actions occur on annotations. Consequently, thanks to the data and GUI facets proposed respectively in Figure 3 and 5, we could retrieve which viewer contains such annotation, its geotype, its geolocation and so on.

Figure 8 presents an RDF/XML sample that corresponds to the user interaction facet described in Figure 7.

---
[14]http://www.w3.org/TR/rdf-mt/

```
<rdf:RDF>
 <rdf:Description rdf:about="&ex;gui.rdf">
  <wm:title>French course</wm:title>
  <wm:viewers>
   <rdf:Bag">
    <rdf:li>
     <rdf:Description rdf:about="&ex;gui.rdf#
         Viewer1">
      <rdf:type rdf:resource="&ex;MapViewer"/>
      <wm:width>400</wm:width>
      <wm:height>300</wm:height>
      <wm:left>300</wm:left>
      <wm:top>20</wm:top>
      <wm:hasData>
       <rdf:Seq">
        <rdf:li rdf:resource="&ex;data.rdf#
            Annotation1"/>
       </rdf:Seq>
      </wm:hasData>
     </rdf:Description>
    </rdf:li>
    ...
   </rdf:Bag>
  </wm:viewers>
 </rdf:Description>
</rdf:RDF>
```

Figure 6: An RDF/XML sample of the GUI facet.

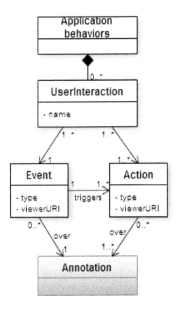

Figure 7: The user interaction facet.

In Figure 8, a potential end-user interaction `I1` is defined. This interaction contains an event `Evt1` and an action `Action1`. This description explains that if a user clicks on any annotations about `Town` in `Viewer2`, the `Viewer1` will zoom in on these annotations. Of course, several interactions could be defined and they may refer several times to the same annotations or the same viewers.

Note that the concept of *Annotation* has an important role, especially for designing a Web-based geographic application, because it allows to link the data, the GUI and the user interactions facets.

```
<rdf:RDF>
 <rdf:Description rdf:about="&ex;hci.rdf">
  <wm:contains>
   <rdf:Bag>
    <rdf:li>
     <rdf:Description rdf:about="&ex;hci.rdf#I1">
      <wm:event>
       <rdf:Description rdf:about="&ex;hci.rdf#Evt1">
        <rdf:type rdf:resource="&ex;Click"/>
        <wm:over rdf:resource="&ex;data.rdf#Town"/>
        <wm:via rdf:resource="&ex;gui.rdf#Viewer2"/>
        <wm:triggers rdf:resource="&ex;hci.rdf#
             Action1"/>
       </rdf:Description>
      </wm:event>
      <wm:action>
       <rdf:Description rdf:about="&ex;hci.rdf#
            Action1">
        <rdf:type rdf:resource="&ex;Zoom"/>
        <wm:over rdf:resource="&ex;data.rdf#Town"/>
        <wm:in rdf:resource="&ex;gui.rdf#Viewer1"/>
       </rdf:Description>
      </wm:action>
     </rdf:Description>
    </rdf:li>
   </rdf:Bag>
  </wm:contains>
 </rdf:Description>
</rdf:RDF>
```

Figure 8: A user interaction encoded in RDF/XML.

6. THE WINDMASH PROTOTYPE

We have implemented our proposal in a prototype named WINDMash[15]. This one allows designers to create and to automatically generate interactive Web-based applications handling geographic information. Figure 9 presents the architecture of our prototype.

As illustrated in Figure 9, our prototype is composed of three modules (i.e., Data, Interface and User interaction) which correspond to our framework proposed in Figure 2. Each module manipulates some RDF/XML descriptions which comply with the model presented in Section 5. Naturally, the merge of all RDF/XML descriptions corresponds to an instance of our unified model which is used to compute and to generate the Web-based geographic application.

In this paper, we only focus on textual contents handled by our prototype. Hence, the data module may invoke several textual indexing services. For instance, the service ① in Figure 9 could directly produce an RDF/XML output complying the data facet proposed in Section 5.1. Besides, other textual indexing services, like the services ② and ③, might be executed. For instance, we have used a service that automatically extracts places from a text and collects places' geolocations [19]. However, the Web service implemented in [19] produces XML outputs that do not correspond to the data facet proposed in Section 5.1. Consequently, we have implemented an XSLT stylesheet in order to translate their XML outputs into some RDF/XML descriptions complying with our data facet.

In order to compute a Web-based geographic application from our unified model, we have implemented a code generator module written in PHP that produces XHTML Web pages containing JavaScript instructions. This module is

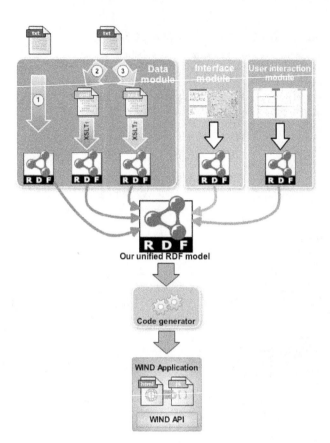

Figure 9: The WINDMash architecture.

using the RAP[16] software package (RDF API for PHP) in order to analyze the RDF descriptions. Indeed, this package allows to query the descriptions with the SPARQL query language [17]. Moreover, we have used our WIND[17] API (Web INteraction Design) which is fully described in [15]. This API enables to describe interactions between components (e.g., map, text) without having to know anything about the underlying Web Mapping Services (IGN Geoportail API, Google Maps API, OpenLayers API, etc.).

In the next section, we will present several tools of our WINDMash prototype which allow designers to instantiate our unified model with its three facets, i.e., data, GUI and user interaction.

7. THE WINDMASH TOOLS FOR DESIGNING WEB-BASED GEOGRAPHIC APPLICATIONS

As illustrated in Figure 9, the WINDMash architecture is composed of three modules which concern, respectively, the data management, the GUI organization and the user interactions specification.

Consequently, three tools have been implemented in our WINDMash prototype for instantiating our unified model presented in Section 5. More precisely, these tools are:

[15]http://erozate.iutbayonne.univ-pau.fr/Nhan/windmash3/

[16]http://www4.wiwiss.fu-berlin.de/bizer/rdfapi/

[17]http://erozate.iutbayonne.univ-pau.fr/Nhan/windapi/wind.html

1. A pipe editor which allows to combine different services and to filter the data manipulated by the application (§7.1);

2. A graphical layout editor which is used to arrange, for instance, mapping components or multimedia contents (§7.2);

3. A UML-like sequence diagram builder which allows to specify potential end-user interactions on the application (§7.3).

In the remainder, each tool listed above is presented. Moreover, these tools are illustrated with screenshots based on the use-case scenario presented in Section 2.1.

7.1 Our pipe editor

In order to manage the data (i.e., contents and annotations) that should be manipulated by the Web-based geographic application, we have developed a pipes editor. This tool has been inspired from the Yahoo! Pipes editor[18] and enables designers to create a processing chain containing different services.

Figure 10 illustrates a possible processing chain which corresponds to the use-case presented in Section 2.1. Actually, from a given text (A), we would like to extract automatically some places (B). For that purpose, we invoke the Place-Extraction Web service which had been detailed in [19]. Thanks to the identified places (B), we have defined two sets of annotations: towns (C) and "Départements" (D). Since the PlaceExtraction service classifies extracted entities, we have implemented some filters (i.e., Town and Department) in order to select the suitable entities. Finally, from the set of towns (C), we needed to get all corresponding "Préfectures" (E). Thanks to available geographical database (e.g., GeoNames, France IGN Database), the Prefecture service computes the appropriate list of "Préfectures" (E) from the list of towns (C).

While constructing the pipes, it is possible to visualize at anytime the computed data by selecting with a double click the B, C, D and E items in Figure 10. Some services may also be customized according to specific needs, e.g., extract places that are situated in a given area. Furthermore, the given text may be updated for computing new datasets.

Each time a dataset is computed, such as the list of extracted places (B), it generates an RDF/XML description which corresponds to the data facet presented in Section 5.1. These descriptions are also accessible at the bottom left of the prototype in Figure 10.

We show in the next subsection, that these descriptions will be used in our graphical layout editor in order to display the data inside viewers.

7.2 Our graphical layout editor

Our graphical layout editor enables a designer to specify the graphical user interface of his/her Web-based geographic application. Indeed, the designer decides which type of viewer he/she wanted in his/her application (e.g., TextViewer, MapViewer, ListViewer) and how these viewers are organized inside the graphical layout.

Figure 11 illustrates how a designer may specify, with our tool, the graphical layout of the use-case presented in Figure 1. The menu on the left hand-side indicates the type

18http://pipes.yahoo.com

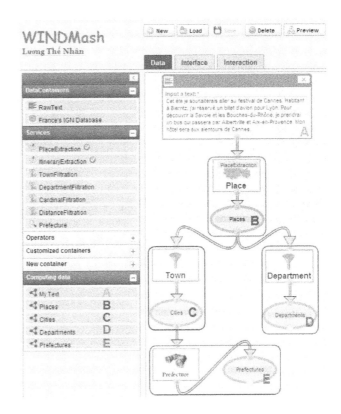

Figure 10: The WINDMash pipes editor.

of viewers that may be manipulated by the designer, the available dataset that have been computed with our pipes editor (Section 7.1) and the viewers that are currently used. In Figure 11, five viewers have been specified: a text viewer (1), a map viewer (2) and three list viewers (i.e., 3, 4 and 5).

Initially, when the designer drags and drops a viewer inside the graphical layout, this viewer is empty, except the map viewer which contains a map. If the designer wants to display some information inside viewers, from the menu, he/she has to drag the computed datasets and to drop them inside a specific viewer. For instance, if the designer wants to see inside the text viewer (1) the text that has been written in Figure 10, he/she has to drag from the menu the My Text element and to drop this element inside the My Text Viewer 1 viewer. Thereafter, if the designer wants to underline the places that have been extracted from this text, he/she has to drag the Places dataset from the menu and to drop it inside the My Text Viewer 1 viewer.

Similarly with the other types of viewers, if the designer wants to see the places on the map, he/she has to drag the Places dataset and to drop it inside the My Map Viewer 2 viewer. Finally, if the designer wants to see the lists of towns, "Départements" and "Préfectures" that have been defined in Figure 10, he/she has to drag the datasets and to drop them inside the list viewers, i.e., My List Viewer 3, My List Viewer 4 and My List Viewer 5.

Of course, it is possible to customize each viewer. For instance, the style of a text inside a text viewer may be modified, such as its font, its size, etc. Furthermore, the type of the lists may also be changed, e.g., with or without

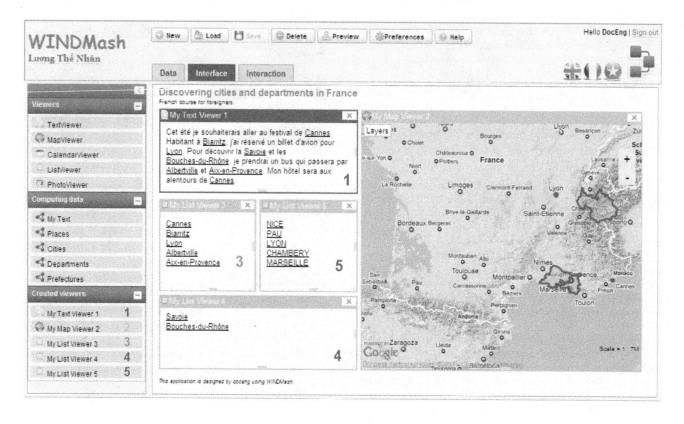

Figure 11: The WINDMash graphical layout editor.

different kinds of bullets. Finally, different types of maps may be used, such as Google Maps, Yahoo! Maps, IGN Maps[19]...

In the next subsection, we present how it is possible to specify potential end-user interactions using the layout components that have been designed in this section.

7.3 A "sequence diagram" builder for specifying end-user interactions

As required in the use-case scenario proposed in Section 2.1, when a user selects a specific place in the lists, the map should focus on this place and, if possible, the corresponding terms in the text should be highlighted. [12] proposed to design this kind of interaction through UML-like sequence diagrams.

Effectively, these diagrams are useful for designers because:

- they are based on few concepts to describe the interactions between a user and the system;

- they clearly distinguish events made by a user on the system (i.e., arrows from the user to the system) and the potential actions of the system to the user (i.e., arrows from the system to the user);

- they can both describe the interactions between the user and the system, as well as the interaction between the components of the system.

Consequently, we have implemented a UML-like sequence diagram builder which is illustrated in Figure 12[20]. On the left hand-side of this figure, the menu is composed of the list of viewers that have been specified in Section 7.2 and the list of datasets that have been defined in Section 7.1.

The sequence diagram example illustrated in Figure 12 describes the following interaction: When a user selects, through a click (see the right arrow), a town contained in the viewer named My list Viewer 3, this town is highlighted (see the left arrow labeled with the term highlight) in the viewer named My Text Viewer 1. Moreover, the viewer My Map Viewer 2 also zooms in on this town (see the left arrow labeled with the term zoom).

When the designer has finished to build the sequence diagram, the WINDMash prototype handles a global RDF/XML description which complies with the unified model presented in Section 5. Consequently, from this description and our code generator module, the designer could preview the Web-based geographic application by selecting the Preview button (see the buttons on top of Figure 12). If the generated application suits his/her needs, the designer may save the application and/or deploy it on a specific client. Otherwise, the designer may come back to the three WINDMash tools presented in this section (i.e., the figures 10, 11 and 12) in order to add, to modify or to remove some application elements.

From the WINDMash tools presented in this section and, especially the examples illustrated in the figures 10, 11 and

---

[19]http://www.geoportail.fr

[20]Currently, this UML-like sequence diagram builder only allows to create lifelines and to specify messages between them via arrows.

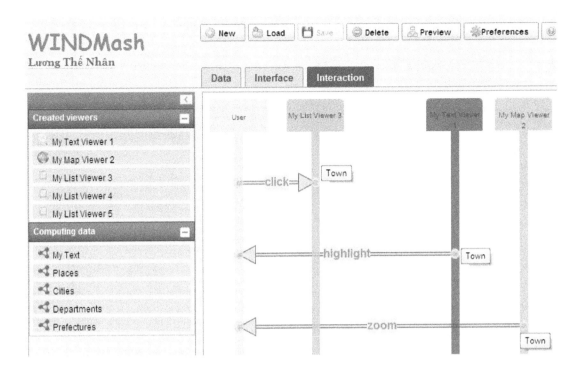

Figure 12: The WINDMash sequence diagram builder.

12, the complete generated Web-based geographic application is available at http://erozate.iutbayonne.univ-pau.fr/Nhan/windmash3/demo/Frenchcourse/app.html. A video screencast of our WINDMash prototype is also available at http://erozate.iutbayonne.univ-pau.fr/Nhan/windmash3/demo/Frenchcourse/video.html

8. CONCLUSION

In this paper, we have presented a framework dedicated to the design of Web-based geographic applications. This framework addresses three complementary tasks which concern the data manipulated by the application, the graphical layout and the user interactions. We have shown through our unified model for describing such kind of applications that annotations are central in the design process. Indeed, they can be used to describe entities, to display information inside viewers and to specify application behaviors. Furthermore, our proposed framework has been implemented in an online prototype named WINDMash. This prototype contains different tools that facilitate the instantiation of our unified model and automatically generates an executable Web-based geographic application.

Currently, our prototype only deals with textual contents. However, the data model presented in this paper is sufficiently generic to be extended in order to deal with multimedia contents, such as videos, audios and images. Furthermore, the geographic information manipulated by our WINDMash prototype can be imported from other repositories, for example from the LinkedGeoData[21] which exploits the spatial information collected by OpenStreetMap[22]. Hence, a future work would consist in developing a mediator be-

tween the imported LinkedGeoData datasets and our unified model. Moreover, we plan to extend our model in order to import non-geographic information, e.g., specific manual annotations.

Finally, we plan to export the generated Web-based geographic applications in XHTML+RDFa [1] instead of XHTML. XHTML+RDFa export will enable WINDMash generated applications to take more advantage of the Semantic Web and, especially the Linking Open Data cloud [11].

9. ACKNOWLEDGMENT

This work has been supported by the French Aquitaine Region (project number 20071104037), the Département of Pyrénées-Atlantiques ("Pyrénées : Itinéraires Éducatifs" project) and the ANR MOANO project (http://moano.liuppa.univ-pau.fr).

10. REFERENCES

[1] ADIDA, B., BIRBECK, M., MCCARRON, S., AND PEMBERTON, S. RDFa in XHTML: Syntax and Processing. Recommendation, W3C, October 2008. http://www.w3.org/TR/rdfa-syntax/.

[2] ALBINOLA, M., BARESI, L., CARCANO, M., AND GUINEA, S. Mashlight: A Lightweight Mashup Framework for Everyone. In *Proceedings of the 2nd Workshop on Mashups, Enterprise Mashups and Lightweight Composition on the Web* (2009).

[3] ALTINEL, M., BROWN, P., CLINE, S., KARTHA, R., LOUIE, E., MARKL, V., MAU, L., NG, Y.-H., SIMMEN, D., AND SINGH, A. Damia: a data mashup fabric for intranet applications. In *Proceedings of the 33rd International Conference on Very Large Data Bases* (2007), VLDB'07, pp. 1370–1373.

---

[21]http://linkedgeodata.org
[22]http://www.openstreetmap.org/

[4] BECKETT, D., AND MCBRIDE, B. RDF/XML Syntax Specification (Revised). Recommendation, W3C, February 2004. http://www.w3.org/TR/rdf-syntax-grammar/.

[5] BELETSKI, O. End user mashup programming environments. Tech. rep., Helsinki University of Technology, Telecommunications Software and Multimedia Laboratory, 2008.

[6] CAPPIELLO, C., DANIEL, F., MATERA, M., PICOZZI, M., AND WEISS, M. Enabling End User Development through Mashups: Requirements, Abstractions and Innovation Toolkits. In *Proceedings of the Third International Symposium on End-User Development* (2011), pp. 9–24.

[7] ENGELS, G., HAUSMANN, J. H., HECKEL, R., AND SAUER, S. Dynamic meta modeling: a graphical approach to the operational semantics of behavioral diagrams in UML. In *Proceedings of the 3rd International Conference on the Unified Modeling Language: advancing the standard* (Berlin, Heidelberg, 2000), UML'00, Springer-Verlag, pp. 323–337.

[8] ENNALS, R. J., AND GAROFALAKIS, M. N. Mashmaker: mashups for the masses. In *Proceedings of the 2007 ACM SIGMOD International Conference on Management of Data* (New York, NY, USA, 2007), SIGMOD'07, ACM, pp. 1116–1118.

[9] GAIO, M., SALLABERRY, C., ETCHEVERRY, P., MARQUESUZAÀ, C., AND LESBEGUERIES, J. A global process to access documents' contents from a geographical point of view. *Journal of Visual Languages and Computing 19* (February 2008), 3–23.

[10] HAKLAY, M., SINGLETON, A., AND PARKER, C. Web Mapping 2.0: The Neogeography of the GeoWeb. *Geography Compass 2*, 6 (2008), 2011–2039.

[11] HEATH, T., AND BIZER, C. *Linked Data: Evolving the Web into a Global Data Space*. Synthesis Lectures on the Semantic Web: Theory and Technology. Morgan & Claypool, 2011.

[12] HENNICKER, R., AND KOCH, N. Modeling the user interface of web applications with UML. In *Practical UML-Based Rigorous Development Methods - Countering or Integrating the eXtremists, Workshop of the pUML-Group held together with the UML 2001* (2001), pp. 158–173.

[13] HUYNH, D. F., KARGER, D. R., AND MILLER, R. C. Exhibit: lightweight structured data publishing. In *Proceedings of the 16th International Conference on World Wide Web* (New York, NY, USA, 2007), WWW'07, ACM, pp. 737–746.

[14] LIN, J., WONG, J., NICHOLS, J., CYPHER, A., AND LAU, T. A. End-user programming of mashups with Vegemite. In *Proceedings of the 14th International Conference on Intelligent User Interfaces* (New York, NY, USA, 2009), IUI'09, ACM, pp. 97–106.

[15] LUONG, T. N., ETCHEVERRY, P., NODENOT, T., AND MARQUESUZAÀ, C. WIND: an Interaction Lightweight Programming Model for Geographical Web Applications. In *Proceedings of the International Opensource Geospatial Research Symposium* (2009), OGRS'09. Available online.

[16] MANOLA, F., AND MILLER, E. RDF Primer. Recommendation, W3C, February 2004. http://www.w3.org/TR/rdf-syntax/.

[17] PRUD'HOMMEAUX, E., AND SEABORNE, A. SPARQL Query Language for RDF. Recommendation, W3C, January 2008. http://www.w3.org/TR/rdf-sparql-query/.

[18] RO, A., XIA, L. S.-Y., PAIK, H.-Y., AND CHON, C. H. Bill Organiser Portal: A Case Study on End-User Composition. In *Proceedings of the 2008 International Workshops on Web Information Systems Engineering* (Berlin, Heidelberg, 2008), WISE'08, Springer-Verlag, pp. 152–161.

[19] SALLABERRY, C., ROYER, A., LOUSTAU, P., GAIO, M., AND JOLIVEAU, T. GeoStream: Spatial Information Indexing Within Textual Documents Supported by a Dynamically Parameterized Web Service. In *Proceedings of the International Opensource Geospatial Research Symposium* (2009), OGRS'09. Available online.

[20] STÜHMER, R., ANICIC, D., SEN, S., MA, J., SCHMIDT, K.-U., AND STOJANOVIC, N. Lifting events in RDF from interactions with annotated web pages. In *Proceedings of the 8th International Semantic Web Conference* (Berlin, Heidelberg, 2009), ISWC'09, Springer-Verlag, pp. 893–908.

[21] TAIVALSAARI, A. Mashware: The future of web applications. Tech. rep., Sun Microsystems Laboratories, 2009.

[22] VIANA, W., FILHO, J. B., GENSEL, J., VILLANOVA-OLIVER, M., AND MARTIN, H. PhotoMap: From Location and Time to Context-Aware Photo Annotations. *Journal of Location Based Services 2* (September 2008), 211–235.

[23] WONG, J., AND HONG, J. I. Making mashups with Marmite: towards end-user programming for the Web. In *Proceedings of the SIGCHI conference on Human Factors in Computing Systems* (New York, NY, USA, 2007), CHI'07, ACM, pp. 1435–1444.

Timesheets.js: When SMIL Meets HTML5 and CSS3

Fabien Cazenave
INRIA
655 avenue de l'Europe
38334 Saint Ismier, France
fabien.cazenave@inria.fr

Vincent Quint
INRIA
655 avenue de l'Europe
38334 Saint Ismier, France
vincent.quint@inria.fr

Cécile Roisin
Grenoble University & INRIA
655 avenue de l'Europe
38334 Saint Ismier, France
cecile.roisin@inria.fr

ABSTRACT

In this paper, we explore different ways to publish multimedia documents on the web. We propose a solution that takes advantage of the new multimedia features of web standards, namely HTML5 and CSS3. While JavaScript is fine for handling timing, synchronization and user interaction in specific multimedia pages, we advocate a more generic, document-oriented alternative relying primarily on declarative standards: HTML5 and CSS3 complemented by SMIL Timesheets. This approach is made possible by a Timesheets scheduler that runs in the browser. Various applications based on this solution illustrate the paper, ranging from media annotations to web documentaries.

Categories and Subject Descriptors

I.7 [**Document and Text Processing**]: Document Preparation—*Markup languages, Multi/mixed media, Standards*

General Terms

Design, Experimentation

Keywords

Declarative languages, Multimedia, Web applications, SMIL, HTML5

1. INTRODUCTION

The multimedia web is evolving very rapidly. With the advent of HTML5, web pages can now integrate graphics, sound and video seamlessly. In addition, scripting languages can animate content such as text, graphics, pictures, as well as continuous media. These multimedia contents can be played on an increasing number of terminals, ranging from the traditional desktop to the smallest pocketable mobile device, and with good performances.

Rendering true multimedia web content is becoming easier. The user experience on web pages embedding video and sound is now much smoother than back in the days when specific plug-ins were required. Modern web browsers can now render natively all sorts of contents embedded in HTML5 pages. In addition, they are supporting graphic formats such as SVG, including its animation feature, as well as the latest properties introduced in CSS to handle transitions and animations, not to mention their ability to manipulate all these contents through powerful script engines.

With these novel multimedia features, new web applications are made possible. They are very diverse, but they all have in common the addition of a time dimension to the usual web document. Time could be present through such continuous content as video or sound, through a time structure added to discrete contents, through animations performed by executing some code (script), or through a combination of these. Concretely, these timed multimedia applications may be slide shows, captioned video clips, annotated audio recordings, graphic animations, augmented recorded conferences, interactive photo albums, web documentaries, and so on.

To implement such web applications, various approaches can be considered, ranging from declarative to imperative:

- The purely declarative approach was taken by SMIL [4], the first multimedia technology specially designed for the web. In SMIL, the time dimension of a document is expressed by a hierarchy of temporal operators. Applications are run by players that interpret the declarative language and play the document (Ambulant [2], RealPlayer, X-Smiles [12]).

- The imperative approach is illustrated by the many ad hoc applications written in JavaScript and ActionScript, where scripts are used to handle user interaction and to make the document change over time.

Obviously, both approaches can be combined; many applications using a declarative language are complemented by some scripting. It is worth noting that scripting languages can also be used to implement players that interpret a declarative language. For instance, the SMIL Timesheets language [17] was implemented in JavaScript (Timesheets JavaScript Engine [16], LimSee3 [10], FakeSmile[1]), but this does not impact application developers, who still work declaratively.

The choice between both approaches should take many criteria into account. What efforts are necessary to make sure the content can be enjoyed on different devices? Is accessibility for people with disabilities granted? What kind

---

[1] http://leunen.d.free.fr/fakesmile/

of user interaction is supported? Can (some parts of) the application be easily reused in another application? How easy is it to maintain applications? Is it possible to reference some particular piece of content or some specific state or date in the document? How can synchronization with continuous content be achieved? Answers to these questions may vary strongly depending on the approach.

In this paper, we try to identify the advantages of each approach. We consequently propose a trade-off that brings the best of each world regarding the questions raised above, and we propose tools that help achieving the best trade-off. In particular, we have chosen to explore a solution built upon HTML5 and SMIL Timing.

The rest of the paper is organized as follows. The next section presents various categories of multimedia web applications. It is followed by a review of the current means available to implement these applications. Sections 4 and 5 propose solutions based on SMIL Timesheets as a new way to improve the current situation. Section 6 provides a few examples to illustrate these solutions. Finally the conclusion summarizes the main contributions of the paper and envisions future work.

2. MULTIMEDIA APPLICATIONS

For the sake of clarity, we divide multimedia web applications in two main categories, depending on the role of the time dimension.

2.1 Media-Driven Applications

We call *media-driven applications* the broad category of multimedia applications where a piece of continuous content plays the role of a backbone for the whole application. In these applications, typically, an audio or video recording constitutes the main content, and various elements (often discrete media such as text or pictures) are associated to parts of this main content to annotate it. Interactive features are also available to the user for moving freely in the main content and to interact with complementary information. This category is exemplified by such applications as:

- a captioned movie: the main content is the movie itself, and the captions constitute the associated content. The display of each caption is precisely synchronized with the video. In addition to the usual VCR controls, a menu allows the user to change caption language or to hide them at any time. A very fine-grained synchronization may be required when the annotation is the transcript of the audio track, like in the MIT150 Infinite History project [11].

- a commented program in an on-line radio archive: the backbone is an audio file, the recorded radio program. Some pictures and textual annotations are associated with specific parts of the program to illustrate them or to provide additional information. Usual controls are provided to pause, play, move forward/backward, and to hide/show additional content.

- a videotaped talk with synchronized slides (see Figure 1): the main content is the video recording of the speaker, which is complemented with the slides (pictures and/or text) the speaker used when giving the talk. Slides are synchronized with the video and displayed next to it. In addition to the usual controls for

a video, an interactive table of contents allows the user to freely navigate through the video and the sequence of slides.

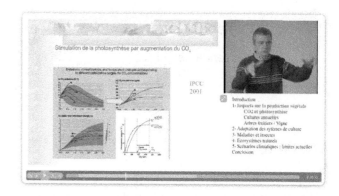

Figure 1: Videotaped talk with synchronized slides

In all these applications, the master media object comes with its intrinsic clock which is used to schedule the whole application. The time dependencies of all other contents are expressed relatively to this clock. In the latest example, each slide is associated with two dates: the time when it must be displayed and the time when it must disappear. Both dates are relative to the master clock. In the same way, items of the table of contents have two dates: the time when the corresponding section starts in the video and the time when it stops. These dates are used to highlight the relevant entry in the table of contents, or to skip to the right position in the video when the user clicks an entry.

Because of this time structure, it is often easy in this kind of application to superimpose one or several sectioning structures on the master media object. Typically, in the example above, both the sequence of slides and the table of contents play this role. The slide titles may be used to associate a series of labels with the timeline, while the headings from the table of contents add a hierarchical structure. This can be reflected through the user interface: when moving the pointing device along the timeline, the user can see various levels of labels, which are helpful to choose a particular part in a long video.

This is for instance the approach taken in the Advene platform where multiple levels of annotations can be associated with a movie [3].

2.2 Event-Driven Applications

As opposed to media-driven applications, *event-driven applications* are not organized around a single continuous media object that provides the main synchronization scheme. Instead, they are made of a collection of multiple media objects related by links such as temporal relations or user interactions.

To illustrate this concept, let us take the example of a slide show. Each slide may be a single picture or a piece of text, but it could also be a small multimedia document itself, with various (possibly continuous) media objects and some interactive features. Slides are linked together to define a preferred sequence for presentation. They also offer the user a way to conveniently move from one slide to the next/previous one in this sequence. A table of contents (or an index) may be available, with links to every slides, offering the user another way to access slides, in any order.

This kind of organization may be used for a photo album for instance. The table of contents contains thumbnails of all pictures, and each slide is constituted by a single photograph with a caption, comments and buttons to go to the next or previous slide, or to the thumbnail index. The same kind of organization is used for the slides displayed on a large screen during a talk with a tool such as HTML Slidy [13].

The category of event-driven applications is broader than the slide show family. It actually includes all applications where there is no dominant time structure, but multiple, different time structures for different parts of the document. Concurrent time structures may occur simultaneously, or some parts may have no intrinsic time dimension, only temporal relations with other pieces of content. For instance, a media object may have to be presented after (or at the same time as) another one. These relations are typically those defined by Allen [1].

3. STATE OF THE ART

3.1 Multimedia Web Authoring

The issue of integrating multimedia content into web pages dates back from the early days of the web. In the initial version of WorldWideWeb, the first web browser ever (1990), even still pictures where displayed in separate windows. They were included in the text later on (1993), when the NCSA-Mosaic browser introduced the `img` tag in HTML. Integration of continuous media objects, in particular video, followed the same way, but much later. For a long time, these objects were handled by separate programs (plug-ins) and were therefore difficult to integrate in the document. Only when SMIL introduced the `video` and `audio` elements could web documents really include continuous media. SMIL was followed in this direction by SVG and more recently by HTML5. All three languages now support audio and video objects natively.

The time dimension of documents was not the priority in the first solutions developed for multimedia web content. Web formats have left aside the issue of synchronization for a long time. Plug-ins did not allow the various components of a web page to be synchronized with continuous media through a standard API. The first step in this direction was made with SMIL, with a rather radical approach: time is the main (and almost the only) dimension for structuring a document. As a consequence, the hierarchical structure offered by other web formats for representing the logical organization of documents can hardly be expressed with SMIL.

Using an XML environment is a natural way to put the focus on the logical dimension when creating multimedia content: the document is structured in XML to encode its logical organization, and time relations are expressed as part of this structure [6]. This option requires some export mechanism to produce documents in a format that can be accepted by a web browser. The main issue is that the authoring and the publishing languages are different. The original XML code has to be converted to SMIL, Flash or HTML5+JavaScript. Authors are prevented from using familiar web languages such as HTML and CSS, and if they want to make adjustments to the final form, they may have to do complex reverse transformations to update the source document [15].

The drawbacks of the conversion is balanced by the advantages of the declarative approach. Authors think in terms of a multimedia *document* instead of a multimedia application they would have to *program* with a scripting language. The document-oriented (declarative) approach thus makes multimedia web authoring available to a broader audience. The declarative approach also provides advantages from an engineering point of view. It makes it easier to maintain and reuse content. Multimedia documents can then be used in a document workflow, for instance.

3.2 Multimedia Web Technologies

There are three main classes of techniques (SMIL, Flash, HTML5) for developing synchronized multimedia documents for the web. Figure 2 reviews these techniques, based on the following criteria:

- Timing and synchronization: This is the most complex part of many advanced multimedia applications. Therefore the language must provide efficient ways to alleviate the task of web developers in this area.

- Scriptability: Whatever the level of declarativeness of a language, it is important to be able to go beyond the limitations of the language by using scripts that extend its capabilities in some particular situations (this requirement has been highlighted by [14]).

- Logical structure: Document formats on the web are typically declarative structured languages. By describing first and foremost the logical organization of a document, they facilitate accessibility, adaptability, reuse, device independence, and processability.

- Content/presentation separation: Keeping this principle, which is widely applied on the web with style sheets, offers flexibility in customizing and adapting content for different contexts of use.

- Temporal links [7]: It is common practice on the web to link to a specific position within a text document, thanks to fragment identifiers in URIs. The equivalent for a timed document is to point to a specific date or state in the execution of the document. This feature allows timed documents to integrate nicely in the web.

- Native rendering in web browsers: The browser is the main tool for accessing the web on all devices. To make sure web users can use multimedia applications, it is important that applications can be run in web browsers.

- Standard compliance: W3C standards are well known to web developers. A solution based on these standards is more likely to be adopted by them. These standards are also widely implemented, in particular in browsers. Applications published according to W3C standards can meet a large audience.

- Device friendliness: Even if web browsers implementing W3C standards are available on all devices, there are significant differences in the capabilities of these devices. The chosen technology should be compatible with as many device classes as possible.

We use these criteria to review the technologies used for developing multimedia web applications (see Figure 2):

	SMIL	Flash	HTML5 + CSS3
Declarative timing and sync.	+	-	-
Scriptability	-	+	+
Logical structure	-	-	+
Content/presentation separation	-	-	+
Temporal links	+	-	-
Native rendering in web browsers	-	-	+
W3C standard compliance	+	-	+
Desktop friendliness	+/-	+	+
Mobile friendliness	+/-	+/-	+

Figure 2: Technology Comparison

- SMIL was designed specifically for multimedia web applications. Its main focus is a rich time structure and content synchronization. Unfortunately, few media players and no web browser support SMIL. Moreover, SMIL has not evolved along with popular web languages such as HTML and CSS. It does not allow to separate presentation from content, and it is impossible to script a SMIL document in a web browser. This makes it difficult to integrate the SMIL language into other web applications. In addition, the structure it represents is the time structure, not the logical structure of a document.

- Flash is currently by far the main technology for multimedia applications. It works well on most desktop browsers, but it is not so widely supported on mobile devices (smartphones, tablets). Since it comes as a binary format, it raises accessibility issues and requires more work to address indexability aspects. More generally, as Flash is not rendered natively by web browsers there are strong limitations regarding the content/presentation separation, and the interactions between a web page and its multimedia (Flash) content are more complex.

- HTML5 was specified for the web with audio/video in mind. HTML5 is complemented by the CSS3 style sheets language. With HTML5 and CSS3, web developers can take advantage of the clean content/ presentation separation, as well as many other web features, but they have to rely on JavaScript to handle timing, synchronization and user interactions. Concerning multimedia applications, HTML5 with its `audio` and `video` elements is supported natively by modern web browsers and, thanks to plug-ins, fallback solutions exist for audio/video contents in legacy browsers (Internet Explorer 6 to 8).

3.3 Declarative Solutions

As seen on Figure 2, Flash suffers from too many limitations to be considered a good web citizen. Web developers are left with SMIL and HTML5+CSS3 (see Figure 3), but neither solution is completely satisfactory.

The latest version of SMIL, namely SMIL 3.0 [4], provides two modules (SMIL State and SMIL Transitions) in addition to the central SMIL Timing module, in order to better cope with advanced multimedia requirements. It is a declarative approach to multimedia contents:

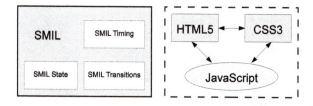

Figure 3: SMIL vs. HTML5

- Timing and Synchronization features are defined by the SMIL Timing module. SMIL Timing brings a simple and very powerful way to describe, in a declarative way, media synchronization and user interaction management for multimedia applications.

- SMIL State [8] and its variant for NCL [14] aim at providing variables inside the declarative time structure (as defined by SMIL or NCL) to cope with the need for controlling document playback, but the resulting syntax is a bit verbose and requires a specific SMIL implementation. Besides, it still cannot be extended to more complex imperative scenarios (e.g. functions, prototypes, objects).

- SMIL Transition Effects can be used to enhance the user experience in pure SMIL documents. CSS3 transitions are the counterpart for HTML5.

The most frequent approach is based on HTML, CSS and JavaScript (see right part of Figure 3):

- Developers of multimedia applications rely on HTML5 to describe the content with its logical structure, and on CSS3 for the presentation. This way, they can separate content from presentation.

- The document playback is controlled in JavaScript by using element APIs and DOM events as defined by W3C recommendations. In particular, the HTMLMediaElement API provides an efficient control for `audio` and `video` elements.

- Developers still have to address most timing and user interaction issues with specific JavaScript code. This is the main difficulty in designing multimedia applications with HTML5: the scripts are designed for a specific DOM structure and use pre-defined CSS classes, which makes such JavaScript developments very difficult to maintain and reuse, unless the relationship between classes and features is carefully documented.

- In many applications, especially in event-driven applications, developing and debugging scripts represent a large part of the development effort.

King *et al.* [9] have proposed XML language extensions to allow multimedia systems to react to dynamic events, and to handle continuous real-time dependencies. In our solution, we take the same implementation approach (a scheduler engine in JavaScript) but we can now take advantage of new features of declarative web languages to partially cover the same needs.

4. PROPOSED SOLUTION

4.1 SMIL Timing and Timesheets

Based on the observations above, it appears that combining HTML5+CSS3 and SMIL Timing would bring a good solution (see Figure 4). SMIL Timing specifies two attributes, `timeContainer` and `timeAction` for integrating timing and synchronization features into HTML and XML documents.[2] Basically, SMIL Timing allows an a-temporal language such as HTML5 to be extended with timing features.

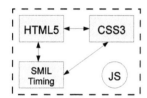

Figure 4: SMIL with HTML5+CSS3

SMIL Timesheets reuses a significant subset of SMIL Timing and allows timing and synchronization to be separated from content and presentation. This can be seen as the counterpart of CSS style sheets in the timing domain: like in CSS, these features are gathered either in an external resource linked to the document, or in a `timesheet` element in the document itself. Like CSS style sheets, timesheets can be associated not only to HTML pages but also to other types of documents such as SVG drawings, for instance, or even to compound documents made of HTML and SVG.

Our approach [5] is based on SMIL Timesheets and can be summed up in three points:

- use HTML5+CSS3 for structuring the content and for rendering it natively in the browser with a clean content/presentation separation;

- rely on SMIL Timing to handle timing, media synchronization and user interaction;

- do not ever redefine what already exists in HTML, SVG and CSS (e.g. animations and transitions), as illustrated in Figure 7.

---

[2]http://www.w3.org/TR/SMIL3/smil-timing.html#Timing-IntegrationAttributes

4.2 Timesheet Engine

As SMIL Timing and Timesheets are not supported natively by web browsers, a JavaScript implementation of these specifications is required. We have developed timesheets.js,[3] which is an open-source, cross-browser, dependency-free library that supports the common subset of the SMIL Timing and SMIL Timesheets specifications.

This still relies on JavaScript, but no specific JavaScript development is required from a web developer for most multimedia applications: the whole application is created using only declarative languages. When such an application is running, some parts of it (HTML and CSS) are executed natively by the browser, some other parts are executed by the browser's JavaScript engine.

Timesheets.js is not the first SMIL Timesheets engine running in the browser. Vuorimaa has developed a Timesheets JavaScript Engine [16] but it has a few limitations for our use cases:

- As it has been developed in 2007, before the raise of SVG 1.2 and HTML 5, it does not support continuous media. In the LimSee3 project [10] (2008), this timesheet engine was adopted to handle the Timing module better and to play continuous media elements through a VLC plug-in, but it still cannot use any event sent by these continuous media, which leads to weak synchronization.

- Both implementations handle only internal timesheets; the W3C Timesheets 1.0 specification does not mention explicitly any other way to use timesheets with a-temporal languages, but we wanted to support both internal and external timesheets, as well as inline SMIL Timing markup, within the same parser.

- Both implementations rely on clock arrays and try to determine begin/end values as soon as possible, which is fine in a fully declarative approach, but is limitative when it comes to adjusting time containers dynamically with JavaScript.

The FakeSmile project has also experimented with SMIL Timesheets, but this implementation is focused on SVG animations and could not be easily reused in a broader scope.

For these reasons, we felt it was preferable to start a new development. After all, our implementation is only about 2000 lines of code, and the whole engine is less than 10 Kbytes in the minified/gzipped version. Technically speaking, the timesheet scheduler is very modular by design:

- Each node declared in the timesheet as a time container has its own clock, methods, properties and event handlers.

- Each time container parses its own descendants (time nodes) and pre-calculates the begin/end time values according to its temporal behavior: `seq`, `par` or `excl`.

- All time containers expose a significant part of the HTMLMediaElement API (which is exposed by the `audio` and `video` elements): web developers can control SMIL time containers with the usual `.play()` / `.pause()` methods, check the time with the `.currentTime` property and register to standard `timeupdate` DOM events.

---

[3]http://wam.inrialpes.fr/timesheets/public/timesheets.js

To put it another way, we wanted the timesheets.js library to be more than just an implementation of the SMIL Timing and Synchronization module. It actually offers a declarative framework for web-based multimedia applications, which can easily be extended to fit specific needs (see Section 5).

It is worth to mention that, as SMIL Timesheets and SMIL Timing are not intended only for HTML documents, the timesheets.js library can be used with SVG content too, and therefore with compound documents. This allows synchronized multimedia applications to include vector graphics in addition to the usual HTML content, and to use time constraints within drawings and between (parts) of drawings and other parts of a HTML document.

4.3 A Basic Example

As an example, here is the very simple case of a rotating banner where three images are displayed one after another:

```
<script type="text/javascript" src="timesheets.js"/>
<div smil:timeContainer = "seq"
     smil:timeAction     = "display"
     smil:repeatCount    = "indefinite">
  <img smil:dur="3s" src="image1.png"/>
  <img smil:dur="3s" src="image2.png"/>
  <img smil:dur="3s" src="image3.png"/>
</div>
```

- The `smil:timeContainer` attribute turns the `div` element into a SMIL time container. Value `seq` defines a sequence in which elements play one after the other.

- The `smil:timeAction` attribute defines *how* the element is to be activated. In this case, the `display` CSS property is set to `block` when the element is active, `none` otherwise. The same mechanism can be used to trigger CSS transitions and animations.

- The `smil:repeatCount` attribute indicates the number of iterations.

- The `smil:dur` attribute specifies the duration of the element.

As a result, the three images are displayed one after the other, each one during 3 seconds, and this is repeated indefinitely, thus creating a rotating banner. The same result may be achieved with an external timesheet, clearly separating timing from content. Here is an equivalent markup:

```
<script type="text/javascript" src="timesheets.js"/>
<link href="banner.smil" rel="timesheet"
      type="application/smil+xml"/>
<div id="banner">
  <img src="image1.png"/>
  <img src="image2.png"/>
  <img src="image3.png"/>
</div>
```

where the external timesheet `banner.smil` contains:

```
<?xml version="1.0" encoding="UTF-8"?>
<timesheet xmlns="http://www.w3.org/ns/SMIL">
  <seq repeatCount="indefinite">
    <item select="#banner img" dur="3s"/>
  </seq>
</timesheet>
```

Attribute `select` of `item` performs a `querySelectorAll()` action: for each DOM node that is matched by the `#banner img` selector, a SMIL item is created. This allows the same timesheet to be reused for several HTML pages: the SMIL markup above always works whatever the number of images in the banner.

4.4 Supported SMIL Features

Our timesheet scheduler supports a significant subset of both SMIL Timing and SMIL Timesheets (see Figure 5). As the same parser is used for both inline timing and timesheets (internal or external), a few SMIL Timing features are also supported in timesheets, and vice-versa.

The `timeContainer` and `timeAction` attributes are the two "integration attributes", as mentioned in the SMIL Timing specification. However, the SMIL Timesheets draft does not mention any `timeAction` attribute. This attribute is essential because it specifies *how* an element is activated in terms of CSS properties.

The `begin` and `end` attributes are supported with two restrictions: only positive values are taken into account; all time formats and event-values (e.g. "button.click") are supported, but there is no support for mixed time *and* event-value yet (e.g. "button.click+5s").

The `item` element and its `select` attribute are very specific to SMIL Timesheets and are thus ignored by our implementation in inline timing markup. On the other hand, the `first`, `prev`, `next`, and `last` attributes, which have been proposed by the SMIL Timesheets specification to control `excl` containers easily (for "lazy user interaction"), do make sense in an inline timing context, and are therefore supported.

4.5 Proposal: Transition Triggers

Timesheets.js fully supports the SMIL `timeAction` attribute, as defined in the SMIL Timing recommendation.[4] To get a sharper control on the way elements are activated, our timesheet scheduler sets a 3-state `smil` custom attribute to targeted elements, containing values `idle | active | done`, before | during | after element activation respectively.

This attribute can be used in CSS selectors to specify asymmetric transitions like in the following example which defines a carousel effect:

```
div[smil=idle] { /* state before transition */
  opacity: 0;
  transform: scale(0.3) translate(+200%);
}
div[smil=done] { /* state after transition */
  opacity: 0;
  transform: scale(0.3) translate(-200%);
}
div[smil=active] { /* state when active */
  opacity: 1;
  /* "transform: none;" is implicit */
}
```

Setting a custom attribute is a working solution, but a more satisfying solution would be to define SMIL-specific pseudo-classes in CSS.

---

[4]http://www.w3.org/TR/SMIL3/smil-timing.html#Timing-timeActionAttribute

Node	SMIL Timing	SMIL Timesheets	timesheets.js
timeContainer	+	+	+
timeAction	+	-	+
begin, end	+	+	+/-
dur	+	+	+
fill, endSync	+	+	+/-
repeatDur, repeatCount	+	+	+/-
item, select	N/A	+	+
first, prev, next, last	-	+	+

Figure 5: SMIL Timing and Timesheets Support in timesheets.js

4.6 Proposal: mediaSync Attribute

The SMIL Timesheets draft does not mention any way to synchronize explicitly a time container with a continuous media. The SMIL Timing recommendation defines a boolean syncMaster attribute on media elements and time containers, that forces other elements in the time container to synchronize their playback to this element.[5] However, as the audio and video elements do not exist in SMIL timesheets, this syncMaster attribute is usable only with inline markup, and not with timesheets.

In order to define the same synchronization feature with timesheets we have introduced the mediaSync attribute, that can be used either with inline markup or within timesheets, which refers to a continuous media element through a CSS selector (in order to be consistent with the timesheet-specific select attribute).

The markup below is an example of a captioned movie that uses the mediaSync attribute and inline timing markup: each caption is an HTML paragraph synchronized with the video element.

```
<video src="myvideo.webm" />
<div smil:timeContainer = "excl"
    smil:timeAction    = "display"
    smil:mediaSync     = "video">
  <p smil:begin="0:00.00" id="intro">
    Title <br />
    by <a href="http://homepage/">Director</a>
  </p>
  <p smil:begin="0:04.93" smil:end="0:10">...</p>
  <p smil:begin="0:11.14">             ...</p>
  ...
    <p smil:begin="1:05.00" id="conclusion">...</p>
</div>
```

The timesheet scheduler parses the value of the mediaSync attribute and performs a querySelector() on its value. As there is only one video element in this page, value video is enough. This could also be done with syncMaster, but the main point is that the proposed markup is suitable for timesheets as well. Another benefit is that the video element does not have to be nested in the time container, which helps separate the content from the timing logic.

The SMIL Timing module includes a note about hyperlink implication on the seq and excl time containers.[6] It allows

links to activate a particular time node in these time containers. URI http://website.tld/page.html#conclusion, for instance, refers to the element with a conclusion id in the above example and sets the video playback to the corresponding time value (1:05.00 in this case). This is an easy way to create temporal pointers.

5. EXTENSIBILITY

A declarative language describing the time structure and user interactions may be enough for most cases, but in more complex scenarios, an imperative language like JavaScript is necessary. Instead of having to choose between JavaScript and SMIL Timing, our implementation allows developers to define the main time structure declaratively and to extend it with JavaScript code when necessary.

5.1 DOM Event Listeners

As mentioned in the SMIL Timing module, SMIL target nodes fire begin and end DOM events when activated and deactivated, respectively. Web authors are used to set DOM event listeners to trigger specific actions. These begin and end events are already well known to SVG authors who use declarative SVG animations. Note that these begin and end events can also be used as event values in the begin and end SMIL attributes – again, like in declarative SVG animations.

5.2 JavaScript API

In our implementation, SMIL time containers can be controlled dynamically through two kinds of JavaScript APIs:

- seq and excl containers expose the same API as the HTML select element, i.e. mainly the selectedIndex property and the onchange DOM event.

- all time containers expose a significant subset of the HTMLMediaElement API (same as audio and video elements in HTML5), e.g.: .currentTime, .duration properties, .play(), .pause() methods, timeupdate, playing, paused DOM events.

We rely on existing web APIs wherever it makes sense. The seq and excl containers can be seen as HTML block level select elements, and all SMIL containers can be seen as general media elements – especially when synchronized with an audio or video element.

Besides, the SMIL Timing implementation comes with a JavaScript API that can be used to retrieve or create SMIL time containers dynamically.

---

[5] http://www.w3.org/TR/SMIL3/smil-timing.html#Timing-ControllingRuntimeSync
[6] http://www.w3.org/TR/SMIL3/smil-timing.html#Timing-HyperlinkImplicationsOnSeqExcl

5.3 Custom Timing Attributes

To keep the benefit of a declarative approach the timesheet scheduler can also be dynamically extended to support custom timing attributes that are too specific to be addressed by the SMIL specification. Timesheets.js provides a way to parse such custom attributes when time containers are initialized. Every custom attribute can be defined by a JavaScript file; we introduce two new attributes, `navigation` and `controls`, which are defined by two libraries.

5.3.1 Example: "navigation"

When using SMIL Timing for interactive presentations, a simple and common case is to ease navigation within the main time container. We are proposing a non-standard `navigation` attribute for this, with the following values:

arrows lets arrow keys control the execution of the time container: left/right to select the previous/next time node, up to reset the current time node, and down to emulate a mouse click on the current time node target;

click detects mouse clicks on the time container: the left/middle buttons select the next/previous time node;

scroll selects the previous/next time node on mouse scroll;

hash updates the fragment identifier (`#id`) in the URI when a time node target has an `id` attribute.

A single `navigation` attribute may have multiple values (which must be separated by semi-colons). For instance, `navigation="arrows; hash;"` activates the arrow-key navigation mode and updates the fragment identifier every time a time node target has an `id` attribute. When this JavaScript extension is loaded, all time containers are checked and mouse/keyboard event listeners are dynamically attached to the target time containers.

Though the navigation extension has proven to be useful for slide shows,[7] it is mainly proposed as a simple code base (less than 150 significant lines of code) for developers intending to write their own custom timing features.

5.3.2 Example: "controls"

Another very frequent need is a user interface for handling time containers: as modern web browsers provide native controls for continuous media, we offer similar but richer controls that take advantage of time container features. Such time controllers typically include (see bottom of Figure 1): a play/pause toggle button; first/prev/next/last buttons (sequential access); a table of contents, i.e. a nested list of links pointing to time nodes (direct access); a graphical timeline, either continuous (like for usual audio/video players) or segmented, where each segment is a link to a time node.

Instead of providing a single UI component, and in order to keep the flexibility of SMIL time containers, these controller elements are defined by a microformat: each UI component has a class name that is used both to define its presentation (in CSS) and its behavior (in JavaScript).

To highlight the current active time node (e.g. a timeline segment or table of contents item) when the time container is running, a time container is created dynamically with a `mediaSync` attribute pointing to the main time container.

A typical use case of such a time controller is a recorded talk[8] where slides are synchronized with the audio track. The timeline may be segmented to display the corresponding slide heading when the mouse hovers over it.

6. USING TIMESHEETS

In this section, we present real applications that were developed using the technology discussed above.

6.1 Media-Driven Applications

A captioned video[9] is a simple example of the category of applications presented in section 2.1. The whole application is implemented in a single HTML5 file, whose content is basically the same as in section 4.6.

In that example, the HTML `div` element is a time container for the video and all captions. Each caption is a HTML `p` element which contains the time when it must be displayed (`smil:begin` attribute) relatively to the beginning of the video. If a caption must disappear exactly when the next one is displayed, that is enough, as the container is `exclusive`, but if it must disappear earlier, its end date has to be explicitly stated.

The full HTML5 language may be used in each caption. The first caption takes advantage of this feature to add a link to the home page of the director and to split the text into two lines. CSS may also be used to select a particular font, its size and color, or to set the position of captions within the `div` element, i.e. over the movie.

6.2 Event-Driven Applications

We have worked with INA, the French national archive of audiovisual, to publish on the web archived radio programs enhanced with associated material.[10] At first glance, the time structure could look similar to the captioned video example discussed above, but the goal here is not only to synchronize pictures or text with the audio content. The objective is really to create an application where the user receives help for moving across the audio recording and is free to choose the associated information s/he wants, which could be multimedia too, with other audio recordings, for example. This is an example of an event-driven application as defined in section 2.2.

Figure 6: Enhanced radio program

[7]see http://wam.inrialpes.fr/timesheets/slideshows/slidy

[8]see http://wam.inrialpes.fr/timesheets/slideshows/audio
[9]see http://wam.inrialpes.fr/timesheets/annotations/video
[10]see http://wam.inrialpes.fr/timesheets/public/webRadio/

In this application (see screenshot on Figure 6), all the content is specified by a HTML5 document, while timing and user interactions are defined in a separate timesheet that refers to elements in the HTML5 file through attributes `select`, `mediaSync` and `controls`, as can be seen in this simplified version:

```
<timesheet xmlns="http://www.w3.org/ns/SMIL">
  <!-- slide show / main section -->
  <excl timeAction="display" mediaSync="#main"
        controls="#timeController" dur="20:47">
    <item select="#section1" begin="00:00.000"/>
    <item select="#section2" begin="01:12.120"/>
    <item select="#section3" begin="04:41.742"/>
  </excl>
  <!-- extra material: multimedia pages -->
  <excl>
    <item select="#extra2"
          begin="open2.click; toc-extra2.click"
          end="close2.click; section2.end"/>
    <item select="#extra3"
          begin="open3.click; toc-extra3.click"
          end="close3.click; section3.end"/>
  </excl>
  <!-- extra material: audio -->
  <par mediaSync="#track2a" controls="#timeline2a"
       dur="2:24.039"/>
  <par mediaSync="#track2b" controls="#timeline2b"
       dur="3:59.928"/>
  <!-- extra material: rotating pictures -->
  <seq timeAction="display"
       repeatCount="indefinite">
    <item select="#extra4 img" dur="3s"/>
  </seq>
</timesheet>
```

In this timesheet, the first `excl` element specifies a slide show synchronized with the audio recording (the `audio` element identified by the `main` id in the HTML5 file). The time structure of this part is similar to the captioned video example. Elements identified as `section`n are divisions in the HTML5 file that contain text and pictures. They define the main slides of the slide show.

The second `excl` element allows the user to display additional slides (identified as `extra`n in the HTML5 file) on request. This is achieved through buttons included in the main slides (ids `open`n refer to `button` elements which are part of the main slides). The right part of Figure 6 shows such a button. Similarly, additional slides contain buttons (ids `close`n) the user can click for closing them.

The element identified as `timeController` in the HTML5 file and referred by the first `excl` element specifies, in addition to the usual controls for an audio stream, a table of contents that the user can display with a button. The items of this table of contents have ids `toc-extra`n. Clicking them not only skips to the corresponding section of the main audio track, but also displays the corresponding additional slide, as specified in the second `excl` element.

The role of the `par` elements is to associate controls with the additional audio tracks, which are part of an additional slide in the HTML5 file. These audio tracks are activated as soon as the user opens the additional slide that contains them. S/he is then free to use the controls for listening to one of these oral comments.

Finally, the `seq` element at the end of the sample code specifies that all images contained in the `extra4` additional slide must be presented one after the other, each during 3 seconds, repeatedly. This automatic picture show starts as soon as the user clicks the button displaying the fourth additional slide.

7. CONCLUSION

Because most web developers are used to the HTML-CSS-JS triple, we have extended its capabilities with timing features borrowed from the SMIL language for enabling multimedia web applications. This approach fully preserves the declarative nature of web formats and their structural model for most applications, while scripts are still available to cover the most complex cases.

Developers do not have to choose between a logical and a temporal structure. Both kinds of structure can co-exist in the same document. In addition, temporal references can be used easily. DOM events, including SMIL time events, are available, which allows complex temporal and interaction behaviors to be defined. Content can be dynamically generated when necessary (e.g. `timeController` structures). Rich media navigation, as well as table of contents navigation, can be provided thanks to additional libraries. Content reusability and multi-device rendering are possible (the videotaped talk of Figure 1, for instance, runs on the iPhone and the iPad too). Components defining common behavior can be used in a declarative way.

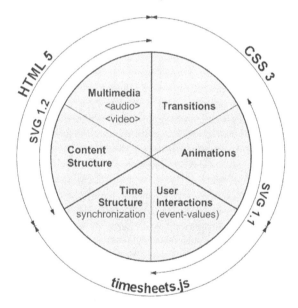

Figure 7: HTML5, CSS3, SVG and timesheets.js

The approach developed here and the timesheets.js library are not restricted to HTML5 documents. They can be used in graphic applications based on SVG, or in compound documents mixing HTML5 and SVG. This is illustrated by Figure 7 (a timed version is also available on-line[11]). The gray disc represents the main features expected from web multimedia languages, which are covered by SMIL to a large extent; the curved arrows show how they have been adopted by (or transposed in) other web languages. Timesheets.js closes

---

[11] http://wam.inrialpes.fr/timesheets/slideshows/svg.html

the loop, thus bringing the full power of SMIL to usual web compound documents that can be rendered natively in the browser.

Several multimedia applications based on timesheets.js are presented in this paper. These are not just examples; most of them are deployed on various web sites. As an example, multiple videotaped talks such as the one of Figure 1 are published by ENS-Lyon.[12]

A further step would be to introduce the features implemented by timesheets.js directly in the document languages of Figure 7. Most timing attributes are already part of the future SVG 2.0 language. Adding three more attributes, `mediaSync`, `timeAction` and `timeContainer`, would enable SVG applications such as slide shows or media annotation. This would also help to put temporal structures on SVG animations. There would be more work for HTML5, but reserving a prefix such as `smil-` could facilitate the integration of timing attributes later. CSS3 already includes transitions and animations, but it would be easier if the three states (idle, active, done) where available as pseudo-classes in selectors.

Because they are so close to usual web documents, multimedia documents based on the timesheets.js library may be developed in the same way as any web page, and can even be hand-coded. But, they could also benefit from specialized tools that would help developers to handle time information in documents. Now that a robust scheduler engine is available, the next step in our work is to develop specialized authoring tools.

8. ACKNOWLEDGEMENTS

The research presented in this paper was conducted in the C2M project, funded by ANR, the French National Research Agency, under its CONTINT 2009 programme.

The authors are grateful to Dominique Saint-Martin from INA-GRM for providing the web radio application and to Gérard Vidal from ENS Lyon for the videotaped talk application. Both have provided real use cases and valuable feedback that were key in testing and validating timesheets.js.

9. REFERENCES

[1] J. F. Allen. Maintaining knowledge about temporal intervals. *Comm. ACM*, 26:832–843, Nov. 1983.

[2] Ambulant. Ambulant open SMIL player, http://www.ambulantplayer.org/.

[3] O. Aubert, P.-A. Champin, Y. Prié, and B. Richard. Canonical processes in active reading and hypervideo production. *Multimedia Systems*, 14:427–433, 2008.

[4] D. Bulterman et al. *Synchronized Multimedia Integration Language (SMIL 3.0)*. W3C Recommendation, Dec. 2008.

[5] F. Cazenave. A declarative approach for HTML Timing using SMIL Timesheets, http://wam.inrialpes.fr/timesheets/, 2011.

[6] R. Deltour and C. Roisin. The LimSee3 multimedia authoring model. In D. Brailsford, editor, *Proceedings of the 2006 ACM Symposium on Document Engineering, DocEng 2006*, pages 173–175. ACM Press, Oct. 2006.

[7] ISO. Iso/iec 10744:1992 - information technology – hypermedia/time-based structuring language (hytime).

[8] J. Jansen and D. Bulterman. SMIL State: an architecture and implementation for adaptive time-based web applications. *Multimedia Tools and Applications*, 43:203–224, 2009.

[9] P. King, P. Schmitz, and S. Thompson. Behavioral reactivity and real time programming in XML: functional programming meets SMIL animation. In *Proceedings of the 2004 ACM Symposium on Document Engineering*, DocEng '04, pages 57–66, New York, NY, USA, 2004. ACM.

[10] J. Mikác, C. Roisin, and B. Le Duc. An export architecture for a multimedia authoring environment. In *Proceeding of the eighth ACM Symposium on Document Engineering*, DocEng '08, pages 28–31, New York, NY, USA, 2008. ACM.

[11] MIT. Infinite history, http://mit150.mit.edu/infinite-history, 2011.

[12] K. Pihkala and P. Vuorimaa. Nine methods to extend SMIL for multimedia applications. *Multimedia Tools and Applications*, 28:51–67, 2006.

[13] D. Raggett. HTML Slidy: Slide shows in HTML and XHTML, http://www.w3.org/talks/tools/slidy2/, 2005.

[14] L. F. Soares, R. F. Rodrigues, R. Cerqueira, and S. D. Barbosa. Variable and state handling in NCL. *Multimedia Tools and Applications*, 50:465–489, Dec. 2010.

[15] L. Villard. Authoring transformations by direct manipulation for adaptable multimedia presentations. In *Proceedings of the 2001 ACM Symposium on Document Engineering*, DocEng '01, pages 125–134, New York, NY, USA, 2001. ACM.

[16] P. Vuorimaa. Timesheets JavaScript Engine, http://www.tml.tkk.fi/~pv/timesheets/, 2007.

[17] P. Vuorimaa, D. Bulterman, and P. Cesar. *SMIL Timesheets 1.0*. W3C Working Draft, Jan. 2008.

---

[12]see http://html5.ens-lyon.fr/

Component-based Hypervideo Model: High-Level Operational Specification of Hypervideos

Madjid Sadallah
DTISI - CERIST
Alger, Algérie
msadallah@mail.cerist.dz

Olivier Aubert
Université de Lyon, CNRS
Université Lyon 1, LIRIS,
UMR5205, France
olivier.aubert@liris.cnrs.fr

Yannick Prié
Université de Lyon, CNRS
Université Lyon 1, LIRIS,
UMR5205, France
yannick.prie@liris.cnrs.fr

ABSTRACT

Hypervideo offers enhanced video-centric experiences. Usually defined from a hypermedia perspective, the lack of a dedicated specification hampers hypervideo domain and concepts from being broadly investigated. This article proposes a specialized hypervideo model that addresses hypervideo specificities.

Following the principles of component-based modeling and annotation-driven content abstracting, the *Component-based Hypervideo Model* (CHM) that we propose is a high level representation of hypervideos that intends to provide a general and dedicated hypervideo data model.

Considered as a video-centric interactive document, the CHM hypervideo presentation and interaction features are expressed through a high level operational specification. Our annotation-driven approach promotes a clear separation of data from video content and document visualizations. The model serves as a basis for a Web-oriented implementation that provides a declarative syntax and accompanying tools for hypervideo document design in a Web standards-compliant manner.

Categories and Subject Descriptors

H.5.1 [**Multimedia Information Systems**]: Video; H.5.4 [**Hypertext/Hypermedia**]: Architectures, Navigation

General Terms

Design, Experimentation

Keywords

Hypervideo, Annotation, component-based modeling, Time and Synchronization, Timeline Reference, CHM

1. INTRODUCTION

Video-based information becomes today an invasive social phenomenon that involves a wide and growing range of digital consumers. Hypervideo is an attractive technology that aims to bring new interaction modalities to video-centric hypermedia documents. Currently, several definitions of hypervideo exist [3], depending on the considered point of view. While some works emphasize the hypermedia aspect of adding information to digital video so that users can activate video hyperlinks and access the added rich content, other authors highlight the storytelling aspects of dynamically creating non-linear and user-defined navigation paths into the document. In this work, we define hypervideo as being *an interactive video-centric hypermedia document that includes an audiovisual content* - a set of video objects - *augmented with several kinds of data in a time synchronized way; the added content supplements the audiovisual part around which the presentation is organized in space and time.* The integration of content-enriched video offers additional interaction and navigation alternatives and additional information levels.

While initial hypervideo concepts can be traced back to the late of nineties [11], the availability of appropriate hardware and software has prevented this kind of documents from being broadly investigated. The need for theoretical foundations for hypervideos motivates the present work. The proposed component-based approach is a means to conceptualize annotation-driven hypervideos structure and behavior. High level components are provided to emphasize the presentation and interaction features while lower level ones allow more complex component building.

2. HISTORY AND RELATED WORK

Following Ted Nelson's hypermedia model extension to include "branching movies" or "hyperfilms" many researchers addressed the field of interactive video. The Aspen Movie Map Project or HyperCafe [11] added interactivity and branching to videos, using specialized software. Basic forms of web-based hypervideos can be achieved for instance through Youtube annotations, but many projects like *VideoClix*[1] and *Popcorn.js*[2] experiment with other approaches.

2.1 Hypervideo Modeling

Hypervideos as General Hypermedia/Multimedia. Models and systems for continuous media integration in hypermedia were discussed since the Amsterdam Hypermedia Model (AHM) [7] proposal, providing mechanisms for structuring,

---

[1] http://www.videoclix.tv/
[2] http://popcornjs.org/

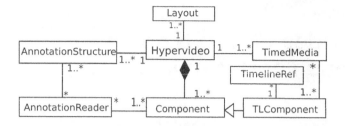

Figure 1: CHM General Overview

linking and synchronizing dynamic multimedia content. Other efforts like NCM (Nested Context Model [10]) also aimed at introducing synchronization in hypermedia.

AHM inspired the definition of SMIL (Synchronized Multimedia Integration Language) which is an ideal candidate for developing general purpose hypervideos. However, hypervideo support has not received much attention [12]. Moreover, SMIL is a huge standard and the specification itself contains no formal model definition [8]. Its concepts are all-purpose and do not precisely describe the hypervideo specific features, even though some proposals [10] have been made to address this issue.

Hypervideo as Annotation-based Multimedia. Video not being intrinsically structured, annotations are needed to add semantic description and content enrichment. They provide the foundation for enriched experience of video content, for instance by breaking the linearity of video [9] or enriching it [4]. We consider here an annotation as a piece of data associated to a spatio-temporal fragment of a video [1].

VideoAnnEx[3] and MediaDiver [9] allow to annotate videos and generate video centric presentations. Hyper-Hitchcock [5] and Advene [1] allow viewers to not only interact with annotated videos but also to create their own annotations and to share them.

The Need for Dedicated Models. Informal models are generally used by existing systems to describe hypervideos. However, the implied representations are mainly technically driven and do not proceed from the definition of hypervideo data models besides the conventional hypermedia/multimedia ones. For instance, hypervideo raises many important rhetorical and aesthetic issues [11, 3] and stresses common hypermedia authoring and reading concerns that risk to overstrain the cognitive capacities of the user and put him under time pressure during navigation [12], provoking a user disorientation [3]. Moreover, hypermedia modeling of such documents results in an inefficiency and a weakness in expressing the logical hypervideo patterns [6] since the underlying concepts are too general to grasp and characterize hypervideo details. Consequently, hypervideo specificities need a dedicated document model able to consider these issues.

3. THE CHM PROPOSAL

3.1 General Overview

The proposed *Component-based Hypervideo Model* (CHM) is based on the presentation of annotations through nested low- or high-level components.

---

[3] http://www.research.ibm.com/VideoAnnEx/

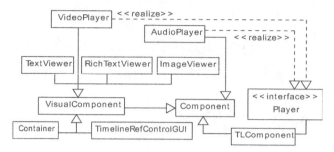

Figure 2: CHM Plain Components

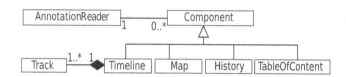

Figure 3: CHM High Level Components

Figure 1 presents a general overview of the model. A hypervideo is formed by a set of low and high level components, building blocks that represent the formal information and composition units. A hypervideo references at least one main audiovisual document accessed through the `TimedMedia` element that addresses at least a temporalized stream, audio or video. A `TimedMedia` played through a player component conveys a timing reference to the document, expressed by the *TimeLine Reference* (TLR) that synchronizes the rendering `TLComponents` related to the played TimedMedia element. Many players (therefore, many TLRs) may be present within the same document, defining different hypervideo sub-documents.

The CHM *annotation* concept expresses the association of any data or resource to a video fragment for enrichment, linking or visualization purposes. An annotation is defined as any information associated to a spatio-temporal video *fragment*. The annotation data may be generated by third party software such as Advene [1] or IBM's VideoAnnEx annotation tool.

CHM Plain Components. Figure 2 presents the basic data components that form a hypervideo. Specific synchronized *display components* offer interactive interfaces for rendering temporalized data, provided as annotations through *AnnotationReader* components.

Document content viewers such as `TextViewer`, `RichTextViewer` and `ImageViewer` allow the display of textual and graphic content, such as a synchronized *Wikipedia* content. The `TimelineRefControlGUI` element defines a graphic user interface for controlling and interacting with TLRs.

CHM High Level Components. For hypervideo document design, we propose an extensible set of high level components shown in figure 3, built upon the plain ones. The *History component* presents a visual representation of the visited nodes, to alleviate user disorientation. The *Document maps and ToCs* (tables of contents) display structuring annotations through interactive graphical or hierarchical layouts. The `Timeline` component is a graphical interactive

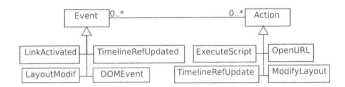

Figure 4: Hypervideo events and actions

representation of annotations through hypervideo time. It displays elements along a timed axis on different tracks.

CHM Spatial Model. Derived from the SMIL spatial model, the CHM spatial model is intended to accommodate the implementation platform specificities. `VisualComponents` are placed within `SpatialRegions`, whose global placement is defined in the root `Layout` element.

CHM Temporal Model. The document temporal specification is achieved through a timeline-based model. The explicit time scale of document components is defined by a *Timeline Reference* (TLR), a virtual time reference attached to a video playback component or to the global document. Time-based components are activated/deactivated when reaching specific timecodes provided or computed by reference to the TLR. The access and control of a TLR is performed through its "state", "position" and "duration" attributes. Any update of the TLR position or state affects all the related components playback.

CHM Event-based Model. In addition to standard hypermedia links, CHM hypervideo links can be defined in space and time and are unidirectional. There is no separate link - like in AHM - within the model; SMIL and HTML also do not use separate link components.

Differently from AHM, CHM does not rely only on a link-based model. The dynamic behavior of a CHM hypervideo is represented by an event-based mechanism, expressed by the `Event` and `Action` elements shown in figure 4. Many actions can be associated to an event, among which: (1) the `OpenURL` action allows the display of a target URL, (2) the `ModifyLayout` action allows to modify the content and placement of elements, (3) the `UpdateTimelineRef` action specifies a state or position change of the timeline reference and (4) the `ExecuteScript` action allows the execution of a user specified set of operations.

3.2 Example

The proposed model is intended to both conceptually analyze common hypervideo documents, and help the design of new ones. We present here a CHM hypervideo document, represented in figure 5 - annotated using Advene[4] and designed through our model implementation[5] - , to give deep insight into the Nosferatu movie[6]. The hypervideo demonstrates some high level components such as data readers, enrichment content viewers, video players, hotspots, timelines, maps and tables of contents. Figure 6 is the abstract representation of the most important components. The example is simplified by defining one spatial region and a unique time-

[4]http://www.advene.org/
[5]http://www.advene.org/chm/API/
[6]http://www.archive.org/details/nosferatu

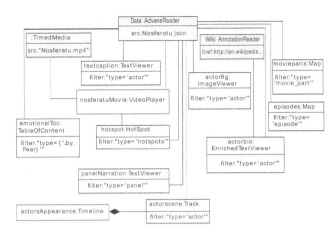

Figure 6: Nosferatu Hypervideo Modeling

line reference, both not represented in the figure. Each class instance is provided with a set of attributes and we present in the figure only the filter used to query the data readers.

4. ARCHITECTURE IMPLEMENTATION

4.1 Rationale

Different programming languages and for various architectures can be used to implement CHM. In order to demonstrate some practical uses of the model, we have developed a publicly accessible Web-based prototype. It relies on a declarative syntax and the development of a set of JavaScript libraries for the model implementation in a Web-standards compliant manner. For rendering the audiovisual content, we make use of the latest HTML5 standard proposal that introduces native browser support for video. The CHM spatial model is supported by the HTML layout model via cascading stylesheets. As there is still no established standard way to add temporal behavior to Web documents, we have used a javascript implementation of SMIL Timesheets [2].

4.2 Web-based Hypervideo Syntax

Based on CHM, a syntax for authoring hypervideos is proposed as an extension above the HTML language. Since HTML does not fully support namespace declarations, we chose to use CHM namespaced attributes associated to standard HTML elements. A javascript-based transformation interprets these attributes and dynamically generates standard HTML5 code. Complex hypervideos can therefore be authored as standard Web documents, styled with CSS, and controlled by scripts. Common Web content is written in standard HTML while hypervideo components are expressed through the CHM attribute-based syntax.

Interested readers can see the code that generates the hypervideo example of figure 5, written in the proposed syntax, by looking at the source of the http://www.advene.org/chm/ document.

4.3 Current State of the Implementation

A first version of the proposed language and tools has been developed, and further developments are underway. We aim to enhance the language support by providing the language DOM API rather than libraries that perform content transformation. The authoring process may be eased by a graphic

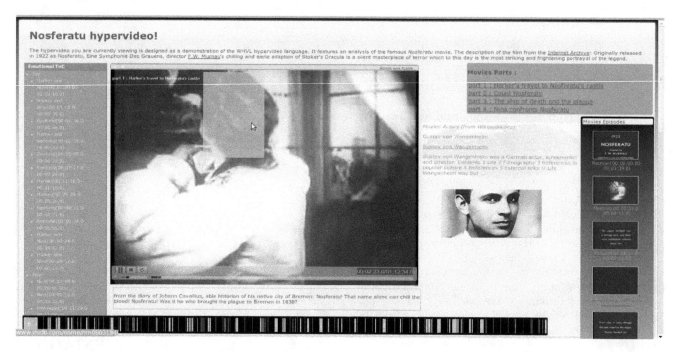

Figure 5: Nosferatu Hypervideo Example

user interface (GUI), planned in future development stages. We also want to rely on cognitive theories to propose more meaningful components. The interaction possibilities can be enriched by more advanced features like offering additional mechanisms to end-users.

5. CONCLUSION

In this paper, we have proposed the *Component-based Hypervideo Model* (CHM), an annotation-based hypervideo document abstraction with a focus on a high level architectural representation. It tries to provide a hypervideo model, building upon conventional hypermedia models. A component-based approach is proposed to apprehend hypervideos and to define their behaviors. The model components allow to use explicit elements to analyse or design hypervideos, providing a high level framework while still offering lower level tools for hypervideo authoring. The model promotes a clear separation of data from the video content, through the notion of annotations. A declarative syntax for defining Web-based hypervideo documents has been proposed and a first implementation has allowed us to validate the principles. We are working on extending and refining it.

Acknowledgements. This work has been partially funded by the French FUI (Fonds Unique Interministériel) - CineCast project and the ANR (Agence Nationale de la Recherche) - Ithaca project. It is supported by the Algerian Research Center on Scientific and Technical Information (CERIST).

6. REFERENCES

[1] O. Aubert and Y. Prié. Advene: active reading through hypervideo. In *ACM Hypertext*, 2005.

[2] F. Cazenave, V. Quint, and C. Roisin. Timesheets.js: When SMIL meets HTML5 and CSS3. In *to appear in DocEng2011*, 2011.

[3] T. Chambel, C. Zahn, and M. Finke. Hypervideo design and support for contextualized learning. In *IEEE Advanced Learning Technologies*, 2004.

[4] R. Fagá, V. G. Motti, R. G. Cattelan, C. A. C. Teixeira, and M. da Graça Campos Pimentel. A social approach to authoring media annotations. In *ACM DocEng 2010*, 2010.

[5] A. Girgensohn, L. Wilcox, F. Shipman, and S. Bly. Designing affordances for the navigation of detail-on-demand hypervideo. In *Proceedings of ACM Advanced Visual Interfaces*, May 2004.

[6] R. I. Hammoud. Introduction to interactive video. In *Interactive Video*, pages 3–24. Springer-Verlag, 2006.

[7] L. Hardman, D. C. A. Bulterman, and G. van Rossum. The Amsterdam Hypermedia Model: adding time and context to the Dexter model. *Commun. ACM*, 37:50–62, February 1994.

[8] J. Jansen, P. Cesar, and D. C. Bulterman. A model for editing operations on active temporal multimedia documents. In *Proc. of DocEng 2010*, 2010.

[9] G. Miller, S. Fels, A. A. Hajri, M. Ilich, Z. Foley-Fisher, M. Fernandez, and D. Jang. Mediadiver: Viewing and annotating multi-view video. In *CHI-Interactivity Session*, May 2011.

[10] L. M. Rodrigues, M. J. Antonacci, R. F. Rodrigues, D. C. Muchaluat-Saade, and L. F. G. Soares. Improving SMIL with NCM facilities. *Multimedia Tools Appl.*, 16:29–54, January 2002.

[11] N. Sawhney, D. Balcom, and I. E. Smith. Hypercafe: Narrative and aesthetic properties of hypervideo. In *UK Conf. on Hypertext*, Bethesda, US, 1996.

[12] C. A. Tiellet, A. G. Pereira, E. B. Reategui, J. V. Lima, and T. Chambel. Design and evaluation of a hypervideo environment to support veterinary surgery learning. In *Proc. of ACM Hypertext*, 2010.

The Art of Mathematics Retrieval

Petr Sojka
Faculty of Informatics, Masaryk University
Botanická 68a, 602 00 Brno, Czech Republic
sojka@fi.muni.cz

Martin Líška
Faculty of Informatics, Masaryk University
Botanická 68a, 602 00 Brno, Czech Republic
255768@mail.muni.cz

ABSTRACT

The design and architecture of MIaS (Math Indexer and Searcher), a system for mathematics retrieval is presented, and design decisions are discussed. We argue for an approach based on Presentation MathML using a similarity of math subformulae. The system was implemented as a math-aware search engine based on the state-of-the-art system Apache Lucene.

Scalability issues were checked against more than 400,000 arXiv documents with 158 million mathematical formulae. Almost three billion MathML subformulae were indexed using a Solr-compatible Lucene.

Categories and Subject Descriptors

H.3.7 [**Information Systems**]: Information storage and Retrieval—*Information Search and Retrieval*; I.7 [**Computing Methodologies**]: Document and text Processing—*Index Generation*

General Terms

Algorithms, Design, Experimentation, Performance

Keywords

MIaS, WebMIaS, digital mathematics libraries, information systems, math indexing and retrieval, mathematical content representation

> There is no abstract art. You must always start with something. Afterward you can remove all traces of reality.
> Pablo Picasso

1. INTRODUCTION

The solution to the problem of mathematical formulae retrieval lies at the heart of building digital mathematical libraries (DML). There have been numerous attempts to solve this problem, but none have found widespread adoption within the wider mathematics community. And as yet, there is no widely accepted agreement on the math search format to be used for mathematical formulae by library systems or by Google Scholar.

MathML standard by W3C has become the standard for mathematics exchange between software tools. Almost no MathML is written directly by authors—they typically prefer a compact notation of some TeX flavour such as LaTeX or AMSLaTeX. The designer of a search system for mathematics is thus faced with the task of converting data to a unifying format, and allowing DML users to use their preferred notation when posing queries. [AMS]LaTeX or similar TeX macropackages are the typical preferences; Presentation MathML or Content MathML are used only when available as outputs of a software system.

During the integration of existing DMLs into larger projects such as EuDML [9], the unsolved math search problem becomes evident—*DML* without math search support is an oxymoron. We have evaluated several systems for our goals: 1) **MathDex**[1] (formerly Math-Find [6]); 2) **EgoMath**[2] developed by Josef Mišutka as an extension of a full text websearch core engine Egothor [5]; 3) **LaTeXSearch**[3], a search tool offered by Springer in SpringerLink; 4) **LeActive-Math**[4] search, developed as part of the ActiveMath-EU project and 5) **MathWebSearch**[5] is an MSE developed in Bremen/Saarbrücken by Kohlhase et al. [1] Our evaluation [7] has lead us to conclude that there is no satisfactory, math-aware and scalable solution. For this reason, we designed and implemented [3, 7] a *new* robust solution for retrieving mathematical formulae.

Section 2 presents our design of this scalable and extensible system for searching mathematics, taking into account not only inherent structure of mathematical formulae but also formula unification and subformulae similarity measures. Our evaluation of a prototypical implementation on the Apache Lucene open source full-featured search engine library is presented in Section 3. The paper concludes with a description of the WebMIaS interface and listing future work directions in Section 4 and with a summary in Section 5.

> Art is never chaste. It ought to be forbidden to ignorant innocents, never allowed into contact with those not sufficiently prepared. Yes, art is dangerous. Where it is chaste, it is not art.
> Pablo Picasso

2. DESIGN OF MIAS

We have developed a math-aware, full-text based search engine called *MIaS* (Math Indexer and Searcher). It processes documents containing mathematical notation in MathML format. MIaS allows users to search for mathematical formulae as well as the textual content of documents.

Since mathematical expressions are highly structured and have no canonical form, our system pre-processes formulae in several steps to facilitate a greater possibility of matching two equal expressions with different notation and/or non-equal, but similar formulae. With

---

[1] www.mathdex.com/
[2] egomath.projekty.ms.mff.cuni.cz/egomath/
[3] www.latexsearch.com/
[4] devdemo.activemath.org/ActiveMath2/
[5] search.mathweb.org/index.xhtml

an analogy to natural language searching, MIaS searches not only for whole sentences (whole formulae), but also for single words and phrases (subformulae down to single variables, symbols, constants, etc.). For calculating the relevance of the matched expressions to the user's query, MIaS uses a heuristic weighting of indexed mathematical terms, which accordingly affects scores of matched documents and thus the order of results.

At the end of all processing methods, formulae are converted from XML nodes to a linear string form, which can be handled by the indexing core.

2.1 Math Indexing Workflow

The top-level indexing scheme with a detailed view of the mathematical part is shown in Figure 1.

MIaS is currently able to index documents in XHTML, HTML and TXT formats. As Figure 1 shows, the input document is first split into textual and mathematical parts. The textual content is indexed in a conventional way, mathematics needs to be processed differently.

2.2 Math preprocessing

Mathematical expressions are pre-analyzed in several steps to facilitate searches not only for exact whole formulae, but also for subparts (tokenization) and for similar expressions (formulae unifications). This addresses the issue of the static character of full-text search engines and creates several representations of each input formula, all of which are indexed.

Tokenization is a straightforward process of obtaining subformulae from an input formula. MIaS makes use of Presentation MathML markup where all logical units are enclosed in XML tags which makes obtaining all subformulae a question of tree traversal.

MIaS performs three types of unification algorithms, the goal of which is to create several more or less generalized representations of all formulae obtained through the tokenization process. These steps allow the system to return similar matches to the user query while preserving the formula structure and α-equality [5]:

1. Ordering—ordering of the operands of the commutative operations;

2. Variables unification—substitution of all variables for unified symbols (ids) while preserving bound variables;

3. Constants unification—substitution of all number constants for one unified symbol (const).

During the searching phase, a query can match several terms in the index. However one match can be more important to the query than another, and the system must consider this information when scoring matched documents. Each indexed mathematical expression has a weight (relevance score) assigned to it describing how far the actual formula is from its original representation. It is computed throughout the whole indexing phase by individual processing steps following this basic rule of thumb—the more modified a formula and the lower the level of a subformula, the less weight is assigned to it.

It is impossible to assemble a weighting function that is exactly right. Such a function needs to consider a document base on which the system will run as well as the established customs in a particular scientific field. We tried to create a complex and robust weighting function that would be appropriate to many fields.

At the end of the preprocessing phase, mathematical formulae are transformed from XML nodes to a compacted string form so it can be handled by a full-text indexing core.

An example of the formula preprocessing is displayed in Figure 2.

2.3 Searching

In the search phase, user input is again split into mathematical and textual parts. Formulae are then preprocessed in the same way as in the indexing phase, except for tokenization—we doubt that users would want to search for subparts of a queried expression rather than the whole.

3. EVALUATION

Art is the elimination of the unnecessary.
Pablo Picasso

For large scale evaluation, we needed an experimental implementation and a corpus of mathematical texts.

3.1 Implementation

The Math Indexer and Searcher is written in Java. The role of full-text indexing and searching core is performed by Apache Lucene 3.1.0. The mathematical part of document processing can be seen as a standalone pluggable extension to any full-text library, however it needs custom integration for each one. In the case of Lucene, a custom Tokenizer (MathTokenizer) has been implemented.

When searching for mathematical formulae, their weights need to be considered in the final score of the document. The scoring function of our MIaS system adds one parameter to the Lucene's standard practical scoring function (described in detail at `http://lucene.apache.org/java/3_1_0/api/all/org/apache/lucene/search/Similarity.html`)—weight w of one matched formula:

$$score(q,d) = coord(q,d) \cdot queryNorm(q) \cdot$$
$$\cdot \sum_{t \, in \, q} \left(tf(t \, in \, d) \cdot avg(w) \cdot idf(t)^2 \cdot t.getBoost() \cdot norm(t,d) \right) \quad (1)$$

If a document contains the same formula more than once (each occurrence can have different weight assigned), the average value of all the weights is taken into consideration ($avg(w)$).

3.2 Corpus of Mathematical Documents

A document corpus MREC (Mathematical REtrieval Corpus) with 439,423 scientific documents was used to evaluate the behavior of the system we modelled. The documents come from the arXMLiv project that is converting document sets from arXiv into XHTML + MathML (both Content and Presentation) [8]. The resulting corpus size was 124 GB uncompressed, 16 GB compressed. This corpus of documents (MREC version 2011.4.439) is available for download at `http://nlp.fi.muni.cz/projekty/eudml/MREC`.

3.3 Results

Math Indexer and Searcher demonstrated the ability to index and search a relatively vast corpus of real scientific documents. Its usability is greatly improved thanks to its preprocessing functions together with the formulae weighting model. The ability to search for exact and similar formulae and subformulae, more so with customizable relevance computation, demonstrates an unquestionable contribution to the whole search experience.

It is very difficult, if not impossible, to completely verify the correctness of the theoretical considerations made in the previous sections and thus correctness of search results. For this purpose, a sufficiently large corpus of documents with fully controlled content would be needed. For any assembled query, there should exist beforehand a complete list of the documents ordered by their relevance to the query to compare the actual results to. On the other hand, the real world relevance of the results being returned needs to be verified by an extensive user study and perhaps several parameters need to be adjusted for the best results.

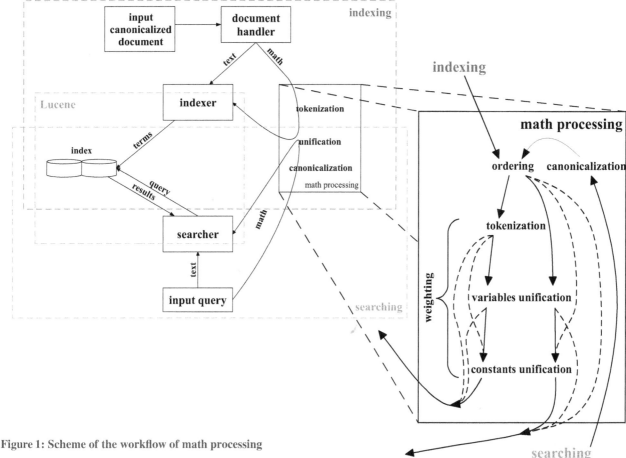

Figure 1: Scheme of the workflow of math processing

3.4 Scalability Testing and Efficiency

We have devised a scalability test to see how the system behaves with an increasing number of documents and formulae indexed. Subsets containing 10,000, 50,000, 100,000 ... and the complete 439,423 documents were gradually indexed and several values were measured: the number of input formulae, the number of indexed formulae, the indexing run-time and indexing CPU-time. Our observation showed that system scales linearly in proportion to the number of documents.

The whole document set contained 158,106,118 formulae, after all the preprocessing was done, our system indexed 2,910,314,146 expressions and the indexing run-time was 1,378 min (almost 23 hours) and the resulting index size was approx. 63 GB. We also measured an average query time by querying the created index with a set of differently complex queries (mixed, non-mixed, more/less complex single/multiple formulae). Resulting average query time was 469 ms.

> You don't make art, you find it.
> Pablo Picasso

4. WEB INTERFACE AND FUTURE WORK

To allow user evaluation, we have created web interface for MIaS—*the WebMIaS system* [4]—available at http://nlp.fi.muni.cz/projekty/eudml/mias/. WebMIaS allows the re-trieval of mathematical expressions written in TEX or MathML. TEX queries are converted on-the-fly into tree representations of Presentation MathML, which is used for indexing. WebMIaS allows complex queries composed of plain text and mathematical formulae. It currently works over our complete mathematical corpus, MREC.

As the semantically same formulae can be represented by different MathML notation, it is evident that some kind of normalization of MathML is necessary. We use Canonical MathML [2] as normalization MathML format and are using the software library UMCL supporting it. Canonicalization (converting to a canonicalized form of MathML) is used both during the indexing and querying phases. It not only increases fairness of similarity ranking, but also helps to match a query against the indexed form of MathML. We plan to extend the effect of the commutative ordering part of the normalization mentioned in 2.2 by arranging a full list of commutative operators and all the operators with their priorities so the ordering can be perfected.

Another area of long-term research planned is supporting Content MathML, in a way similar to the current handling of Presentation MathML. The architectural design is open to it, but as most of the math available is in Presentation MathML taken from PDFs, this is not currently a high priority.

> Art is the lie that enables us to realize the truth.
> Pablo Picasso

5. CONCLUSIONS

We have presented an approach to mathematics searching and indexing—the architecture and design of the MIaS system, and

We have applied an empirical approach to the evaluation so far. For these purposes we created a demo web interface, see Section 4.

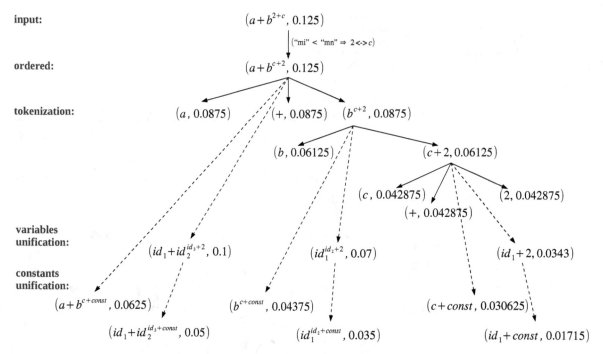

input: $(a+b^{2+c}, 0.125)$

$\big(\text{"mi"} < \text{"mn"} \Rightarrow 2 <\text{-}> c\big)$

ordered: $(a+b^{c+2}, 0.125)$

tokenization: $(a, 0.0875)$ $(+, 0.0875)$ $(b^{c+2}, 0.0875)$

$(b, 0.06125)$ $(c+2, 0.06125)$

$(c, 0.042875)$ $(2, 0.042875)$

$(+, 0.042875)$

variables
unification: $(id_1+id_2^{id_3+2}, 0.1)$ $(id_1^{id_2+2}, 0.07)$ $(id_1+2, 0.0343)$

constants
unification: $(a+b^{c+const}, 0.0625)$ $(b^{c+const}, 0.04375)$ $(c+const, 0.030625)$

$(id_1+id_2^{id_3+const}, 0.05)$ $(id_1^{id_2+const}, 0.035)$ $(id_1+const, 0.01715)$

Figure 2: Example of formula preprocessing. Ordered pairs are (<expression written naturally>, <it's weight>). All expressions as shown are indexed, except for the original one.

its WebMIas interface. The feasibility of our approach has been verified on large corpora of real mathematical papers from arXMLiv. Scalability tests have confirmed that the computing power needed for fine math similarity computations is readily available and allows the use of this technology for projects on world-wide scale.

Acknowledgements. This work has been partially supported by the Ministry of Education of CR within the Center of Basic Research LC536 and by the European Union through its Competitiveness and Innovation Programme (Policy Support Programme, 'Open access to scientific information", Grant Agreement No. 250503). We thank Michal Růžička for help with figure drawings and web form of MIaS interface.

Bad artists copy. Good artists steal.
Pablo Picasso

6. REFERENCES

[1] Ş. Anca. Natural Language and Mathematics Processing for Applicable Theorem Search. Master's thesis, Jacobs University, Bremen, Aug. 2009.
https://svn.eecs.jacobs-university.de/svn/eecs/archive/msc-2009/aanca.pdf.

[2] D. Archambault and V. Moço. Canonical MathML to Simplify Conversion of MathML to Braille Mathematical Notations. In K. Miesenberger, J. Klaus, W. Zagler, and A. Karshmer, editors, *Computers Helping People with Special Needs*, volume 4061 of *Lecture Notes in Computer Science*, pages 1191–1198. Springer Berlin / Heidelberg, 2006.
http://dx.doi.org/10.1007/11788713_172.

[3] M. Líška. Vyhledávání v matematickém textu (in Slovak), Searching Mathematical Texts, 2010. Bachelor Thesis, Masaryk University, Brno, Faculty of Informatics (advisor: Petr Sojka), https://is.muni.cz/th/255768/fi_b/?lang=en.

[4] M. Líška, P. Sojka, M. Růžička, and P. Mravec. Web Interface and Collection for Mathematical Retrieval. In P. Sojka and T. Bouche, editors, *Proceedings of DML 2011*, pages 77–84, Bertinoro, Italy, July 2011. Masaryk University. http://www.fi.muni.cz/~sojka/dml-2011-program.html.

[5] J. Mišutka and L. Galamboš. Extending Full Text Search Engine for Mathematical Content. In P. Sojka, editor, *Proceedings of DML 2008*, pages 55–67, Birmingham, UK, July 2008. Masaryk University. http://dml.cz/dmlcz/702546.

[6] R. Munavalli and R. Miner. MathFind: A Math-Aware Search Engine. In *Proceedings of the 29th annual international ACM SIGIR conference on Research and development in information retrieval*, SIGIR '06, pages 735–735, New York, NY, USA, 2006. ACM.
http://doi.acm.org/10.1145/1148170.1148348.

[7] P. Sojka and M. Líška. Indexing and Searching Mathematics in Digital Libraries – Architecture, Design and Scalability Issues. In J. H. Davenport, W.M. Farmer, J. Urban and F. Rabe, editors, *Proceedings of CICM Conference 2011 (Calculemus/MKM)*, volume 6824 of *Lecture Notes in Artificial Intelligence, LNAI*, pages 228–243, Berlin, Germany, July 2011. Springer-Verlag. http://dx.doi.org/10.1007/978-3-642-22673-1_16.

[8] H. Stamerjohanns, M. Kohlhase, D. Ginev, C. David, and B. Miller. Transforming Large Collections of Scientific Publications to XML. *Mathematics in Computer Science*, 3:299–307, 2010.
http://dx.doi.org/10.1007/s11786-010-0024-7.

[9] W. Sylwestrzak, J. Borbinha, T. Bouche, A. Nowiński, and P. Sojka. EuDML—Towards the European Digital Mathematics Library. In P. Sojka, editor, *Proceedings of DML 2010*, pages 11–24, Paris, France, July 2010. Masaryk University. http://dml.cz/dmlcz/702569.

Automated Conversion of Web-based Marriage Register Data into a Printed Format with Predefined Layout

David F. Brailsford
Document Engineering Research Group
School of Computer Science
University of Nottingham
Nottingham NG8 1BB, UK
dfb@cs.nott.ac.uk

ABSTRACT

The Phillimore Marriage Registers for England were published in the period 1896 to 1922 and have defined a standard layout format for the typesetting of marriage data. However, not all English parish churches had their marriage registers analysed and printed by the Phillimore organisation within this time period.

This paper tells the story of Wirksworth, a town in Derbyshire with a large church, licensed for marriages, yet whose marriage data was not released to the Phillimore organisation. Hence there is no printed Phillimore Marriages volume for Wirksworth. However, in recent years, a Wirksworth web site, created by John Palmer, has become famous as being probably the most comprehensive record of a parish's activities anywhere on the Web.

Within a total of 120 MB of data on the web site, covering events in Wirksworth from medieval times to the present, is a set of data recording births, marriages and deaths transcribed from the original hand-written church register volumes.

The work described here covers the software tools and techniques that were used in creating a set of *awk* scripts to extract all the marriage records from the Wirksworth web site data. The extracted material was then automatically re-processed, typeset and indexed to form an entirely new Phillimore-style volume for Wirksworth marriages.

Categories and Subject Descriptors

D.2.3 [**Software Engineering**]: Coding Tools and Techniques; I.7.5 [Document and Text Processing]: Document Capture–*document analysis*; I.7.2 [Document and Text Processing]: Document Preparation –*Markup languages*

General terms

Documentation, Languages

Keywords

Re-typesetting, Web-to-Print, *troff*, genealogy, hyperlinking, indexing

DocEng'11, September 19–22, 2011, Mountain View, California, USA.
Copyright 2011 ACM 978-1-4503-0863-2/11/09...$10.00.

1. INTRODUCTION

In previous work [1,2] a set of software tools was developed which enabled scanned pages of printed English Marriage Registers to be re-typeset, and indexed.

The printed originals were published in the early years of the 20th century by the Phillimore company as an aid to genealogical researchers. These printed registers were, in turn, transcribed from the original hand-written registers, kept in local parish churches throughout England and Wales. For the purposes of the previous work the 15 Phillimore volumes relating to the county of Derbyshire were used as the test corpus. As a proof of concept the first two of these volumes have been retypeset and indexed. They have now been published and are available through Amazon (UK).

The idea of using a pipelined sequence of simple software tools, for the task of retypesetting whole Phillimore volumes from OCR scans of bitmap pages, was in the hope that the various software components could be re-used if other sources of marriage raw material came to light.

For reference, the original processing chain from scanned bitmap pages to re-typeset volumes is shown starting at the upper left of Figure 1. In this present paper we investigate how a new processing chain, shown in the lower half of Figure 1, was developed to cope with a Web-based source of information for an extra Derbyshire parish, not included in the original Phillimore set. This new source of information is able to feed into the existing processing chain along a common central spine (shown in bold type in the centre of Figure 1). Also included in that Figure is the indexing process common to both input chains.

2. The parish of Wirksworth

Wirksworth is a town in the county of Derbyshire, with a population of about 5000 in the town itself, rising to more than 9000 if immediately surrounding villages are included. It has a substantial parish church, set in its own grounds, with very much the air of a small cathedral.

The Wirksworth parish was not included in the original Phillimore Derbyshire series of registers, Of course, there are numerous other English parishes that were also not included, either because permission to copy the registers was not given, or simply due to lack of volunteers to carry out the transcription work.

A remarkable Web site for the Wirksworth parish was started in 1991, by John Palmer, and is available at http://www.wirksworth.org.uk. It is believed to

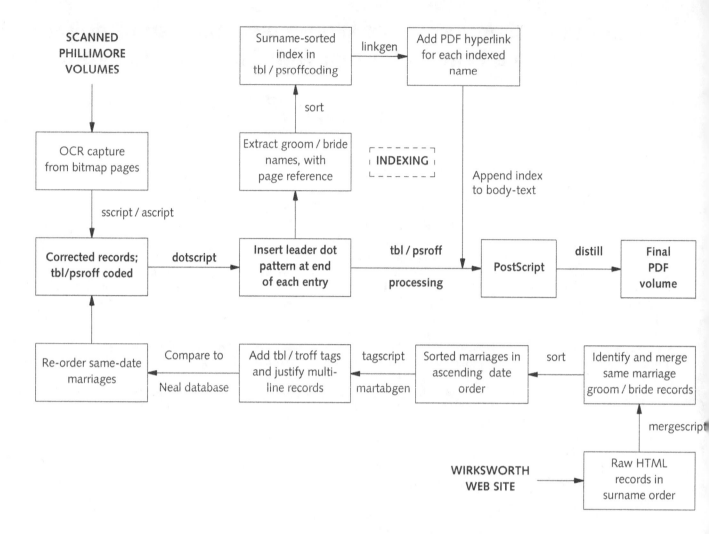

Figure 1. Marriage register processing chains for scanned volume and Web site starting points

be the most comprehensive historical and contemporary record of a single parish, anywhere on the Web. Among approximately 128 MB of material on the Wirksworth site is a complete transcription, into hand-coded HTML, of the Wirksworth marriage registers, from 1608 to 1899.

It should be explained that, in the years since the printed Phillimore volumes of the early 20th century, parish registers have generally become much more accessible to researchers. Since the enactment of the Parochial Registers and Records Measure, of 1987, most church registers have been moved to their respective County Record Offices. Those for Wirksworth now reside at the Derbyshire Records Office (DRO) in Matlock. Thus the transcription of the Wirksworth registers, by John Palmer, was conducted using microfilm copies made available by the Genealogical Society of Utah (GSU) supplemented by several visits to DRO to cross-check faded or near-illegible entries against the original hand-written registers.

3. The Web-based marriage entries

Register material on the Wirksworth site was initially generated from an MPRO database, but unfortunately this particular software dates back to the days of the MS DOS operating system. Although it now runs, without crashing, on more modern systems such as Windows XP this has been at the expense of losing some of its ability to generate subsets of the database (e.g. all the marriages) in various sorted orders and in various output formats.

For these reasons, when it came to data gathering, the only way forward was to deal with the raw HTML coding on the Wirksworth Web site itself. The presentation of register information for marriages, christenings and burials, on this site, is understandably tied to searches based on family surnames. To increase the likelihood of success in the case of marriages, when details may be known for only one of the persons involved, a system of double entry is used whereby a marriage on a given date is indexed under both bride and the groom surnames.

3.1. First steps — merging and sorting

The infrastructure behind the Wirksworth web site registers is a very simple HTML tabular format. The various records comprising the table entries are all date tagged, but they make little or no use of more advanced HTML metanotation. On any given date there may be records for Burials (B), christenings (C) and Marriages (M), all intermixed. For example a search for records relating to 27th April 1704 gives:

```
C 1704apr27 BLOUNT Thomas=(son)Thomas/(Wirksworth)
M 1704apr27 BRUKSHALL Mary(Biggin)/WOOD John
M 1704apr27 MADDOCK George(Wirksworth)/MELLOR Mary
M 1704apr27 MELLOR Mary(Wirksworth)/MADDOCK George
M 1704apr27 WOOD John(Biggin)/BRUKSHALL Mary
```

Now, it is a simple matter to reject any records beginning with a B or a C, but then comes the trickier task of merging pairs of groom-first and bride-first marriage (M) entries into a single record. To make things worse the above example shows that it is common to have two or more marriages on the same day and it will not always be the case that bride and groom entries for a given marriage will appear on successive lines, especially if the respective surnames are far apart in the alphabet (as shown in the WOOD-BRUKSHALL wedding, above).

It is the job of mergescript, seen at the lower right of Figure 1, to identify and merge pairs of records relating to the same marriage, though whether the single merged record ends up as groom-first, or bride-first, is arbitrary; it depends entirely on which of the surnames is earlier in the alphabet. The towns, or villages, of origin for bride and groom appear in parentheses in the Web site data and are carried over into the output of mergescript, which for the current example would now look like:

```
17040427 BRUKSHALL Mary(Biggin)/WOOD
              John(Biggin)!!@
17040427 MADDOCK George(Wirksworth)/MELLOR
              Mary(Wirksworth)!!@
```

Note that the date of the marriage, at the start of each record, has been converted into a totally numeric format to enable date-based sorting to take place at a later stage. The trailing !!@ is inserted as an end-of-record marker to assist in the processing of multi-line records when, as above, records are split over one or more lines. A standard technique for joining lines with just such end-markers is given in the classic AWK text [3].

It now only remains to sort these pseudo-single-line records into ascending date order via:

```
sort +2n -3 mar-merged.txt > marsorted.txt
```

yielding a large (approximately 2MB) file of all 9,380 marriages at Wirksworth from 1608–1899

3.2. Tagging and justifying

The bulk of the processing for Wirksworth marriages is done by yet another *awk* script called tagscript. This takes on the task of identifying whether a given merged record is groom-first, or bride-first, by consulting a pre-clared set of strings, within tagscript itself, for male and female first names. Every time an unknown first name is encountered it is flagged up on the controlling terminal window, so that it can be incorporated into an updated version of tagscript.

A second *awk* script, called martabgen takes the output from tagscript and takes the opportunity to parse out, from each merged input record, the parenthesized places of origin, for bride and groom, and to convert these into

standard Phillimore format. Other less common information (usually contained in Comment: fields within the original raw records) relates to previous marital status or to occupation (e.g. 'widow', or 'priest') and is also carried over.

Another major task undertaken by martabgen is to take what may now potentially be a multi-line output record and to split it according to the width of the first column in a Phillimore marriage table entry (2.8 inches), which accommodates approximately 40 characters of Caslon Roman text at the point size in use (10 pt.) Each generated line needs to be justified by adding or reducing inter-word spacing. The default is to set a line slightly too tight rather than over-stretching it. Every line after the first is initialised with six en-space characters to allow for the fact that the second, and subsequent, lines are always indented in Phillimore layout.

The martabgen script is set up to report an error if the interleaved spacing on any line is too wide or too narrow. Rather than incorporating a hyphenation algorithm into this *awk* script, any necessary hyphenation is imposed by hand for the tiny number of records (less than one in every 1000) that require it.

Each output record from the two scripts just described is also invested with all of the *tbl* and *troff* coding necessary for formatting the register pages (similarly to the job performed by ascript in the earlier work described in [1]).

3.3. Final checking

The output from tagscript and martabgen looked perfectly plausible as potential input to the common, horizontal, processing chain of Figure 1, which begins by invoking a script to add leader dots. However one small further editing step was still needed.

The Palmer HTML database, extracted from the Wirksworth web site, is originally surname-sorted. Therefore once it has been re-sorted into ascending date order it is impossible to tell the ordering of the marriage services themselves, if several of them took place on a single give day. By default same-day marriages are sorted alphabetically on grooms' surnames. However, the precise actual ordering is of great interest to local historians and is certainly made manifest in the registers themselves (and in any microfilm copies of them). Fortunately, an independent, but unpublished, transcript of the Wirksworth Marriage registers was discovered. This was made by Ivor Neal of Middleton, a village near to Wirksworth.

The Neal database is sorted on a primary key of the marriage date and a secondary key of the register ordering, for any dates with multiple marriages. At the moment any corrections to marriage ordering for the new Phillimore-style Wirksworth volume are done by hand but consideration is being given to generating this re-ordering automatically from a comparison of the Neal database against the tagscript/martabgen output.

Figure 2 shows the output of a typeset sample page from the Wirksworth directory, generated by the processing chain starting at the lower right of Figure 1. By sheer chance the three marriages on 13th January 1705 really did occur in accordance with the alphabetical order of groom surnames. But the three on 15th April did not, and the ordering has been corrected using the Neal database. Figure 2 also shows, on line 9, that it was necessary to hyphenate the place-name of 'Alderwasley'.

1706]   *Wirksworth Marriages.*   55

Thomas Dud & Elizabeth Wetton, both of Idereshay	13 Jan.	1705
Cornelius Roper & Sarah Beardesley, both of Cromford	13 Jan.	,,
William Smith & Mary Clouse, both of Ireton	13 Jan.	,,
Robert Hill & Alice Buxton, both of this p. ...	17 Jan.	,,
George Hodgkinson & Mary Bateman, both of Cromford	15 Apr.	,,
Joseph Fox & Ann Needham, both of Alderwasley	15 Apr.	,,
John Gregory & Sarah Stear, both of this p. ...	15 Apr.	,,
Thomas Allen & Martha Kemp, both of this p.	19 Apr.	,,
Thomas Wagstaff & Atlinah Gregory, both of Cromford	20 Apr.	,,
Edmund Vallence & Elizabeth Smith, both of this p.	29 Apr.	,,
William Short & Ruth Hops, both of Ashlehay	6 May	,,
Stephen Wall & Ann Wood, both of this p. ...	5 June	,,
Michael Clay & Elizabeth Barlow, both of this p.	1 July	,,
John Simms & Dorothy Wean, both of Ashlehay	22 July	,,
Thomas Wigley & Mary Keeling, both of Cromford	29 Aug.	,,
William Valence & Elizabeth Fern, both of this p.	9 Oct.	,,
John Tatom & Ann Clay, both of this p. ...	12 Oct.	,,
John Mather & Elizabeth Wingfield, both of this p.	1 Nov.	,,
Ralph Gell & Jane Hallworth, both of Middleton	7 Nov.	,,
Robert Gothard & Grace Cawdale, both of Alderwasley	1 Dec.	,,
John Smith & Mary Smith, both of Tisington	31 Dec.	,,
Anthony Wilson & Alice Fone, both of this p.	1 Jan.	1706
Thomas Godbere & Mary Collinson, both of Middleton	9 Jan.	,,
Robert Spencer & Elizabeth Hole, both of this p.	15 Jan.	,,
Samuel Spencer & Ann Blackwell, both of this p.	15 Jan.	,,

Figure 2: A sample typeset page of Wirksworth marriages

4. FINAL TESTING

At this stage the output from the processing of Wirksworth marriage records was ready to be fed into the common part of the processing chain in the centre of Figure 1. As related in [1] the precise leader dot patterns needed are extraordinarily sensitive to small variations in the set widths of characters on a given line. The crucial test was whether `dotscript` would cope adequately with input from a completely new processing chain in the lower part of Figure 1, and it certainly did so, as can be seen in Figure 2.

That being said the Wirksworth data did throw up some string widths that revealed, yet again, the need for sensitive fine-tuning when switching from one set of padding

and leader patterns to another. A refined and revised version of `dotscript`, called `wirkdots`, was developed as a result of this new source of data. This refined version has now become the standard for all future work (as well as being retro-fitted into the source code for the previously produced Phillimore Register volumes).

Checks on the operation of the common indexing software, at the upper right of Figure 1, show that it copes flawlessly with the task of indexing Wirksworth entries.

5. CONCLUSIONS

In the real world, when transforming already-available material into another format, one seldom has any control of the way that the source material was created or published. That has certainly been the case for the current work. The test of success can only be whether the approach of pipelined software tools worked well for the new material and the extent to which existing tools from the previous work could be adapted and re-used. In these respects this latest work, which serves as a coda to the original investigations, can be accounted a success.

At present the Phillimore Registers for Derbyshire run to just 15 volumes, of which two have been republished as a result of the work described in [1]. The current work has been tentatively designated as Volume XVI to mark the fact that it is an unexpected extension of the original Phillimore Derbyshire series. When equipped with an index this new volume is likely to run to 500 pages or more. At present the first cut at typesetting and pagination is complete. The limiting step on progress is the need for patient cross-checking of the input data. It is hoped to publish Volume XVI in mid-2012

Acknowledgements

Grateful thanks are due to John Palmer whose patient and meticulous transcripts of Wirksworth marriage data, on his Web site, made the present project possible. The availability of Ivor and Valerie Neal's independent transcript of Wirksworth marriages greatly eased the cross-checking of missing entries, and precise entry ordering, within the original registers. Finally, thanks are due to Margaret O'Sullivan and James Davies, of DRO Matlock, who acted as a backup service for checking details of missing, or near-illegible, entries.

References

[1] David F. Brailsford, "Automated Re-typesetting, Indexing and Content Enhancement for Scanned Marriage Registers", pp. 29–38 in *Proceedings of the ACM Symposium on Document Engineering (DocEng09)*, ACM Press (September 2009).

[2] David F. Brailsford, "Reconstituting typeset Marriage Registers using simple software tools", in *Computer Science — Research and Development (Online first)*, Springer (22 December 2010). DOI: 010.1007/s00450-010-0145-x

[3] Alfred V. Aho, Brian W. Kernighan, and Peter J. Weinberger, *The AWK Programming Language*, Addison-Wesley (1988).

A Cloud-based and Social Authoring Tool for Video

Naimdjon Takhirov
Dept. of Computer and Information Science
Norwegian University of Science and Technology
NO-7491 Trondheim, Norway
takhirov@idi.ntnu.no

Fabien Duchateau*
Dept. of Computer and Information Science
Norwegian University of Science and Technology
NO-7491 Trondheim, Norway
fabiend@idi.ntnu.no

ABSTRACT

In this paper, we present a cloud-based collaborative authoring tool called Creaza VideoCloud. This authoring tool offers an extensive set of features for document-based video authoring in the cloud.

Categories and Subject Descriptors

I.7.2 [**Document and Text Processing**]: Document PreparationHypertext/hypermedia; H.5.1 [**Information Interfaces and Presentations**]: Multimedia Information SystemsVideo

General Terms

Standardization, Documentation, Design, Human Factors.

Keywords

Authoring Tool, Cloud Computing, Collaborative work, Social Video, SMIL

1. INTRODUCTION

Cloud-based services are a relatively recent trend in cloud computing that have attracted a lot of interest lately [4, 2]. The main idea behind cloud computing is that on one hand, virtualized computing resources are used on-demand and on the other hand, user's computer mainly serves as display for processes being executed on the cloud. It implies a greater flexibility, a service-oriented architecture, reduced information technology overhead for the end-user, reduced total cost of ownership, and on demand services [9]. Parallel to cloud computing, various cloud-based services have appeared [6]. A few examples of this kind of services are Dropbox[1], which serves as backup and storage of files backend on the cloud,

or Amazon's Elastic Compute Cloud (EC2)[2], which provides computing power in the form of virtual machines.

The Web is a common environment where social dissemination of information is easily achieved by current innovations [3]. The growth in multimedia documents such as collections of photos, music and videos has been remarkable along with social websites such as Flickr, del.icio.us and Youtube [1]. This rapid development promoted the widespread usage of multimedia systems, i.e., images, sound and animation and these systems can be used in combination with texts and time-based annotations in a variety of settings, providing a platform for a range of story formats combining literary and video elements [5].

Thus, the Web has become a major point of sharing multimedia content including video. More and more devices are increasingly equipped with the HD-quality capability and this results in a significantly increased size of the final content. This increase does not facilitate collaborative video production, which also requires a shared location to version control files and manual change propagation. Additionally, there is a need to publish specific versions for each type of device, i.e. PC, handheld devices, tablet, etc. These limitations make the process of authoring a creative video unnecessarily challenging. We present the system called *Creaza VideoCloud* [3] that addresses all of these issues. Creaza VideoCloud includes a set of tools to help users create, edit, reuse and share their productions. The collection of resources that can be used to author videos includes professional and user-generated content. All tools only require a flash plugin installed on the browser and an Internet connection. The content is stored on the cloud and it can be accessed and shared on the Web any time. One of the important features supported by Creaza VideoCloud is social collaboration, which enables members of a project to interact with each other and share their multimedia files.

2. CREAZA VIDEOCLOUD

Creaza VideoCloud is a cloud-based online video service designed to enable users to collaboratively produce, stream, share and store user-generated video. By combining the power of broadcast-quality HD and the vast reach of social media, Creaza VideoCloud is at the vanguard of a rapidly emerging market in which millions of users in thousands of communities are beginning to continually interact via user-generated online video.

---

*The author carried out this work during the tenure of an ERCIM "Alain Bensoussan" Fellowship Programme.

[1]http://www.dropbox.com

[2]http://aws.amazon.com/ec2/
[3]http://creaza.com

Figure 1: Creaza VideoCloud Dashboard.

The vision of Creaza VideoCloud is to become the premium cloud video service of choice. For instance, football fans want to take their self-generated video and create broadcast-quality produced pieces in HD and then share them with other fans and their social networks.

2.1 The Platform

The Creaza VideoCloud platform consists of a dashboard that includes a rich set of features. The dashboard is the main point of entry to any other functionality. Figure 1 depicts an example of dashboard page. On the top of the menu, there are three tabs, each reminding one visionary aspect of the platform:

- *Projects* tab(collaborative aspect) is a workspace where all projects and project details are accessible. In this tab, the user can create a new timeline, i.e. a set of audiovisual content as well as (still/floating) texts and effects which are put together either in a sequence or in an overlapped manner. The order decides which layer is shown on top of the other layers. A sample timeline is depicted in Figure 2 showing a production about a ski weekend. In the timeline, a user can additionally create a custom audio or video layer. The platform supports direct export of final productions to either Youtube[4] or as locally exported MPEG-4 file.

- *Media* tab (cloud-based storage aspect) enables the management of the different content (video, images sounds and other file types). The multimedia content can be arranged in various folders for easy management.

- *Account* tab (social aspect) is where the user can change his/her profile information as well as connect his Creaza VideoCloud account to other social networking sites. This tab includes the integration with other social networking sites such as Facebook[5], Twitter[6].

The dashboard is based on the latest RIA application framework, the Flex platform. These platforms use BlazeDS[8] which is the server-based Java remoting and web messaging technology that enables to easily connect to back-end distributed data and push data in real-time to the client side of the Creaza VideoCloud platform. Consequently, this results in a rich interactive feature, directly from the user's browser.

Creaza VideoCloud is powered by Inspera Platform[7] which together with BlazeDS open source technology facilitates high performance data transfer making the platform's client side a more responsive application. The client side pushes data to the server over standard HTTP protocol, in real-time. Furthermore, the full-scale publish/subscribe messaging approach used in the platform provides a basis for a scalable infrastructure where real-time data delivery is of significant importance. The platform also supports Webdav [10] based communication but BlazeDS integration is far more efficient both in terms of transfer and latency.

No local installation of the application is necessary. The content is stored on the server side and the storage servers support double backup. This redundancy gives possibility for a safe failover and a faster disaster recovery.

---

[4]http://youtube.com

[5]http://facebook.com
[6]http://twitter.com
[7]http://inspera.com

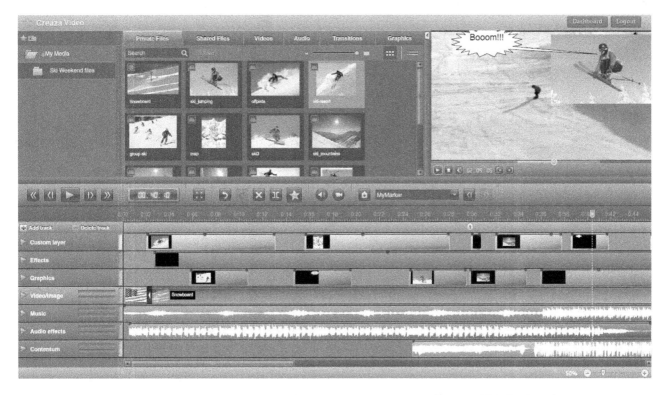

Figure 2: Sample Timeline Showing a Production about a Ski Weekend.

2.2 Document Representation

Creaza VideoCloud features multi-layer tracks for audio content as well as graphics, effects (e.g. transitions, watermark) and videos. An interesting issue addressed by the platform is that it enables multiple videos to be played in different parts of the screen, all at the same time. Furthermore, the texts appearing in the productions such as watermark, greeting texts or any other type of informative text are rendered on-the-fly and across the layers. The same applies to the audiovisual content. A *production* is the outcome of a set of audiovisual content along with effects represented in the timeline. The document representing productions is transferred to the cloud-based storage facility as a *SMIL* file. Synchronized Multimedia Integration Language (SMIL) [7] is an XML-based markup language endorsed by World Wide Web Consortium[8]. SMIL facilitates simpler authoring of interactive audiovisual documents. The W3C recommendation specifies the definition of timing markup, animation and layout, visual transition, embedding, etc. Creaza VideoCloud platform makes an extensive use of the latest version of SMIL (v3.0) for documents representing audiovisual content, i.e., the creative content collaboratively authored by users. Figure 3 depicts an example of how these documents are represented as SMIL. An important aspect to mention is that since the actual multimedia content may be stored anywhere in the cloud, there is no fixed absolute reference to the media. For instance, our example has an audio object with reference "A0690016-D12F-3BA3-6E26-5DC9DCDA444F". The actual address of this object is resolved at runtime as requested. The different *GraphPoints* are objects that appear in the timeline

with the specified start and stop positions. In the above example, audio object with identifier "A0690016-D12F-3BA3-6E26-5DC9DCDA444F" has a position from 1 to 12600 milliseconds in the timeline.

3. DEMONSTRATIONS

In this section, we describe two simple applications of Creaza VideoCloud. Currently, Creaza VideoCloud is in closed beta and only invited users can use the service and provide feedback. An account has been created for demonstration purpose at the following url: `http://www.creaza.com/doceng2011.jsp`. This account will remain active until the Creaza VideoCloud is finally launched, late summer 2011.

3.1 Use Case 1: DocEng'2011 Experience

During a conference, many participants take pictures and record movies of the presentations and conference dinner. As conferences aim at exchanging ideas and promoting collaboration, the participants are interested in sharing their videos and photos with others, listening to presentations that they might have missed or using one or more media during their own teaching courses. A social platform is therefore helpful to fulfill these goals. Let us imagine what could happen at DocEng'2011 using the free Creaza VideoCloud platform for sharing videos. The organizers can propose to participants to upload their media related to the conference in a Creaza VideoCloud project that they have created. A participant needs to login either by creating a Creaza Video-Cloud account or by using an existing authentication service (Google, Facebook, Yahoo and Twitter are currently supported). Once logged in, the participant is invited or added as project member and she has access to the shared media

---

[8]`http://www.w3.org/AudioVideo/`

```
<smil ...>
  <head>...</head>
  <body>
    <par>
      <par ins:layerLock="false" ins:layerColor="10581326"
           ins:layerTitle="Effekte" ins:layerType="SVG_LAYER"/>
      <par ins:layerLock="false" ins:layerColor="5953842" ins:layerTitle="Grafik"
           ins:layerType="SVG_LAYER"/>
      <seq ins:layerVolume="100" ins:layerMute="false" ins:layerSolo="false"
           ins:layerLock="false" ins:layerColor="3776464" ins:layerTitle="Musik"
           ins:layerType="AUDIO_LAYER">
        <audio ins:objectType="content_sound" ins:logicalName="fused"
               ins:uid="A0690016-D12F-3BA3-6E26-5DC9DCDA444F"
ins:volume="1.00"
               ins:mute="false" dur="00:07.640">
          <ins:GraphPoints>
            <ins:GraphPoint ins:time="0" ins:value="1"/>
            <ins:GraphPoint ins:time="7640" ins:value="1"/>
          </ins:GraphPoints>
        </audio>
        <audio ins:objectType="content_sound" ins:logicalName="cool_kille_loop"
               ins:uid="317B6B18-FFDE-96FA-E298-5DCA143605A5"
ins:volume="1.00"
               ins:mute="false" dur="00:12.600">
          <ins:GraphPoints>
            <ins:GraphPoint ins:time="0" ins:value="1"/>
            <ins:GraphPoint ins:time="12600" ins:value="1"/>
          </ins:GraphPoints>
        </audio>
      </seq>
      ....
      ....
    </par>
  </body>
  <markers/>
</smil>
```

Figure 3: Sample SMIL File.

files. When uploading a new content in this shared project, all multimedia metadata (e.g., number of frames, duration) are automatically filled in and the participant should only provide descriptive metadata (e.g., tags, annotation). Anyone can then download, annotate or edit these media. Similarly to other social websites, Creaza VideoCloud enables discussions about a project or media. Furthermore, a video can be directly uploaded to YouTube for easy sharing with people that do not have access to the project. This feature is especially meaningful to release a collaborative work, as described in the next demonstration.

3.2 Use Case 2: Ski Weekend

We now illustrate the cloud-based video authoring features of Creaza. After the DocEng'2011 conference, some participants decide to spend a week-end at the *Bear Mountain Resort* close to San Bernardino CA, for discussing future collaboration and enjoying skiing. These inspiring, fruitful days are generally productive in terms of creating videos. When the participants are back home, they create a new project in Creaza VideoCloud platform to share creatively document and enjoy virtually once again their experience in an easy way. In addition, they want to produce a summary of this week-end to show to their absent friends and colleagues the fun moments they have missed. Creaza Video-Cloud is clearly dedicated to this task since it enables a collaborative authoring. The participants choose to produce a new video which contains the most interesting slices of the videos that they have recorded. Because of the use of multiple (overlapping) layers, this task can be divided as follows: one participant is in charge of mixing the video slices, an-

other one selects the various background musics while a last one could add some texts to describe the scenes. Once all participants have agreed on the editing, the resulting video is finally either exported to YouTube or as a local standalone production file.

4. CONCLUDING REMARKS

In this paper, we presented a cloud-based collaborative authoring tool, Creaza VideoCloud. This tool offers an extensive set of features for document-based social video authoring in the cloud. The combination of cloud technology and video authoring simplifies the collaborative work and provides a seamless range of services for the heavy process of video production. Thus, the Creaza VideoCloud platform is a realization of a "making video social" vision and a demonstration of the fact that video authoring can be made smooth, simple and collaborative.

5. REFERENCES

[1] E. Bertino, A. K. Elmagarmid, and M.-S. Hacid. Quality of service in multimedia digital libraries. *SIGMOD Rec.*, 30(1):35–40, 2001.

[2] R. Buyya, C. S. Yeo, S. Venugopal, J. Broberg, and I. Br. Cloud Computing and Emerging IT Platforms: Vision, Hype, and Reality for Delivering Computing as the 5th Utility, 2009.

[3] M. J. Halvey and M. T. Keane. Analysis of online video search and sharing. In *Proceedings of the eighteenth conference on Hypertext and hypermedia*, HT '07, pages 217–226, New York, NY, USA, 2007. ACM.

[4] B. Hayes. Cloud computing. *Commun. ACM*, 51:9–11, July 2008.

[5] R. Laiola Guimarães, P. Cesar, and D. C. Bulterman. Creating and sharing personalized time-based annotations of videos on the web. In *Proceedings of the 10th ACM symposium on Document engineering*, DocEng '10, pages 27–36, Manchester, United Kingdom, 2010. ACM, New York, NY, USA.

[6] M. Miller. *Cloud Computing: Web-Based Applications That Change the Way You Work and Collaborate Online.* Que Publishing Company, 1 edition, 2008.

[7] L. Rutledge, L. Hardman, and J. V. Ossenbruggen. The use of smil: Multimedia research currently applied on a global scale. In *Modeling Multimedia Information and System Conference*, pages 1–17. World Scientific, 1999.

[8] S. Tiwari. *Professional BlazeDS: Creating Rich Internet Applications with Flex and Java.* Wrox Press Ltd., Birmingham, UK, 2009.

[9] M. A. Vouk. Cloud computing: Issues, research and implementations. In *Information Technology Interfaces, 2008. ITI 2008. 30th International Conference on*, pages 31–40, jun 2008.

[10] E. J. Whitehead, Jr. and Y. Y. Goland. WebDAV: a network protocol for remote collaborative authoring on the Web. In *Proceedings of the sixth conference on European Conference on Computer Supported Cooperative Work*, ECSCW'99, pages 291–310, Norwell, MA, USA, 1999. Kluwer Academic Publishers.

Collaborative editing of Multimodal Annotation Data

Demo

Stephan Wieschebrink
CRC 673 – Alignment in Communication
Bielefeld University,
Postfach 10 01 31, D-33501 Bielefeld
swieschebrink@uni-bielefeld.de

ABSTRACT

The annotation of multimodal speech corpora is a particularly tedious task, since annotatable events can be composed of smaller events that span across several modalities (e.g. speech and gesture), which imposes the need to operate on the same data, using a wide range of different tools in order to cover all the different modalities and layers of abstraction within multimodal data. MonadicDom4J has been developed as a highly generic general purpose java-based Rich Client framework that opens the possibility to simultaneously operate on any kind of XML data through several different views and from several remote locations. It allows for the dynamic allocation of plugins, needed to render a given type of XML markup, and takes care of the concurrency between different sites viewing the same data, by means of differential synchronization. The demonstration will involve several different applications ranging from general textual hyperdocument editing to multimodal annotation tools, whose contents can be freely intermixed, interlinked and transcluded into different contexts, using drag and drop interaction. The audience will have the opportunity to try collaborative editing on the presented examples from their own devices.

Categories and Subject Descriptors

C2.2 [**Network Protocols**]: protocol architecture (OSI model), applications; C.2.4 [**Distributed Systems**]: client/server, distributed databases, distributed applications; D.2.11 [**Software Architectures**]: patterns; D.2.13 [**Reusable Software**]: reuse models; E.1 [**DATA STRUCTURES**]: tree, distributed data structures; E.2 [**DATA STORAGE REPRESENTATIONS**]: linked representations, object representation; H.5.3 [**Group and Organization Interfaces**]: collaborative computing, computer-supported cooperative work, synchronous interaction, web-based interaction, WEB; H.5.4 [**Hypertext/Hypermedia**]: architectures; I.7 [**DOCUMENT AND TEXT PROCESSING**]: desktop publishing, hypertext/hypermedia, markup languages, standards

General Terms

Design, Standardization

Keywords

Hyperdocument Systems, XML data binding, software engineering, online collaboration, CSCW

1. INTRODUCTION

One of the hot topics in web technology is the emerging field of collaborative games [4,5], which are also referred to *games with a purpose*. The predominant purpose of these games is to create annotation corpora, which are sufficiently large enough, to be employed for various machine learning tasks. One of the major difficulties, concerning such web-based real-time collaboration and content creation scenarios, is the fact that WWW-architecture does not natively support in-place editing and live-synchronization of rich contents: The AJAX-Framework that is credited much of the success of recent developments in collaborative web technology, known as Web2.0, examined under the hood, seems much like a huge and fancy ship-in-a-bottle which is crafted through the narrow neck of JavaScript-animated HTML+CSS-contents that are communicated by http-protocol.

The driving motive in the creation of the monadicDom4J framework was to revive some of the concepts that the original visionaries of hypertext had foreseen [1,2,3]. Hence, monadicDom4J consolidates another internet-based hyperdocument system on its own, using its own protocol (xttp) and its own serialization format (RAXML).

The projected field of application for monadicDom4J was to serve as a platform for editing- and visualization tools, that operate on XML documents of the *Speech and Gesture Alignment Corpus* (SaGA) Corpus [10].

There are certain special requirements that such a tool-platform has to meet in order to be able to be adequate for some of the more advanced goals of our working group (namely the implementation of real time collaborative annotation scenarios and collaborative games):

First of all the framework needs to be performant and clear cut enough in order to serve as a platform for rather sophisticated applications, like multimodal annotation editors, that are capable of playing multimedia web-based contents from various online sources, synchronously. Additionally, it needs to support the parallel viewing of the same documents through several different views, which have to remain up to

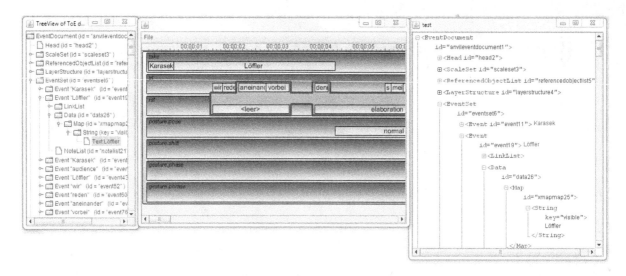

Figure 1: Several views on the same document: All visualizations are monadic documents themselves.

date on any change of the document. Finally, in order to support the notion of internet-based real-time collaboration, it has to be capable of synchronizing different locations that are accessing the same contents, without delay. These considerations have led to the creation of an extremely generic and abstract java-based Rich Client Platform that is based on entirely different paradigms than conventional Web architecture. Here, two pivotal aspects are (1) the use of so called Document Factories in the front end, and (2) the use of a new XML serialization format called RAXML, in the back end. Together they make up a new paradigm, we refer to as monadic XML representation. The core notion of this paradigm is that any type of XML content or representation (i.e. Serialized, memory loaded or in the internet) is encapsulated by the same XML object representation. Changes to that object representation are implicitly synchronized with any other representation of the same content (i.e. a different client viewing the same data. Hence, the dichotomy of several different XML representations of the same data is virtually lifted from the users' as well as the designers' perspective. From the developers'-perspective, the immediate advantage of monadic XML representation is, that developers of new document formats only have to implement the Object representation (i.e. A Document Factory) of their new format in order to profit from all the subsequent features, which the underlying framework provides (e.g. Real time collaboration and implicit versioning). In addition, since all parts of any document share the same XML representation, contents of different XML types can still be mashed up. Together, this adheres to the notion of an Open Notebook space, which enables researchers to work, collaborate, publish and track the evolution of their work immediately within the same workspace [14].

2. THE ARCHITECTURE

The most central element in the overall architecture of monadicDom4J is its new XML serialization format: In RAXML, XML nodes are not represented in textual form, but as fixed-length data nodes, similar to those in file systems (e.g. hard disk partition formats). Other than in file

systems, deleted nodes are not physically deleted from the disc. Instead, only the addresses of their according parent and sibling nodes are updated, in order to take them out of the content. The nodes themselves are preserved as they are (including the references to their parents and siblings). This imposes the circumstance that all nodes appear in the sequence they have been added to the document since any new node has to be appended to the end of the current document. For any deletion, a surrogate node that holds record of only the address of the deleted node is added to the end of file. That way, the entire change history of a document becomes tractable and undoable at the level of individual edits.

A memory loaded XML document of RAXML data consists of DOM nodes which are created by a custom *document factory* which is associated with the *namespace* of that node. Each DOM node holds a reference to its corresponding RAXML node within the RAXML file. As the resulting DOM composite is edited, the according RAXML nodes within the file are updated implicitly.

This rather simple front-end back-end design patter enables all the subsequent key-features of a real-time collaborative Hyperdocument system, which are:

1. *Fine grained addressability* of individual nodes: Each node accounts for the offsets of its neighboring nodes (children, parents, siblings), which makes it possible, to directly allocate individual branches of an XML document, traversing the node structure selectively.

2. *Proper hyperlinking*: That same format also enables the linking of individual nodes and text ranges. Links are a special type of node within the RAXML format that any node type (including links themselves) can have as a child node. Hence, it is possible to create an arbitrary number overlapping and bidirectional links on the same content. This also allows for transclusions. E.g., the embedding of a branch of content from one location within another context.

3. *Gapless versioning*: The overwriting of space occupied by deleted RAXML nodes is omitted in order to allow

for the gapless reconstruction of any state of the editing history of a RAXML document. Additionally, each node is furnished with its own version number, which allows for the restoration of any version any individual branch has been in.

4. *Real-time online collaboration.* The above versioning mechanism allows for the lag-free synchronization of remote instances of the same document, using differential updating, which (along with (1)) implements the possibility of live collaboration.

5. *Mashing up of arbitrary types of XML markup.* Any new node of a monadic document is instantiated by a so called Document Factory. Document Factories can be associated with the namespace of a certain XML dialect. They are implemented as .jar-plugins that can be side-loaded and allocated on demand and at runtime, which makes it possible to dynamically extend the functionality of a Rich Client instance by new Document Factories for previously unknown markup languages. Since all nodes of any document factory implement dom4J's DOM interface, it is still possible to intermix any of these nodes within the same document.

3. THE DEMO

The demonstration will display all of the above mentioned major features of monadicDom4J. As of the time of this paper, several plugins, that provide for the visualization and interaction on multimodal data and XML data in general are yet being developed. However, the demo session will definitely include following document factories:

- An editor for multimodal events that allows for the time-aligned replay of multimedia streams, called *MMEViz*.

- A general purpose treelike XML visualization, that allows for the editing of any XML tree.

- A MathML formula editor.

- A general Hypertext editor, that allows for overlapping link layers, and the visual navigation of such.

- A SWING-Components Document Factory, that is based on the SwiXml[12] markup language.

All of the visible elements of any of these applications can be freely intermixed by means of drag-and-drop interaction or using copy-and-paste (as a special action, the *clude*-command will allow for the transclusion of any item within another context). I.e., a mathematical formula can be embedded in a multimodal event as annotation, which in turn can be transcluded in a hypertext document etc.. Editing of any of the transcluded items will implicitly update any other location of the same content (E.g., locally between different views, or over the internet at remote sites). In the case that new (foreign) markup is inserted into a document, the appropriate Document Factory is also side-loaded from the source of that edit event and relayed to any other location viewing the same content. The individual steps that are incorporated to provide for the

differential synchronization during collaborative editing, can be exemplified using eclipse's debugger feature and a a network protocol analyzer.
The audience may to join a collaborative editing session, using their own devices.

4. REFERENCES

[1] Nelson, T.H. 1974,1987 *Computer Lib: You can and must understand computers now / Dream Machines: New freedoms through computer screens - a minority report* (1974), Microsoft Press, rev. edition 1987: ISBN 0-914845-49-7

[2] Nelson, T.H. 1980-1993. *Literary Machines: The report on, and of, Project Xanadu concerning word processing, electronic publishing, hypertext, thinkertoys, tomorrow's intellectual revolution, and certain other topics including knowledge, education and freedom.* Mindful Press

[3] Vannevar Bush, 1945. *As we may think.* Atlantic Monthly 176 y Eastgate Press, Watertown MA, USA.

[4] Ahn, L., 2006. *Games with a purpose.* IEEE Computer Magazine, 39(6), 92-94.

[5] Ahn, L., 2008. Human Computation. IEEE 24th International Conference on Data Engineering (ICDE 2008), Cancun, Mexico (pp. 1-2)

[6] V. Quint and I. Vatton, 2004. *Techniques for authoring complex XML documents.* In Proc. of DocEng '04, pages 115-123. ACM.

[7] O. Beaudoux and M. Beaudouin-Lafon, 2001. *DPI: A conceptual model based on documents and interaction instruments.* In Proc. of IHM-HCIŠ01, pages 247-263. Springer Verlag.

[8] O. Beaudoux, 2005. *XML active transformation (eXAcT): Transforming documents within interactive systems.* In Proc. of DocEng'05, pages 146-148. ACM.

[9] CRC673: *Alignmet In Communication: The X1 Project.* http://www.sfb673.org/projects/X1

[10] Lücking, A., Bergmann, K., Hahn, F., Kopp, S., Rieser, H. (2010). *The Bielefeld Speech and Gesture Alignment Corpus (SaGA).* In M. Kipp, J.-C. Martin, P. Paggio D. Heylen (Eds.), LREC 2010 Workshop: Multimodal Corpora - Advances in Capturing, Coding and Analyzing Multimodality.

[11] *The dom4J Project.* http://dom4j.sourceforge.net/

[12] *The SwiXml Project.* http://www.swixml.org/

[13] Qinyi Wu and Calton Pu, 2010. Modeling and Implementing Collaborative Text Editing Systems with Transactional Techniques. In Proceedings of IEEE Conference on Collaborative Computing: Networking, Applications and Work Sharing (CollaborateCom'10).

[14] Bradley, Jean-Claude, Owens, Kevin, and Williams, Antony, (2008). *Chemistry crowdsourcing and Open Notebook Science.* Nature Precedings. 10 Jan 2008.

Developer-Friendly Annotation-Based HTML-to-XML Transformation Technology

Lendle Chun-Hsiung Tseng

Department of Computer Information and Network Engineering, Lunghwa University of Science and Technology

lendle_tseng@seed.net.tw

ABSTRACT

Nowadays, the amount of information accessible on the web is huge. Although web users today expect a more integrated way to access information on the web, it is still rather difficult to "integrate" information from different web sites since most web pages are authored in HTML format, which is actually a presentation-oriented language and is usually considered unstructured. Today, there are many research works aiming at extracting information from web pages. Existing works typically transform the extracting results into structured or semi-structured data formats, thus other applications can further process the results to discover more useful information. Nevertheless, the unstructured nature of HTML makes the transformation process complex and can hardly be widely adopted. In this paper, an annotation-based HTML-to-XML transformation technology is proposed. The mechanism is developed with both usability and simplicity in mind. With the proposed mechanism, ordinary web site developers simply add annotations to their web pages. Annotated web pages can then be processed by our software libraries and transformed into XML documents, which are machine-understandable. Software agents thus can be developed based on our technology.

Categories and Subject Descriptors

H.3.3 [**Information Search and Retrieval**]: Retrieval models

General Terms

Design

Keywords

XML, Mashup, Information Extraction, Annotation

1. INTRODUCTION

It is without doubt that adopting XML data format in electronic data exchange has advantages. XML data format is machine-understandable, which means an XML document can be "understood" by software agents. Furthermore, since an XML document can be associated with XML schemas, strict data type information is available for preventing misunderstanding. Hence, integration among several XML information data sources is

possible. However, compared with the number of information sources on the web, the number of XML-enabled ones is much smaller. Most web sites deliver information in a data format that is usually considered as unstructured and not machine-understandable, say, the HTML data format. The original purpose of the HTML data format is for layout/presentation only. As a result, the data format lacks of some key concept for modeling data, such as type information. The situation reduces the usability of the huge amount of information.

To overcome these issues, several information extraction technologies have been developed. Basically, existing technologies provide means for locating desired regions of information from HTML documents and then transform them into more structured representations. Conceptually, existing technologies can be categorized as automatic ones and manual ones. The former ones are convenient, but with the cost of less accuracy and flexibility. The latter ones are accurate and flexible but require developers to define detailed rules and even write programs.

In this paper, the author proposes an alternative mechanism for extracting information from HTML documents. With the proposed mechanism, web site developers at first annotate their HTML pages with the provided set of annotations. These annotations are used for declaring the relationships between HTML nodes and the target XML nodes as well as the positioning rules for these XML nodes. After the transformation, HTML documents will be converted to XML data conforming to some specified XML schemas. As a result, the extracted information can be "understood" by software agents that have knowledge about the corresponding XML schemas. A software library is also provided for performing the transformation. The proposed technology is developer-friendly because of the following criterion:

1. The required annotations are easy to understand and easy to use.

2. Adopting these annotations will not affect the normal functionality of a web page.

3. The degree of flexibility that can be achieved by the proposed mechanism is close to which can be achieved by other programming-based approaches such as XSLT [9].

The resulting software library is named as H2X, which stands for "Html to XML transformation". H2X is a java-based software library and can be adopted in both standalone applications and server-side applications.

2. RELATED WORK

Rather than simple web surfing, web users nowadays request for an integrated view of information on the web. Roughly speaking, web information extraction systems can be divided into four different types: manually constructed systems, supervised systems, semi-supervised systems, and unsupervised systems [2]. Manually constructed systems such as WebOQL [1] require users to use provided programming or query languages for extraction and thus become extremely expensive. Supervised systems such as DEByE [5] require users to provide a set of pre-labeled web pages for training and then will automatically infer rules for extracting data from similar web pages. Semi-supervised systems such as OLERA [3] require no pre-labeled training pages, but post-efforts from the user is required to choose the target pattern and indicate the data to be extracted. Unsupervised systems, on the other hand, require no user interactions. However, they may only be applicable for data-rich regions, e.g. the table part, of a web page. The work proposed by Wang and Lochovsky [10] is an unsupervised web information extraction system.

Generating XML documents from information sources is not an uncommon request for web information extraction. There are several technologies and research works targeting at providing such functionality. For example, XSLT [9] is a popular technology aiming at transforming an existing XML document (probably an XHTML document) into a new one. Besides programming-based transformation, some technologies may rely on pre-specified mappings to generate XML documents from user inputs on GUI components. Examples of applications/software libraries adopting such technology are XForms [8], FormsXML [4], InfoPath [7], and XUIB [6].

Compared with existing works, the proposed mechanism appears to be both feature-rich and user-friendly. XSLT-based approaches require programming skills. XForms-based technologies need special browser support and thus have not been widely adopted. Most mapping-based technologies lack of flexibility due to the constraints caused by XML schemas. XUIB is flexible, but probably too complex for ordinary developers. On the other hand, the proposed mechanism requires developers to add only needed annotation and can determine some transformation rules automatically.

3. Annotation-Based HTML-to-XML Transformation

The proposed mechanism is aimed at providing a transformation framework for generating XML documents from annotation-augmented HTML documents. Here, it is assumed that the HTML documents to be processed are always XHTML-compliant. If not, a tidy process will be invoked in advance to ensure XHTML-compliance. An annotation-augmented HTML document is an HTML document with HTML nodes augmented by annotations. The augmenting information is then used as rules for generating the target XML document. Two types of rules are needed during the generation process:

1. node construction rules that are used for providing the specification of the target XML nodes

2. positioning rules that are used for determining the final positions of generated XML nodes

That is, for all nodes (including element nodes, attribute nodes, and text nodes) within the HTML document to be processed, a *(node, node construction rule, positioning rule)* tuple is maintained. The transformation process is listed below:

1. walk through the HTML DOM tree and construct a meta tree, the transforming tree, which reflects the original tree structure; nodes on the transforming tree, the transforming tree nodes, maintain the original HTML DOM nodes and associated annotation information

2. for each HTML node, create the corresponding XML node according to its associated construction rules and store the generated XML node into the meta tree

3. traverse the meta tree and position XML node contained in each tree node according to its explicit or implicit (will be explained later) positioning rule

The proposed mechanism defines the following extended attributes for specifying the augmenting information:

1. *as_root*: an element node annotated with this attribute is regarded to as a root node in the resulting XML document; if there are more than one element HTML nodes annotated with this attribute, more than one XML documents will be generated from this HTML document

2. *root*: specify the root XML element of the resulting XML node generated from the annotated target

3. *def*: specify how to generate a resulting XML node from the annotated target

4. *relation*: define the relationship among the resulting XML node and other XML nodes; allowable relationships are *child* and *next*; these relationships are used for positioning resulting XML nodes

Most extended attributes expect values in the form: (*target_exp, parameter*). *target_exp* is used for locating the target nodes to apply this annotation. The design of *target_exp* allows developers to annotate HTML nodes more freely. For example,

<input type="text" value="book1" h2x:def="@{}value, @{}bookName"></input>

indicates that the attribute node "value" is associated with (and will be transformed to) the "bookName" attribute on the target XML page. Multiple rules separated by ";" can be defined within one annotation if they have different *target_exp*. For instance, *h2x:def="@{}value, @{}bookName; ., {}bookMetaData"* defines an additional rule that will map the annotated HTML node itself to the "bookMetaData" element on the target XML page.

Furthermore, two types of positioning rules, the explicit and the implicit ones, are used for determining the final position of a generated XML node. Explicit positioning rules are specified with the *relation* annotation and have the highest priority. Implicit rules are adopted when no explicit rule is specified. Two types of implicit rules will be adopted when no explicit rule declared (listed from highest to lowest priority):

1. schema position: to position a transforming tree node according to the restrictions imposed by the associated XML schema

2. natural position: to position a transforming tree node according to the relative position of nodes with *def* annotations

Schema position needs more explanation. For an XML element with complex content type, the allowable sequential order of its child nodes has to fit the constraint imposed by the corresponding schema. Without loss of generality, one can construct a state machine to represent the constraint. Each transition arc of the state machine is labeled with the qualified name of a legal child node. By traversing along the transition path from the start state to the end state, one or more legal child node sequences can be determined. The proposed mechanism utilizes an approach based on this to discover a legal child node sequence. As a result, developers do not have to explicitly position every node and hence their efforts can be greatly reduced.

4. AN EXAMPLE

In this section, an example is used for demonstrating the concept of the proposed mechanism. At first, image an online book-shopping agent service that is designed as a portal for browsing/buying books from several other online book stores. The agent may want to provide more advanced services such as price comparison. To achieve the goal, the agent has to extract information form several online book stores. A commonly-used approach is web page scraping. This can be difficult and unstable since web pages of different book stores will organize information in different ways. Note that although different web sites have different layouts, it is still possible that they adopt the same data model. Assume the following (simplified) XML schema is used as the common data model:

<xs:schema targetNamespace="test1"

xmlns:editix="test1"
xmlns:xs="http://www.w3.org/2001/XMLSchema">

<xs:element name="books">

<xs:complexType>

<xs:sequence maxOccurs="unbounded">

<xs:element ref="editix:book"/>

</xs:sequence>

</xs:complexType>

</xs:element>

<xs:element name="book">

<xs:complexType>

<xs:sequence>

<xs:element ref="editix:title"/>

<xs:element ref="editix:author"/>

<xs:element ref="editix:price"/>

</xs:sequence>

</xs:complexType>

</xs:element>

<xs:element name="title" type="xs:string"/>

<xs:element name="author" type="xs:string"/>

<xs:element name="price" type="xs:double"/>

</xs:schema>

Instead of relying on web page scraping for extracting information from web pages, here, we demonstrate how to associate a normal XHTML web page with the above data model through the proposed approach. Below is an annotation-augmented web page:

<html xmlns:h2x="h2x">

<head>

<title>TODO supply a title</title>

</head>

<body h2x:def="{test1}books"

h2x:as_root="books.xsd">

*Books:
*

<ul>

<li h2x:def="{test1}book" id="book1">

<table border="1">

<tbody>

<tr h2x:def="{test1}price" id="price1">

<td>Price</td>

<td h2x:def="text(),text()">100</td>

</tr>

<tr h2x:def="{test1}author" id="author1"
h2x:relation="id(title1),.,next">

<td>Author</td>

<td h2x:def="text(),text()">Author1</td>

</tr>

<tr h2x:def="{test1}title" id="title1">

<td>Title</td>

<td h2x:def="text(),text()">Book1</td>

</tr>

</tbody>

</table>

</li>

</ul>

</body>

</html>

As shown in the above example, the <body> element is associated with the {test1}books schema element via the *def* annotation and is defined as a root node. Note there are several different syntaxes of the *def* annotation shown in the example. For instance, the <td> element with "Book1" as its text content is annotated with "*h2x:def="text(),text()"*". The annotation specifies that the text content of the <td> element is mapped to a text node in the target XML document. Another example is the <tr> element with id "*author1*". According to the annotation, "*h2x:def="{test1}author"*", this <tr> element is mapped to an element node with "*{test1}author*" as the qualified name.

It is possible that, the hierarchical structure of the XHTML document is different from the target XML document. The "*relation*" annotation is used for dealing with this scenario. For instance, the <tr> element with id *author1* is annotated with "*h2x:relation="id(title1),.,next"*". The annotation states that the

XML node generated by this XHTML node is the next sibling node of the XML node generated by the XHTML node with id *title1*. Furthermore, XHTML nodes not associated with the *relation* annotation will be positioned according to implicit positioning rules, so there is no need to specify positioning rules for all XHTML nodes. For example, XHTML node with id *price1* and *title1* will be automatically positioned according to the associated XML schema.

Below is the XML document generated from the sample XHTML document:

```
<books xmlns="test1">
 <book>
   <title>Book1</title>
   <author>Author1</author>
   <price>100</price>
 </book>
</books>
```

Now, consider another web page with a different layout:

```
<html xmlns="http://www.w3.org/1999/xhtml"
xmlns:h2x="h2x">
   <head>
     <title>TODO supply a title</title>
   </head>
   <body h2x:def="{test1}books" h2x:as_root="books.xsd">
     Books:<br/>
     <ol>
       <li h2x:def="{test1}book" id="book1">
         <ul>
           <a h2x:def="{test1}title" id="title1"></a><li
h2x:relation="id(title1),text(),child"
h2x:def="text(),text()">Book1</li>
           <a h2x:def="{test1}price" id="price1"></a><li
h2x:relation="id(price1),text(),child"
h2x:def="text(),text()">100</li>
           <a h2x:def="{test1}author" id="author1"></a><li
h2x:relation="id(author1),text(),child"
h2x:def="text(),text()">author1</li>
         </ul>
       </li>
     </ol>
   </body>
</html>
```

Note that in the above web page, dummy <a> tags are introduced as the container node holding the information for the generation of some XML elements. The resulting XML document is compatible with the previous one since they adhere to the same XML schema. Hence, software agents can understand information extracted from both web sites. As a result, instead of performing web page scraping, the agent service mentioned above can simply extract information from the resulting XML documents. This will be simpler and more stable.

5. CONCLUSION AND FUTURE WORK

In this paper, an annotation-based technology for HTML to XHTML transformation is proposed. Utilizing the provided set of annotations, web site developers can specify transformation rules directly on HTML documents. These annotations can configure not only the generation of XML nodes but also the positioning rules of the generated XML nodes. Despite of the richness in functionalities, the proposed mechanism is carefully designed to reduce the usage complexity. By employing schema information associated with the HTML document, the mechanism is capable of determining the position of most generated XML nodes. Hence, web site developers do not have to specify positioning rules for all generated XML nodes. Furthermore, additional tool support is not needed for adopting the proposed technology. Developers can utilize the proposed technology with their acquainted HTML tools.

In the near future, it is scheduled to extend the current work to provide scripting support. By allowing scripting in the value part of annotations, flexibility can be further enhanced; take the "id(*#id*)" style target expression as an example, with scripting support, the "*#id*" can be determined by evaluating an scripting function.

Moreover, it is planned to implement the web database concept based on H2X. With H2X, it will be easier to extract information from the web; as a result, it appears reasonable to regard the web as a huge database of machine-understandable information; by extending current design with ontology concept and query functionalities, it will be possible to perform more advanced queries than current search engines can support. For example, type information can be included in a query thus the result will be more accurate.

6. REFERENCES

[1] Arocena, G.O. and Mendelzon, A.O. 1998. WebOQL: Restructuring documents, databases, and webs. In Proceedings of the 14th IEEE international conference on Data engineering, 1998, 24-33

[2] Chang, C.H. and Girgis, M. 2006. A survey of web information extraction systems. IEEE Transactions on Knowledge and Data Engineering 18, 1411-1428

[3] Chang, C.H. and Kuo, S.C. 2004. OLERA: A semisupervised approach for web data extraction with visual support. IEEE Intelligent Systems 19, 56-64

[4] Kuo, Y.S., Shih, N.C., Tseng, L., and Hu, H.C.: Generating form-based user interfaces for xml vocabularies. In Proceedings of the 2005 ACM symposium on Document engineering, pp. 58–60. ACM, New York, NY, USA, 2005

[5] Laender, A.H.F., Ribeiro, B., and Silva, A.S. 2002. DEByE--- Data extraction by example. Data and Knowledge 40, 121-154.

[6] Lendle Tseng, Y.S. Kuo, Hsiu-Hui Lee, and Chuen-Liang Chen: XUIB: XML to User Interface Binding. In Proceedings of the 2010 ACM Symposium on Document Engineering, 2010, page 51-60

[7] Microsoft: Infopath. http://office.microsoft.com/en-us/default.aspx

[8] W3C: XForms. http://www.w3.org/MarkUp/Forms/

[9] W3C: XSLT. http://www.w3.org/TR/xslt

[10] Wang, J. and Lochovsky, F.H. 2003. Data extraction and label assignment for web databases. In Proceedings of the 12th international workshop on World wide web, 2003, 187-196

EDITEC: Hypermedia Composite Template Graphical Editor for Interactive TV Authoring *

Jean Ribeiro Damasceno*   Joel dos Santos†   Débora Muchaluat-Saade†

MídiaCom Lab
Telecommunications Engineering Department;*   Computer Science Department†
Universidade Federal Fluminense
R. Passo da Pátria, 156 - Bloco E - Sala 408 - Niterói, RJ - Brazil
(damascon, joel, debora)@midiacom.uff.br

ABSTRACT

This paper presents EDITEC, a graphical editor for hypermedia composite templates that can be used for authoring interactive TV programs. EDITEC templates are based on the XTemplate 3.0 language. EDITEC was designed for offering a user-friendly visual graphical approach. It provides several options for representing iteration structures. The editor provides a multiple-view environment, giving the user a complete control of the composite template during the authoring process. Composite templates can be used in NCL programs for embedding spatio-temporal semantics into NCL contexts. NCL is the standard declarative language used for the production of interactive applications in the Brazilian digital TV system and ITU H.761 IPTV services.

Categories and Subject Descriptors

H.5.2 [**Information Interfaces and Presentation**]: User Interfaces – graphical user interfaces (GUI), interaction styles

General Terms

Design, Languages

Keywords

EDITEC, NCL, XTemplate, XPath, graphical editor, hypermedia composite templates, authoring tool

1. INTRODUCTION

Ginga-NCL is the declarative middleware of the Brazilian digital TV system and ITU international standard for IPTV services [4]. Ginga uses NCL (Nested Context Language) [1] for the development of declarative applications.

As the use of NCL becomes widespread, the number of applications tends to grow, so it is important to provide tools for facilitating application development.

The authoring of more elaborated digital TV declarative applications produces bigger hypermedia documents. The need for defining a great number of relationships (represented as links in NCL) among document parts (nodes), as well as presentation specifications (regions and descriptors in NCL), introduces more difficulty in the authoring process. Bigger hypermedia documents increase the probability of specification errors. In this scenario, reuse of generic program specifications and use of graphical editors make the authoring process easier.

The XTemplate 3.0 language [2] provides the definition of generic document structures through the concept of hypermedia composite templates. Composite templates allow the reuse of document structural specifications in distinct documents that have a common structural characteristic. The template user only has to define the specific media nodes to be used by the final document. Spatial layout and temporal relations, as well as document interactivity specifications, are automatically generated by the template processing. Thus, final document authors do not need to worry about those definitions.

A remaining issue about composite template creation, when using XTemplate XML-based textual approach, is that it is not an easy task for non-expert authors. The template author needs a good knowledge of XTemplate syntax and semantics as well as a good abstraction capacity. In addition, the understanding of the document structure becomes more difficult as the quantity of information grows. Although textual edition brings great flexibility and expressiveness, it consumes much time. For some users, the time spent defining documents in a textual approach makes the authoring process very tiring.

To help non-expert users creating composite templates, this work proposes EDITEC, a hypermedia composite template graphical editor based on XTemplate 3.0. The editor allows the creation of templates in an interactive way, making the user authoring effort using the editor smaller than using XTemplate textual approach, as evidenced by usability tests. The editor offers a friendly graphical environment with multiple views/windows that help the user creating templates efficiently. EDITEC's graphical notation is based on the notation used by both NCL and XTemplate documentation to represent their language elements.

The remaining parts of the paper are organized in the

---

*This work was partially supported by CNPq, FAPERJ and CAPES.

following way. Section 2 discusses related work. Section 3 presents EDITEC, its main views and functionalities. Section 4 presents final remarks and future work.

2. RELATED WORK

LimSee2 [5] is a SMIL [10] document authoring tool. It offers several integrated views (temporal, spatial, textual, etc). Users can define media synchronization by moving media nodes in the temporal view.

LimSee3 [6] is a multimedia authoring tool based on document templates. Although the names may suggest different versions of the same editor, LimSee3 is not a newer version of LimSee2. In LimSee3, the user creates a document by instantiating a template, in a kind of guided application. A template may be seen as a "document with holes", called zones. The instantiation process consists of filling those "holes" with multimedia content.

Composer [3] is an authoring tool that supports multiple integrated views (structural, temporal, textual and layout) for creating NCL documents. The structural view abstracts NCL main entities [1]. Composer does not offer the use of templates for authoring.

SCO Creator Tool [7] is a tool used for creating learning self-adaptive content through digital TV systems. It is used for creating t-learning self-adaptive objects, called SCO (Sharable Content Object.) The main characteristic of those objects is to be able to change their behavior according to the characteristics of the student (viewer), making the learning process through iTV more enjoyable.

Similar to almost all tools mentioned in this section, EDITEC also has different views that facilitate the better understanding of the template result during the authoring process. LimSee3 uses XSLT standard elements [9] in order to model templates. In EDITEC, XSLT elements are used only for building iteration structures. SCO Creator Tool has an interface for creating adaptation rules that has some characteristics similar to EDITEC graphical interface for building basic XPath expressions [8]. The user creates logical expressions, interactively, which are converted into XML code. Finally, Composer is used for building NCL programs. In an NCL program, there are not generic components (used to represent a set of document elements), so it is not necessary to define iteration structures. On the other hand, in a composite template construction, those structures are often used, which makes EDITEC design more complex.

3. EDITEC: COMPOSITE TEMPLATE GRAPHICAL EDITOR

Figure 1 presents EDITEC. It provides several integrated views in order to facilitate document understanding, giving the user a complete control during authoring.

EDITEC was designed to be used by authors with great knowledge of NCL, once they need to identify application parts that can be generalized with templates; and a basic knowledge of XTemplate, once they need to understand XTemplate concepts in order to create a template. Besides NCL and XTemplate, the author should at least understand the construction of XPath expressions. Note that only the understanding of those languages are required, since EDITEC provides abstractions for those XML language elements.

In order to facilitate authoring, EDITEC presents a minia-

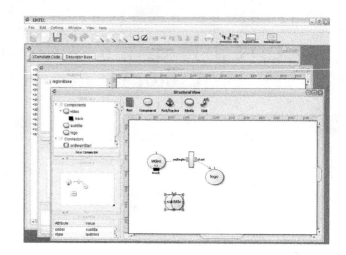

Figure 1: EDITEC graphical environment.

ture overview of the workspace in both structural and layout views. The overview presents the entire workspace, visible or not, highlighting the visible workspace. So elements out of the visible area remain accessible to the author.

EDITEC's main view is the structural view. Graphical elements represent main language entities, indicating media nodes, links, connectors, ports and contexts. Components and ports created in the structural view are automatically presented in the vocabulary area. Connectors used in the template, on the other hand, are created in the vocabulary area through a "new connector" button. In the structural view, spatio-temporal relationships among template components are defined through the creation of links. When creating a link, the author chooses one of the connectors available in the vocabulary area.

In order to create a new element (port, component node, or link), the user selects the icon that represents the desired element and drags it into the workspace in the desired position. When the mouse button is released, the editor presents a popup window where the user can insert complementary information about the created object. The objects may be moved and resized, in order to obtain a spatial disposition that helps the template visualization and creation. The user may graphically create, edit or delete ports, components and links. When an element is selected, its properties are presented in the window lower left corner.

Specific NCL media nodes can be defined selecting the media icon and dragging it over an already created component. Thus, that media node will receive the same component logic, after processing the template.

The layout view presents the screen regions where document media nodes will be presented. In this view, the author may graphically create, edit and delete regions. The layout view also provides options for aligning a group of regions and creating NCL descriptors, with the option to refer to an existing region.

The textual view is divided in two parts. One presents the XML code of the template and another, related to the layout view, presents the XML code of the template descriptor, region and rule bases. Any changes made in EDITEC's structural and layout views are automatically made in the XML code exhibited in the textual view.

Table 1: Iteration structure options.

Option	Definition
i	Associates a connector role or creates a port mapping, at each iteration, to the i-th element of an element set. Where 'i' is incremented at each iteration until it refers to all the positions of the set.
All	Associates a connector role or creates a port mapping to each element of an element set.
All-i	Associates a connector role to each element of an element set, excluding the i-th element of the set.
Prev/Next	Associates a connector role, at each iteration, to the previous/following element of an element set, regarding the current position.

3.1 Iteration Options

XTemplate provides XML iteration structures to indicate the creation of multiple elements by the template processor. In order to graphically represent those structures, EDITEC translates those iteration structure capabilities into iteration options. EDITEC analyses the template graphical iteration options, processes them and generates the correspondent XTemplate code. The iteration options can be specified in link binds and port mappings when those entities are attached to generic components that represent a set of nodes/ports. Table 1 presents the iteration structure options. The following sections present those options in more details.

3.1.1 The "i" Option

Figure 2 presents an example of XTemplate link, created in the EDITEC structural view, that illustrates the use of the "i" option in the creation of a spatio-temporal relationship. That XTemplate link has two components that represent two distinct node sets. The "i" option is used in both binds associated to components "x" and "y".

Figure 2: "i to i" example.

An NCL context that uses a template containing that link, after processed, will have a set of links. For each node in that context with xlabel equal to "x", the processor will apply the same logic defined for the template generic component "x", that is, for each node it creates a link. The first link created has as source the first node with xlabel equal to "x", which is attached to the *onBegin* condition role. It also has, as destination, the first node with xlabel equal to "y", which is attached to the *start* action role. This iteration goes on until the node set ends. Figure 3 graphically represents the resulting NCL links after the template processing.

3.1.2 The "All" Option

In order to illustrate the use of the "All" option, suppose an XTemplate link connecting two node sets, one as the source and another as the destination. That link uses the "i" option for the bind referring to the component representing the source node set and the "All" option for the bind referring to the component representing the destination node set. Figure 4 illustrates this example.

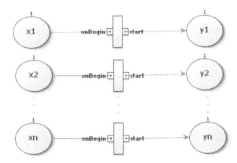

Figure 3: Graphical representation of the final NCL links for "i to i" example.

Figure 4: "i to All" example.

An NCL context that uses a template containing that link, after processed, will have a set of links. For each node in that context with xlabel equal to "x", the processor will apply the same logic defined for the template generic component "x", that is, for each node it creates a link. The first link created has as source the first node with xlabel equal to "x", which is attached to the *onBegin* condition role. It also has, as destination, all the nodes with xlabel equal to "y", presenting several binds attaching them to the *start* action role. This iteration goes on until the node set ends. Figure 5 graphically represents the resulting NCL links after the template processing.

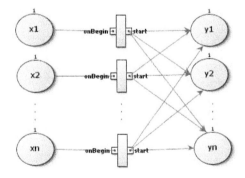

Figure 5: Graphical representation of the final NCL links for "i to All" example.

In the example presented in Figure 4, one possibility is to use the "All" option also in the bind referring to the *onBegin* role, which would generate only one link. That link would have several binds associated to the *onBegin* role, one for each node with xlabel equal to "x" and several binds associated to the *start* role, one for each node with xlabel equal to "y". The "All" option may also be used for creating ports.

In this case, one port is created for each node with xlabel equal to the port mapping target component. The "All-i" option is a particular case of the "All" primitive. For example, this option can be used in an application where 'N' buttons (or images) appear on the screen and, after selecting one of them, the others stop their presentation, except the selected one.

3.1.3   The "Next" Option

A relationship using the "Next" option usually has the same type of component as source and destination, as presented in Figure 6. This type of component represents a node set where each node, except the first and the last ones, will be attached to a condition role in a link and to an action role in another link. However, the first node will be attached only to a condition role and the last node only to an action role.

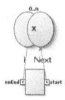

Figure 6: "Next" example.

An NCL context that uses a template containing this link, after processed, will have a set of links with two binds. As shown in Figure 7, the first link created has as source the first node with xlabel equal to "x", which is attached to the *onEnd* condition role, and as destination the second one, which is attached to the *start* action role. The second link creates a relationship among the second and the third nodes and so on, until the node set ends. However, the last node will be attached only to the (*start*) action role and not to a condition role. This example illustrates a generic sequential presentation for all nodes of a specific node set (identified by xlabel "x").

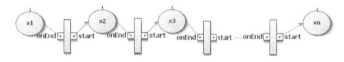

Figure 7: Graphical representation of the final NCL links for "Next" example.

The "Prev" (*previous*) option is similar to the "Next" option. The difference is that there is an inversion among the link source and destination. Usually, the use of the "All-i", "Prev" and "Next" options is when a component type representing a node set is the source and the destination of the link. Those options could also be used when the source and destination components are different, but only in specific cases. Therefore, the use of those options requires author's extra attention when creating a template. During the processing of those iteration graphical options, EDITEC takes into account some special cases, for example, where it is possible to define, in a single link, options "All" and "Next" in different binds.

4.   CONCLUSION

This paper presented EDITEC, a graphical editor for creating composite templates based on the XTemplate 3.0 language. It also presented EDITEC model to represent iteration structures allowing an interactive way of creating templates with a great variety of spatio-temporal semantics.

There may be situations where the graphical authoring tool may not be capable of providing all the functionalities desired by the author. In those cases, the use of a textual approach is recommended.

As a future work, the implementation of the editor temporal view is expected. EDITEC was developed as part of an integrated edition environment. In such an environment it will be possible to create documents from a template created with EDITEC and, consequently, verify the behavior of an application created using that template. The creation of an integrated environment, however, is an ongoing work and its discussion is out of the scope of this article.

5.   REFERENCES

[1] ABNT NBR. Digital Terrestrial Television — Data Coding and Transmission Specification for Digital Broadcasting — Part 2: Ginga-NCL for Fixed and Mobile Receivers — XML Application Language for Application Coding, 2007.

[2] J. A. F. dos Santos and D. C. Muchaluat-Saade. XTemplate 3.0: spatio-temporal semantics and structure reuse for hypermedia compositions. *Multimedia Tools and Applications*, 2011. available at http://www.springerlink.com/content/m3932258853567j0/.

[3] R. L. Guimarães, R. M. R. Costa, and L. F. G. Soares. Composer: Authoring Tool for iTV Programs. In *Proceedings of the 6th EITV*, July 2008.

[4] ITU. Nested Context Language (NCL) and Ginga-NCL for IPTV services. www.itu.int/rec/T-REC-H.761-200904-P, 2009. ITU-T Recommendation H.761.

[5] LimSee2. limsee2.gforge.inria.fr/.

[6] LimSee3. limsee3.gforge.inria.fr/.

[7] López, M. R. et al. T-MAESTRO and its authoring tool: using adaptation to integrate entertainment into personalized t-learning. *Multimedia Tools and Applications*, 40(3):409–451, 2008.

[8] W3C. XML Path Language (XPath) Version 1.0, 1999.

[9] W3C. XSL Transformations (XSLT) Version 1.0, 1999.

[10] W3C. Synchronized Multimedia Integration Language - SMIL 3.0 Specification, 2008.

An Exploratory Analysis of Mind Maps

Joeran Beel

Docear & OvGU/FIN/DKE
Magdeburg, Germany

beel@docear.org

Stefan Langer

Docear & OvGU/FIN/DKE
Magdeburg, Germany

langer@docear.org

ABSTRACT

The results presented in this paper come from an exploratory study of 19,379 mind maps created by 11,179 users from the mind mapping applications 'Docear' and 'MindMeister'. The objective was to find out how mind maps are structured and which information they contain. Results include: A typical mind map is rather small, with 31 nodes on average (median), whereas each node usually contains between one to three words. In 66.12% of cases there are few notes, if any, and the number of hyperlinks tends to be rather low, too, but depends upon the mind mapping application. Most mind maps are edited only on one (60.76%) or two days (18.41%). A typical user creates around 2.7 mind maps (mean) a year. However, there are exceptions which create a long tail. One user created 243 mind maps, the largest mind map contained 52,182 nodes, one node contained 7,497 words and one mind map was edited on 142 days.

Categories and Subject Descriptors

I.7.5 [**Document and Text Processing**]: Document Capture – *document analysis*

General Terms

Measurement, Design.

Keywords

mind maps, concept maps, content analysis, document analysis, mind mapping software, information retrieval

1. INTRODUCTION

Millions of people are using mind maps for brainstorming, note taking, document drafting, project planning and other tasks that require hierarchical structuring of information. Figure 1 shows a mind map which was created as draft for this paper. As all mind maps, it has a central node (the root) which represents the main topic the mind map is about. From this root node, child-nodes branch out, in order to describe sub-topics. Each node may contain an arbitrary number of words. This way, a mind map is comparable to an outline but with stronger focus on the graphical representation. Mind maps created on a computer may also contain links to files, hyperlinks to websites (in Figure 1 indicated by red arrows), pictures, and notes (indicated by yellow note icons).

In this paper we present the initial results of an exploratory study of 19,379 mind maps. The overall research objective was to find out how mind maps are structured and what information they

contain. To our knowledge this is the first study of its kind. We therefore aimed at a broad overview to determine further areas of interesting research.

2. RELATED WORK

There is lots of research on content and structure of other documents: Web pages, emails, academic articles, etc. have all been analyzed thoroughly in the past (e.g. [1-3]). With respect to mind maps there is mostly research about the effectiveness as learning tool (e.g. [4]).

The lack of analyses of mind maps is not surprising. Emails, web pages, etc. had to be thoroughly researched to make information retrieval tasks, for instance, indexing and spam detection, effectively possible. Such information retrieval tasks have never been applied to mind maps, and therefore the need for knowledge about mind map content and structure was low.

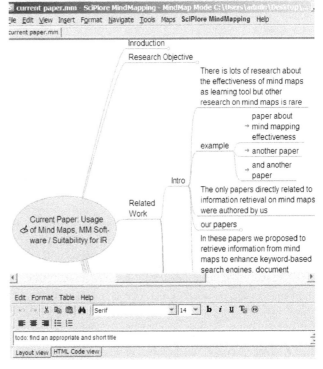

Figure 1: Screenshot of a mind mapping software

However, recently we proposed to apply information retrieval tasks to mind maps to enhance keyword-based search engines, document recommender systems, and user profile generation [5]. To do this effectively, knowledge about the content and structure of mind maps is required.

There was only one paper we found that is somewhat related: a survey from the *Mind Mapping Software Blog* [6]. For this survey 334 participants answered questions about their use of mind mapping software. However, the survey was based on 334 self-selected participants from a single source (readers of the *Mind Mapping Software Blog*). Accordingly, it seems likely that predominantly very active mind mapping users participated in the survey and results are not representative. In addition, the survey focused on the usage of mind mapping software rather than the content and structure of mind maps.

3. METHDOLOGY

We conducted an exploratory study on 19,379 mind maps created by 11,179 users from the two mind mapping applications *Docear*[1] and *MindMeister*[2] (the latter one is abbreviated as 'MM' in figures and tables).

Docear is a mind mapping application for Windows, Linux and Mac, focusing on academic literature management, and developed by ourselves [7]. 2,779 users agreed to have their mind maps analyzed. They created 7,506 mind maps between April 1, 2010 and March 31, 2011.

MindMeister is a web based mind mapping application. 8,400 users published 11,873 mind maps in MindMeister's public mind map gallery[3] between February 2007 and October 2010. For our study these public mind maps were downloaded in XML format via MindMeister's API[4], parsed, and analyzed.

Numbers include only mind maps containing six or more nodes[5], and that were not being edited between April 1, 2011 and the day of the analysis (June 2, 2011). This way it is ensured that mind maps in the beginning of their life-cycle do not spoil the results but only "mature" mind maps were analyzed.

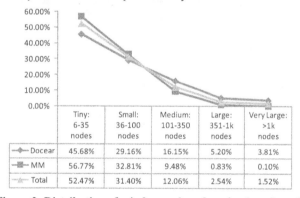

	Tiny: 6-35 nodes	Small: 36-100 nodes	Medium: 101-350 nodes	Large: 351-1k nodes	Very Large: >1k nodes
Docear	45.68%	29.16%	16.15%	5.20%	3.81%
MM	56.77%	32.81%	9.48%	0.83%	0.10%
Total	52.47%	31.40%	12.06%	2.54%	1.52%

Figure 2: Distribution of mind maps based on size (number of nodes)

We were particularly interested in finding out whether differences existed for different types of mind maps and between the two mind mapping applications. Therefore, mind maps were grouped based on their size, measured by the number of nodes. Mind maps with 6 to 35 nodes were considered as 'tiny', with 36 to 100 nodes as 'small', with 101 to 350 nodes as 'medium', with 351 to 1000 nodes as 'large' and with more than 1000 nodes as 'very large'. In the data set, the majority of mind maps were tiny (52.47%) or small (31.40%) as shown in Figure 2.

---

[1] http://docear.org

[2] http://mindmeister.com

[3] http://mindmeister.com/maps/public

[4] http://mindmeister.com/services/api

[5] A random sample of 50 mind maps showed that the vast majority of mind maps with five or fewer nodes were created for testing purposes and did not contain valuable content.

4. RESULTS & INTERPRETATION

4.1 Mind Maps per User

Figure 3 shows the number of mind maps users created. The majority of MindMeister users created, or we should say published, exactly one mind map (81.26%). Only 2.32% of MindMeister users published five or more mind maps. In contrast, 56.75% of Docear users created one mind map and 11.36% created five or more mind maps. On average (mean), users created 2.7 mind maps (Docear) during the 12 month period of data collection, respectively 1.4 (MindMeister) during ~3.5 years. The highest number of mind maps created by one user was 243 for Docear and 73 for MindMeister.

It has to be noted that numbers of MindMeister and Docear are only limitedly comparable, as we did only analyze MindMeister mind maps that were published by their users. It can be assumed that most users who published mind maps on the Web, created further private mind maps that were not publicly available.

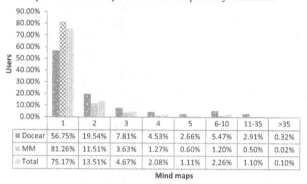

	1	2	3	4	5	6-10	11-35	>35
Docear	56.75%	19.54%	7.81%	4.53%	2.66%	5.47%	2.91%	0.32%
MM	81.26%	11.51%	3.63%	1.27%	0.60%	1.20%	0.50%	0.02%
Total	75.17%	13.51%	4.67%	2.08%	1.11%	2.26%	1.10%	0.10%

Figure 3: Number of created mind maps per user

4.2 Nodes per mind map

As mentioned in the methodology and shown in Figure 2, most mind maps were rather small. On average, Docear mind maps contained 232 nodes (mean), respectively 41 nodes (median). MindMeister mind maps contained 51 nodes (mean), respectively 31 (median). Docear mind maps tended to be larger than MindMeister mind maps. For instance, while only 0.10% of MindMeister mind maps were 'very large', 3.81% of Docear mind maps were. The largest Docear mind map contained 52,182 nodes (and there are several more mind maps containing 10,000+ nodes); the largest MindMeister mind map contained 2,318 nodes.

4.3 File Links

In a mind map, users may link to files on their hard drive. Figure 4 shows the distribution of mind maps containing a certain number of links (for Docear mind maps only since MindMeister does not provide this feature). Well over half of mind maps do not contain any links to files (63.88%).

Table 1: File types linked in mind maps

PDFs	Images	Documents	HTML	Excel/CSV	PowerPoint	MP3s	Other
89.58%	1.26%	0.53%	0.47%	0.42%	0.34%	0.27%	7.14%

However, some users make heavy use of the feature. 2.94% of mind maps contained more than 1,000 links to files and 2.97% of mind maps contained between 351 and 1,000 links. The highest number of links in a mind map was 52,138 and all 7,506 Docear mind maps together contained 1,184,547 links to files on the users' hard drives. This does not mean that 1,184,547 different files were linked. Most users linked the same file multiple times in a mind map.

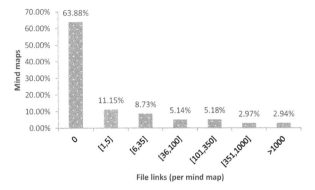

Figure 4: Number of file-links in Docear mind maps

From all links, 89.58% pointed to PDF files (see Table 1). Other files being linked included images (.gif, .png, .jpeg, .tiff), MP3s and text documents (.doc, .docx, .odt, .rtf, .txt), but with much smaller frequency.

4.4 Hyperlinks

Looking at all mind maps, 81.57% do not contain a single hyperlink to a website (see Figure 5). However, there are differences between Docear and MindMeister. While 92.37% of Docear mind maps do not contain hyperlinks at all, only 75.27% of MindMeister mind maps do not contain any hyperlinks. In other words: 7.63% of Docear mind maps and 24.73% of MindMeister mind maps contain at least one hyperlink.

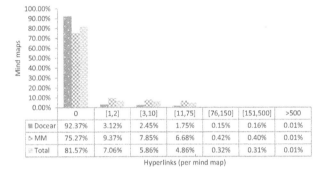

	0	[1,2]	[3,10]	[11,75]	[76,150]	[151,500]	>500
Docear	92.37%	3.12%	2.45%	1.75%	0.15%	0.16%	0.01%
MM	75.27%	9.37%	7.85%	6.68%	0.42%	0.40%	0.01%
Total	81.57%	7.06%	5.86%	4.86%	0.32%	0.31%	0.01%

Hyperlinks (per mind map)

Figure 5: Number of hyperlinks in mind maps

Larger mind maps more often contain hyperlinks when compared to smaller mind maps. For instance, around 20% of Docear's (very) large mind maps but only 3.94% of tiny mind maps contain hyperlinks. Similarly, around 40% of MindMeister's (very) large mind maps but only 22% of tiny mind maps contain hyperlinks[6].

4.5 Notes

Most mind mapping software tools (such as Docear and MindMeister) allow users to add notes to a node.

Table 2: Number of notes in mind maps

		0	[1,2]	[3,10]	[11,75]	[76,150]	[151,500]	>500
		\multicolumn{7}{c}{Amount of Notes}						
Mind Maps Size	Tiny	68.66%	19.53%	8.00%	3.81%	0.00%	0.00%	0.00%
	Small	65.72%	15.02%	9.58%	9.53%	0.15%	0.00%	0.00%
	Medium	59.58%	13.73%	9.92%	14.37%	1.97%	0.43%	0.00%
	Large	52.15%	11.86%	11.66%	17.59%	4.29%	2.25%	0.20%
	Very large	61.74%	6.04%	7.38%	16.44%	3.69%	3.36%	1.34%
	Total	66.12%	17.01%	8.81%	7.42%	0.45%	0.16%	0.03%

Many users do not use this feature – 66.12% of mind maps do not contain any notes (see Table 2). Results are similar for both, MindMeister and Docear mind maps[6].

4.6 Words per node

Figure 6 shows the distribution of words per node (everything separated by whitespace characters was assumed to be a word). Nodes in mind maps generally contain few words. Nearly 1/3 of all 2,352,584 nodes contained a single word (29.91%). Only 8.25% of nodes contained more than ten words.

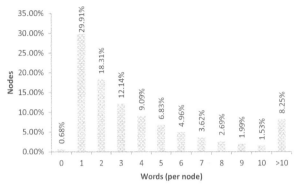

Figure 6: Number of words per node

However, there is a long tail in the distribution – the maximum word count for a node was 7,497 for Docear and 1,184 for MindMeister. Although the most frequent word count per node is one, mean is 4.80 words per node and median is 3. There is a slight tendency that the larger mind maps are, the more words their nodes contain. Details are provided in Table 3.

Table 3: Number of words per node by mind map size

	\multicolumn{4}{c}{Word count per node}			
	Mean	Median	Modal	Max
Tiny maps	4.67	2	1	1,874
Small maps	4.45	2	1	687
Medium maps	5.07	2	1	1,463
Large maps	5.76	3	1	2,723
Very large maps	4.60	3	1	7,497

Also, the deeper a node is in a mind map (further out on the branch), the more words it tends to contain. While root nodes (level 0) contain 3.03 words on average (mean), respectively 2 (median), nodes in level 5 contain 5.11 words on average (mean), or 3 (median) respectively (see also Figure 7).

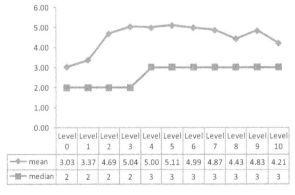

	Level 0	Level 1	Level 2	Level 3	Level 4	Level 5	Level 6	Level 7	Level 8	Level 9	Level 10
mean	3.03	3.37	4.69	5.04	5.00	5.11	4.99	4.87	4.43	4.83	4.21
median	2	2	2	2	3	3	3	3	3	3	3

Figure 7: Number of words per node based on node level

---

[6] Detailed results are not provided due to space restrictions.

Results are similar for both, Docear and MindMeister mind maps[6]. Except, the median word count for Docear is three, and for MindMeister two.

4.7 Days Edited

The majority of mind maps seem to be used for rather short term activities such as brainstorming or maybe taking meeting-minutes. Figure 8 shows on how many days mind maps were edited[7]. 60.76% of mind maps were edited only during a single day[8]. However, also a large proportion of mind maps were edited on several days, and a small fraction (0.55%) even on more than 25 days. On average, mind maps were edited on one day (median), respectively 2.36 days (mean). The maximum was 142 days.

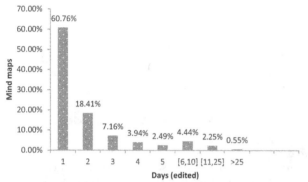

Figure 8: Number of days mind maps were edited

5. INTERPRETATION & SUMMARY

For some features, there appear to be significant differences between mind maps created with Docear and those created with MindMeister. However, most of the differences can be attributed to the special functionality of the corresponding software. For instance, Docear offers special features for literature management such as automatically importing PDF bookmarks as new nodes to a mind map. Accordingly, it was expected that Docear mind maps would be larger, in terms of number of nodes. Concerning this case, probably MindMeister numbers are more representative for other mind maps than Docear's are.

On the other hand, when estimating the number of mind maps per user, Docear's numbers are probably more suitable for generalizations, as we could only analyze public mind maps of MindMeister users.

The study showed that a 'typical' (average) mind map is rather small, with a few dozen nodes (31 was the median for MindMeister mind maps), whereas each node contains probably between one to three words (more for large mind maps or nodes deeper in a mind map). The mind map probably contains few if any notes (66.12%). The number of hyperlinks depends on the mind mapping application and tends to be rather low, too. Probably the mind map was edited only on one (60.76%) or two days (18.41%) and it is expected that a typical user creates around 2.7 mind maps a year (mean, Docear).

However, these are only averages. Most results followed a power-law distribution with a long tail. There was one user who created 243 mind maps (and several users more created 10+ mind maps).

The largest mind map in the data set contained 52,182 nodes (and several more with 10,000+ nodes existed), there was one node containing 7,497 words (and several more nodes with 100+ words existed), one mind map was edited on 142 days (and several more were edited a few dozen times) and several mind maps contained a few hundred notes.

6. OUTLOOK

For future research, analysis of the evolvement of mind maps could be interesting. Maybe there are different patterns how mind maps evolve and are used by users. Also, differences between user types should be analyzed. In addition, the content of mind maps has only been analyzed superficially, yet. It would be interesting to know what exactly the content is and what mind maps are used for exactly (brainstorming, literature management, etc.). A more detailed analysis should also look at the extremes and outliers (e.g. the node with 7,497 words).

Most importantly, mind maps need to be compared to other types of documents and consequences for information retrieval needs to be drawn. What does it mean when nodes usually contain one to three words? Are they comparable to search queries which usually consist of a similar number of terms? If so, can approaches for search query recommender easily be adopted to create a 'node recommender'? Are mind maps with a few dozen nodes comparable to a user's collection of social tags which usually also consist of a few dozen tags each with one or two words? If so, can approaches for user modeling based on social tags easily be applied to model the interests of mind map users? And are mind maps, which contain a few thousands nodes or words, comparable to web pages, academic articles, or emails? If so, what does this mean for the ability to apply information retrieval on mind maps? All these questions need to be answered in further research.

7. ACKNOWLEDGEMENT

This work was supported by grant No. 03EGSST033 of the 'German Federal Ministry of Economics and Technology', 'German Federal Ministry of Education and Research', and 'European Social Fund' (ESF).

REFERENCES

[1] Alexandros Ntoulas, Marc Najork, Mark Manasse, and Dennis Fetterly. Detecting spam web pages through content analysis. In *15th International Conference on World Wide Web*, pages 83–92, 2006.

[2] Dean F Sittig. Results of a content analysis of electronic messages (email) sent between patients and their physicians. *BMC Medical Informatics and Decision Making*, 3 (11), October 2003..

[3] Ken Hyland and Polly Tse. "i would like to thank my supervisor". acknowledgements in graduate dissertations. *International Journal of Applied Linguistics*, 14 (2): 259–275, July 2004.

[4] Glennis Edge Cunningham. *Mindmapping: Its Effects on Student Achievement in High School Biology*. PhD thesis, The University of Texas at Austin, 2005.

[5] Joeran Beel, Bela Gipp, and Jan-Olaf Stiller. Information Retrieval on Mind Maps - What could it be good for? In *Proceedings of the 5th International Conference on Collaborative Computing: Networking, Applications and Worksharing (CollaborateCom'09)*, 2009. IEEE.

[6] Chuck Frey. Mind mapping software user survey, March 2010. URL http://mindmappingsoftwareblog.com/2010-survey-form/.

[7] Joeran Beel, Bela Gipp, Stefan Langer and Marcel Genzmehr. Docear: An academic literature suite for searching, organizing and creating academic literature. *In Proceedings of the 11th ACM/IEEE Joint Conference on Digital Libraries (JCDL'11)*, 2011.

---

[7] Data was available for Docear mind maps only.

[8] Creation of a mind map was counted as one edit. All edits made during one day were combined.

Models for Video Enrichment

Benoît Encelle
Université de Lyon,
CNRS Université Lyon 1, LIRIS,
UMR5205, F-69622, France
bencelle@liris.cnrs.fr

Pierre-Antoine Champin
Université de Lyon,
CNRS Université Lyon 1, LIRIS,
UMR5205, F-69622, France
pchampin@liris.cnrs.fr

Yannick Prié
Université de Lyon,
CNRS Université Lyon 1, LIRIS,
UMR5205, F-69622, France
yprie@liris.cnrs.fr

Olivier Aubert
Université de Lyon,
CNRS Université Lyon 1, LIRIS,
UMR5205, F-69622, France
oaubert@liris.cnrs.fr

ABSTRACT

Videos are commonly being augmented with additional content such as captions, images, audio, hyperlinks, etc., which are rendered while the video is being played. We call the result of this rendering "enriched videos". This article details an annotation-based approach for producing enriched videos: enrichment is mainly composed of textual annotations associated to temporal parts of the video that are rendered while playing it. The key notion of enriched video and associated concepts is first introduced and we second expose the models we have developed for annotating videos and for presenting annotations during the playing of the videos. Finally, an overview of a general workflow for producing/viewing enriched videos is presented. This workflow particularly illustrates the usage of the proposed models in order to improve the accessibility of videos for sensory disabled people.

Categories and Subject Descriptors

I.7.2 [**Document Preparation**]: Multi/mixed media.

General Terms

Design, Languages.

Keywords

Models for videos enrichment, video enrichment, hypervideo

1. INTRODUCTION

Videos are commonly being augmented with additional content such as captions, images, audio, hyperlinks, etc., which are rendered while the video is being played. We call the result of this rendering "enriched videos". The goal of video enrichment can be either to make parts of the video content available to people that cannot fully perceive its visual or audio content, or for complementing it with additional information so as to enhance the watching experience.

This article presents several contributions regarding the production of enriched videos. Two models are first detailed: the first one is for representing the content of enrichments as temporally situated structured annotations, and the other is for describing the presentation modalities of these annotation contents. These models are illustrated with an example corresponding to the ACAV (Collaborative Annotation for Video Accessibility) project general workflow. ACAV explores how enriching videos can improve their accessibility for sensory disabled people.

We first introduce the key notion of video enrichment and associated concepts (section 2). We then present in section 3 the models we have developed for video enrichment: the first one is for annotating the video and the second is for rendering annotations during the playing of the video. Next, in order to illustrate a possible usage of these models, we present an overview of the general ACAV workflow for producing/viewing enriched videos. The related works (section 4) focuses on existing approaches for producing enriched videos before concluding and presenting future work.

2. VIDEOS ENRICHMENT USAGES

Utility of video enrichment is twofold. Video can be enriched either to *translate* parts of its content or to *complement* it with additional information in order to enhance the watching experience. Concerning *translation-based video enrichment*, the objective is that people who cannot fully understand the video visually or aurally can apprehend it. For instance, subtitling and superimposed dubbing have been two common means of enriching a video to translate its dialogs in a foreign language. For sensory impaired people, the objective is to present the key audio information or key visual information of the video using respectively either some visual presentation modalities or some audio, tactile (Braille) presentation modalities. For instance, the *audio description* of a video concerns visually impaired people and consists in adding verbal information to the audio track of the video in order to describe the visual content of the video [6]. For deaf and hearing-impaired users, *teletext* is for instance a digital service that allows a television channel to broadcast closed-captions that describe the audio track of a program (dialogs, sounds, music), as are subtitles for hearing-impaired on DVDs.

Complement-based video enrichment is different from *translation-based video enrichment*: the objective is not anymore to ensure that viewers will apprehend the video as intended by its creators, but to offer new experiences. For instance, chat messages rendered as subtitles can comment on a TV program, video meaning can be changed (*e.g.* from tragedy to comedy) by added sounds, added visual elements (e.g. arrows) can underline important elements in a scenery, *etc*. As another example of complement-based video enrichment, Díaz Cintas [4] emphasizes some new subtitling practices of professionals or hobbyist annotators. He stresses out subtitling activities that add precision to little-known terms (*e.g.* specialized vocabulary) by using explanations in brackets or texts placed at the top of the screen, which he calls *headnotes* or *topnotes*. One step further, videos can become part of hypermedia, as video enrichment paves the way for new interaction possibilities [7]. Indeed, a hyperlinked video, or hypervideo, is a hypermedia document into which video streams are enriched with embedded, clickable anchors. Clicking these anchors results in navigating to other places in the same video, or to other videos, or to others information elements. Such combination of video with non-linear information structure can be used in various domains: storytelling (*HyperCafe* [9]), e-learning (adding slides, references, links, *etc.* to video content).

Considering the technological point of view on video enrichment, our approach considers *annotation-based video enrichment*. A video annotation is here defined as *any information associated to a fragment of a video* (e.g a textual transcription of a dialog associated to a temporal fragment, defined by two timecodes) [2]. Annotation data can be *rendered* so as to enrich a video – i.e. presenting its content using an adequate modality (*e.g.* visual enrichment with textual captions, images, video fragments, *etc.* or auditory enrichment with voice, music or sounds). As a result, the general process for enriching videos is made up of two main steps: an annotation step and a rendering step (*cf.* Figure 1).

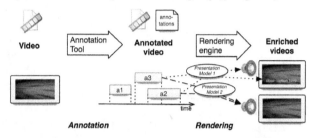

Fig. 1. The general process of annotation-based video enrichment.

In our opinion, this separation –similar to the structure/presentation separation in document engineering– has good properties: annotations and their renderings are independently defined. This can foster innovation by allowing different people to create content and content rendering, for example in a collaborative process. It also allows performing "live" video enrichment according to end-user preferences that can change during the rendering itself, paving the way to real-time adapted enrichments. The following section deals with models we have developed first for annotating videos and second for specifying presentation intents of the content of annotations.

3. MODELS FOR VIDEO ENRICHMENT

3.1 Annotation Model

We have previously proposed [1] a general model for video annotation. This model has been implemented in the Advene application[1], and experimented within different contexts, including multimodal presentations of annotated videos. We borrow from that general model the main elements of our **annotation model** (cf. Figure 3, annotation package):

- *Annotations* are the main elements of our model. Basically, an annotation has a unique id, a content and is associated to a temporal fragment (two timecodes addressing the original video).

- *Annotation Types* are a way to structure annotations as every annotation has exactly one type (e.g. annotations of type *Character*, of type *Setting*, etc.). They define the semantics of annotations and constrain their content.

- *Annotation Tags* are a more flexible way to categorize annotations. Every annotation can be associated to one or more tags.

- An *Annotation Schema* embodies a particular annotation practice as a set of annotation types. For example, one could define a schema for describing the dialogues of a video, another schema for the musical part, etc.

[1] http://www.advene.org

As an example related to improving the accessibility of a movie for blind people, a schema called "VisualBase" that could contain textual annotations of type "*Character*", "*Action*", "*Setting*" can be created to describe key visual elements of the movie (*Character* annotations for describing characters appearance/role and their interrelations, *Settings* annotations for describing the different settings of the movie, etc.)[2].

3.2 Annotation Presentation Model

This section exposes the main elements of our annotation presentation model (*cf.* Figure 3, presentation package) as a specialization of the notion of views defined in [1].

Presentation rules. A *Presentation Rule* R specifies the presentation of a subset of annotations according to one or several presentation modalities (e.g. a text-to-speech engine, a subtitle displayer, etc.). A presentation rule is composed of an annotation selector S associated to a set of presentation actions A_i.

$$R = <S, \{A_1,..., A_q\} > \text{ with } q > 0$$

An *Annotation Selector* S selects a subset of annotations from an annotation set according to a set of constraints. Constraint types can be:

- Structural: upon structural elements of our annotation model (schemas, types, tags);
- Intrinsic: upon annotation id or content.

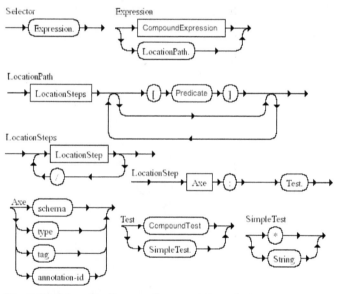

Fig. 2. Partial syntax diagram of an annotation selector.

A syntax similar to XPath [12] was used to specify selectors (cf. Figure 2). In fact we reuse the XPath concepts of "*Location Path*" and "*Location Step*" for selecting a subset of annotations using "constraints". A *location step* is first expressed according to an *axe* that indicates the nature of the constraint and then according to a *test* that filters the annotations subset according to the axe. For instance, the location step "*schema:VisualBase*" filters a set of annotations selecting only annotations associated to the *schema* "*VisualBase*". Others possible axes are: *type, tag, annotation-id*.

[2] This way of describing a movie actually corresponds to an existing description practice that is called "audio-description".

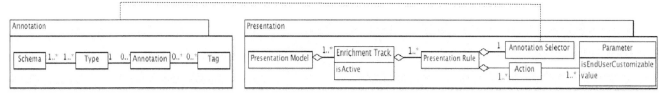

Fig. 3. Main elements of the models for video enrichment

Annotation selection also makes use of *predicates* that add intrinsic constraints by filtering the previously selected annotation subset (e.g. with the location steps) mainly by using conditions on annotation content. A predicate can be modeled as a regular expression. For instance the predicate "[cC]at | [dD]og" selects the annotations in the subset that have their content equal to "cat", "Cat", "dog" or "Dog".

Actions. An *action* A corresponds to an *action type* T and specifies the presentation of annotation content. Suggested action types are: speech synthesizing, Braille displaying, subtitling, close-captioning, audio icon playing, etc.

The author of a presentation model can parameterize actions. Some actions might also be customizable by the end-users watching the enriched videos. Most parameters and their possible values are taken from CSS [10].

$$A = < T, \{P_1,\dots, P_r\} > \text{ with } r \geq 0$$

$$P_i = < parameterName, value, isUserModifiable >$$

For each parameter, the author of the model can give to the end-user the permission for changing its value (boolean *isUserModifiable*). For instance, some parameters associated to an action of type "speech synthesizing" are: *voice-family* (with possible values: male, female, child), *defaultPlaybackRate* (with possible values x0.5, x1, x1.5, x2), *volume* (percentage), etc.

We have proposed in [5] several families of parameters depending on properties associated to each action type. For instance, temporal actions (i.e. actions that present messages that generally evolve during time) parameters include *defaultPlayBackRate*, *minPlaybackRate*, *maxPlaybackRate*. Parameters related to audio actions include *volume*, *panning*.

In our example, the list of parameters of the action "speech synthesizing" would be completed with those parameters of both temporal actions and audio actions.

Enrichment tracks. An *Enrichment Track* T is made up of one or several presentation rules R_i and can be activated/deactivated during the playing of the video.

$$T = \{R_1,\dots, R_n\} \text{ with } n > 0$$

Presentation model. Finally, a *Presentation Model* M, aims to specify the presentation of a considered annotation set. It is made up of m enrichment tracks T_i.

$$M = \{T_1,\dots, T_m\} \text{ with } m > 0$$

Some integrity constraints apply to our model (we only describe them informally because of space limitations):

- an action A cannot contain two parameters with the same *parameterName*,

- a rule R cannot have two actions with the same action type,

- in a track T, two actions with the same action type (hence in two different rules) cannot have different values for the same *parameterName*.

Going on with our example of annotations of types "*Character*", "*Action*", "*Setting*" corresponding to the "*VisualBase*" schema, we propose an example presentation model for producing audio enrichments in order to have an audio description of a movie. This presentation model is composed of two enrichment tracks, each one containing one presentation rule.

The presentation rule of the first enrichment track selects annotations of types *Character* and *Action* and presents the content of these annotations using a speech synthesizer with a male voice:

R_1 = < "*type:Character or type:Action*", {<*speechSynthesizing*, {*voice-family, male, false*}>}>

The presentation rule of the second enrichment track selects *Setting* annotations and presents each annotation by first playing a short sound (that indicates a set change) and second speech synthesizing the annotation content (i.e. the set description) with a female voice:

R_2 = < "*type:Setting*", {<*soundPlaying*, {*file, bip.mp3, false*}>}, <*speechSynthesizing*, {*voice-family, female, false*}>}>

Note that, according to the third integrity constraint, those two rules have to belong to different tracks, as they use two different values for the *voice-family* parameter of *speechSynthesis*.

3.3 An Example Workflow For Producing Enriched Videos Integrating Proposed Models

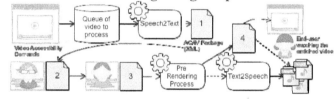

Fig. 4. ACAV general architecture for producing enriched videos

The ACAV project general workflow for producing enriched videos (cf. Figure 4) illustrates a usage of the suggested video enrichment models. The format of documents 1 to 3, and associated workflow steps (e.g. annotation and presentation model authoring steps) are a specialization of the document format and workflow steps used in the Advene project [1]. According to the ACAV workflow, the first output document mainly contains annotations that come from a *Speech2Text* process used for transcribing the dialogs of the video. The second document is made up of the content of the first one and of supplementary annotations added by "annotator" users (e.g. for describing key visual elements of the video). These two documents represent, in an XML syntax, elements of the annotation model (i.e. schemas, types, tags, annotations) created by annotator users and by the Speech2Text process. The third document expands the second one with the XML description of the presentation model. According to these specifications and some annotation contents, enriching contents can be created, such as *mp3* files containing generated audio description. Finally, the sensory disabled end-user watches the resulting enriched video and may customize it in adjusting some track action parameters (e.g. volume, defaultPlayBackRate for speechSynthesizing): document 4 that contains annotations and information about their rendering (e.g. links to *mp3* files) is interpreted during the playing of the video, producing an accessible "audio described" video.

An important feature of that workflow, illustrated by Figure 4, is that different users can contribute to different elements of the packages (e.g. schemas, annotations, presentation models), as those can be defined independently. Furthermore, generic schemas or presentation models can be pre-defined, shared and reused with multiple videos.

4. RELATED WORKS

Concerning approaches for video enrichment, some work has been done for annotating multimedia content (i.e. including video content) [2, 3]. In comparison with our approach, the main difference in our opinion is that these approaches do not separate the annotation structure (content) from its presentation. To our mind, they do not foster, as much as our approach does, reusability and sharing. Indeed, with our approach, the same annotation package can be reused for making several different kinds of enrichments in using different presentation models (e.g. audio enrichments, visual enrichments, audio + tactile enrichments, etc.). In the same way, a successful presentation model can be applied on several videos through different annotation packages. Moreover, our workflow for producing enriched videos is collaborative by essence: it allows the involvement of several people for annotating videos and/or for designing presentation models, paving the way to the emergence of video enrichment practices.

Concerning technical solutions for publishing enriched videos, SMIL (Synchronized Multimedia Integration Language) can be used for synchronizing different multimedia contents (e.g. a video synchronized with an audio file containing an audio-description and with a subtitle file). SMIL can be also used to annotate some SMIL content [2]. Several technical recommendations, initiatives and formats have emerged from the community working on accessibility. The Web Accessibility Initiative (WAI) advocates in its recommendation entitled "Web Content Accessibility Guidelines (WCAG)" [11] the development of different versions of a given temporal content (audio and visual versions for sensory disabled people). Concerning formats, the Mozilla Foundation [8] advocates the usage of the Ogg format with multiplexed specialized tracks for video accessibility. In the same way, the HTML accessibility task force suggests adding several tracks to a video content to improve its accessibility: e.g. a subtitle track, an audio-description track, etc. These "enrichment" tracks would be represented as HTML 5 *Track* elements inside a *Media* element (i.e. *Video* or *Audio* element). The notion of enrichment track in our model is very closed from this HTML Track element. In comparison with the HTML 5 Track notion, our concept of enrichment track permits a deeper end-user customization of enrichment, by adjusting some parameters of actions (cf. 3.2).

5. CONCLUSION AND FUTURE WORK

We have proposed an annotation-based approach to produce enriched videos. Two models are presented: the annotation model permits the association of typed annotations to fragments of a video; the annotation presentation model allows to describe how the video is to be enriched with the rendering of annotation contents using various modalities (*e.g.* textual captions, images, video fragments, spoken texts, music or sounds). Concerning video enrichment for improving video accessibility, several experiments involving people with disabilities are currently being conducted in order to evaluate

the consistency of these models. The annotation presentation model is inspired from XPath for defining annotation selectors, and from CSS for specifying presentation intents for selected annotations. The selector specification model should probably be extended in order to support the definition of temporal constraints upon annotations, e.g. for selecting annotations that start before the beginning of an annotation X and that end before the beginning of an annotation Y. We will also study mechanisms for checking the "consistency" of annotation presentation, as a presentation model can indicate that several time-overlapping annotations have to be presented using the same presentation action, resulting in hardly perceptible information for the end-user (e.g. two or more overlapping subtitles, etc.).

6. ACKNOWLEDGEMENTS

This paper was partly supported by the French Ministry of Industry (*Innovative Web call*) under contract 09.2.93.0966, "Collaborative Annotation for Video Accessibility" (ACAV).

7. REFERENCES

[1] O. Aubert and Y. Prié. 2007. Advene: an open-source framework for integrating and visualising audiovisual metadata. In *15th international conference on Multimedia* (MULTIMEDIA '07). ACM, 1005-1008.

[2] D.C.A. Bulterman. 2003. Using SMIL to encode interactive, peer-level multimedia annotations. In *2003 ACM symposium on Document engineering* (DocEng '03). ACM, 32-41.

[3] R.G. Cattelan, C. Teixeira, R. Goularte, and Maria Da Graça C. Pimentel. 2008. Watch-and-comment as a paradigm toward ubiquitous interactive video editing. ACM Trans. Multimedia Comput. Commun. Appl. 4, 4.

[4] J. Díaz Cintas. 2005. Back to the Future in Subtitling. In Marie Curie Euroconferences MuTra: Challenges of Multidimensional Translation, University of Saarland, 2005

[5] B. Encelle. Modèle pour la spécification de modèles de présentation d'annotations associées à des videos. Online : http://liris.cnrs.fr/Documents/Liris-5147.pdf. Accessed 06/14/11.

[6] L. Gagnon, S. Foucher, M. Heritier, M. Lalonde, D. Byrns, C. Chapdelaine, J. Turner, S. Mathieu, D. Laurendeau, N. T. Nguyen, and D. Ouellet. 2009. Towards computer-vision software tools to increase production and accessibility of video description for people with vision loss. *Univers. Access Inf. Soc.* 8, 3 (July 2009), 199-218.

[7] J. Geißler. Surfing the movie space: advanced navigation in movie-only hypermedia. In *3th international conference on Multimedia* (MULTIMEDIA '95). ACM, 391-400.

[8] S. Pfeiffer and C. Parker. Accessibility for the HTML5 <video> element. In *6th International Cross-Disciplinary Conference on Web Accessibilty* (W4A '09). ACM, 98-100.

[9] N. Sawhney, D. Balcom, and I. Smith. 1996. HyperCafe: narrative and aesthetic properties of hypervideo. In *7th ACM conference on Hypertext* (HYPERTEXT '96). ACM, 1-10.

[10] W3C, Cascading Style Sheets Level 2 Revision 1 (CSS 2.1) Specification. Online : http://www.w3.org/TR/CSS2/. Accessed 06/14/11.

[11] W3C, Web Content Accessibility Guidelines (WCAG) 2.0. Online : http://www.w3.org/TR/WCAG20/. Accessed 06/14/11.

[12] W3C, XML Path Language (XPath) 2.0. Online : http://www.w3.org/TR/xpath20/. Accessed 06/14/11.

Print-Friendly Page Extraction for Web Printing Service

Sam Liu
HP Labs
1501 Page Mill Rd
Palo Alto, CA 94304 USA
sam.liu@hp.com

Cong-Lei Yao
HP Labs - China
No. 1 Zhong Guan Cun East Rd
Beijing 100084, China
conglei.yao@hp.com

ABSTRACT

Printing Web pages from browsers usually results in unsatisfactory printouts because the pages are typically ill formatted and contain non-informative content such as navigation menu and ads. Thus, print-worthy Web pages such as articles generally contain hyperlinks (or links) that lead to print-friendly pages containing the salient content. For a more desirable Web printing experience, the main Web content should be extracted to produce well formatted pages. This paper describes a cloud service based on automatic content extraction and repurposing from print-friendly pages for Web printing. Content extraction from print-friendly pages is simpler and more reliable than from the original pages, but there are many variations of the print-link representations in HTML that make robust print-link detection more difficult than it first appears. First, the link can be text-based, image-based, or both. For example, there is a lexicon of phrases used to indicate print-friendly pages, such as *"print"*, *"print article"*, *"print-friendly version"*, etc. In addition, some links use printer-resembling image icons with or without a print phrase present. To complicate matter further, not all of the links contain a valid URL, but instead the pages are dynamically generated either by the client Javascript or by the server, so that no URL is present. Experimental results suggest that our solution is capable of achieving over 99% precision and 97% recall performance measures for print-friendly link extraction.

Categories and Subject Descriptors

H.3.5 [**Information Storage and Retrieval**]: Online Information Service – *commercial services, web-based services.*

General Terms

Algorithms, Design, Experimentation.

Keywords

Web content extraction, Web printing, HTML, DOM.

1. INTRODUCTION

The World Wide Web (WWW) has become the platform of choice for publishers to distribute content to their audiences. A Web page, however, typically includes auxiliary information that has little association with the main content, such as navigation

menu and ads. The auxiliary content is generally considered "noise" by information retrieval (IR) systems, and should be removed for Web data mining [2-4]. Recently, there is work in content extraction to provide a better Web printing experience, mainly motivated by trying to improve the poor quality prints rendered by the Web browser [5]. A desired printout would contain only the main content and aesthetically layout in a popular file format, e.g. PDF. A robust solution for accurate Web content extraction, however, is still elusive given the complexity of modern Web pages. Recognizing the Web printing problem, many publishers are now providing print-friendly pages via print-links, especially for pages that are more likely to be printed, such as articles. These print-friendly pages contain essentially the same main content as the original page but in a much cleaner and simpler format, as illustrated in Figure 1. Thus far, all the effort on content extraction is devoted to the original page, which in general is a difficult problem. We propose instead using the print-friendly page as an alternative for content extraction when it is available. This requires the detection of the print-link and the extraction of the print-friendly URL. The key challenge here is to find a solution that can achieve very high extraction accuracy since an incorrect link would lead to a page with the wrong content, which is a "catastrophic" mistake. The target application for this technology is a cloud-based Web printing service as illustrated in Figure 2, but it can also be used for other applications such as search and indexing.

Figure 1a: Example of a news article page

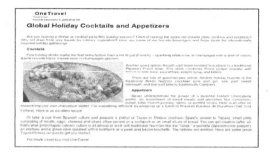

Figure 1b: Example of the print-friendly page

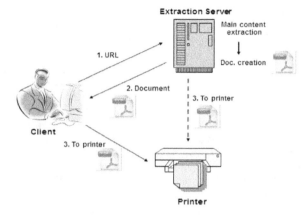

Figure 2: Cloud based Web printing service

As shown in Figure 3, there are many variations of the print-link expressions, making the detection problem more difficult than initially thought. First, the link content between the *"<a>...</a>"* (hyperlink) tags can be text-based, image-based, or both, and the text expression can be a lexicon of phrases of many possibilities, such as *"print"*, *"print article"*, *"print-friendly version"*, etc. Furthermore, some pages use printer-resembling image icons with or without a print phrase present. It turns out that not all the links contain valid URLs because the print-friendly pages can be dynamically generated either by the client Javascript or by the server upon request.

As mentioned earlier, the goal of the system is to provide high extraction accuracy or precision (% of extracted URLs that are correct). In general, systems that require high precision usually have to relax the recall (% of URLs correctly extracted relative to all the valid URLs in the dataset), but experiments show that our solution can maintain a precision of over 99% while achieving a recall of over 97%, a very high performance measure.

Figure 3: Examples of print-links from various Web pages

2. SYSTEM DESCRIPTION

2.1 Overall Architecture

The processing pipeline of a Web article extraction and reformatting system for Web printing is illustrated in Figure 4. To summarize, the first stage of the system uses the method described in reference [5] to extract the article components such as the title, text body, the associated image and caption. The input to the system is a Web HTML page from which we first attempt to acquire the print-friendly page URL, but if it is not available, content extraction would resort back to the original page. After the article components have been extracted, they are reassembled to create the final aesthetic document in PDF format.

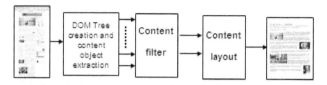

Figure 4: Block diagram of print-link detection

Our solution to the print-link extraction problem places particular importance on the correctness (precision) of the extracted URL, thus we impose a precision target of over 99% while maximizing the recall. The print-link detection strategy uses a print-phrase dictionary to find an exact match with the link text content or the link/image attribute values (described in more details in Section 2.3). It is important to populate the print phrase dictionary with an appropriate set of phrases since it impacts both the precision and recall. We can control the precision performance by using only print-phrases encountered in actual Web pages, and maximize the recall by having a comprehensive dictionary. By using exact print-phrases that only appear on Web pages, there is little chance that the match would result in a false alarm. To further improve the precision, we also compare the extracted (a real URL exists) and original URL domain names. We found that the print-friendly page is also typically archived in the same server as the original page and share the same domain name. It turns out, however, that this is not always the case, but the exceptions are very rare and do not impact the precision (only lower the recall). If a print-friendly URL cannot be extracted, either because there is no print-link on the page or the print-friendly URL is not valid, content extraction would resort back to the original page.

The print-link detection and extraction system basically takes on two stages: (1) detection of the print-link using the DOM (Document Object Model), as illustrated in Figure 5, and (2) print-friendly URL retrieval from the link attributes and the test for its validity, as illustrated in Figure 6. In the following sections, we will provide more details of the various components of the system.

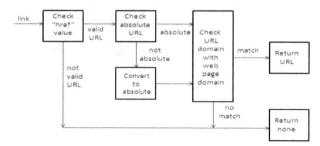

Figure 5: Block diagram of print-link detection

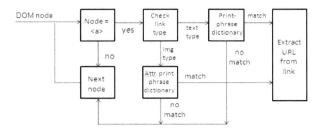

Figure 6: Block diagram of URL extraction and validation

2.2 DOM – Document Object Model

To find the print-link of a Web page, we need to examine the content of the link tag *"<a>"* in the HTML file for a print-phrase. One method of finding the *"<a>"* tags is to simply use text matching, but we instead choose the DOM approach because the DOM is a data structure that captures all the HTML information while avoiding all non-HTML syntaxes such as Javascript [1]. Every HTML file can be mapped to a DOM, which given the embedding nature of HTML, the DOM is a hierarchical data structure representing the organization of the various content objects in the HTML file. Thus the DOM is a data structure that is much cleaner and suited for efficient analysis and manipulation of the Web content than the original HTML file. Each node of the DOM tree is an HTML-Tag element, so we simply just walk the tree to search and examine the *"<a>"* nodes.

2.3 Print-Link Types

We found that all print links in our dataset belong to one of the following types in the HTML representation: (1) *"<a>print-phrase</a>"*, (2) *"<a><img></a>"*, (3) *"<a><img>print-phrase</a>"*, and (4) *"<a><img></a>print-phrase"*, where the *"<a>"* and *"<img>"* are the HTML (hyper)link and image tags, respectively (note the attributes are not shown). Examples of the print-phrases are *"print"*, *"print this article"*, *"print story"*, etc. There can be many variations of the print-phrases but they all convey the notion of a print-friendly page. Note that in type 2, there is no print-phrase (only an image icon), but a print-phrase is typically embedded in the attributes of the *"<img>"* or *"<a>"* tag, such as the *"alt"* or *"title"* attribute. Our print-link detection strategy is to simply use a print-phrase dictionary to find an exact match with the link text content or the image attribute values. Again, it is very important to populate this dictionary with the appropriate set of print-phrases since it impacts both the precision and recall. In the following sections, we will provide the details of the four types of print-link expressions found in Web pages.

2.3.1 Type1: <a>print phrase</a>

This is the simplest and the most common type of the four print-link expressions. The link is just represented by a print-phrase, e.g. *"<a> print-phrase </a>"*, as illustrated in following HTML code fragment from a real Web page.

"<a id="printaction" class="actionlink" rel="nofollow" href="/story/story/print?guid=60B21A5A-015E-11E0-B254-00212804637C">Print</a>"

The print-phrase in this example is simply *"Print"*, but as mentioned earlier it can take on many expressions. *"Print"* is the most common print-phrase, used by over 80% of the Web pages containing a print-link. A print-phrase is also typically found in the attributes of "<a>" tag, but for this type, we only examine the link text content for the print-phrase.

2.3.2 Type1: <a><img></a>

The Type 2 print-link expression contains only an image between the *"<a>...</a>"* tags, as illustrated in the following real HTML code fragment.

"<a href="/print/story/10939358.html" title="Print This Article"><img src="http://i.thestreet-static.com/files/tsc/v2008/css/images/print_14x13.gif" /></a>"

From this example, we see that there is no text content to indicate that this is a print-link (the icon image is providing that information visually). Notice, however, that the *"title"* attribute value of the *"<img>"* tag does contain a print phrase, *"Print This Article"*, that can be used to determine this link is indeed a print-link. We also found that the *"alt"* is another attribute that commonly contains a print-phrase. Thus, for image-only print-links, we need to examine the *"title"* and *"alt"* attributes for print-link detection. Our solution, however, is not capable of detecting the print-link if none of these attributes contain a print-phrase, but the miss only impacts recall (precision is not changed).

2.3.3 Type1: <a><img>print phrase</a>

Type 3 print-link expressions simply contain both an image and text print phrase between the *"<a>...</a>"* tags, as illustrated below.

"<a href="/news/religion/ct-met-stroger-chaplain-ordination-20101205,0,7578136,print.story" rel="nofollow" ><img src="/hive/images/icons/print_icon.png" alt="print" class="icon" />Print</a>"

Note that this example suggests that we can search for a print-phrase in the link text content or in the attributes of the *"<img>"* tag. For this type, however, we just test for the print-phrase in the link text content, similar to Type 1.

2.3.4 Type1: <a><img></a> print phrase

The last type is when an image is the link, but the print-phrase is outside of the *"<a>...</a>"* tags, as shown below.

"<a href="index.php?fa=PAGE.printable&pageId=117444" target="_new"><img src="/images/print_friendly.gif" width="29" height="15" border="0"></a> Printer-friendly version</font>"

As in type 2, if the attributes of the *"<img>"* tag contain a print-phrase, then the print-link can be detected. If not, we need to examine the text node adjacent to the *"<a>"* node of the DOM tree. This case, however, is extremely rare. Out of the two 2000 sample Web pages, we only found one such case, so we just apply the Type 2 detection scheme to this type.

2.4 Print Phrase Dictionary

To test for print-link, we simply search for a print-phrase in the text content of the *"<a>"* tag for Type 1 and 3, or the attribute values of the *"<a>"* and *"<img>"* tags for Type 2 and 4. We found that the *"alt"* and *"title"* attributes are the best candidates containing a print-phrase, so no other attribute is examined. Based on empirical studies of over 2000 Web pages, we found that a solution is not robust if it just searches for the word or word fragment *"print"* as the print-phrase. It turns out that there are links with text content such as *"printers"*, *"print run"*, or *"reprint"* that are not true print-links. As mentioned earlier, achieving very high precision is the key objective since any incorrectly extracted URL would lead to a page with the wrong content. Thus to ensure high precision, we use a dictionary containing the exact print-phrases we collected from the dataset (over 2000 pages from the top 50 popular news sites), and requires an exact match between the link text content (or attribute value) and a print-phrase in the dictionary. To maximize the chances of capturing all of the print-links (recall), the dictionary should contain a comprehensive set of print-phrases from a large enough dataset. Currently, the dictionary size is greater than 20, including the most popular phrases such as *"print"*, *"print article"*, *"printable version"*, and the lesser used phrases such as *"print it"* and *"print story"*. Since the print phrases are simply to convey the notion of print-friendly pages, we do not expect the dictionary to evolve over time. Also note that it is straight forward to internationalize the dictionary to support other languages. With this dictionary, we are able to achieve very high precision and recall performance measures.

2.5 Print-Link URL

The *"href"* attribute value of the *"<a>"* contains the print-link URL (absolute or relative), but not all them are valid. Some of the pages are actually dynamically generated either by the client Javascript or by the server upon request, so no pre-generated print-friendly HTML page is available for content extraction. The following HTML code fragments are examples of print-links with no valid URLs.

"<a href="#">Print</a>" or *"<a rel="canonical" href="javascript:window.print()">print this story</a>"*

From examining various Web pages with dynamically generated print-friendly pages, the value of the *"href"* attribute usually contains these text symbols or words: *"#"*, *"javascript"*, or *"funct(...)"*, where *"funct"* denotes a Javascript function call. To determine whether a URL is valid, we can simply test for the following text symbols in the in the *"href"* attribute value: *"#"*, *"javascript"*, *"("*, or *")"*.

2.6 URL Domain Validation

Since the goal of print-link detection is to achieve the highest precision possible, we require the extracted URL (non-dynamically generated print-friendly pages) to have the same domain name as the URL of the original Web page. The assumption is that if the extracted URL is indeed the print-friendly page, this page should be pre-generated and archived in the same server, sharing the same domain name as the original page. For example, the following URL is that of an original Web page, *"http://www.foxnews.com/leisure/2010/12/06/gift-guide-gourmands/"*, and the print-friendly page URL is *"http://www.foxnews.com/leisure/2010/12/06/gift-guide-gourmands/print"*, which has the same domain name *"foxnews"* as the original page. By imposing this requirement, it can hurt the recall performance, but based on the evaluation of the dataset, it is very rare (1 out of 2000) that the two pages do not share the same domain name. It is also good to know that even though the recall might be impacted, the precision is not. Also note that the URL in the *"href"* attribute can be relative, which by default it is in the same domain. The overall URL extraction and validation flow diagram is illustrated in Figure 4.

3. EXPERIMENTAL RESULTS

To evaluate our print-link detection and extraction scheme, we collected roughly 2000 Web pages from the top 50 popular news sites. The ground-truths of the print-links and the print-friendly URLs are manually labeled using multiple subjects. As mentioned earlier, for quantitative performance measures, we use the

traditional definition of precision (% of extracted URLs that are correct) and recall (% of URLs correctly extracted relative to all the valid URLs). An evaluation system has been build based on Java, using the Java *swing* package for HTML parsing as an efficient method to create the DOM. Our experiment shows that to achieve high precision and recall it is critical that the print-phrase dictionary be comprehensive, containing specific print-phrases for full phrase matching. The dictionary size is currently greater than 20, and we do not anticipate it to grow much, if at all, as we continue to collect more Web pages. With this system, we are able to achieve a precision of over 99% and recall of over 97%. All the false alarms in the precision are caused by miss-classification of invalid URLs as real URLs. With such high precision and recall measures, the system is highly accurate and robust for print-link detection and extraction.

4. SUMMARY

This paper presents a highly robust solution to print-link detection and print-friendly URL extraction from Web pages. The targeted application is a Web printing service, but the technology is applicable to general Web content extraction for search and indexing, data management, targeted advertisement, and more. The print-friendly pages, which are pages created for better print outputs, can be used as a cleaner and simpler alternative for main content extraction. We estimate that over 35% of the Web articles contains usable print-friendly URLs (pre-generated pages). The majority of the articles, however, dynamically create the print-friendly pages upon request, and thus our method cannot retrieve any usable URL for these pages (content extraction is resort back to the original pages). The key goal of the solution is to achieve very high precision while maximizing the recall since any incorrectly extracted print-friendly URL would lead to a page with the wrong content. We have identified there are basically four HTML expressions to represent print-links, and used a dictionary populated with print-phrases collected from over 2000 Web pages from the 50 most popular news sites for print-phrase matching. The print-phrase dictionary is expected to stay fairly stable, and can easily internationalize to support other languages. To further improve the precision, we also impose the requirement that both the extracted print-friendly and original page URLs must share the same domain name. This requirement has only a small negative impact on the recall, but no impact on the precision. With this system, we are able to achieve over 99% precision and 97% recall performance measures.

5. REFERENCES

[1] Le Hégaret, Philippe (2002). "The W3C Document Object Model (DOM)". World Wide Web Consortium. http://www.w3.org/2002/07/26-dom-article.html.

[2] J. Pasternack and D. Roth. "Extracting article text from the web with maximum subsequence segmentation". In Proceedings *of the 18th WWW*, 2009.

[3] Gupta, Suhit et al. "Automating Content Extraction of HTML Documents". World Wide Web: Internet and Web Information System, 8, 2005, 179-224.

[4] Reis, D. et al. "Automatic Web News Extraction using Tree Edit Distance". In Proceedings of the 13th International Conference on World Wide Web, 2004, New York.

[5] Luo, Ping et al. 2009. "Web Article Extraction for Web Printing: a DOM+Visual based Approach". In Proceedings of the 9th ACM Symposium on Document Engineering. DocEng 2009, New York, NY, 66-69.

Skeleton Comparisons

The Junction Neighbourhood Histogram

Jannis Stoppe
University of Bremen
Am Fallturm 1
D-28359 Bremen
jannis@informatik.uni-bremen.de

Björn Gottfried
University of Bremen
Am Fallturm 1
D-28359 Bremen
bg@informatik.uni-bremen.de

ABSTRACT

For analysing and comparing characters, using skeletons is a promising approach due to their topology-preserving nature and the resemblance of the skeleton to the original writing movement. We suggest a novel qualitative approach to skeleton comparison that is based on the adjacency of junctions and end points and the steps of a preceding skeleton simplification. By using a multi-dimensional histogram that contains information about the adjacency and the degree of joints, we gain high comparison speeds which, when combined with the multi-step approach, can be used for a generic topology distance metric.

Categories and Subject Descriptors

I.4.7 [**Image Processing and Computer Vision**]: [Feature Measurement - Size and shape]

General Terms

Algorithms

Keywords

Skeletons, Characters, Comparison, Similarity

1. INTRODUCTION

Skeletons are well-suited for the analysis of characters and text, as they focus on representing topological features, which are the major type of feature to classify characters. However, there are no straightforward ways to process both, raster based skeletons and graph based representations, concerning the topology and their comparison with other skeletons, which are suitable for fast processing of vast datasets of characters. In this paper, we present a method to efficiently compare skeleton topologies and make this comparison more robust to small changes in the topology based on the chosen skeletonisation algorithm or slight shape differences.

2. SKELETON SIMPLIFICATION

Skeletons consist of those points in a shape that are equidistant to two or more borders. As such, they represent a shape's topol-ogy. We do believe that such a topology based information is well suited for the analysis of text of all kinds, especially handwritings. Several factors (such as pen tip shape) that are not part of the actual information to be extracted are inherently removed during the skeletonisation process.

However, as raster based skeletonisation approaches usually result in a too detailed skeleton, we use the approach suggested in [11] to simplify the given raster skeletons first. As the suggested approach simplifies the skeleton iteratively, the question remains when to stop the process and use the current skeleton for further analysis or comparison. As each simplification step also results in an error level that indicates the appearance's difference between the simplified skeleton and the original one, this value can be used to specify a threshold at which the simplification process should be stopped. However, two similar shapes do not neccessarily have the same simplified skeletons when using the same simplification threshold (see fig. 1).

Our approach to solve this problem is to use *all* resulting simplification stages instead of just one predefined final stage.

In addition, we added a second error metric to the calculation of skeleton errors. In [11], the only measure suggested was the distance from the original pixels to the skeleton bones representing them. In some cases, this leads to very long bones which run parallel to each other, distorting the topology. Our solution is to add a penalty for bones that are longer or shorter than the pixels they represent. In general, the "length" of a pixel could be anything between 1 and $\sqrt{2}$. If a bone exceeds or falls below this range (with its lower and upper bound being multiplied by the amount of pixels the bone represents), the difference between the bone's length and this range is added to the skeleton's error level (see eq. 1). Consider P the set of points $p_0 \ldots p_n$ that are assigned to a bone. The error level of a bone then receives an additional penalty of p (see eq. 1). The *weight* is an additional factor that allows coordinates to be assigned to several bones: as our algorithm creates a graph that has its nodes places directly on the coordinates of pixels, the original coordinates usually cannot be attached to a bone unambiguously. Instead of picking one bone, we assigned the coordinate to all adjacent bones, with the sum of all weights being 1, resulting in a more predictable and consistend behaviour of the simplification.

Figure 1: Two similar shapes, simplified with given threshold 1, resulting in two different skeleton topologies.

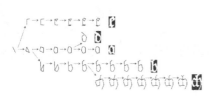

Figure 2: Using a tree structure to differentiate between skeleton shapes

Figure 3: The topology of similar shapes is not certain to remain the same throughout the inverse simplification history.

$$
\begin{aligned}
l_{\min} &= \sum_{i=0}^{|P|} weight(p_i) \\
l_{\max} &= |P| \cdot \sqrt{2} \\
p &= \begin{cases} 0, & \text{iff } l_{min} \leq length \leq l_{max} \\ min(|length - l_{min}|, |length - l_{max}|), & \text{else} \end{cases}
\end{aligned} \quad (1)
$$

3. USING THE SIMPLIFICATION STEPS

Each simplification step yields a skeleton by merging two adjacent joints. The simplification process stops either at a predefined threshold or if only one bone remains. That means that if the threshold is not used, all skeletons are simplified down to a single bone, sharing the same final result.

This sequence of simplification steps can also be seen in reverse. In this case, the skeleton simplification history would instead be a sequence of successively added details, with more significant information added first. A tree structure that represents changes in the skeleton topology could then be used to differentiate between different types of characters (see fig. 2).

However, due to the nature of the simplification algorithm, similar shapes may well have a different topology at a certain state of their inverse simplification history while returning to the same topology later in the given list of steps (see fig. 3). If a tree structure was used to represent similarities, this would result in an early split of similar shapes with low chances of a correct classification. A more appropriate comparison would therefore include all steps of the simplification history.

Our approach to this problem is not to use a tree to differentiate between different shape classes but to compare all available states and use the sum of all differences and similarities as the basis for the distance between two skeletons. This in turn results in a distance metric between any two shapes based on the amount of common skeleton topologies at the same simplification levels. The most straightforward way to calculate this would be to sum up all steps in which the topologies are equal among the objects (see eq. 2). In this case, with $\sigma(simplificationHistory)$ being defined as the amount of simplification steps that are stored for a skeleton and s_1 and s_2 being the skeleton simplification histories that are to be compared (each consisting of a set S of skeletons, starting at the most simplified level), the similarity function $sim(s_1, s_2)$ is defined as

$$
n = \min(\sigma(s_1), \sigma(s_2))
$$
$$
sim_1(s_1, s_2) = \sum_{i=0}^{n} \begin{cases} 1, & \text{iff } s_1(i) = s_2(i) \\ 0, & \text{else} \end{cases} \quad (2)
$$

Figure 4: The last few elements of a simplification chain and their respective ranges of error levels, which can be used for an error level based equality check. Error levels between two steps are considered to belong to the more simplified step. Previous steps are not displayed to avoid a cluttered display but have been calculated.

However, this would also result in a lack of penalty for complex, big shapes that share some stages but have much more stages that are different from each other. We therefore prefer a solution like eq. 3 or eq. 4 which also take the amount of different stages into account, either absolute or relative.

$$
sim_2(s_1, s_2) = -|\sigma(s_1) - \sigma(s_2)| + \sum_{i=0}^{n} \begin{cases} 1, & \text{iff } s_1(i) = s_2(i) \\ -1, & \text{else} \end{cases} \quad (3)
$$

$$
sim_3(s_1, s_2) = \frac{\sum_{i=0}^{n} \begin{cases} 1, & \text{if } s_1(i) = s_2(i) \\ 0, & \text{else} \end{cases}}{max(\sigma(s_1), \sigma(s_2))} \quad (4)
$$

The error level of a skeleton (which is used during the simplification process to determine the next simplification step by punishing a difference in the simplified skeleton's appearance compared to the original one) can also be used to further classify the skeletons that should be compared. Instead of comparing each step one by one (which implies that stage $steps_1 - n$ of one simplification chain should usually match stage $steps_2 - n$ of a second chain), the error levels of a step and its successor can be used to assign a span to each skeleton of the simplification history within which it is valid. When comparing histograms, the amount of overlap between those spans indicates the weight of this comparison. This results in complex skeletons (e.g. that of an "@") with high error levels in their final states not matching less complex skeletons (e.g. that of an "l") despite identical final states.

4. THE JUNCTION NEIGHBOURHOOD HISTOGRAM

The question how to compare the resulting skeletons remains. Current approaches [12] usually rely on graph matching algorithms. There are a variety of different approaches available for applying this graph matching problem to shape retrieval:

- matching finite rooted trees [8, 9], which is in turn based on transforming the skeletons to such trees by breaking loops apart into two disjoint paths [10]

- matching paths within the skeleton between two endpoints [2] and/or junctions [13]

- finding a many-to-many match between the two graphs that are compared [3] by mapping both graphs into a geometric space

- breaking up the skeleton parts and comparing the disconnected sets of branches using detected symmetry axes [1]

While this works well for a variety of shapes, letters in particular are often defined by few features, with sparse information apart from an occuring loop (or none) and maybe one or two elements branching off this structure. Another major problem concerning traditional graph matching approaches is their computational complexity: "Also, the complexity of the descriptions or the data structures leads to computationally expensive matching and recognition algorithms" [1].

It has already been suggested to use graph histograms as a feature to more efficiently compare graphs [7, 4]. However, the suggested approach uses only a node's own degree to be added to a histogram. This in turn does not contain much information about the topological structure of the shape.

The idea of matching two skeleton graphs usually implies that for comparing one shape's features to another shape's features the shapes themselves are compared. Our approach instead uses a histogram based representation of the shapes' features to be able to efficiently compare the shapes' structure. This histogram is based on the neighbourhood relations in a skeleton: starting from each end point or junction in a skeleton, we examine the degree of the next neighbouring node n in the graph with $deg(n) \neq 2$. For each node, we thus construct a neighbourhood degree histogram, storing the amount of neighbouring nodes by their degree. This means that each node can be positioned in a k-dimensional space, with k being the maximum degree of each of its neighbours and its position on each axis a being the amount of neighbours sharing $deg(n) = a$. These k-dimensional positions can then be accumulated for all end points and junctions of a skeleton, the result being a k-dimensional junction neighbourhood histogram. Instead of leaving the histogram's second dimension empty, we use it to store information about a node's self-neighbourhood, as a circle that points at the same junction is usually a topologically important feature of a letter.

These histograms contain sufficient topological information in order to compare two skeletons. Note that these only have to be computed once per skeleton. The histogram computation's worst case complexity is $O(n^2)$, with n being the amount of nodes of the graph the histogram is computed for if the given graph is a clique, which is, however, rarely the case for the skeletons of letters. The comparison of the histograms then depends on the amount of entries in the histogram, the worst case for comparing two skeletons $s1$ and $s2$ with n_{s1} and n_{s2} being the number of nodes with $deg(n) \neq 2$, being $min(n_{s1}, n_{s2})$.

Generally, our approach bypasses the computational complexity of mapping all important nodes of one graph to another but instead builds a histogram that contains information about specific features for each skeleton, which in turn can be compared rapidly. This is sufficient in our case and this is what distinguishes our approach from the computationally hard graph matching problem.

The next step is to combine the histogram based comparison with the simplification steps suggested in section 2. This is done by generating a histogram for each of the simplification stages, resulting in a chain of histograms which can either be compared one by one (see eq. 2 to 4) or by matching those histograms that represent skeletons of the same error level.

5. EVALUATION

For evaluating our approach, we use the Bull's Eye test to determine its accuracy [5]. Due to the nature of the discrete steps, the equality of distances becomes a major problem of this test for the given scenario though: When searching for the correct hits within the 40 closest distances, the test specifications do not include any information concerning the occurance of several shapes sharing the same distance (possible any combination of correct and incorrect ones), all being placed across the 40th place, thus having more than

$$acc = \frac{hits}{(\text{\# shapes ranked} \leq 2 \times \text{\# instances per class}) \times 0.5} \quad (5)$$

40 shapes in the first 40 hits. We therefore modify the calculation to what we consider was the intended meaning: The amount of correctly identified shapes divided by half the amount of shapes that are checked (see eq. 5).

As test sets, we use the MPEG Core Experiment CE-Shape-1 B test set and a collection of letters taken from the *Grenzboten* which were separated and binarized, 20 classes with 20 instances each (see fig. 5) [6].

For the *Grenzboten* test set, we performed the analysis using the absolute error levels given in eq. 3, the relative error levels given in eq. 4, both compared one by one as shown in fig. 3, and the relative error levels compared using the overlapping error levels approach shwon in fig. 4. The 20×20 instances would result in an accuracy of 10 per cent for randomly drawn results. The results were $\approx$ 24.83 per cent accuracy for the absolute levels, $\approx$ 43.50 per cent accuracy for the relative levels and $\approx$ 45.96 per cent accuracy for the error level based comparison.

For the MPEG test set, we added a resize step to the processing chain that normalized the image sizes to 64 pixels on the image's long side. The relative error levels performed the best, with $\sim$ 26.21 per cent accuracy, the error level based comparison with $\sim$ 15.88 per cent and the absolute levels with $\sim$ 14.83 per cent being far behind.

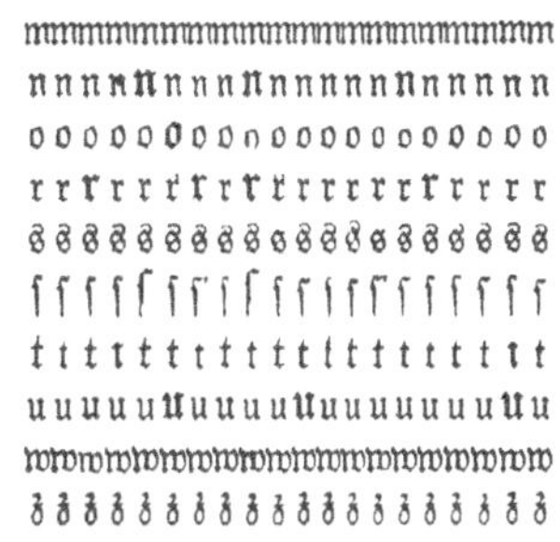

6. CONCLUSION

The proposed method works fast and compares skeletons with sufficiently different topologies well. The histogram that contains neighbourhood information seems a valid approach to compare skeletons based on their structure without having to solve a computationally expensive graph matching problem.

The less well results when using the proposed algorithm on the MPEG test are probably due to topological differences between similar objects: Especially those objects that contain holes (due to some darker parts in the original objects, like the wings of butterflies or the cows' black patches) that are not actually topological features but a result of the way the shape was created from the original picture can hardly be compared using a measure that focuses only on topological features of the shape.

For the *Grenzboten* test set, there are several conclusions to be drawn from the results. While there are some classes that could be detected quite well, there are others with some rather low detection rates. For example, the "w" class had an average detection rate of 68.75 per cent on the error level based comparison, with the major drawback being sample #13 (ɯ) due it not being connected properly (and only one correct match apart from itself). The "ch" class scored 79 per cent accuracy on the error level based comparison, with the major drawbacks being samples #1 (ɸ), #2 (ɸ) and

Figure 5: *Grenzboten* test set.

#16 (ch); the first due to having too many instances ranked among 40 closest samples, the latter ones due to their significantly shorter arm on the right.

In addition, there might be room for significant improvements by combining the given comparison methods. Although the accuracy for the "ch" class goes down to $\approx$ 59 per cent when using the relative calculation, the most problematic cases are then #17 (ch) and #9 (ch), the first due to a missing connection between the arc and the straight line at the bottom. The other instances performed rather well, although some scored lower than they usually would have due to the *ch* being one of the bigger shapes, which results in lots of the same similarity values due to the denominator from eq. 4 remaining the same for most calculations (and in turn to more than 40 instances on the first 40 places, which results in a penalty due to the denominator from eq. 5 rising).

Problematic for most classes is the transformation invariance of the approach. Although the combination of skeleton comparisons as such with the simplification steps resulted in more robust results concerning variances in the size and shape of particular features (as e.g. a bigger circle that is part of a shape would be erased later than a smaller circle that is part of another shape), certain shapes are just too similar concerning their topological features: a, b, d, c and g are all characters that basically consist of a closed circle with an additional line being attached to it, which leads to similar histograms being used to describe each of them (and accordingly false results). This means that the current approach's coarse topology description contains too many invariances for reliable character recognition.

Also, the skeleton simplification algorithm used currently only focuses on merging adjacent joints to approximate a given skeleton. While this significantly reduces the computational complexity, this means that noise is only removed from the skeleton if all parts are connected correctly. As soon as a skeleton is actually broken apart, the simplification will keep the parts seperate throughout all approximation steps, in turn resulting in incorrect histogram information and therefore also in incorrect classifications.

Still, we have come up with a novel and working technique to rapidly compare skeletons. While the skeleton simplification takes some time, especially for skeletons that are generated from bigger pictures (due to the pixel based nature of raster based skeletons and a consequentially high amount of nodes that are generated from the pixel data), this calculation has to be done only once. The comparison itself can be done rapidly using the histograms, which do not need to be re-computed once they were created from the simplification steps. The recognition rate for topologically similar shapes can be improved upon, but shapes with a unique topology are recognized reliably.

7. FUTURE WORK

For comparing characters, it is often important to differentiate between similar features that are arranged in different ways relative to each other or in relation to a global coordinate system. One way to achieve less invariances in our skeleton comparisons without affecting the overall performance of the system would be to incorporate that kind of information directly into the histogram.

Another step forward would be to improve upon the skeleton simplification. Aspects of the Gestalt laws of organization could be used to connect or merge nodes that are not necessserily adjacent to each other to achieve less sensitivity concerning unwanted breaks in the skeleton structure.

Finally, the histograms are currently only checked for equality. While this works in our case due to the relatively small size of the shapes, this might not be appropriate for all tasks that the skeletons and their topology analysis can be used for. Due to the adjacency based nature of the histogram, changes in small parts of bigger shapes should only affect a small part of the histogram as well, which in turn would mean that a similarity between two histograms could go well beyond a simple check for equality.

8. REFERENCES

[1] C. Aslan and S. Tari. An axis-based representation for recognition. In *Proceedings of the Tenth IEEE International Conference on Computer Vision - Volume 2*, ICCV '05, pages 1339–1346, Washington, DC, USA, 2005. IEEE Computer Society.

[2] X. Bai and L. J. Latecki. Path similarity skeleton graph matching. *IEEE Transactions on Pattern Analysis and Machine Intelligence*, 30:1282–1292, 2008.

[3] M. F. Demirci, A. Shokoufandeh, Y. Keselman, L. Bretzner, and S. Dickinson. Object recognition as many-to-many feature matching. *Int. J. Comput. Vision*, 69:203–222, August 2006.

[4] K. Kailing, H. Kriegel, S. Schönauer, and T. Seidl. Efficient similarity search for hierarchical data in large databases. In E. Bertino, S. Christodoulakis, D. Plexousakis, V. Christophides, M. Koubarakis, K. Böhm, and E. Ferrari, editors, *Advances in Database Technology - EDBT 2004*, volume 2992 of *Lecture Notes in Computer Science*, pages 643–644. Springer Berlin / Heidelberg, 2004.

[5] L. Latecki, R. Lakamper, and T. Eckhardt. Shape descriptors for non-rigid shapes with a single closed contour. In *Computer Vision and Pattern Recognition, 2000. Proceedings. IEEE Conference on*, volume 1, pages 424 –429 vol.1, 2000.

[6] Miscellaneous. Die mecklenburgischen domainenbauern und die mecklenburgische verfassung. *Die Grenzboten*, 28.2.1:20 ff, 1869.

[7] A. Papadopoulos and Y. Manolopoulos. Structure-based similarity search with graph histograms. In *Database and Expert Systems Applications, 1999. Proceedings. Tenth International Workshop on*, pages 174 –178, 1999.

[8] A. Shokoufandeh and S. Dickinson. Applications of bipartite matching to problems in object recognition. In *Proceedings of the IEEE Workshop on Graph Algorithms and Computer Vision*, 1999.

[9] A. Shokoufandeh and S. Dickinson. A unified framework for indexing and matching hierarchical shape structures. In C. Arcelli, L. Cordella, and G. di Baja, editors, *Visual Form 2001*, volume 2059 of *Lecture Notes in Computer Science*, pages 67–84. Springer Berlin / Heidelberg, 2001.

[10] K. Siddiqi, A. Shokoufandeh, S. Dickenson, and S. Zucker. Shock graphs and shape matching. In *Computer Vision, 1998. Sixth International Conference on*, pages 222 –229, Jan. 1998.

[11] J. Stoppe and B. Gottfried. Down to the bone: simplifying skeletons. In *Proceedings of the 10th ACM symposium on Document engineering*, DocEng '10, pages 215–218, New York, NY, USA, 2010. ACM.

[12] H. Sundar, D. Silver, N. Gagvani, and S. Dickinson. Skeleton based shape matching and retrieval. In *Shape Modeling International, 2003*, pages 130 – 139, May 2003.

[13] Y. Xu, B. Wang, W. Liu, and X. Bai. Skeleton graph matching based on critical points using path similarity. In H. Zha, R. ichiro Taniguchi, and S. Maybank, editors, *Computer Vision âĂŞ ACCV 2009*, volume 5996 of *Lecture Notes in Computer Science*, pages 456–465. Springer Berlin / Heidelberg, 2010.

Version-Aware XML Documents

Cheng Thao
Department of EECS
University of Wisconsin-Milwaukee
Milwaukee, WI 53201-0784, USA
chengt@uwm.edu

Ethan V. Munson
Department of EECS
University of Wisconsin-Milwaukee
Milwaukee, WI 53201-0784, USA
munson@uwm.edu

ABSTRACT

A document often goes through many revisions before it is finalized. In the normal document creation process, newer revisions overwrite older ones and only the final revision is kept. At any stage of document creation, it might be desirable to see how the document came to its current form or to revert back to a previous revision. Conventional version control tools such as CVS could help authors do exactly this. However, these tools are unlikely to be adopted by non-technical document authors due to the overhead of managing a repository and the tools' learning curves.

This paper presents an approach called *version-aware documents* that embeds versioning data within the document thus making version control for single documents a seamless part of the authoring process.

Categories and Subject Descriptors

I.7.1 [**Document and Text Editing**]: Document and Text Editing—*Document management, Version control*

General Terms

Management

Keywords

Version control, Collaborative editing, XML document

1. INTRODUCTION

A conventional document, such as a user manual, a technical report or a software requirement document, will go through multiple revisions. Creating such a document may require a collaboration among multiple persons who will create and edit content in parallel. While the document is evolving, authors may want to see how and why it has changed and they may want revert back to a previous revision if the current revision is beyond repair. The ability to capture the evolution of a document is also important to

organizations such as pharmaceutical and aerospace companies that must stay compliant with government regulations where traceability or auditability of documents is important. Also, the ability to retrieve arbitrary revisions frees authors to experiment with the current revision without fear of losing it.

A naive approach to version control is to use a manual process of copying and naming files to indicate revision or version. This solves the problem of version identification but requires the overhead of managing the files and manually selecting the correct files from among many. Collaboration is still possible and merging may be done by hand or by 3-way merge tools such as diff3 if the files are text-based. When multiple authors want to have access to the change history, files for every revision must be copied and shared. When there are too many files and revisions, the overhead and confusion can be overwhelming.

A more attractive approach would be to use conventional version control tools such RCS [23], CVS [4], Subversion [21], Git [8], or Mercurial [16]. These tools often require access to a central repository or a shared file system where the version data is stored to support both versioning and collaboration. This may require setting up the infrastructure needed, and in organizations that must be in compliance with government rules, gaining approval could be cumbersome. These tools are excellent for collaboration in large groups where there's a centralized repository and access to a network. But non-technical authors who do not have the skills or the will to learn the tools are unlikely to adopt them. Some cloud storage systems offer versioning and collaboration support for documents stored but that makes the author dependent on the particular system [6, 9].

We propose an approach called *version-aware documents* that does not require a repository or network to collaborate and makes version control as seamless as possible. Currently our approach is limited to documents that can be represented as XML.

2. BACKGROUND AND RELATED WORK

XML (Extensible Markup Language) [27] is a markup language that is used for many purposes ranging from representation of data in databases and remote procedure calls, to configuration files such as Ant build scripts, to human oriented documents found in publications, documentation, and marketing materials. An XML document can be composed from multiple *markup vocabularies* from different domains which may have the same element and attribute names. XML *namespaces* can be used to expand names in order to prevent name collisions between vocabularies with overlapping element and attribute names. The XML Security standards [28] provide support for maintaining document data confidentiality and integrity. Our approach, version-aware documents, makes use of namespaces and XML signature from the XML Security standards for version management and integrity checking. The use of namespaces allows version data to co-exist peacefully with the primary content of an XML document while XML signature is used to prevent users from altering the version data.

Conventional version control tools [4, 8, 16, 21] require the user to maintain a repository either locally or remotely and rely on the diff3 merge tool to support merging. Some researchers are interested in representing *multiple structures* (or variants) of a document [5] in an XML model but not the versions of each variant. If we view a variant as a version, then we have a set of versions but not a tree of versions. Some researchers use XML to encode data with a time dimension [7, 15, 29, 30] known as *temporal XML* to achieve a time based retrieval functionality found in *temporal databases* [17, 20], where the emphasis is on the ability to query the data for a given time. These approaches extend XML and introduce query techniques and tools over the XML model.

There are several XML differencing techniques [1, 12, 13, 14, 18] which can be used as basis for XML version control. Several approaches to XML document versioning use structural differencing [2, 3, 19, 24, 26]. These versioning approaches store the document and the change information (the *delta*) separately on the file system or in a database. Vion-Dury proposed encoding XML deltas and version history as a standalone XML file [25].

Microsoft Office and OpenOfffice have limited versioning capabilities. Microsoft Word's change tracking system is essentially version control with only two revisions: previous and current. OpenOffice documents support linear versioning. In OpenOffice, each version is stored in full form (ZIP archive) in the current version's ZIP archive. Both Microsoft Office and OpenOffice support collaboration by user accepting or rejecting new changes but lack advance collaboration support such as merging and branching.

3. APPROACH AND UNIQUENESS

Our *version-aware XML documents* are created via three simple extensions to an application's native XML data format using vocabularies and namespace from our versioning XML framework. The namespace in our framework is simply refer to as *molhado*. The extensions do not change the document's semantics and thus the document can be rendered and edited by an editor as if it were a native document. For example, an extended SVG document can be read, rendered, and edited as if it were a normal SVG document. The extensions are:

- New document elements representing reverse deltas;

- An XML signature on the version data to ensure its integrity; and

- The addition of an identifier attribute to each element, which eases identification of changes.

As mentioned before, our version-aware documents use reverse deltas (edit operations needed to compute a previous

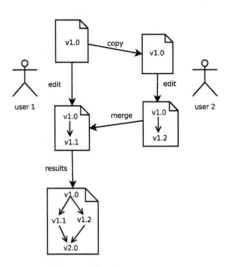

Figure 1: Version-aware XML document collaboration workflow example.

version from the current version) to encode the version history of a document. With reverse deltas, the latest version is stored in full form, while all other versions are stored as deltas. Figure 2 is an example of an extended Inkscape SVG document. Each version is encoded as an XML element, and each has a parents attribute which points to the previous version elements. The sub-elements of the versions represent edit operations. Possible edit operations are attribute value update, node child sequence update, node deletion, node addition, and node name update. The latest version is the complete document content ignoring the version data. A version other than the latest version is retrieved by applying a chain of deltas to the latest version. An editor, whether version-aware or not, can modify the document directly as if it were a normal document. An editor without version support can simply ignore the version data and renders the document normally, though the system works best with editors that preserve the delta elements and UID attributes that have been added to the base document representation.

To support branching and merging, each version is labelled with a 32 hex character unique ID and each version has zero or more parents. The unique IDs are generated using the Java Unique ID Generator (JUG) library [11]. The IDs are generated using the system clock and the host MAC address to prevent possible collisions when versions are created in different machines.

Branching occurs when two copied and modified documents are merged. A merge creates branches composed of new changes from both documents. Note that this is a 3-way merge because they have a common ancestor version. The branches start from the common version and join at the merged version forming a diamond. The branches contain all changes made by both users and both changes can be retrieved. Figure 1 illustrates this process.

Version-aware documents contain the entire document history without requiring users to interact with a version repository. To collaborate, an author simply gives a copy of his working file to another author. This could be done in any manner, such as via email or using a flash drive. Once the other author has finished his work, he gives the first author his copy of the file. The first author can then perform a

```
❶ <svg xmlns="http://www.w3.org/2000/svg" xmlns:cc="http://creativecommons.org/ns#"
   xmlns:dc="http://purl.org/dc/elements/1.1/" xmlns:inkscape="http://www.inkscape.org/namespaces/inkscape"
❷ xmlns:molhado="http://www.cs.uwm.edu/molhado" xmlns:rdf="http://www.w3.org/1999/02/22-rdf-syntax-ns#"
   xmlns:sodipodi="http://sodipodi.sourceforge.net/DTD/sodipodi-0.dtd" xmlns:svg="http://www.w3.org/2000/svg"
   xmlns:xlink="http://www.w3.org/1999/xlink" height="1052.3622047" id="svg2" inkscape:version="0.48.0 r9654"
   molhado:id="0" sodipodi:docname="drawing.svg" version="1.1" width="744.09448819">

❸    <molhado:revision-history id="revision-history">

❹      <molhado:revision id="311d0dba-93a1-11e0-9d6b-001b6393b591" name="0">
          <molhado:attr-update attr="sodipodi:docname" newvalue="New document 1" nodeid="0"/>
          <molhado:attr-del attr="xmlns:molhado" nodeid="0" value="http://www.cs.uwm.edu/molhado"/>
          <molhado:attr-del attr="xmlns:xlink" nodeid="0" value="http://www.w3.org/1999/xlink"/>
          <molhado:children-update newchildren="[]" nodeid="1"/>
          <molhado:children-update newchildren="[10]" nodeid="9" />
       </molhado:revision>
       <molhado:revision id="8e4d8ed4-93a1-11e0-a21c-001b6393b591" parents="311d0dba-93a1-11e0-9d6b-001b6393b591" name="1">
          <molhado:attr-del attr="xmlns:xmldsig" nodeid="0" value="http://www.w3.org/2000/09/xmldsig#"/>
          <molhado:attr-update attr="inkscape:current-layer" newvalue="layer1" nodeid="2"/>
          <molhado:children-update newchildren="[10, 60, 74]" nodeid="9"/>
          <molhado:children-update newchildren="[12, 13, 14, 15, 16, 17, 18, 19, 20, 21, 22, 23, 24, 25, 26, 27, 28, 29, 30, 31, 37, 43]"
   nodeid="11" />
          <molhado:attr-update attr="style" newvalue="fill:#80ff80;fill-rule:evenodd;stroke:#000000;
   stroke-width:1.08585px;stroke-linecap:butt;stroke-linejoin:miter;stroke-opacity:1" nodeid="12" />
          <molhado:attr-update attr="d" newvalue="m 326.31384,304.8995 c 15.29596,54.01718 -42.60436,106.27838 -70.15238,138.91691 -26.84874,
   31.81 -73.69085,27.29921 -91.98482,27.29921 -61.12266,0 -97.068173,-26.78221 -117.733017,-56.87141 C 25.77878,384.15502 20.394602,
   350.75883 20.394602,337.53088 20.394602,269.23747 8.9028782,194.2 41.88246,163.28374 c 49.73638,-46.62469 95.8792,-31.59793 114.13633,
   -32.75822 64.18186,-4.07892 97.82282,47.99734 121.04715,81.34495 28.17395,40.45473 49.2479,79.80108 49.2479,93.02903 z" nodeid="15" />
          <molhado:attr-update attr="style" newvalue="fill:#40ed48;fill-opacity:1;fill-rule:evenodd;stroke:#000000;stroke-width:1px;
   stroke-linecap:butt;stroke-linejoin:miter;stroke-opacity:1"
   nodeid="15" />
       </molhado:revision>

❺    <Signature xmlns="http://www.w3.org/2000/09/xmldsig#"><SignedInfo>
          <CanonicalizationMethod Algorithm="http://www.w3.org/TR/2001/REC-xml-c14n-20010315"/>
          <SignatureMethod Algorithm="http://www.w3.org/2000/09/xmldsig#rsa-sha1"/><Reference URI="#revision-history"><Transforms>
          <Transform Algorithm="http://www.w3.org/2000/09/xmldsig#enveloped-signature"/></Transforms>
          <DigestMethod Algorithm="http://www.w3.org/2000/09/xmldsig#sha1"/><DigestValue>LlpISOZCGlaS2/L7vMb920tvMds=</DigestValue>
          </Reference></SignedInfo><SignatureValue>VAxbyxmW0+T9W3pzrz0d/wt2WDRMRJIN7luXyNp7vMQbO9MjOQxgLTaURg9hleQHXW9YPukefwzL
          3A6a534uRA==</SignatureValue><KeyInfo><KeyValue><RSAKeyValue>
          <Modulus>qSDvU7kLsj9H9niakcQuToqWDdjvo85EWTzBNdcZITB5Py3+0NYIAGfm+1uxbpfJpv5JyE1aWyt+YFQcxeGbgQ==</Modulus><Exponent>AQAB
          </Exponent></RSAKeyValue></KeyValue></KeyInfo>
       </Signature>

      </molhado:revision-history>

❻    Inkscape SVG document content (latest version)

   </svg>
```

Figure 2: This version-aware Inkscape document contains three revisions. Box 1 represents the entire version-aware document with molhado namespace defined in box 2. Box 3 contains the version data (box 4) and the version data signature (box 5). Box 6 contains the Inkscape document content which represents the latest revision.

3-way merge. Merge rules and conflicts detection are described in [22]. Our versioning framework can be supported by native application to make versioning more seamless or by third party tools to enable an application to support versioning. We show in the next section how an editing application can be "wrapped" by a version-aware application so that an end user can remain mostly unaware of how the versioning system works.

4. EVALUATION

To demonstrate that our approach is feasible, we implemented the XML versioning framework for XML documents and the tools to support the framework. We tested them by Inkscape [10], a popular open-source SVG editor and its native SVG file format. We sought to add versioning support to Inkscape without modifying Inkscape in any way. In order to capture the delta each time Inkscape changes the document, we created a Java program called *vinkscape* that is called to open an Inkscape SVG file, instead of calling Inkscape directly. When *vinkscape* is called, it starts Inkscape with the file specified by the user. Once Inkscape has modified the file and exited, *vinkscape* computes the delta between the modified content and the previous content and updates the version data within the Inkscape file. Note that the rendering and editing are done with Inkscape with the versioning data intact. The delta is then signed to ensure the integrity of the version data. Any version can be retrieved and viewed by calling *vinkscape <file> -rev <revision>*. If the revision argument is not provided by the user, *vinkscape* will load the latest version. Figure 3 demonstrates *vinkscape*. The version-aware SVG document *inkscape.logo.svg* has three versions. Each of the Inkscape window shows each of the versions contained in *inkscape.logo.svg*. The *vinkscape* command is executed in the terminal window which in turn loads the selected version in Inkscape.

5. CONCLUSION

We have described an approach, *version-aware XML documents*, in which the version data is stored within the working document. This introduces the notion that versioning is part of the document and collaborative editing does not have to revolve around shared storage or a centralized repository. Our framework can be used to extend a document type to

Figure 3: vinkscape: retrieval of revision 1, 2 and 3 from a version-aware Inkscape SVG document.

incorporate versioning. We have shown that this is feasible by extending Inkscape SVG files.

There are two ways to use the versioning framework. One is by modifying an application to support the versioning framework. A second is to create external tools to support an application that is not version-aware by capturing the changes as demonstrated in the Inkscape example.

For future work, we will include features allowing the user to delete versions, delete branches, and clear the version history. To optimize version retrieval performance, we plan to use full document snapshots to represent certain versions, instead of using deltas.

6. ACKNOWLEDGMENTS

This research is supported by a grant from the HP Labs Innovation Research Program.

7. REFERENCES

[1] R. Al-Ekram, A. Adma, and O. Baysal. diffx: an algorithm to detect changes in multi-version XML documents. In *Proceedings of the 2005 conference of the Centre for Advanced Studies on Collaborative research*, CASCON '05, pages 1–11. IBM Press, 2005.

[2] S. Y. Chien, V. J. Tsotras, and C. Zaniolo. Xml document versioning. *SIGMOD Rec.*, 30(3):46–53, September 2001.

[3] S.-Y. Chien, V. J. Tsotras, C. Zaniolo, and D. Zhang. Storing and querying multiversion XML documents using durable node numbers. In *Web Information Systems Engineering, 2001. Proceedings of the Second International Conference on*, volume 1, pages 232–241, dec. 2001.

[4] CVS-concurrent versions system. http://www.nongnu.org/cvs.

[5] K. Djemal, C. Soule-Dupuy, and N. Valles-Parlangeau. Management of document multistructurality: Case of document versions. In *Research Challenges in Information Science, 2009. RCIS 2009. Third International Conference on*, pages 325–332, april 2009.

[6] Dropbox - online backup, file synch, and sharing made easy. http://www.dropbox.com.

[7] N. Fousteris, Y. Stavrakas, and M. Gergatsoulis. Multidimensional xpath. In *Proceedings of the 10th International Conference on Information Integration and Web-based Applications & Services*, iiWAS '08, pages 162–169, New York, NY, USA, 2008. ACM.

[8] Git - fast version control system. http://git-scm.com.

[9] Google docs. http://www.google.com/google-d-s/documents.

[10] Inkscape. draw freely. http://inkscape.org.

[11] Java uuid generator (JUG) home page. http://jug.safehaus.org.

[12] K.-H. Lee, Y.-C. Choy, and S.-B. Cho. An efficient algorithm to compute differences between structured documents. *Knowledge and Data Engineering, IEEE Transactions on*, 16(8):965–979, aug. 2004.

[13] T. Lindholm, J. Kangasharju, and S. Tarkoma. Fast and simple XML tree differencing by sequence alignment. In *Proceedings of the 2006 ACM symposium on Document engineering*, DocEng '06, pages 75–84, New York, NY, USA, 2006. ACM.

[14] A. Marian. Detecting changes in XML documents. In *Proceedings of the 18th International Conference on Data Engineering*, ICDE '02, page 41, Washington, DC, USA, 2002. IEEE Computer Society.

[15] A. O. Mendelzon, F. Rizzolo, and A. Vaisman. Indexing temporal XML documents. In *Proceedings of the Thirtieth international conference on Very large data bases - Volume 30*, VLDB '04, pages 216–227. VLDB Endowment, 2004.

[16] Mercurial SCM. http://mercurial.selenic.com.

[17] G. Ozsoyoglu and R. T. Snodgrass. Temporal and real-time databases: A survey. *IEEE Trans. on Knowl. and Data Eng.*, 7(4):513–532, August 1995.

[18] S. Rönnau, G. Philipp, and U. M. Borghoff. Efficient change control of XML documents. In *Proceedings of the 9th ACM symposium on Document engineering*, DocEng '09, pages 3–12, New York, NY, USA, 2009. ACM.

[19] L. A. Rosado, A. P. Marquez, and M. S. Sanchez. A data model for versioned XML documents using xquery. In *Digital Information Management, 2008. ICDIM 2008. Third International Conference on*, pages 931–933, nov. 2008.

[20] R. T. Snodgrass. *The TSQL2 Temporal Query Language.* Kluwer Academic Publishers, Norwell, MA, USA, 1995.

[21] Subversion. http://subversion.tigris.org.

[22] C. Thao and E. V. Munson. Using versioned tree data structure, change detection and node identity for three-way XML merging. In *Proceedings of the 10th ACM symposium on Document engineering*, DocEng '10, pages 77–86, New York, NY, USA, 2010. ACM.

[23] W. F. Tichy. Rcs—a system for version control. *Softw. Pract. Exper.*, 15(7):637–654, July 1985.

[24] Z. Vagena, M. M. Moro, and V. J. Tsotras. Supporting branched versions on XML documents. In *Research Issues on Data Engineering: Web Services for e-Commerce and e-Government Applications, 2004. Proceedings. 14th International Workshop on*, pages 137–144, march 2004.

[25] J.-Y. Vion-Dury. Stand-alone encoding of document history(or one step beyond XML diff). In *Proceedings of Balisage: The Markup Conference 2010*, volume 5, 2010.

[26] R. K. Wong and N. Lam. Managing and querying multi-version XML data with update logging. In *Proceedings of the 2002 ACM symposium on Document engineering*, DocEng '02, pages 74–81, New York, NY, USA, 2002. ACM.

[27] Extensible markup language (XML). http://www.w3.org/XML/.

[28] XML security standard. http://www.w3.org/standards/xml/security.

[29] F. Zhang, X. Wang, and S. Ma. Temporal XML indexing based on suffix tree. In *Software Engineering Research, Management and Applications, 2009. SERA '09. 7th ACIS International Conference on*, pages 140–144, dec. 2009.

[30] Y. Zhang, X. Wang, and Y. Zhang. A labeling scheme for temporal XML. In *Web Information Systems and Mining, 2009. WISM 2009. International Conference on*, pages 277–279, nov. 2009.

Google's International Bloopers...
and How We Fixed One

Mark Davis
Google, Inc.
1600 Amphitheatre Parkway
Mountain View, CA 94043
markdavis@google.com

Abstract

Google has millions of users around the world...but occasionally we mess up. We'll explore some of Google's international "bloopers" and show how to avoid similar mistakes in other applications.

We'll also highlight how we solved a persistent blooper, namely having UI strings like "Alice has added 3 contact to his address book." Our Plural/Gender API allows complicated UI strings to change appropriately, based on numbers and personal gender. We'll look at the document formats used to express plural and gendered messages, and explore the impact on the translation process.

Categories & Subject Descriptors: I.2.7 [**Artificial Intelligence**]: Natural Language Processing—*Language models, Machine translation, Text analysis*; I.7.5 [**Document and Text Processing**]: Document Capture—*Document Analysis.*

General Terms: Design.

Keywords: Internationalization, Number agreement, Gender agreement.

Copyright is held by the author/owner(s).
DocEng'11, September 19–22, 2011, Mountain View, California, USA.
ACM 978-1-4503-0863-2/11/09.

Evaluating CRDTs for Real-time Document Editing*

Mehdi Ahmed-Nacer
Université de Lorraine
LORIA
54506 Vandœuvre-lès-Nancy,
France
mahmedna@loria.fr

Claudia-Lavinia Ignat
INRIA Nancy - Grand Est
LORIA
54600 Villers-lès-Nancy,
France
ignatcla@loria.fr

Gérald Oster
Université de Lorraine
LORIA
54506 Vandœuvre-lès-Nancy,
France
oster@loria.fr

Hyun-Gul Roh
INRIA Nancy - Grand Est
LORIA
54600 Villers-lès-Nancy,
France
roh@loria.fr

Pascal Urso
Université de Lorraine
LORIA
54506 Vandœuvre-lès-Nancy,
France
urso@loria.fr

ABSTRACT

Nowadays, real-time editing systems are catching on. Tools such as Etherpad or Google Docs enable multiple authors at dispersed locations to collaboratively write shared documents. In such systems, a replication mechanism is required to ensure consistency when merging concurrent changes performed on the same document. Current editing systems make use of operational transformation (OT), a traditional replication mechanism for concurrent document editing.

Recently, Commutative Replicated Data Types (CRDTs) were introduced as a new class of replication mechanisms whose concurrent operations are designed to be natively commutative. CRDTs, such as WOOT, Logoot, Treedoc, and RGAs, are expected to be substitutes of replication mechanisms in collaborative editing systems.

This paper demonstrates the suitability of CRDTs for real-time collaborative editing. To reflect the tendency of decentralised collaboration, which can resist censorship, tolerate failures, and let users have control over documents, we collected editing logs from real-time peer-to-peer collaborations. We present our experiment results obtained by replaying those editing logs on various CRDTs and an OT algorithm implemented in the same environment.

Categories and Subject Descriptors

I.7.1 [**Document and Text Processing**]: Document and Text Editing; D.2.8 [**Software Engineering**]: Metrics—*complexity measures, performance measures*; C.2.4 [**Computer-Communication Networks**]: Distributed Systems—*Distributed applications*

General Terms

Algorithms, Experimentation, Performance

Keywords

Real-time Editing, Collaboration, Benchmark, Commutative Replicated Data Types

1. INTRODUCTION

Collaboration is a very important aspect of any team activity and hence of importance to any organisation such as business, science, education, administration, political or social institutes. Central to collaboration is editing of shared documents. Collaborative editing has to take into account the geographical distribution of team members, possibly across a wide range of time zones, together with the mobility of individuals. Collaborative systems require that users be able to edit shared documents as easily as one edits a single-author document. The major benefits of collaborative editing include reducing task completion time, reducing errors, getting different viewpoints and skills, and obtaining an accurate text [13, 27].

The nature of collaboration varies extensively in terms of strategies and proximity of writing groups [27]. For instance, there exist various writing strategies; users can jointly write a document by working closely together; or they can work separately, and afterwards their works are subject to review by other group members. According to proximity, all members of some groups can work in the same location and on the same time schedule, while other groups work on different schedules and may be located thousands of miles apart.

Depending on such strategies and proximities, the collaboration between users can be synchronous or asynchronous. In asynchronous collaboration, members of the group modify the copies of the documents in isolation, synchronising afterwards their copies to re-establish a common view of the data.

*This work is partially funded by the french national research programs ConcoRDanT (ANR-10-BLAN-0208) and STREAMS (ANR-10-SEGI-010).

Synchronous or real-time collaboration means that modifications by a member are immediately seen by other members of the group. Such real-time and synchronous features could be very helpful; for example, when a paper deadline is imminent, authors of a paper can avoid the time consumed for synchronisation of the copied documents and can edit their parts, being aware of the changes made by the other members.

Recently, real-time collaborative editing gained much attention as Google Docs and Google Wave support such features. It is also a main subject of investigation by recent user studies [20] that focus on user behaviour during real-time collaboration. This paper focuses on consistency maintenance algorithms for real-time collaborative editing.

Data replication is necessary in collaborative editing to achieve high responsiveness. Many mechanisms dealing with data replication have been introduced. In the domains of replicated databases, pessimistic approaches are usually used, which gives users the illusion of a single copy [1]. High availability can be obtained by allowing read operations on any replica, while update operations have to be atomically applied on all replicas. Pessimistic approaches cannot ensure high responsiveness for real-time collaboration because the initiator of an update should acquire an exclusive access, and an update cannot be applied immediately before resolving conflicts.

Optimistic replication [19] is a family of approaches suitable for real-time collaboration. Compared to pessimistic approaches, optimistic replication needs no atomic update. Though replicas are allowed to diverge temporarily, they are expected to converge eventually.

Optimistic synchronisation algorithms are classified into state-based and operation-based. State-based approaches use only information about the different states of the documents and no information about the evolution of one state into another. Examples of state-based approaches are three-way merges adopted by version control systems such as Subversion [3] and differential synchronisation approaches [5]. On the other hand, operation-based merging approaches keep the information about the evolution of one state of the document into another in a buffer. The merging is done by executing the operations performed on a copy of the document onto the other copy of the document to be merged. State-based approaches are generally not suitable to be used in the real-time communication as the difference between versions has to be computed each time an operation is performed resulting in an increased time complexity. In this paper we study operation-based optimistic synchronisation approaches.

Operational transformation (OT) approach has been identified as an appropriate operation-based synchronisation mechanism for real-time collaboration [4, 17, 25, 12, 24, 29, 26]. Modifications are represented by means of operations such as insertion and deletion of a character in the case of textual documents. OT changes the index of an operation based on the history of operations in order to take into account the effects of concurrent operations and to eventually achieve consistency. A few commercial document editing systems such as Google Wave and Google Docs adopt centralised OT algorithms, such as Jupiter [12]. However, the centralised approaches may store not only shared documents and all changes, but also some personal information, which could

be a privacy threat, in the hands of a single large corporation.

In order to overcome the disadvantages of central authority, i.e., to let users have control over their documents, to resist censorship, and to tolerate failures, a tendency is to move towards decentralised peer-to-peer collaboration [2]. This solution is suitable for user communities that rely on sharing their infrastructures and administration costs. An example of this tendency in the domain of real-time collaborative editing is Microsoft SharePoint Workspace 2010 [22] (previously known as Groove). This software allows documents of desktop office applications to be synchronised through a peer-to-peer network of SharePoint servers while managing security of local copy. However, most decentralised OT algorithms use version vectors [24, 26] or central timestamps [29] to detect concurrent operations which do not scale well in cloud and peer-to-peer environments with dynamic groups.

Performance is a key factor of the success of real-time collaborative applications. If these applications cannot quickly respond to user actions, users may get frustrated and will quit the application. Studies in [23, 8] found that people can comfortably notice changes which respond to local and remote actions within $50\,ms$.

Very recently, a new class of algorithms called CRDT (commutative replicated data types) [15, 31, 16, 18] that ensures consistency of highly dynamic contents on peer-to-peer networks emerged. Unlike OT algorithms, CRDTs require no history of operations, and no detection of concurrency in order to ensure consistency. Instead, they are designed for concurrent operations to be natively commutative by actively using the characteristics of abstract data types such as lists or ordered trees. However, CRDT algorithms have not yet been applied for real-time collaborative editing.

In this paper, we investigate the suitability of CRDT algorithms by performing evaluations of CRDT algorithms against real editing logs of real-time peer-to-peer collaboration. The results we obtained show that CRDT algorithms are suitable for real-time collaboration and that they outperform a representative class of OT approaches.

The paper is structured as follows. In the next section, we start by giving an overview of main CRDTs and OT algorithms, and study their theoretical complexity. Then, we describe how we designed real-time peer-to-peer collaborations from which the editing logs are collected, and provide some details on the logs. We then provide evaluation results of the analysed CRDTs and OT algorithms by replaying the collected logs. We also compare our work with existing experimental results of related work. In the last section we provide concluding remarks and directions of future work.

2. STUDIED REAL-TIME DOCUMENT EDITING ALGORITHMS

In the nineties, various OT algorithms have been proposed [4, 17, 25, 12, 24, 29, 26, 10]. They constitute the groundwork of CRDT approaches that emerged in the last decade: WOOT [15], Treedoc[16], Logoot[31], RGA [18], partial persistent data structure (PPDS) [33] and causal tree model [6] were presented.

In the following subsections, we briefly describe some representative algorithms belonging to the families of CRDT and OT approaches. We also provide their theoretical complexities.

2.1 WOOT Algorithm

WOOT [15] is the first CRDT algorithm which was proposed. Operations used by this algorithm are insertion and deletion of elements in a linear structure. Elements are uniquely identified. An insertion is defined by specifying the new element identifier, the element content and the identifiers of the preceding and following elements. Concurrent operations determine partial orders between elements. The merging mechanism can be seen as a linearisation of the partial order to obtain a total order. To obtain convergence, the total order has to be the same at all peers. As operations are integrated in any order at a site, merging has to be computed incrementally and independently of the order of arrival. WOOTO [30] is an optimisation of WOOT that uses element degree to compare unordered elements.

The advantage of these algorithms is their suitability for open user groups where users often join and leave the network. Moreover they do not require a causal delivery of operations. A disadvantage is that the algorithms use tombstones, i.e. elements are not physically deleted but only marked as deleted. Since tombstones cannot be removed without compromising consistency, performance degrades during time.

We designed a new version of WOOTO, called WOOTH, inspired by RGA approach. A hash table and a linked list are used for optimising retrieval, update and insertion of an element.

2.2 Logoot Algorithm

Logoot [31] is another CRDT approach that ensures consistency of linear structures. Logoot associates to the list of elements of the structure, an ordered list of identifiers. Identifiers are composed by a list of positions. Positions are 3-tuples formed with a digit in specific numeric base, a unique site identifier and a clock value. When inserting an element, Logoot generates a new identifier. Identifiers have unbounded lengths and are totally ordered by a lexicographic order. So a new identifier can always be generated between two consecutive elements. Different strategies can be adopted to produce the new identifier [32], all of them using randomness to prevent different replicas to produce concurrently close identifiers.

The advantage of Logoot is the absence of tombstones and an improvement of the algorithmic complexity compared to WOOT/WOOTO/WOOTH. A disadvantage is the size of the identifier that can grow unbounded. However, experiments made on collaboration traces produced on Wikipedia shown that the size of identifiers stays low even for largest collaborative contributions [32].

2.3 Replicated Growable Array (RGA)

Replicated growable array [18] is a CRDT that supports not only insertion and deletion but also update operations which replace the content of an element without changing the size of the document. In this paper, we evaluate only insertion and deletion operations. RGAs maintain a linked list of elements, via which local operations find their target elements with integer indexes. Meanwhile, a remote operation retrieves its target element via a hash table with a unique index of the target.

As a unique index, the paper [18] introduces s4vector consisting of four integers. A unique s4vector is issued with every operation, and the oldness or newness of multiple s4vectors can be determined transitively, respecting causality. Using the properties of uniqueness and transitivity, the s4vector associated with the insertion that creates an element is used as a unique index of the element, which is also used to resolve conflicts between concurrent insertions at the same position. An insertion compares its s4vector with the s4vector identifiers of elements next to its target element, and adds its new element in front of the first encountered element that has an older s4vector. That is, a newer insertion inserts an element closely to its target position with higher precedence than relatively order concurrent insertions. Such precedence transitivity, realised with s4vectors, ensures consistency of concurrent insertions. As some other CRDTs, RGAs also uses tombstones for deleted elements. Tombstones should be preserved as long as they can be accessed by other remote operations.

2.4 Operational Transformation Algorithms

Most representative OT algorithms that do not make any assumption on using a central server for a total order broadcast of operations are SOCT2 [24] and GOTO [25] algorithms. The principle of this class of algorithms is that when a causally ready operation is integrated at a site, the whole log of operations is traversed and reordered. After reordering, causally preceding operations come before concurrent ones in the history buffer. Finally, the remote operation has to be transformed according to the sequence of concurrent operations. In order to reduce the complexity of the integration mechanism of a remote operation, history buffer is pruned by using a garbage collection mechanism as proposed in [9, 26]. Replicas use this mechanism to remove operations they know to be received by all other replicas. However, the garbage collection mechanism has no effect in open peer-to-peer networks were users often join and leave.

These algorithms require transformation functions satisfying conditions C_1 and C_2 [17, 24]. Satisfying C_1 allows executing in any order two concurrent operations defined on the same document state while ensuring convergence of the document. C_1 is sufficient with only two sites or in client-server applications. C_2 expresses the equality between an operation transformed against two equivalent sequences of operations. These two conditions ensure that transforming any operation with any two sequences of the same set of concurrent operations in different execution orders always yields the same result. C_1 and C_2 are sufficient for ensuring convergence in a peer-to-peer architecture. In [7], it was shown that many proposed transformation functions fail to satisfy these conditions. The only existing transformation functions that satisfy conditions C_1 and C_2 are the ones proposed by the TTF (Tombstone Transformation Functions) approach [14]. In the TTF approach when a deletion of a character is performed, the character is not physically removed from the document, but just marked for deletion, i.e. deleted characters are replaced by tombstones.

2.5 Theoretical Evaluation

The worst case complexity for each of the above described algorithms is presented in Table 1. We consider the time complexity of generation of a local user operation (single character insert or delete) and for the execution of a remote operation. We denote by R the number of replicas and by H the number of operations that had affected the document. We consider constant time for accessing an element in a hash table. In the worst case scenario for the approaches that use tombstones, the document size including tombstones equals H. For the approaches that use state vectors we took into account the complexity of state vector creation, i.e. $O(R)$, associated with the operation at the moment of its generation.

ALGORITHM	LOCAL		REMOTE	
	INS	DEL	INS	DEL
WOOT	$O(H^3)$	$O(H)$	$O(H^3)$	$O(H)$
WOOTO	$O(H^2)$	$O(H)$	$O(H^2)$	$O(H)$
WOOTH	$O(H^2)$	$O(H)$	$O(H^2)$	$O(log(H))$
Logoot	$O(H)$	$O(1)$	$O(H.log(H))$	$O(H.log(H))$
RGA	$O(H)$	$O(H)$	$O(H)$	$O(log(H))$
SOCT2/TTF	$O(H+R)$	$O(H+R)$	$O(H^2)$	$O(H^2)$

Table 1: Worst-case time-complexity analysis

The average complexity of each of the above described algorithms is presented in Table 2. We denote by:

c the average number of operations concurrent to a given one,

n the size of the document (non deleted characters),

N the total number of inserted characters (including the ones that were deleted called tombstones),

k the average size of Logoot identifier[1].

$t = N - n$ the number of tombstones,

$d = \lceil (t+c)/n \rceil$ the average number of elements (tombstones and concurrently inserted elements) found between to successive document elements.

Algorithms using tombstones (WOOTs, RGA and TTF) have a complexity depending on N for retrieving an element or a document position in their model. WOOT algorithms have a complexity proportional to d^2 since they call a recursive algorithm to place a newly inserted element between these $O(d)$ elements. RGA algorithm compares a remote inserted element only with the elements inserted concurrently at the same position (c/n in average). The SOCT2 algorithm reorders each operation of the log against $O(c)$ concurrent ones. With an efficient garbage collection mechanism, there are $O(c)$ operations in the SOCT2 log.

ALGORITHM	AVG. LOCAL		AVG. REMOTE	
	INS	DEL	INS	DEL
WOOT	$O(N.d^2)$	$O(N)$	$O(N.d^2)$	$O(N)$
WOOTO	$O(N.d^2)$	$O(N)$	$O(N+d^2)$	$O(N)$
WOOTH	$O(N+d^2)$	$O(N)$	$O(d^2)$	$O(1)$
Logoot	$O(k)$	$O(1)$	$O(k.log(n))$	$O(k.log(n))$
RGA	$O(N)$	$O(N)$	$O(1+c/n)$	$O(1)$
SOCT2/TTF	$O(N+R)$	$O(N+R)$	$O(H.c)$	$O(H.c)$
with g.c.	$O(N+R)$	$O(N+R)$	$O(c^2)$	$O(c^2)$

Table 2: Average time-complexity analysis

The space complexity of meta-data used by each replica is presented in Table 3. In average, algorithms using tomb-

[1] Theoretically, the size of a Logoot identifier is only bounded by H, but due to stochastic nature of Logoot identifier generation, it has only an infinitesimal chance to be proportional to H.

stones need to store N elements in their model. Logoot stores n identifiers with an average size of $O(k)$. SOCT2 additionally stores a log of operations, each one containing a version vector with size of $O(R)$.

ALGORITHM	SPACE COMPLEXITY	
	WORST	AVERAGE
WOOT-WOOTO-WOOTH	$O(H)$	$O(N)$
Logoot	$O(H^2)$	$O(k.n)$
RGA	$O(H)$	$O(N)$
SOCT2/TTF	$O(H.R)$	$O(H.R)$
SOCT2/TTF with g.c.	$O(H.R)$	$O(N+c.R)$

Table 3: Space complexity analysis of meta-data

3. OBTAINING LOGS FROM REAL-TIME P2P COLLABORATION

Currently, some commercial real-time collaboration systems such as Google Docs are on service, but their logs are not complete and freely available. For example, the revision log provided by Google server is a serialisation of user operations transformed by the Jupiter algorithm [12]. Therefore, the revision logs open to the public do not include the information needed for replaying the real-time peer-to-peer collaboration, such as version vectors.

Due to unavailability of logs of the real-time peer-to-peer collaboration, we set up an experiment where we asked students to collaboratively write documents by using a collaborative editor and logged a number of operations generated during this experiment.

3.1 Design of Real-time P2P Collaborations

We designed real-time peer-to-peer collaborations in order to obtain their logs. In this section, we describe the collaboration design and some features of the obtained operation logs such as their lengths and operation types. TeamEdit [28] is the real-time collaborative editor used in our collaborations. We modified it in order to log user operations performed during the collaboration. TeamEdit software uses a central server, but only to establish communication between the different sites. It does not serialise the operations, nor uses the server as a merge mechanism for concurrent operations. Students could use the undo/redo feature we added to TeamEdit, but, however, these operations were transformed into the corresponding insert/delete operations.

We performed two collaborations with groups of students: report and series. The first collaboration (report) was performed with 13 master students divided into three groups: two groups composed of 4 students and one group composed of 5 students. Each group was asked to collaboratively write a report of its semester project in the Software Engineering lecture. Each student in a group worked on the report from a private computer, and was not allowed to use any other communication tools except TeamEdit. The collaboration lasted for one and a half hours. Collaboration was encouraged by noticing that each student will be evaluated according not only to the content of the report but also to the size of his/her contribution. Students were allowed to copy-paste text blocks from some other documents.

In our second collaboration (series) that lasted for about one and a half hours, we asked 18 students to watch an episode of the series "The Big Bang Theory" and to produce a transcription of the episode while watching it. The transcription of the episode was edited into a shared document with TeamEdit editor. Students were divided into nine groups of two, and each group was asked to make a transcription of a certain hero or to describe the environment and actions that happened during the episode. During this collaboration, students were allowed to communicate mutually. Each student had his/her own computer for editing the shared document. The same episode was played twice, and we assigned different groups to each task in order to obtain more operations from the collaboration.

To minimise internal threats to validity, the whole experiment was conducted in the same period (during one morning), with the same working stations, and with students not aware of our research and non experts in real-time collaborative editing. Of course, this experiment only captures a subset of all the behaviors observable when users collaborate. This selection bias can only be reduced with access to the internal logs of a public widely used real-time editor. Another threat to validity is that the experiment was conducted on a local area network. Thus, propagation time between user desktops was shorter than in a wide area network, leading to a slightly lower concurrency degree between operations. Thus, one can expect sightly lower performance for all studied algorithms in wide area networks.

3.2 Description of Collaboration Logs

TeamEdit was modified to log the following user operations: insertions of a text block and deletions of a range of characters. Text blocks and ranges have a size of one character when a user types on the keyboard. They have a larger size when a user copy-pastes a text block or deletes a selected block. When an operation is generated it has associated a version vector that indicates the number of operations received by the generating site from each of the other sites. In order to apply the studied algorithms on the generated logs, user operations must be transformed into character operations. In the Table 4, the total numbers of user operations and character operations are specified for all the collaborations, the report and the series.

	Report			Series	
	Group 1	Group 2	Group 3	Doc 1	Doc 2
No. user operations	11 211	11 066	13 702	9 042	9 828
No. char. operations	26 956	47 992	42 443	29 882	10 268
% of del	12	12	12	9	5

Table 4: Total number of user/character operations

The proportion of delete operations is smaller in the series experiments due to the difficulty for non-specialists to type a transcript as quickly as actors talk. So the students had less time to make corrections on their document.

We also observed that, without instructions from us, some students disconnected and then reconnected to the real-time collaboration session during almost all collaborations.

4. EXPERIMENTAL EVALUATION

To evaluate algorithms performance, we designed a framework called ReplicationBenchmark in Java, and reveal the source on GitHub platform[2] under the terms of the GPL license.

The framework provides base classes for common elements of real-time document editing algorithms, such as document, operation, generation and integration algorithms, and version vector, so that each algorithm can be implemented by inheriting them. The framework lets replicas of every algorithm generate character operations in its own formats for the given user operations in the logs; we first measured this generation and local execution time of user operations. The framework also provides a dispatcher that enables each replica to receive generated character operations in the same order as that in the logs. A replica, therefore, can execute character operations, enabling measurement of the net execution time of character operations in each algorithm. The framework uses java.lang.System.nanoTime() for the measurement of execution time of each user operation and each character operation.

To obtain the presented results, we ran each algorithm on each log twenty times on the same JVM execution. The heap size was 1GB. We compute the average execution time per replica for each user operation and for each character operation. From the obtained results, we remove the aberrant values due to java garbage collection, i.e. values more than twice the average for a given operation.

We ran our experiments on a dual processor machine with Intel(R) Xeon(R) 5160 dual-core processor (4Mb Cache, 3.00 GHz, 1333 MHz FSB), that has installed GNU/Linux 2.6.9-5. During the experiment, only one processor was used. We present no results for the WOOT algorithm since it is obviously outperformed by its optimised versions WOOTO and WOOTH algorithms.

4.1 Execution times

4.1.1 User operations

The execution times, in microseconds, of user operations including generation of the corresponding character operations are presented in Table 5. We present the average and the maximum response time, and the standard deviation σ[3].

During execution of an operation the user interface of a document editor is frozen and the user is prevented from typing. If this value is greater than the $50\,ms$, users will notice the bad performance of the collaborative application. The average values measured and presented in Table 5 give the impression that all algorithms performed very well since no average value is greater than $0.2\,ms$. However, if we consider $50ms$ as a limit for the maximum response time, the algorithms belonging to the WOOT family cannot be used safely to build a real-time collaborative applications since the maximum value, confirmed by a high standard deviation, is often greater than $50\,ms$.

The maximum execution time, for every experiment and algorithm, is almost always due to an operation inserting or suppressing a block of hundreds, or thousands, of characters. The only experiment where all algorithms perform very well is the second series since there is no such user operation.

4.1.2 Character operations

The average execution time of character operations are

---

[2] http://github.com/PascalUrso/ReplicationBenchmark
[3] σ computed using the classical formula $\sqrt{\frac{\sum_{i=1}^{N}(x_i-x_{av})^2}{N}}$.

		REPORT			SERIES	
		GROUP 1	GROUP 2	GROUP 3	DOC 1	DOC 2
LOGOOT	AVG	6	7	7	5	5
	MAX	751	901	2 322	2 267	77
	σ	10	18	26	30	3
WOOTH	AVG	26	43	49	46	16
	MAX	3 623	40 042	**156 407**	164 934	453
	σ	44	464	1 396	1 735	16
WOOTO	AVG	43	112	96	110	23
	MAX	13 489	**208 985**	340 068	494 030	162
	σ	207	2 948	3 331	5 388	16
SOCT2	AVG	21	40	30	27	19
	MAX	5 753	24 741	15 312	8 912	147
	σ	85	389	190	119	16
RGA	AVG	27	32	32	20	17
	MAX	998	2 082	1 971	2 671	550
	σ	32	46	38	52	19

Table 5: User operation execution times (in μs)

presented in Table 6. This value represents the computation time needed to integrate a remote incoming character operation into the current document.

		REPORT			SERIES	
		GROUP 1	GROUP 2	GROUP 3	DOC 1	DOC 2
LOGOOT	AVG	5	6	5	3	4
	MAX	91	127	110	80	36
	σ	2	3	3	2	3
WOOTH	AVG	2	4	7	8	2
	MAX	694	44 190	4 330	567	200
	σ	8	211	24	17	3
WOOTO	AVG	54	97	99	84	25
	MAX	2 027	**72 937**	8 903	1 334	133
	σ	51	352	108	71	18
SOCT2	AVG	80	573	130	**1 383**	305
	MAX	5 286	13 278	5 832	20 727	2 848
	σ	133	1 087	175	1 974	397
RGA	AVG	2	2	2	1	2
	MAX	750	1 295	1 002	403	252
	σ	5	7	7	3	3

Table 6: Character operation execution times (in μs)

Since one user operation corresponds to one or more character operations (up to 5000 characters for the biggest copy-paste), we expect that each algorithm performs better for character operations. This is actually the case for almost all maximum execution times. However, the WOOTO algorithm still has a maximum execution time higher than $50\,ms$.

The average character operation execution times are much better than user ones for WOOTH and RGA algorithms (due to hash table usage) and similar for Logoot and WOOTO. The case of SOCT2 is different. SOCT2 has a low maximum execution time but an average execution time that exceeds $1\,ms$. Let us consider a user that copy-pastes a block of 5000 characters, as seen in our experiments. This user operation is translated into 5000 character operations that will be executed one by one on the document replicas of other users. It means that the other users will see the characters

of the inserted block appearing one by one in a total duration of five seconds. This is not acceptable from a user's point of view. This performance issues are mainly due to the transformation algorithm coupled with the inefficiency of the garbage collection mechanism as seen in the behavior study described in the next subsection.

Finally, concerning character operations execution times, Logoot algorithm has the best maximum execution times (MAX values in Table 6) and RGA has the best average for every experiment.

4.1.3 Consistency with theoretical evaluation

The average results obtained for user and character operations are consistent with the average time-complexities presented in table 2. For local user operations, Logoot has the best results with $O(k)/O(1)$, WOOTO has the worst results with $O(N.d^2)/O(N)$ while RGA, WOOTH and SOCT2 have medium results with complexities around $O(N)$. For remote character operations, RGA and WOOTH have the best complexities and best results, followed by Logoot.

The worst case complexities are also consistent with the results obtained. WOOTO and WOOTH have the worst maximum result for user operations with $O(H^2)/O(H)$. SOCT2 and WOOTO have the worst maximum results for character operations with $O(H^2)$. These correlations validate our implementation of the algorithms for average and worst cases.

4.2 Behaviour

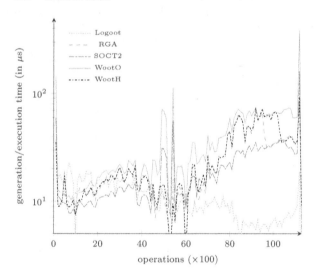

Figure 1: User operation execution times - 1st group report

Since performance of all studied algorithms may degrade over time due to tombstones or growing identifiers, we present in this subsection performance behaviour over time.

The observed behaviour was approximately the same within the two collaboration categories (report and series). We therefore present here only a selection of the comparisons we obtained for certain collaboration logs.

In order to obtain meaningful representations of algorithms behaviour we computed an average of the local generation and respectively execution times for every hundred user operations and every three hundred character operations. The horizontal axis uses a linear scale representing the number

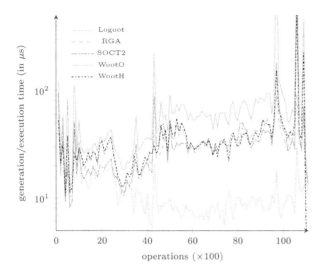

Figure 2: User operation execution times - 2nd group report

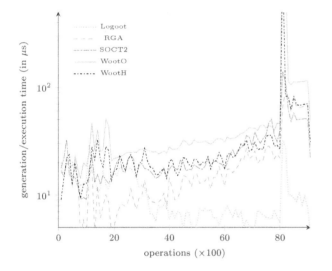

Figure 3: User operation execution times - 1st series

of elapsed operations. The vertical axis uses a logarithmic scale and represents the average time, in microseconds, required to execute operations.

4.2.1 User operations

In Figure 1, 2 and 3 we present algorithms' behaviour for execution times of user operation for the first and second project report and respectively for the first series.

Accordingly to their average time-complexity, the performances of the algorithms using tombstones (WOOTs, RGA and SOCT2) eventually degrade over time. Indeed, all these algorithms have to count the number of tombstones and characters present before an inserted or suppressed string. Temporary improvement of the performance of these algorithms, for instance in first report between the 40th and 60th block of operations, are due to a period where users mostly edit at the beginning of the document. This is also the case at the very end of the three report experiments due to instructions given to students to sign the document by

typing their names at the very beginning of the document. The performance of Logoot remains good during all experiments. Indeed, the average size of Logoot identifier stays very low even for large documents as demonstrated in [32].

For every experiment, peaks of low performance common to all algorithms exist, for instance in second report for the 43th block of operations. Such peaks are all due to operation inserting a block of hundreds, or thousands, of characters.

Finally, the global performance behaviors of RGA, WOOTH and SOCT2 algorithms are quite similar (especially for 2nd report), even if they are very different algorithms. This similarity is less obvious in table 5. Such an observation, leads to state the conjecture that any algorithm counting tombstones will have, at best, similar performances.

4.2.2 Character operations

In Figure 4, 5 and 6 we present execution time behaviours for character operations for the second and third project report and respectively for the second series.

During these experiments, their is no peak of low performance common to all the algorithm as for user operation. So, their is no character operation that represent the worst case for all the algorithm. Only WOOTH and WOOTO have such common peaks, for instance in the second report for the 126th block of operation, due to the similar nature of these algorithms.

We can notice that the performance remains stable for Logoot. The performance of RGA and WOOTH are, in average, better than Logoot but have a more erratic behavior for the reports. The behavior of RGA and WOOTH is composed of a base line at 1 μs and some lower performance periods due to more frequent concurrent editing. RGA overperform WOOTH in case of concurrent delete operation.

The performances of WOOTO and SOCT2 degrade over time since they are around ten times slower at the end of the experiments than at the beginning. SOCT2 have the most erratic behavior and the worst average performance.

The behavior of SOCT2 performance is mainly due to its garbage collection mechanism. In our experiments, some users had a period of inactivity between performing two successive modifications. Their inactivity implies that other users do not receive any information regarding the progress of the document state of this inactive user and therefore the garbage mechanism cannot purge the history log. The same situation happens when users left the editing session without notification. It is well-known that pruning history in peer-to-peer networks by using a garbage collection is impossible in these situations. One the over hand, when a user inactive since long produces an operation the performance of SOCT2 temporally improves, for instance in the third report for the 87th block of operation.

The experiments on the series have a slightly different behavior since they contain few deletes and a lot of concurrent operations. Indeed, the students don't have time to type as fast as actor talks.

5. RELATED WORK

Although OT algorithms have been studied since 1989, the first performance report was published in 2006 on the analysis of SDT and ABT algorithms [10]. Performance of the improved versions of these algorithms was published in [11] [21]. In [18] an evaluation of RGA approach was provided. However, these algorithms were not compared with

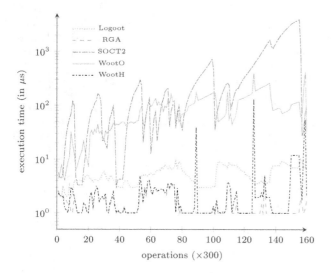

Figure 4: **Character operation execution times - 2nd group report**

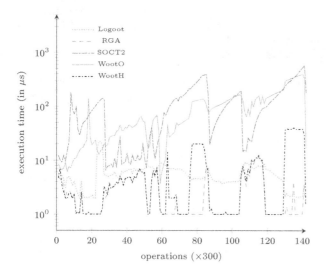

Figure 5: **Character operation execution times - 3rd group report**

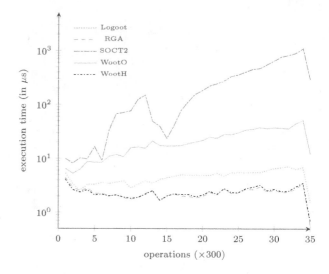

Figure 6: **Character operation execution times - 2nd series**

other existing ones. Moreover, performance was measured by using simulated data and not real collaboration traces.

CRDTs such as Treedoc [16] and Logoot[31, 32] presented experimental results, but they were focused not on performance, but on the overhead incurred by tombstones or meta data. PPDS approach [33] presented performance evaluation on their own algorithm without comparing it with other algorithms. All the above mentioned CRDT approaches [16, 31, 32, 33] were evaluated by using Wikipedia and/or Subversion collaboration traces. However, Wikipedia and Subversion traces represent a serialisation of user operations where conflicts between concurrent changes were already resolved by the users.

In this regard, this is the first paper that presents the comparison of real-time document editing algorithms written in the same language and environment. Moreover, this is the first work that evaluates algorithms by using collaboration traces including concurrency and generated during real-time collaborative editing.

6. CONCLUSION AND FUTURE WORK

In this paper, we evaluated representative consistency maintenance algorithms for real-time collaboration. We provided a theoretical evaluation as well as an experimental one against traces of real-time collaborative editing. We found out that CRDT algorithms initially designed for peer-to-peer asynchronous collaboration are suitable for real-time collaboration. Moreover, they outperform some representative operational transformation approaches that were well established for real-time collaboration in terms of local generation time and remote integration time. As an example, in average case, Logoot and RGA algorithms outperform between 25 and 1000 times faster than SOCT2 OT algorithm. We can also notice that the results we obtained are conforming to the worst case and average theoretical complexities. Best overall performances are obtained by RGA and Logoot algorithms.

One of our directions for future work is to extend our study to other operational transformation and CRDT algorithms and to study other evaluation criteria such as memory occupation, communication complexity and convergence latency. We also plan to obtain larger size traces of real-time collaborations and to generate automatically traces that have the same characteristics as real traces. In this paper we considered and compared decentralised algorithms for real-time collaborative editing. We plan to extend our study to centralised real-time collaborative editing and to analyse suitability of CRDT approaches for this kind of collaboration.

7. REFERENCES

[1] P. A. Bernstein, V. Hadzilacos, and N. Goodman. *Concurrency Control and Recovery in Database Systems.* Addison-Wesley, Boston, MA, USA, 1987.
[2] S. Buchegger, D. Schiöberg, L. H. Vu, and A. Datta. PeerSoN: P2P Social Networking - Early Experiences and Insights. In *Proceedings of the Second ACM*

EuroSys Workshop on Social Network Systems - SNS 2009, pages 46–52, Nürnberg, Germany, March 2009. ACM Press.

[3] B. Collins-Sussman, B. W. Fitzpatrick, and C. M. Pilato. *Version control with Subversion*. O'Reilly & Associates, Inc., 2004.

[4] C. A. Ellis and S. J. Gibbs. Concurrency Control in Groupware Systems. *SIGMOD Record : Proceedings of the ACM SIGMOD Conference on the Management of Data - SIGMOD '89*, 18(2):399–407, May 1989.

[5] N. Fraser. Differential Synchronization. In *Proceedings of the 9th ACM Symposium on Document engineering - DocEng 2009*, pages 13–20, Munich, Germany, September 2009. ACM Press.

[6] V. Grishchenko. Deep Hypertext with Embedded Revision Control Implemented in Regular Expressions. In *Proceedings of the 6th International Symposium on Wikis and Open Collaboration - WikiSym 2010*, pages 1–10, Gdańsk, Poland, July 2010. ACM Press.

[7] A. Imine, P. Molli, G. Oster, and M. Rusinowitch. Proving Correctness of Transformation Functions in Real-Time Groupware. In *Proceedings of the European Conference on Computer-Supported Cooperative Work - ECSCW 2003*, pages 277–293, Helsinki, Finland, September 2003. Kluwer Academic Publishers.

[8] C. Jay, M. Glencross, and R. Hubbold. Modeling the Effects of Delayed Haptic and Visual Feedback in a Collaborative Virtual Environment. *ACM Transactions on Computer-Human Interaction*, 14(2), August 2007.

[9] P. R. Johnson and R. H. Thomas. Maintenance of Duplicate Databases. RFC 677, Internet Engineering Task Force, January 1975. http://www.ietf.org/rfc/rfc677.txt.

[10] D. Li and R. Li. A Performance Study of Group Editing Algorithms. In *Proceedings of the International Conference on Parallel and Distributed Systems - ICPADS 2006*, pages 300–307, Minneapolis, MN, USA, July 2006. IEEE Computer Society.

[11] D. Li and R. Li. An Operational Transformation Algorithm and Performance Evaluation. *Computer Supported Cooperative Work*, 17(5-6):469–508, December 2008.

[12] D. A. Nichols, P. Curtis, M. Dixon, and J. Lamping. High-latency, Low-bandwidth Windowing in the Jupiter Collaboration System. In *Proceedings of the 8th Annual ACM Symposium on User interface and Software Technology - UIST '95*, pages 111–120, Pittsburgh, PA, USA, November 1995. ACM Press.

[13] S. Noël and J.-M. Robert. Empirical Study on Collaborative Writing: What Do Co-authors Do, Use, and Like? *Computer Supported Cooperative Work*, 13(1):63–89, 2004.

[14] G. Oster, P. Molli, P. Urso, and A. Imine. Tombstone Transformation Functions for Ensuring Consistency in Collaborative Editing Systems. In *Proceedings of the International Conference on Collaborative Computing: Networking, Applications and Worksharing - CollaborateCom 2006*, pages 1–10, Atlanta, GA, USA, November 2006. IEEE Computer Society.

[15] G. Oster, P. Urso, P. Molli, and A. Imine. Data Consistency for P2P Collaborative Editing. In

Proceedings of the ACM Conference on Computer-Supported Cooperative Work - CSCW 2006, pages 259–267, Banff, AB, Canada, November 2006. ACM Press.

[16] N. Preguiça, J. M. Marquès, M. Shapiro, and M. Letia. A Commutative Replicated Data Type for Cooperative Editing. In *Proceedings of the 29th International Conference on Distributed Computing Systems - ICDCS 2009*, pages 395–403, Montreal, QC, Canada, June 2009. IEEE Computer Society.

[17] M. Ressel, D. Nitsche-Ruhland, and R. Gunzenhäuser. An Integrating, Transformation-Oriented Approach to Concurrency Control and Undo in Group Editors. In *Proceedings of the ACM Conference on Computer-Supported Cooperative Work - CSCW '96*, pages 288–297, Boston, MA, USA, November 1996. ACM Press.

[18] H.-G. Roh, M. Jeon, J. Kim, and J. Lee. Replicated Abstract Data Types: Building Blocks for Collaborative Applications. *Journal of Parallel and Distributed Computing*, 71(3):354–368, 2011.

[19] Y. Saito and M. Shapiro. Optimistic Replication. *ACM Computing Surveys*, 37(1):42–81, March 2005.

[20] L. Scissors, N. S. Shami, T. Ishihara, S. Rohall, and S. Saito. Real-Time Collaborative Editing Behavior in USA and Japanese Distributed Teams. In *Proceedings of the ACM International Conference on Human Factors in Computing Systems - CHI 2011*, pages 1119–1128, Vancouver, BC, Canada, May 2011. ACM Press.

[21] B. Shao, D. Li, and N. Gu. A Fast Operational Transformation Algorithm for Mobile and Asynchronous Collaboration. *IEEE Transactions on Parallel and Distributed Systems*, 21(12):1707–1720, December 2010.

[22] Microsoft sharepoint workspace 2010. http://office.microsoft.com/en-us/sharepoint-workspace/.

[23] B. Shneiderman. Response Time and Display Rate in Human Performance with Computers. *ACM Computing Surveys*, 16(3):265–285, September 1984.

[24] M. Suleiman, M. Cart, and J. Ferrié. Serialization of Concurrent Operations in a Distributed Collaborative Environment. In *Proceedings of the ACM SIGGROUP Conference on Supporting Group Work - GROUP '97*, pages 435–445, Phoenix, AZ, USA, November 1997. ACM Press.

[25] C. Sun and C. Ellis. Operational Transformation in Real-Time Group Editors: Issues, Algorithms and Achievements. In *Proceedings of the ACM Conference on Computer-Supported Cooperative Work - CSCW '98*, pages 59–68, Seattle, WA, USA, November 1998. ACM Press.

[26] C. Sun, X. Jia, Y. Zhang, Y. Yang, and D. Chen. Achieving Convergence, Causality Preservation, and Intention Preservation in Real-Time Cooperative Editing Systems. *ACM Transactions on Computer-Human Interaction*, 5(1):63–108, March 1998.

[27] S. G. Tammaro, J. N. Mosier, N. C. Goodwin, and G. Spitz. Collaborative Writing Is Hard to Support: A Field Study of Collaborative Writing.

Computer-Supported Cooperative Work, 6(1):19–51, 1997.

[28] TeamEdit. *A collaborative text editor.* http://teamedit.sourceforge.net/.

[29] N. Vidot, M. Cart, J. Ferrié, and M. Suleiman. Copies Convergence in a Distributed Real-Time Collaborative Environment. In *Proceedings of the ACM Conference on Computer-Supported Cooperative Work - CSCW 2000*, pages 171–180, Philadelphia, PA, USA, December 2000. ACM Press.

[30] S. Weiss, P. Urso, and P. Molli. Wooki: A P2P Wiki-based Collaborative Writing Tool. In *Proceedings of the International Conference on Web Information Systems Engineering - WISE 2007*, pages 503–512, Nancy, France, December 2007. Springer-Verlag.

[31] S. Weiss, P. Urso, and P. Molli. Logoot : A Scalable Optimistic Replication Algorithm for Collaborative Editing on P2P Networks. In *Proceedings of the 29th International Conference on Distributed Computing Systems - ICDCS 2009*, pages 404–412, Montreal, QC, Canada, June 2009. IEEE Computer Society.

[32] S. Weiss, P. Urso, and P. Molli. Logoot-Undo: Distributed Collaborative Editing System on P2P Networks. *IEEE Transactions on Parallel and Distributed Systems*, 21(8):1162–1174, August 2010.

[33] Q. Wu, C. Pu, and J. E. Ferreira. A Partial Persistent Data Structure to Support Consistency in Real-Time Collaborative Editing. In *Proceedings of the 26th IEEE International Conference on Data Engineering - ICDE 2010*, pages 776–779, Long Beach, CA, USA, March 2010. IEEE Computer Society.

A Generic Calculus of XML Editing Deltas

Jean-Yves Vion-Dury
Xerox Research Centre Europe
6 chemin de Maupertuis, 38330 Meylan, France
viondury@xeroxlabs.com

ABSTRACT

In [18, 17], we outlined a mathematical model of the so-called *XML editing deltas* and proposed a first study of their formal properties. We expected at least three outputs from this theoretical work: a common basis to compare performances of the various algorithms through a structural normalization of deltas, a universal and flexible patch application model and a clearer separation of patch and merge engine performance from delta generation performance. This paper presents the full calculus and reports significant progresses with respect to formalizing a normalization procedure. Such method is key to defining an equivalence relation between editing scripts and eventually designing optimizers compiler back-ends, new patch specification languages and execution models.

Categories and Subject Descriptors

E.1 [**Data Structures**]: Trees; F.2 [**Theory of Computation**]: Analysis of Algorithms and Problem Complexity; H.4 [**Information Systems Applications**]: Miscellaneous; I.7.1 [**Document and Text Editing**]: [Document Management, version control]

General Terms

Theory, Algorithms, Management, Performance, Reliability

Keywords

Formal Model, XML, distance metric, tree transformation, tree edit distance, tree-to-tree correction, version control

1. INTRODUCTION AND STATE OF THE ART

XML diff operations are basically used to realize two fundamental operations on structured documents: comparison and merging. Therefore tree based diff algorithms are central to solving various important problems related to XML document management and editing processes ([2]), among which:

- checking modifications with respect to a reference document(version management [3])

- synchronizing variants([8])

- merging variants, possibly imply solving potential conflicts if the underlying document management system applies weak control policies

- recovering anterior state(s) of a document

An abundant literature exists on this topic, including recent synthetic work on an old topic, linear differencing ([7]). Most work covers :

1. *Algorithmic complexity of the differencing operation, either for ordered or unordered tree models.* The problem was first approached as a particular case of string oriented diff computation (see [7] for a point on most recent algorithms in this area), implying computing an adequate linearization of the XML trees. Zhang and Shasha described a fast algorithm in 1989 for computing tree edit distance [21] improving previous work conducted in 1979. In [19] an algorithm for unordered tree was presented with a polynomial complexity for so-called "optimal differences" (although this notion is not defined), whereas previous works established a NP-complete complexity for general cases. A survey (still of interest 14 years later) and a comparative study on the topic can be found in [1] and [4]. Most recent algorithms for ordered diff computation are described in [10], [13, 14].

2. *Optimality of editing scripts (also called deltas).* The very notion of optimality is highly controversial. From a pure theoretical perspective, it means measuring the number of atomic operations (this is a simple transposition of the so called Levenstein distance on strings), however, this method lacks two qualities to be really relevant: (*i*) a normal form for editing scripts has never been defined and (*ii*) different algorithms can not be compared using a common vocabulary of deltas. From engineering perspectives, measuring performance depends on the implementation of the delta application engine (far from being objective as depending on languages and implementer's skills), and the code is usually optimized toward particular document profiles and some implicit applicative expectations. In [6], authors

focus on analyzing commutativity property of XML update primitives, but have to deal with fundamental undecidability due to XPath and XQuery Update expressive power. They proposed a restrictive subset of the language in order to shape conditions for which commutativity is guaranteed. The underlying operational model is also restricted to "strict evaluation order and immediate update application". Based on anterior work ([18]) a first step is proposed in [17] to abstract over basic delta operation and their combination through an exploratory mathematical model suited to XML tree models. However, this formalism was not characterized.

3. *Delta models and use of diff operations inside storage architectures (typically databases)*. In [12], a versioning graph is proposed based on XLink designation mechanism. In [15], an application is proposed that offers online versioning services for XML documents. However, the history itself cannot be exported or organized into an explicit form, and the delta model is borrowed from jXYDiff and its underlying model [5]. A recent paper [13] proposes reversible deltas using a straightforward mechanism as deletion operation conveys explicitly the deleted subtree (this could quickly lead to enormous overhead, especially in the perspective of storing the whole history). The authors focus on algorithmic and speed performance issues of a novel diff algorithm using a bottom-up approach (and also precomputed hash codes on a sliding window of adjacent nodes.) Another approach from DB community explored a diff model based on so-called *completed deltas* with strong properties among which reversibility and reduction of delta composition [11]. However, the designation mechanism entirely relies on maintaining persistent identifiers for each node of the tree. Thus, according to this framework, no notion of path is needed (as a designation mean) and the problem of comparing nodes is reduced to numerical equality (of IDs). However, this approach confines the application field to document management systems based on a dedicated indexation system. Last year, [16] proposed a very efficient model for a three-way merge tool, in which computation is driven by node identity and preexisting versioning information stored in a particular tree data-structure. The use of such infrastructure, if quite relevant from the applicative point of view, does not bring a generic light on the fundamental issue of designating tree regions (no designation mechanism since nodes are uniquely identified via a static ID), and its impact on diff specifications ; similarly, the delta model is based on some standard operators directly derived from the model's peculiarities (*insert*, *delete move* and *update* primitives).

As a conclusion, we observe that most work conducted in this area focused on related algorithm performance, time and space complexity of the diff operation and also optimality of generated deltas, but little attention has been paid to tools and models to make use of these in efficient ways. From the practical point of view, no work considered the necessity and applicative interest of abstracting over deltas and related operation models.

2. THE FORMALISM: DEFINITIONS

2.1 Overview

We first define a tree model really close to XML. Then, we define the syntax of *paths*, objects used to designate subtrees of any given reference tree. A well-formedness condition allows us to assert that a path is indeed applicable to a given tree, i.e. that a dedicated **get** primitive will actually be able to "extract" this subtree from the reference tree. Paths, by construction, are subject to various logical ordering relations, which abstractly express important properties inherent to the geometry of trees: embedding, sibling dependencies, independence of various parts of the tree in term of node insertion or deletion. Together with specific operations on paths (addition $p \oplus p'$, subtraction $p \ominus p'$, fingerprint extraction $\zeta(p)$ - a kind of signature that reflects the depth structure of the path), those relations allow us reasoning on the locality of tree modification, and defining shift operations able to propagate these modifications in accordance with the topological constraints of trees.

The syntax and semantics of the so-called *delta operators* is studied in section 2.4. The more basic operators (atomic deltas) are insertion and deletion, plus a neutral operation that do not change the tree. These operators are semantically defined through their abstract property regarding the **get** primitive. This approach allows focusing on logical properties rather than on operational behavior [1]. More complex operations are specified by mean of composition operators. The most common is the sequence ; however, we discovered that two subtle but recurrent and significant phenomenons were not adequately captured by sequential construct: the full independence of some sub-transformations (concurrency) and the use of a static referential to specify several tree diff operations (snapshot). Making these more explicit inside the calculus opens the way to commutativity analysis, and consequently, to new possibilities for designing deep reorganization of delta based transformations modeled through the calculus.

2.2 XML model and basic operations

We propose to use a simplifying data model for capturing XML particularities: total order over nodes including attributes (direct translation of the linear semantics of document) but order over attributes not significant in tree comparison. Moreover, attributes (noted $@n : t$) can be attached to element nodes only, and are uniquely defined through a name n. We denote by d the document fragments (or subtrees), t the terminal nodes (text, attribute value) and n the names of elements or attributes. The following syntax captures the the positional information of nodes thanks to the exponents. Sequences of trees are captured by the f non terminal (*forest*), with k exponent denoting the rank in that sequence. Empty element nodes are captured through $n^i[\epsilon]$, just noted n^i. The following attributed tree grammar capture our tree model together with most structural constraint of XML.

DEFINITION 1. *document and tree model*

$$
\begin{array}{llll}
d^i & ::= & n^i[f^k] \mid t^i \mid @n^i{:}t & (\forall i > 0) \\
f^k & ::= & f^{k-1}d^k & (\forall k > 0) \\
f^0 & ::= & \epsilon &
\end{array}
$$

---

[1]It would be interesting also to define these operator through a recursive function, and then establish the soundness of abstract semantics definition

At this abstraction level, we assume that the lexical constraints of XML are always satisfied, and so are well-formedness constraints, leading to always successful parsing and identification of documents with trees resulting from parsing. In our proposal, trees are operands of abstract operations (thus, similar to *abstract types* [20]): **diff**, **get**, **patch** and **merge**.

DEFINITION 2. *Abstract modeling of basics functions*

$$\forall d_1, d_2, \exists \Delta \quad \textbf{diff}(d_1, d_2) = \Delta \quad [\text{a-diff}]$$

$$\forall d_1, d_2, \exists \Delta \quad \textbf{patch}(d_1, \Delta) = d_2 \quad [\text{a-patch}]$$

$$\frac{\textbf{diff}(d_1, d_2) = \Delta}{\textbf{patch}(d_1, \Delta) = d_2} \quad [\text{p-patch}]$$

$$\frac{\textbf{diff}(d, d_1) = \Delta_1 \quad \textbf{diff}(d, d_2) = \Delta_2 \quad \Delta_1 \perp \Delta_2}{\textbf{patch}(d, \Delta_1; \Delta_2) = d'} \quad [\text{a-merge}]$$

Note that **diff** and **patch** are far from being defined *stricto sensu*: the formulas a-diff and a-path just state that the operations shall work on any pair of well-formed documents, producing a result Δ which syntax will be detailed in the sequel. Patch and diff relate through an invariance property *p-patch*. Finally, the (conflict-free) three way merge requires a logical condition, we called it orthogonality ($\perp$, formally defined later), over the deltas issued from applying diff on d_1 and d_2 with respect to a reference document d.

2.3 Paths

Paths are closely related to trees and serve as designation mechanisms. The definition (3) below shows three types of paths to designate the elements, attributes and text nodes in this order, and is similar to many previous works (e.g. [9, 13, 14]), but starting counting children nodes at value 1, and being always relative to the context tree (defined by the parent node). These two points are not minor, as they enables algebraic operations on paths (see (3)) and also compositional concatenation.

DEFINITION 3. *Syntax of paths*

$$p \quad ::= \quad i \mid @n \mid i/p$$

This syntax distinguishes paths p designating all kind of node including terminal text and attributes, and paths pp designating only element nodes. A path p is well-formed with respect to a document d ($d \models p$) if it indeed allows accessing part of d:

DEFINITION 4. *well-formed paths and tree access*

$$d^i \models i \qquad @n{:}t \models @n \qquad \frac{d^j \models p}{n^i[\cdots d^j \cdots] \models i/p}$$

DEFINITION 5. *tree access*

$$\frac{d = n^i[f^j]}{\textbf{get}(d, i) = d} \qquad \frac{\textbf{get}(d^j, p) = d'}{\textbf{get}(n^i[\cdots d^j \cdots], i/p) = d'}$$

$$\textbf{get}(@n{:}t, @n) = t$$

PROPOSITION 1. *path well-formedness*

$$d \models p \quad \Leftrightarrow \quad \exists d' \mid \textbf{get}(d, p) = d' \qquad (1)$$

$$\models p \quad \Leftrightarrow \quad \exists d \mid d \models p \qquad (2)$$

The proposed paths can be compared (useful for asserting precedence over subtrees) and added (useful to propagate changes in the structure of the tree). Our first comparison expresses subtree embedding. Intuitively, 1/2/3 designates a subtree of 1/2, which is directly captured by $1/2/3 \subset 1/2$. For reader accustomed to XPath language, this relation is logically similar to **ancestor** :: *. In the rest of the paper, paths are assumed to be well-formed.

DEFINITION 6. *prefix path ordering*

$$\forall p, p' \quad \models p/p' \ \Rightarrow \ p/p' \subset p$$

The following relation is larger, and defines a strict path ordering that reflects the total document node ordering defined in the XML data model (corresponds to a leftmost innermost tree traversal, but attribute nodes are not ordered):

DEFINITION 7. Strict path ordering.

$$\frac{i < j}{i \prec j} \ [\prec\text{-1}] \qquad \frac{i < j}{i/p \prec j/p'} \ [\prec\text{-2}] \qquad \frac{i < j}{i \prec j/p} \ [\prec\text{-3}]$$

$$\frac{p \prec p'}{i/p \prec i/p'} \ [\prec\text{-4}] \qquad \frac{}{i \prec i/p} \ [\prec\text{-5}]$$

$$\frac{i \prec p}{i/@n \prec p} \ [\prec\text{-6}] \qquad \frac{p \prec i}{p \prec i/@n} \ [\prec\text{-7}]$$

Note that by construction, this relation is irreflexive. The next relation, smaller, captures ordering between sibling nodes, and turned out to be central in studying dependencies among delta operations. For reader accustomed to XPath language, this relation is logically similar to **following-sibling** :: *//*.

DEFINITION 8. *Sibling path ordering*

$$\frac{i < j}{i \ll j} \ [\ll\text{-1}] \qquad \frac{i < j}{i \ll j/p} \ [\ll\text{-2}] \qquad \frac{p \ll p'}{i/p \ll i/p'} \ [\ll\text{-4}]$$

$$\frac{i \ll p}{i/@n \ll p} \ [\ll\text{-6}] \qquad \frac{p \ll i}{p \ll i/@n} \ [\ll\text{-7}]$$

As an illustration, the next proof tree illustrates how to use these definitions to assess $1/2/1 \ll 1/2/4/1$.

$$\frac{\dfrac{\dfrac{\dfrac{1 < 4}{1 \ll 4/1} \ [\ll\text{-2}]}{2/1 \ll 2/4/1} \ [\ll\text{-4}]}{1/2/1 \ll 1/2/4/1} \ [\ll\text{-4}]}{}$$

These relations lead us to define a more abstract relation, the orthogonality of paths, which application is to assess the independence of atomic deltas, and subsequently, of composite deltas. The idea captured by the definition 9 is to discriminate subtrees that cannot be impacted by deletions or insertions, thanks to a branching topological property, reflected by paths structure.

DEFINITION 9. *Path orthogonality*

$$\frac{p \perp p'}{i/p \perp i/p'} \ [\perp\text{-1}] \qquad \frac{i \neq j}{i/p \perp j/p'} \ [\perp\text{-2}]$$

$$\frac{i < j}{i/p \perp j} \ [\perp\text{-3}] \qquad \frac{j < i}{i \perp j/p} \ [\perp\text{-4}]$$

The proof tree below illustrates how to use these definitions to assess $1/2/1 \perp 1/4$.

$$\dfrac{\dfrac{2 < 4}{2/1 \perp 4}^{[\perp\text{-}3]}}{1/2/1 \perp 1/4}^{[\perp\text{-}1]}$$

It is easy to observe that the $\perp$ relation is commutative (see properties (63) in annex).

Now we introduce a partial path addition (operands cannot be path selecting attributes) as the relation satisfying the following properties:

DEFINITION 10. *path addition*

$$\begin{array}{rcll}
i/p_1 & \oplus & j/p_2 & = & (i+j)/(p_1 \oplus p_2) \\
i/p & \oplus & j & = & (i+j)/p \\
i & \oplus & j/p & = & (i+j)/p \\
i & \oplus & j & = & i+j
\end{array} \tag{3}$$

Path addition is useful to propagate structural information related to tree modification. Together with what we call *fingerprints*, it allows to relocate atomic deltas thus permitting swaps in modification sequences. We define quite similarly a path subtraction, noted $\ominus$, using subtraction of integer instead addition.

DEFINITION 11. *path fingerprint*

$$\begin{array}{rcl}
\zeta(i) & = & 1 \\
\zeta(i/p) & = & 0/(\zeta(p)) \\
\zeta(i/@n) & = & 0
\end{array} \tag{4}$$

The two operation, combined, allows modeling shifting phenomenons that arise when deleting or inserting a node. As an example, inserting a subtree at $1/2/3$ will shift "to the right" the subtree located at $1/2/5$. The new position of the latter can be computed as

$$\begin{array}{rcl}
1/2/5 \oplus \zeta(1/2/3) & = & 1/2/5 \oplus 1/\zeta(2/3) \\
 & = & 1/2/5 \oplus 0/\zeta(2/3) \\
 & = & 1/2/5 \oplus 0/0/\zeta(3) \\
 & = & 1/2/5 \oplus 0/0/1 \\
 & = & 1/(2/5 \oplus 0/1) \\
 & = & 1/2/(5 \oplus 1) \\
 & = & 1/2/6
\end{array}$$

2.4 The syntax and semantics of delta-based transformations

We propose to express a clear distinction between atomic operations (noted δ) and transformations of tree using combinations of atomic deltas (noted Δ).

2.4.1 Atomic Transformations

At this first stage, we define three atomic operators, that we will slightly extend afterward in order to gain expressiveness.

DEFINITION 12. *Syntax of Atomic Transformations*

$$\delta \quad ::= \quad \boldsymbol{ins}(p,d) \mid \boldsymbol{del}(p) \mid \boldsymbol{nop} \tag{5}$$

Their semantic is expressed by definition below forward. The first denote the insertion of a tree d at position p. Note that according to the defined syntactic structure the subtree can be also a textual leave or an attribute as well (e.g. $\boldsymbol{ins}(1/2/@id,' test')$).

DEFINITION 13. *Semantic of insertion*

$$\forall d, d', p, p', \quad d \models p, \ d \models p'$$

$$\boldsymbol{patch}(d, \boldsymbol{ins}(p, dd)) = d'$$
$$\Leftrightarrow$$

$$\left\{
\begin{array}{llll}
 & & \boldsymbol{get}(d', p) = dd & (a) \\
p \perp p' & \Rightarrow & \boldsymbol{get}(d, p') = \boldsymbol{get}(d', p') & (b) \\
p \ll p' \vee p = p' & \Rightarrow & \boldsymbol{get}(d, p) = \boldsymbol{get}(d', p' \oplus \zeta(p)) & (c) \\
p' = p/pp & \Rightarrow & \boldsymbol{get}(dd, 1/pp) = \boldsymbol{get}(d', p') & (d) \\
p = p'/pp & \Rightarrow & \boldsymbol{patch}(\boldsymbol{get}(d, p'), \boldsymbol{ins}(1/pp, dd)) & \\
 & & = \boldsymbol{get}(d', p) & (e)
\end{array}
\right.$$

The line (a) says that insertion is done "in place", which means that after insertion, the **get** primitive applied to the same path will return the operand.

This property is however not enough to fully define the semantics of insertion. We need to express the locality of the action, i.e. the way other parts of the tree are affected (or not) by the operation.

Hence, (b) says that independent parts of the tree are not affected by the change, whereas (c) says that following sibling trees are shifted by one place to the right and (d) says that the inserted subtree becomes fully accessible afterward while (e) asserts that containing subtrees are impacted by the operation in the way we expect; in other words, (d, e) say that integration is done in an homogeneous way if we consider tree embedding relation.

The definition of deletion is similar while simpler. Following sibling subtrees are shifted too, but on left side (see $(6.c)$ below).

DEFINITION 14. *Semantic of deletion*

$$\forall d, d', p, p', \quad d \models p, \ d \models p'$$

$$\boldsymbol{patch}(d, \boldsymbol{del}(p)) = d'$$
$$\Leftrightarrow$$

$$\left\{
\begin{array}{llll}
p \perp p' & \Rightarrow & \boldsymbol{get}(d, p') = \boldsymbol{get}(d', p') & (b) \\
p \ll p' & \Rightarrow & \boldsymbol{get}(d, p) = \boldsymbol{get}(d', p' \ominus \zeta(p)) & (c) \\
p = p'/pp & \Rightarrow & \boldsymbol{patch}(\boldsymbol{get}(d, p'), \boldsymbol{del}(1/pp)) & \\
 & & = \boldsymbol{get}(d', p) & (e)
\end{array}
\right. \tag{6}$$

The last atomic operation is neutral, which is captured by:

DEFINITION 15. *Semantic of neutral operation*

$$\forall d \quad \boldsymbol{patch}(d, \boldsymbol{nop}) = d \tag{7}$$

This axiomatic semantic leads to some complexity due on the nature of objects (trees); however, it turned to be quite handful in further proving equational properties of operations.

Well-formedness properties assert some constraints on the geometry on paths and trees. For instance, the root cannot be deleted, an insertion cannot transform a root tree in forest, insertion cannot attach a subtree to a leave or an attribute. They are captured by:

DEFINITION 16. *Well-formedness of atomic operations*

$$\dfrac{d^j \models \boldsymbol{ins}(p, d)}{d^i[\cdots d^j \cdots] \models \boldsymbol{ins}(i/p, d)} \qquad \dfrac{1 \le j \le k+1}{d^i[f^k] \models \boldsymbol{ins}(i/j, d)}$$

$$\overline{@n : t \models \boldsymbol{ins}(@n, t')}$$

We consider in addition that the neutral operation is always well-formed: $d \models \boldsymbol{nop}$.

116

2.4.2 Composite transformation

Realistic transformations based on tree editing deltas need to articulate many primitive operations. In the literature, they are built on the basis of atomic deltas and, up to our knowledge, on a unique sequential composition. We propose to use three binary operators: one for sequence, one for concurrent operations and one for snapshot. These notions will be detailed further in the sequel.

DEFINITION 17. *Syntax of Composite Transformations*

$$\Delta \quad ::= \quad \delta \mid \Delta_1;\Delta_2 \mid \Delta_1 \diamond \Delta_2 \mid \Delta_1 \bowtie \Delta_2 \qquad (8)$$

Sequential composition. The $\Delta_1;\Delta_2$ notation expresses sequences of transformations, understood that Δ_2 is applied on the tree resulting from applying the Δ_1 transformation. Thus, Δ_1 and Δ_2 necessarily refer to different trees. This is the most common compositional operation found in literature.

DEFINITION 18. *Well-formedness and Semantics of sequential composition*

$$\frac{d \models \Delta_1 \quad d \models \Delta_2}{d \models \Delta_1;\Delta_2} \qquad (9)$$

$$\frac{\models \Delta_1;\Delta_2}{patch(d, \Delta_1;\Delta_2) = patch(patch(d, \Delta_1), \Delta_2)} \qquad (10)$$

The first property we need here is associativity.

PROPOSITION 2. *associativity of* ;

$$\Delta_1;(\Delta_2;\Delta_3) = (\Delta_1;\Delta_2);\Delta_3$$

The proof comes easily from the definition by developing both sides.

Concurrent composition. On the contrary, $\Delta_1 \diamond \Delta_2$ describes transformations that can be done concurrently without any risk of interfering each others. At this end, well-formed composition must ensure that deltas are orthogonal.

DEFINITION 19. *Well-formedness of parallel composition*

$$\frac{d \models \Delta_1 \quad d \models \Delta_2 \quad \Delta_1 \perp \Delta_2}{d \models \Delta_1 \diamond \Delta_2} \qquad (11)$$

The "diamond" operator is finally defined by underlying its relation with sequential operations: order is not significant.

DEFINITION 20. *Semantics of parallel composition*

$$\frac{\models \Delta_1 \diamond \Delta_2}{\Delta_1 \diamond \Delta_2 = \Delta_1;\Delta_2 = \Delta_2;\Delta_1} \qquad (12)$$

This operator allows describing concurrent transformations, in that sense that whatever sub-transformation is applied first, the result is always the same. Moreover, as a corollary effect, both sub-transformations use the same tree operand as reference (or equivalently, the structural changes induced by one operation can not modify the references involved in the other operation). As an example,

$$patch(a^1[b^1[c^1]d^2], \mathbf{ins}(1/1/2, e) \diamond \mathbf{del}(1/2)) =$$
$$patch(a^1[b^1[c^1]d^2], \mathbf{ins}(1/1/2, e);\mathbf{del}(1/2)) =$$
$$patch(a^1[b^1[c^1\ e^2]d^2], \mathbf{del}(1/2)) =$$
$$a^1[b^1[c^1\ e^2]]$$

and however

$$\mathbf{patch}(a^1[b^1[c^1]d^2], \mathbf{del}(1/2);\mathbf{ins}(1/1/2, e)) =$$
$$\mathbf{patch}(a^1[b^1[c^1]], \mathbf{ins}(1/1/2, e)) =$$
$$a^1[b^1[c^1\ e^2]]$$

Snapshot composition. The third operator, $\Delta_1 \bowtie \Delta_2$ lies in between sequential and concurrent compositions. Indeed, it may involve dependent operations, but using a common referential to interpret the paths. This referential is the tree before applying the transformation. To illustrate this concept, let us combine two different operations on a tree $a[b\ c\ d]$: the first one adds an attribute $n :''\ 1''$ to subtree b (that is $\mathbf{ins}(1/1, @n :''\ 1'')$), and the other one adds the same attribute to subtree d ($\mathbf{ins}(1/3, @n :''\ 1'')$). Note that both operations are specified independently, with the same original tree as reference for paths. However, the two operations are sibling dependent (for illustration purposes, we produce the proof tree):

$$\frac{\dfrac{\dfrac{1 < 3}{1 \ll 3}^{[\ll\text{-}1]}}{1/1 \ll 1/3}^{[\ll\text{-}4]}}{\mathbf{ins}(1/1, @n :''\ 1'') \ll \mathbf{ins}(1/3, @n :''\ 1'')}^{[\text{see 25, annex}]}$$

Conditions are met to define a snapshot as

$$\mathbf{ins}(1/1, @n :''\ 1'') \bowtie \mathbf{ins}(1/3, @n :''\ 1'')$$

Hence, to be well-formed, such operation must only be applied on operands verifying certain conditions.

DEFINITION 21. *Well-formedness of snapshot composition*

$$\frac{d \models \Delta_1 \quad d \models \Delta_2 \quad (\Delta_1 \ll \Delta_2 \lor \Delta_2 \ll \Delta_1 \lor \Delta_2 \perp \Delta_1)}{d \models \Delta_1 \bowtie \Delta_2} (13)$$

The definition below makes use of sequential composition, like for the diamond composition, but potentially requires a transformation of one operand to express the shift due to dependencies.

DEFINITION 22. *Semantics of snapshot composition*

$$\frac{\Delta_1 \ll \Delta_2}{\Delta_1 \bowtie \Delta_2 = \Delta_2;\Delta_1 = \Delta_1;\lceil\Delta_2\rceil^{\Delta_1}} \qquad (14)$$

$$\frac{\Delta_2 \ll \Delta_1}{\Delta_1 \bowtie \Delta_2 = \Delta_1;\Delta_2 = \Delta_2;\lceil\Delta_1\rceil^{\Delta_2}} \qquad (15)$$

$$\frac{\Delta_2 \perp \Delta_1}{\Delta_1 \bowtie \Delta_2 = \Delta_1;\Delta_2 = \Delta_2;\Delta_1} \qquad (16)$$

$$\models \Delta_1 \bowtie \Delta_2 \text{ is assumed in 14, 15, 16}$$

This definition says that for this operator, order is not significant. However, to be translated into a sequence of transformation, any potential sibling dependency must be interpreted in term of transposition. This transposition is inductively defined (the notation δ^p abstracts over $\mathbf{ins}(p, d)$ and $\mathbf{del}(p)$):

DEFINITION 23. *Semantic of transposition*

$$\lceil \delta^p \rceil^{ins(p', d)} = \delta^{p \oplus \zeta(p')} \text{ if } p' \ll p, \ \delta^p \text{ otherwise} \quad (17)$$
$$\lceil \delta^p \rceil^{del(p')} = \delta^{p \ominus \zeta(p')} \text{ if } p' \ll p, \ \delta^p \text{ otherwise} \quad (18)$$
$$\lceil \Delta \rceil^{nop} = \Delta \qquad (19)$$
$$\lceil \Delta_1;\Delta_2 \rceil^{\Delta} = \lceil \Delta_1 \rceil^{\Delta};\lceil \Delta_2 \rceil^{\Delta} \qquad (20)$$
$$\lceil \Delta \rceil^{\Delta_1;\Delta_2} = \lceil\lceil \Delta \rceil^{\Delta_1}\rceil^{\Delta_2} \qquad (21)$$

The inverse composition, $\lfloor\Delta_1\rfloor^{\Delta_2}$, used to reverse dependent sequences, is defined quite similarly in annex (70) with similar properties.

LEMMA 1.

$$\models\Delta_1\ \wedge\ \models\Delta_2\ \implies\ \models\lfloor\Delta_1\rfloor^{\Delta_2} \quad (22)$$

proof: by structural induction. Terminal cases are handled by a reasoning on shifting mechanism, in case of dependent paths.

PROPOSITION 3.

$$\Delta_1\perp\Delta_2\ \implies\ \lfloor\Delta_1\rfloor^{\Delta_2}=\Delta_1 \quad (23)$$
$$\models\Delta_1\diamond\Delta_2\ \implies\ \lfloor\Delta_1\diamond\Delta_2\rfloor^{\Delta}=\lfloor\Delta_1\rfloor^{\Delta}\diamond\lfloor\Delta_2\rfloor^{\Delta} \quad (24)$$
$$\models\Delta_1\bowtie\Delta_2\ \implies\ \lfloor\Delta_1\bowtie\Delta_2\rfloor^{\Delta}=\lfloor\Delta_1\rfloor^{\Delta}\bowtie\lfloor\Delta_2\rfloor^{\Delta} \quad (25)$$

proofs: Hypotheses allow substituting composite operators by sequential variant, thanks to definitions. Than for each case, hypothesis is refined, and after applying again the definition, a reverse substitution is done to obtain the result. We produce below the proof for (24):

Hypothesis: $\quad\models\Delta_1\diamond\Delta_2$

$\models\Delta_1$ from hypothesis by (11)
$\models\Delta_2$ from hypothesis by (11)

$$\begin{aligned}\lfloor\Delta_1\diamond\Delta_2\rfloor^{\Delta} &= \lfloor\Delta_1;\Delta_2\rfloor^{\Delta} &\text{by (12)}\\ &= \lfloor\Delta_1\rfloor^{\Delta};\lfloor\Delta_2\rfloor^{\Delta} &\text{by (17)}\\ &= \lfloor\Delta_1\rfloor^{\Delta}\diamond\lfloor\Delta_2\rfloor^{\Delta} &\text{by (12,22)}\end{aligned}$$

3. EQUATIONAL PROPERTIES OF DELTAS

Now we address the heart of the calculus, the equational properties, tools enabling deep transformations of expressions and therefore, revealing the deepest nature of the objects we study.

PROPOSITION 4. *Equational laws for sequence*

$$\begin{aligned}\Delta_1\perp\Delta_2 &\Rightarrow\ \Delta_1;\Delta_2 = \Delta_2;\Delta_1 &(26)\\ \Delta_2\ll\Delta_1 &\Rightarrow\ \Delta_1;\Delta_2 = \Delta_2;\Delta_1 &(27)\\ \Delta_1\ll\Delta_2 &\Rightarrow\ \Delta_1;\Delta_2 = \lfloor\Delta_2\rfloor^{\Delta_1};\Delta_1 &(28)\\ &nop;\Delta = \Delta &(29)\\ &\Delta;nop = \Delta &(30)\\ &ins(p,d);del(p) = nop &(31)\\ &del(p/p');del(p) = del(p) &(32)\\ &ins(p,d);ins(p/p',d') = ins(p,d'') &(33)\\ &\text{with } d'' = patch(d,ins(1/p',d'))\\ &ins(p,d);del(p/p') = ins(p,d') &(34)\\ &\text{with } d' = patch(d,del(1/p')d)\end{aligned}$$

PROPOSITION 5. *Associativity*

$$\begin{aligned}\Delta_1\diamond(\Delta_2\diamond\Delta_3) &= (\Delta_1\diamond\Delta_2)\diamond\Delta_3 &(35)\\ \Delta_1\bowtie(\Delta_2\bowtie\Delta_3) &= (\Delta_1\bowtie\Delta_2)\bowtie\Delta_3 &(36)\end{aligned}$$

PROPOSITION 6. *Distributivity - part 1*

$$\begin{aligned}\Delta_1\perp\Delta_3 &\Rightarrow (\Delta_1\diamond\Delta_2);\Delta_3 = \Delta_1\diamond(\Delta_2;\Delta_3) &(37)\\ \Delta_2\perp\Delta_3 &\Rightarrow (\Delta_1\diamond\Delta_2);\Delta_3 = (\Delta_1;\Delta_3)\diamond\Delta_2 &(38)\\ \Delta_1\perp\Delta_3 &\Rightarrow \Delta_1;(\Delta_2\diamond\Delta_3) = (\Delta_1;\Delta_2)\diamond\Delta_3 &(39)\\ \Delta_2\perp\Delta_1 &\Rightarrow \Delta_1;(\Delta_2\diamond\Delta_3) = \Delta_2\diamond(\Delta_1;\Delta_3) &(40)\end{aligned}$$

PROPOSITION 7. *Distributivity - part 2*

$$\begin{aligned}\Delta_1\perp\Delta_3 &\Rightarrow (\Delta_1\diamond\Delta_2)\bowtie\Delta_3 = \Delta_1\diamond(\Delta_2\bowtie\Delta_3) &(41)\\ \Delta_2\perp\Delta_3 &\Rightarrow (\Delta_1\diamond\Delta_2)\bowtie\Delta_3 = (\Delta_1\bowtie\Delta_3)\diamond\Delta_2 &(42)\\ \Delta_1\perp\Delta_3 &\Rightarrow \Delta_1\bowtie(\Delta_2\diamond\Delta_3) = (\Delta_1\bowtie\Delta_2)\diamond\Delta_3 &(43)\\ \Delta_2\perp\Delta_1 &\Rightarrow \Delta_1\bowtie(\Delta_2\diamond\Delta_3) = \Delta_2\diamond(\Delta_1\bowtie\Delta_3) &(44)\end{aligned}$$

PROPOSITION 8. *Distributivity - part 3*

$$\begin{aligned}\models\Delta_1\bowtie\Delta_3 &\Rightarrow (\Delta_1\bowtie\Delta_2);\Delta_3 = \Delta_1\bowtie(\Delta_2;\Delta_3) &(45)\\ \models\Delta_2\bowtie\Delta_3 &\Rightarrow (\Delta_1\bowtie\Delta_2);\Delta_3 = (\Delta_1;\Delta_3)\bowtie\Delta_2 &(46)\\ \models\Delta_1\bowtie\Delta_3 &\Rightarrow \Delta_1;(\Delta_2\bowtie\Delta_3) = (\Delta_1;\Delta_2)\bowtie\Delta_3 &(47)\\ \models\Delta_2\bowtie\Delta_1 &\Rightarrow \Delta_1;(\Delta_2\bowtie\Delta_3) = \Delta_2\bowtie(\Delta_1;\Delta_3) &(48)\end{aligned}$$

PROPOSITION 9. *Distributivity - part 4*

$$\begin{aligned}\models\Delta_1\bowtie\Delta_3 &\Rightarrow (\Delta_1\bowtie\Delta_2)\diamond\Delta_3 = \Delta_1\bowtie(\Delta_2\diamond\Delta_3) &(49)\\ \models\Delta_2\bowtie\Delta_3 &\Rightarrow (\Delta_1\bowtie\Delta_2)\diamond\Delta_3 = (\Delta_1\diamond\Delta_3)\bowtie\Delta_2 &(50)\\ \models\Delta_1\bowtie\Delta_3 &\Rightarrow \Delta_1\diamond(\Delta_2\bowtie\Delta_3) = (\Delta_1\diamond\Delta_2)\bowtie\Delta_3 &(51)\\ \models\Delta_2\bowtie\Delta_1 &\Rightarrow \Delta_1\diamond(\Delta_2\bowtie\Delta_3) = \Delta_2\bowtie(\Delta_1\diamond\Delta_3) &(52)\end{aligned}$$

PROPOSITION 10. *Conversion*

$$\begin{aligned}\Delta_1\perp\Delta_3 &\Rightarrow \Delta_1\bowtie\Delta_2 = \Delta_1\diamond\Delta_2 &(53)\\ \Delta_1\ll\Delta_3 &\Rightarrow \Delta_1;\Delta_2 = \Delta_1\bowtie\lfloor\Delta_2\rfloor^{\Delta_1} &(54)\end{aligned}$$

4. SYNTHESIS AND CONCLUSION

Our formalism defined the following elements:
- a simple tree model still capturing basic XML peculiarities (in particular, mixed node ordering policies)
- a simple designation mechanism to specify tree areas and various items of the tree
- a syntax for building diff operations
- an axiomatic semantics of basic delta operations ("patch" primitive)
- an axiomatic semantics of composite delta operations taking in account operation dependencies
- an equational semantics of a transposition operation and its inverse
- a set of equational laws allowing for script transformation and equivalence checking

Our contribution showed that this calculus exhibits unexpected commutative properties, appearing as soon as we discriminate the many path relationships reflecting topological properties of trees. One can summarize those properties through the following illuminating commutative diagrams:

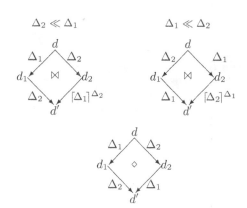

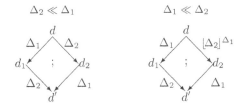

The basis of the calculus are settled, and derivative work can be envisioned, such as normalization algorithms using equational laws as backbone for soundness assertion. Still, a *completeness theorem* is missing. Such tool would guaranty that no equation is missing, or, using different words, that all actual equivalence can be established by using the equational system and only it. Ideally, one may expect another property, well known in logic, stating that the equational system is *minimal*, in that sense that no formula can be deduced from others. It is yet unclear to the author how those fundamental theorem can be reached (there no indication this would be particularly hard neither), but this goal was not in the scope of this paper.

5. REFERENCES

[1] P. Bille. A survey on tree edit distance and related problems. *Theoretical computer science*, 337(1-3):217–239, 2005.

[2] S. Chawathe, A. Rajaraman, H. Garcia-Molina, and J. Widom. Change detection in hierarchically structured information. In *Proceedings of the 1996 ACM SIGMOD international conference on Management of data*, pages 493–504. ACM, 1996.

[3] S. Chien, V. Tsotras, and C. Zaniolo. A comparative study of version management schemes for XML documents. Technical Report TR-51, TimeCenter technical Report, September 2001.

[4] G. Cobéna, T. Abdessalem, and Y. Hinnach. A comparative study for XML change detection. *Research Report, INRIA Rocquencourt, France*, 2002.

[5] G. Cobena, S. Abiteboul, and A. Marian. Detecting changes in XML documents. In *ICDE*, page 0041. Published by the IEEE Computer Society, 2002.

[6] G. Ghelli, K. Rose, and J. Siméon. Commutativity analysis for xml updates. *ACM Trans. Database Syst.*, 33(4):1–47, 2008.

[7] S. Khanna, K. Kunal, and B. Pierce. A Formal Investigation of diff3. *FSTTCS 2007: Foundations of Software Technology and Theoretical Computer Science*, pages 485–496, 2007.

[8] R. La Fontaine. Merging XML files: a new approach providing intelligent merge of XML data sets. In *XML Europe*, pages 03–03. Citeseer, 2002.

[9] T. Lindholm. A 3-way merging algorithm for synchronizing ordered trees. Master's thesis, Helsinki University of Technology, 2001.

[10] T. Lindholm, J. Kangasharju, and S. Tarkoma. Fast and simple xml tree differencing by sequence alignment. In *DocEng '06: Proceedings of the 2006 ACM symposium on Document engineering*, pages 75–84, New York, NY, USA, 2006. ACM.

[11] A. Marian, S. Abiteboul, G. Cobena, and L. Mignet. Change-centric management of versions in an XML warehouse. In *Proceedings of VLDB*, pages 581–590, 2001.

[12] M. Martínez, J.-C. Derniame, and P. de la Fuente. A method for the dynamic generation of virtual versions of evolving documents. In *SAC '02: Proceedings of the 2002 ACM symposium on Applied computing*, pages 476–482, New York, NY, USA, 2002. ACM.

[13] S. Rönnau, C. Pauli, and U. M. Borghoff. Merging changes in xml documents using reliable context fingerprints. In *DocEng '08: Proceeding of the eighth ACM symposium on Document engineering*, pages 52–61, New York, NY, USA, 2008. ACM.

[14] S. Rönnau, G. Philipp, and U. M. Borghoff. Efficient change control of xml documents. In *DocEng '09: Proceedings of the 9th ACM symposium on Document engineering*, pages 3–12, New York, NY, USA, 2009. ACM.

[15] A. Rosado, A. Márquez, and M. Sánchez. A web-based version editor for XML documents. In *Proceedings of the 9th ACM symposium on Document engineering*, pages 249–250. ACM, 2009.

[16] C. Thao and E. Munson. Using versioned tree data structure, change detection and node identity for three-way xml merging. In *DocEng '10: Proceedings of the 2010 ACM symposium on Document Engineering*, pages 77–86, New York, NY, USA, 2010. ACM.

[17] J.-Y. Vion-Dury. Diffing, patching and merging xml documents: toward a generic calculus of editing deltas. In *DocEng '10: Proceedings of the 2010 ACM symposium on Document Engineering*, pages 191–194, New York, NY, USA, 2010. ACM.

[18] J.-Y. Vion-Dury. Stand-alone encoding of document history. In *Balisage: The Markup Conference 2010*, August 3 - 6 2010.

[19] Y. Wang, D. DeWitt, and J. Cai. X-Diff: A fast change detection algorithm for XML documents. In *International Conference on Data Engineering (ICDE'03)*. Citeseer, 2003.

[20] Wikipedia. Abstract data type. *en.wikipedia.org*, 2010.

[21] K. Zhang and D. Shasha. Simple fast algorithms for the editing distance between trees and related problems. *SIAM J. Comput.*, 18(6):1245–1262, 1989.

APPENDIX

DEFINITION 24. *path subtraction*

$$\begin{aligned}
i/p_1 \ominus j/p_2 &= (i-j)/(p_1 \ominus p_2) \\
i/p \ominus j &= (i-j)/p \\
i \ominus j/p &= (i-j)/p \\
i \ominus j &= i-j
\end{aligned} \tag{55}$$

DEFINITION 25. *Sibling dependencies($\star \in \{\diamond, ;, \bowtie\}$)*

$$\frac{p \ll p' \ \vee \ p = p'}{ins(p,d) \ll \delta^{p'}} \qquad \frac{p \ll p'}{del(p) \ll \delta^{p'}}$$

$$\frac{\Delta \ll \Delta_1 \quad \Delta \ll \Delta_2}{\Delta \ \ll \ \Delta_1 \star \Delta_2} \qquad \frac{\Delta_1 \ll \Delta \ \vee \ \Delta_2 \ll \Delta}{\Delta_1 \star \Delta_2 \ \ll \ \Delta}$$

PROPOSITION 11. *Anti-symmetry*

$$p \prec p' \ \Rightarrow \ \neg(p' \prec p) \tag{56}$$

$$p \ll p' \ \Rightarrow \ \neg(p' \ll p) \tag{57}$$

$$p \subset p' \ \Rightarrow \ \neg(p' \subset p) \tag{58}$$

PROPOSITION 12. *Containment*

$$p \ll p' \quad \Rightarrow \quad p \prec p' \tag{59}$$

$$p \subset p' \quad \Rightarrow \quad p' \prec p \tag{60}$$

$$p \perp p' \quad \Rightarrow \quad p \prec p' \vee p' \prec p \tag{61}$$

$$p \perp p' \quad \Rightarrow \quad \begin{cases} \neg(p \subset p') \wedge \neg(p' \subset p) \\ \neg(p \ll p') \wedge \neg(p' \ll p) \end{cases} \tag{62}$$

PROPOSITION 13. *Commutativity*

$$p \perp p' \quad \Leftrightarrow \quad p' \perp p \tag{63}$$

PROPOSITION 14. *Strict total order*

$$p \neq p' \quad \Rightarrow \quad (p' \prec p) \vee (p \prec p') \tag{64}$$

PROPOSITION 15. *Irreflexivity*

$$p \perp p' \quad \Rightarrow \quad p \neq p' \tag{65}$$

$$p = p' \quad \Rightarrow \quad \neg(p \prec p') \wedge \neg(p' \prec p) \tag{66}$$

$$p = p' \quad \Rightarrow \quad \neg(p \ll p') \wedge \neg(p' \ll p) \tag{67}$$

$$p = p' \quad \Rightarrow \quad \neg(p \subset p') \wedge \neg(p' \subset p) \tag{68}$$

$$p = p' \quad \Rightarrow \quad \neg(p \perp p') \wedge \neg(p' \perp p) \tag{69}$$

DEFINITION 26. *Orthogonality of composite deltas ($\star \in \{\diamond, ;, \bowtie\}$)*

$$\frac{p \perp p'}{\delta^p \perp \delta^{p'}} \qquad \overline{del(p) \perp del(p)} \qquad \overline{nop \perp \Delta}$$

$$\frac{\Delta \perp \Delta_1 \quad \Delta \perp \Delta_2}{\Delta \perp \Delta_1 \star \Delta_2} \qquad \frac{\Delta_1 \perp \Delta \quad \Delta_2 \perp \Delta}{\Delta_1 \star \Delta_2 \perp \Delta}$$

DEFINITION 27. *Semantic of inverse transposition*

$$\lfloor \delta^p \rfloor^{ins(p',d)} \quad = \quad \delta^{(p \ominus \zeta(p'))} \text{ if } p' \ll p, \delta^p \text{ otherwise}$$

$$\lfloor \delta^p \rfloor^{del(p')} \quad = \quad \delta^{(p \oplus \zeta(p'))} \text{ if } p' \ll p, \delta^p \text{ otherwise}$$

$$\lfloor \Delta \rfloor^{nop} \quad = \quad \Delta \tag{70}$$

$$\lfloor \Delta_1 ; \Delta_2 \rfloor^{\Delta} \quad = \quad \lfloor \Delta_1 \rfloor^{\Delta} ; \lfloor \Delta_2 \rfloor^{\Delta}$$

$$\lfloor \Delta \rfloor^{\Delta_1 ; \Delta_2} \quad = \quad \lfloor \lfloor \Delta \rfloor^{\Delta_1} \rfloor^{\Delta_2}$$

REMARK 1.

$$\lfloor \delta^p \rfloor^{ins(p',d)} \quad = \quad \lfloor \delta^p \rfloor^{del(p')}$$

$$\lfloor \delta^p \rfloor^{del(p')} \quad = \quad \lfloor \delta^p \rfloor^{ins(p',d)}$$

$$\lfloor \Delta \rfloor^{nop} \quad = \quad \lfloor \Delta \rfloor^{nop}$$

An Efficient Language-Independent Method to Extract Content from News Webpages

Eduardo Cardoso
ecardoso@inf.puc-rio.br

Iam Jabour
ijabour@inf.puc-rio.br

Eduardo Laber
laber@inf.puc-rio.br

Departamento de Informática, PUC-Rio
Mq. de São Vicente 225, RDC, Rio de Janeiro, RJ, Brazil

Rogério Rodrigues
roger@microsoft.com

Pedro Cardoso
pedrolazera@gmail.com

Microsoft Corporation
Avenida Rio Branco 1, 1611,
Rio de Janeiro, RJ, Brazil

Departamento de Informática,
PUC-Rio
Mq. de São Vicente 225,
RDC, Rio de Janeiro, RJ,
Brazil

ABSTRACT

We tackle the task of news webpage segmentation, specifically identifying the news title, publication date and story body. While there are very good results in the literature, most of them rely on webpage rendering, which is a very time-consuming step. We focus on scenarios with a high volume of documents, where performance is a must. The chosen approach extends our previous work in the area, combining structural properties with hints of visual presentation styles, computed with a quicker method than regular rendering, and machine learning algorithms. In our experiments, we took special attention to some aspects that are often overlooked in the literature, such as processing time and the generalization of the extraction results for unseen domains. Our approach has shown to be about an order of magnitude faster than an equivalent full rendering alternative while retaining a good quality of extraction.

Categories and Subject Descriptors

H.3.3 [**Information Storage and Retrieval**]: Information Search and Retrieval—*Information filtering*; I.7.m [**Document and Text Processing**]: Miscellaneous; I.2.6 [**Artificial Intelligence**]: Learning—*Concept learning*

General Terms

Algorithms, Experimentation, Performance

Keywords

News segmentation, webpage rendering

1. INTRODUCTION

We face an ever growing amount of content being produced every day. In this scenario, identifying and extracting the contents of a webpage, discarding templates and similar non-relevant parts of the page, is useful for several applications. To name a few, screen reading software for the visually impaired may focus on the content and skip templates and other irrelevant content; search engines can more cleanly store the page's data, which provides more accurate search results; and small screen devices, such as modern mobile phones, can use it to increase readability.

The task of news segmentation consists in identifying the key regions of the webpage. These regions might have a smaller or bigger role depending on the application. For our purposes, we consider the title, publication date and story body as regions of interest. Very good results have already been reported for this task: using the F-score metric, described in section 3.5, we observe works reaching 97% of F1 for title detection [16, 18], 87% of F1 for publication date detection [18] and 94% of F1 for news body detection [18] for an exact DOM node metric. We believe that these results are satisfactory for some applications and there's not room for large improvements other than refining current approaches. However, most approaches rely on rendering the web page, which demands lots of processing and, consequently, time.

Our focus is on large-scale document processing, specifically the page processing step of search engine's web crawlers, extending the previous work of [8]. Search engines keep local versions of the webpages for indexing purposes. With a cleaner copy of a page, search results tend to be much more relevant as terms that would otherwise be part of the page, but not part of the relevant content (templates, advertisements, etc.), are discarded. In addition, the knowledge of the news title and its publication date can be useful for ranking results, either by relevance (since the title is a general summary of the text) or by date. Some works in the literature have shown how information extraction can improve document retrieval, such as [6] and [28].

For this specific scenario, we feel that there is a lack of suitable approaches. Rendering webpages is not an option because it is a time-consuming task, as shown by [8, 10], and it would slow down the throughput of these systems. Thus, we propose a new approach that more closely keeps up with the high volume of documents in this scenario, while still producing satisfactory results. First, our method locates a DOM node that includes the page's relevant content in its subtree. Then, we proceed to remove noise from this subtree. Finally, we use machine learning models that identify the title and publication date. For these models, we use structural features and visual presentation information computed by a simplified CSS parser. The reduced subset of the page in which we apply these models, along with the simplified CSS implementation, provides us with the necessary performance we were looking for.

To test our approach, we constructed a corpus consisting of 200 news webpages in English, Portuguese and Spanish. A total of 20 websites were crawled and each contributed with exactly 10 pages. These pages have been manually annotated for news title, publication date and article text. We have carefully observed the often overlooked aspects of processing time and results quality when applying our model to a website not previously seen in training, as well as those already seen. Our approach has shown to be about an order of magnitude faster than an equivalent full rendering alternative while retaining a good quality of extraction. The results we obtained in a cross-validation for seen websites using this corpus, measured with a text-based metric, are 92% of F1 for title detection, 84% of F1 for date detection and 88% of F1 for body detection. For unseen websites, the results are 91% of F1 for title detection, 77% of F1 for date detection and 88% of F1 for body detection. Slightly better results are achieved when testing against other corpora such as the NEWS600 [17], which consists of 604 pages from 177 distinct domains, all written in English.

2. RELATED WORK

Related works in the area may be classified in various ways. It is common to differentiate by their scope, which creates the notions of *site-level* and *page-level approaches*, and by the requirements to solve the problem, which range from *strictly structural* properties to a *full rendering* of the page, which provides the geometric positioning of elements and allows the use of computer vision algorithms.

Site-level approaches require some mass of example data to build a model or rules that are specific for its pages. As it is tailored for a specific group, the results are generally better, but come at the cost of high maintenance, high setup costs and limited usability due to the wrappers that are created to exploit particularities of each site's design. A good example of this approach is [22], which identifies site templates using tree edit distance.

On the opposite side of the spectrum lies the page-level approaches. These are devised to work on virtually any webpage, including those from websites never seen before. Its generality comes at the cost of slightly worse results, but requires little maintenance, has low setup costs and broad usability since the approach works independently of the site's design. Examples of these are [6, 28], which identify titles in generic webpages and [23, 24], which train a model from a single website and apply it to 11 others to extract the news article content.

The structural approaches depend on features directly extracted from the HTML file, which may or may not be con-

verted into a DOM tree. Information such as number of nodes, link density, word tokens, among others, are considered structural features. Methods that make good use of them include [11], which describes one of the winning approaches used in the CleanEval shared task [2], and [15], which uses a token-based local classifier to identify the boundaries of article text.

The rendering approaches may include all features available structurally, but have access to other information such as geometric positioning, bounding box size, font size and font color of various elements in the webpage, commonly available in web browsers. Works making use of this information generally achieve better results than strictly structural ones, from which we may cite [16, 17, 18], which perform segmentation of news pages' content in several classes, and [10], which takes the approach of [15] and applies it to visual features with good results.

3. OUR APPROACH

In this work, we design a page-level approach that lies in between the strictly structural vs. full rendering spectrum. It uses structural properties of news web pages and visual presentation information from cascading style sheets, such as font size and color, which may be calculated without a full rendering of the web page, providing us with good execution times. With this information available, we train a machine learning model to classify DOM nodes and employ some post-processing afterwards. Also, we strive to keep it language independent. Since we don't rely on textual cues, we can apply our approach to virtually any document and observe little variation in the results obtained.

An outline of our execution pipeline is displayed in Figure 1.

3.1 Document Object Model (DOM)

The Document Object Model (DOM) [13] is a standard created by the World Wide Web Consortium (W3C) [27] to represent HTML, XHTML and XML documents such as web pages.

A document is seen as a rooted tree where each tag or element represents an *element node*. Text contained in these elements is placed in *text nodes*, which are leaves of this tree.

Other types of nodes exist, but won't have any particular impact on our approach. We will be focusing on the text nodes, as that is where the content lies.

3.2 Cascading Style Sheets (CSS)

Cascading Style Sheets [1] is a language used to describe presentation attributes of HTML and XML documents, such as font face, color and size; absolute and relative positioning; etc. It allows the separation of content and presentation in a web page and the use of several style sheet definitions to be combined in a predictable (or cascade) manner.

It was created in 1996, with the latest revisions dating from 2008. Over the years, it has seen wide adoption by most web sites as the browsers started providing support for it and presentational elements were deprecated from the HTML specification [7, 21]. All current major browsers have a complete implementation of CSS Level 1 and a near-complete implementation of CSS Level 2.1 [3, 14, 20, 26].

Simplified CSS parser: We have developed a simplified CSS parser that exempts us from using a rendering engine

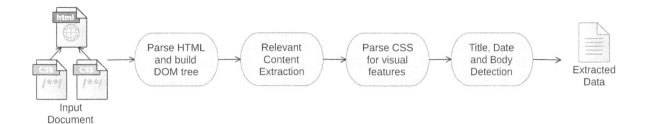

Figure 1: Outline of our execution pipeline

such as Mozilla Firefox's Gecko [4], Microsoft Internet Explorer's MSHTML [12] or Apple Safari and Google Chrome's WebKit [25]. These engines are capable of a much more accurate parsing, but end up computing much more information than we require, raising execution time. With our parser we have access to all information we need much faster, making our approach ideal for use in time-sensitive tasks such as crawling the web. A time comparison between rendering engines and our approach is made in the next section.

3.3 Attributes

For each text node, we use 9 attributes computed from the parsed web page as features for our machine learning models. These are:

1. the text length, measured in characters;

2. amount of digit characters present in the text;

3. percentage of digit characters in the text;

4. font size, normalized to pixels;

5. font size relative to the biggest font size seen, measured in percentage;

6. whether the node is presented in bold letters;

7. the amount of similarly styled nodes in the document;

8. a measure of distance between the current node and the document's node that contains the title;

9. edit distance between the node text and the document's <title> tag contents.

The first six attributes are very natural. The last three, however, deserve some further explanation.

Attribute 7 attempts to capture the uniqueness of a presentation style. This is motivated by the observation that the title is often easily recognized by humans because it stands out from the rest of the text in the page, usually because of a combination of style attributes. Two presentation styles are considered similar if the font size, font color, font family (serif, sans-serif font, etc.) and bold text attributes are the same.

Attribute 8 is discarded for title detection. It is only used for date detection, measuring the distance to the detected title node, as described in more detail in section 3.4.2. Similarly, for date detection, attribute 9 is not used.

Attribute 9 measures edit distance as

$$EditDistance(node\ text,\ title\ tag)\ /\ length(title\ tag)$$

where $EditDistance(\cdot, \cdot)$ is the classic Levenshtein metric [9] with weights adjusted to penalize deletions from the node text much more than insertions. This is done to smooth out the impact of the web site identification that usually comes in the document title while penalizing the loss of the node's information.

3.4 Extraction process

3.4.1 Extraction of relevant content

The first step in our approach is the extraction of relevant content from the news page. We define as *relevant* the body of the news story, its title and associated metadata, such as author and publication date. That is, the relevant content is what is left after removing all templates associated with the webpage.

To detect the relevant content, we employ the NCE algorithm described in [8]. This algorithm is quite fast and provides a very good starting point for the news body, which will later be improved by the next steps of our approach. It is based on the assumption that a *separator node* exists in the webpage; defined as a node for which the textual content of the subtree rooted on it contains most or all of the relevant content of the page. We proceed by refining the selection of relevant content with the identification and removal of subtrees that present a high link density or repeated textual patterns. The former is indicative of navigational links (menus), advertisements, related stories, etc.; and the latter often indicates a comments section at the end of the news body. As a definition of link density we use the number of characters in <a> tags in a subtree over the number of characters in the textual content of that subtree. The remaining subtrees combined should provide high values of precision and recall for the relevant content. We suggest the reader to refer to [8] for more details on the implementation and the different steps of the algorithm.

3.4.2 Title and date detection

We apply a binary classification model at each step, separating the title from all the rest, then the date from all the rest. We use an ensemble of decision tree models for this task [19]. For this, we used the Weka tool [5] for training and model settings were kept with default values, with no specific optimizations.

First, we classify the document title. Second, we classify the publication date. Our premise is that dates are generally close to the document title [24], so we use the classified title as a feature for the date classification. For each node, we calculate its distance to the classified title and use that to guide our model. This constitutes attribute 8, mentioned in section 3.3.

It's worth considering what happens with the date classification model when no title node is detected. We make some considerations on this regard in the next section.

3.4.3 Post-processing

Each of the previously mentioned steps receive some post-processing.

For titles, we only allow one node per document to be detected. In case two or more are detected (excluding the `<title>` tag), we always pick the one that comes first on the document. However, if no title nodes are detected, we take the `<title>` tag as the title.

For dates, we cluster all detected nodes by their proximity in the document and select the cluster closest to the detected title. In case no title was detected (and thus the `<title>` tag is used), no clustering takes place and all nodes are considered. Next, nodes which would add large amounts of text are discarded. This is usually a sign of noise being added as our observations show that publication dates are well separated from the document's content.

Finally, we redefine as body every non-title and non-date relevant node visited in a depth-first search that starts at the detected title and ends at the last node from the extracted relevant content.

3.5 Metrics

We have been using two different metrics to assess our results: the first is node-based and the second uses bag of words. A discussion on the advantages and disadvantages of each of them is provided in the following subsections.

The final user application should dictate which metric to use. In this paper, unless otherwise noted, all results reported use the bag of words metric.

For evaluating both methods, we use the concepts of *precision*, *recall* and *F-measure*. A high value for precision indicates few wrong classifications, while a high value for recall indicates most of the annotated data is retrieved. They are calculated for each class (title, date, etc.) as follows:

$$\text{precision} = \frac{\#correctly\ classified\ elements}{\#total\ classified\ elements}$$

$$\text{recall} = \frac{\#correctly\ classified\ elements}{\#total\ elements\ in\ class}$$

It can be tricky to improve the results on these metrics: to improve recall, more elements may be classified; however, the more elements classified, the smaller the precision obtained.

As these two measures are often reported together, it is common to use the harmonic mean of both measurements, called F-measure or F-score. Here, we will be working with the *F1-score*, which means precision and recall are given the same weight. To calculate the F1-score, we use the following formula:

$$\text{F1-score} = \frac{2 \times precision \times recall}{precision + recall}$$

3.5.1 Node-based

Since we are dealing with a DOM tree, it is natural and convenient to think of a node-based metric. However, this may not always be the best solution.

One issue we found is that all nodes have the same weight when calculating precision and recall, regardless of their content. This means that small nodes with little information are as valuable as large nodes with lots of information. While they could be weighted by character size or word count, small nodes might contain crucial information such as the publication date or the news title. Consider the case of news titles: detecting a node that does not represent the exact news title would be a miss, but its contents might highly resemble the title, which would make it valuable in a search engine setting. Likewise for dates: detecting a node that only partially contains the date, for instance the day and month of publication, but not the exact time, would have a high impact on the precision and recall metrics, but may do no harm for a given application that disregards such time information.

3.5.2 Bag of Words

We employ a metric based on bag of words for measuring the results of our extraction. In this approach, text is represented as a collection of unordered tokens (referred to as words) which may appear repeatedly. Words are often defined as a contiguous sequence of either non-whitespace characters or characters from a given set (for instance, letters and numbers, but no punctuation). Neither approach for tokenization requires linguistic information for this task, which does not invalidate the language independence of our method for languages that make use of the Latin alphabet. Also, special care has been taken to reduce the impacts of eventual encoding errors in the text.

We prefer the bag of words approach because we have found that it more accurately represents what a human would consider as a correct or incorrect extraction. The issues present on the node-based approach are not completely absent, but smoothed out in a way that they cause less negative impact for an imperfect match.

4. EXPERIMENTS

We constructed the RCD4, a corpora of 200 news pages that we employed in our experiments. Its pages were manually annotated for news title, publication date and story body. The pages in the corpus are written in English, Portuguese and Spanish. All objects referred by the pages are also downloaded, such as images, style sheets, script files, etc., in order to keep our copies as close as possible to the originals.

Special care has been taken to collect the same amount of pages from each domain: a total of 20 websites were used, and 10 pages were downloaded from each of them. This allows us to perform balanced cross-validation tests to evaluate how well our models generalize the problem for unseen domains and how results can be improved by adding examples of pages from the same domains. We denote by *extra-site cross-validation* an experiment that trains a model on a different set of domains from the testing documents; that is, folds are domain-disjoint. Similarly, we denote by *intra-site cross-validation* an experiment that trains and tests on pages from the same domain; that is, folds are representative of the domain distribution in the corpus.

As previously mentioned, we start by extracting the relevant content of the webpage with the NCE algorithm described in [8], using it as a base for the next steps. This reduces the amount of nodes that we need to classify in order to find the news title and publication date.

Our experiments show that a 35 to 40% speedup in total execution time is obtained by discarding the nodes that are not considered part of the relevant content. These discarded nodes have affected the quality of our results in at most a 1% decrease in F1.

Title detection: We conducted a series of experiments to evaluate the impact of visual presentation attributes. As a baseline for comparison, we use a classifier that always determines the document's `<title>` tag as the document title. The results for strictly structural features and our combined approach can be seen in Tables 1 and 2.

Method	Prec.	Recall	F1
Baseline (`<title>` tag)	0.61	0.89	0.72
3-fold CV (extra-site)	0.72	0.90	0.80
3-fold CV (intra-site)	0.73	0.91	0.81

Table 1: Title extraction results on the RCD4 corpus for strictly structural attributes

Method	Prec.	Recall	F1
Baseline (`<title>` tag)	0.61	0.89	0.72
3-fold CV (extra-site)	0.88	0.95	0.91
3-fold CV (intra-site)	0.90	0.94	0.92

Table 2: Title extraction results on the RCD4 corpus for both structural and visual attributes

We observe that the use of visual presentation attributes significantly improves our results, with over 10% increase in F1.

A time comparison of these approaches is given in Table 3, measured relatively to the structural approach. As an example, the value 1.77 for the Structural + Visual method indicates that it is 77% slower than the strictly structural approach. The table also includes timings for the approach that skips the relevant content detection step, thus classifying every node in the DOM tree, which we identify as "whole tree", and time estimates for an equivalent approach that uses full rendering instead of our simplified CSS parser. The rendering engine used was WebKit [25], with scripts and plugins disabled. We then proceeded to add to the rendering time of the pages the average time it took for our algorithm to run once every feature is computed. We toggled the use of images, as they might be of interest to preserve the appearance of the webpage in case geometric positioning of nodes is needed.

Date detection: Our model for date detection is title-dependent. That is, it depends on a correct classification of the title because this information is used for attribute 8 (see section 3.3) during date detection. However, when a title node is not classified, this dependency will most likely prevent our models from obtaining a correct classification.

We then employed two different models, depending on the title detection outcome. If some node is classified as title in the previous step, we proceed with the title-dependent model for dates. Otherwise, a title-independent model is used. Results for the conditional and title-dependent approaches are shown, respectively, in Tables 4 and 5.

Method	Time taken
Baseline (`<title>` tag)	0.40
Strictly structural	1.00
Structural + Visual	1.77
Structural + Visual, whole tree	3.01
WebKit rendering, no images	10.80
WebKit rendering	39.16

Table 3: Title extraction execution times on the RCD4 corpus, relative to strictly structural approach

Method	Prec.	Recall	F1
3-fold CV (extra-site)	0.88	0.66	0.75
3-fold CV (intra-site)	0.88	0.82	0.85

Table 4: Date extraction results on the RCD4 corpus for the conditional approach with two models, after post-processing

Surprisingly, we observe that the title-dependent model still seems to be the best choice as it produces more balanced results for precision and recall, specially for the extra-site cross-validation.

Body detection: For body detection, we consider every node returned from the NCE algorithm of [8] used at the start of the pipeline, excluding only those that were detected as title or publication date. As only a few nodes were excluded from the returned set, the results were unaffected, as shown in Table 6.

We experimented with a post-processing stage that would also consider all nodes between the title and the first detected body node, but the results weren't very revealing: very little have changed, and often times for the worse.

4.1 Results comparison

For the sake of comparison, we have applied our methods to the NEWS600 corpus [17], which consists of 604 pages from 177 distinct domains, all written in English. Despite the small size of our corpus, the results obtained in NEWS600 are consistent with the numbers obtained in our experiments, which reinforces the stability of our results. The results obtained are shown in Table 7, along with the best published results we have found for the same corpus.

It is worth noting that we require considerably less computational resources. Our small set of less than a dozen features behaves quite well considering that listed results tend to use tens of thousands of features and depend on a full rendering for geometric positioning.

Method	Prec.	Recall	F1
3-fold CV (extra-site)	0.79	0.75	0.77
3-fold CV (intra-site)	0.85	0.83	0.84

Table 5: Date extraction results on the RCD4 corpus for the title-dependent model, after post-processing

Method	Prec.	Recall	F1
NCE of [8]	0.82	0.92	0.87
After our pipeline	0.82	0.92	0.87

Table 6: Body extraction results on the RCD4 corpus

Approach	Label	Prec.	Recall	F1
Baseline (`<title>`)	Title	0.63	0.93	0.75
Our approach	Title	0.90	0.96	0.93
SVM of [16]	Title	0.99	0.96	0.97
Our approach	Date	0.87	0.85	0.86
CRF of [18]	Date	0.83	0.92	0.87
Our approach	Body	0.82	0.95	0.88
SVM of [16]	Body	0.90	0.98	0.94

Table 7: Results on the NEWS600 corpus, ordered by F1. The baseline and our results are measured using bag of words.

5. CONCLUSION

We presented an approach that makes use of some visual features of webpages while being approximately an order of magnitude faster than a full rendering approach.

We applied it to the task of segmenting a news webpage, obtaining satisfactory results with a small number of features and regardless of the language of the story and website where it has been published. In our experiments, we achieve at least 91% of F1 for title detection, 77% of F1 for date detection and 88% of F1 for body detection.

Future work: We would like to extend our approach to other relevant metadata in the news domain, such as author and news agency, when available, and other tasks and domains in the web where visual features might be useful, such as e-commerce and price comparison.

6. REFERENCES

[1] ÇELIK, T., BOS, B., HICKSON, I., AND LIE, H. W. Cascading style sheets level 2 revision 1 (CSS 2.1) specification. Candidate recommendation, W3C, Sept. 2009. http://www.w3.org/TR/2009/CR-CSS2-20090908.

[2] Cleaneval home page. http://cleaneval.sigwac.org.uk. [Online; accessed 18-January-2011].

[3] DOM CSS Properties – MDC Doc Center. https://developer.mozilla.org/en/DOM/CSS. [Online; accessed 13-April-2011].

[4] Gecko. https://developer.mozilla.org/en/Gecko. [Online; accessed 01-September-2010].

[5] HALL, M., FRANK, E., HOLMES, G., PFAHRINGER, B., REUTEMANN, P., AND WITTEN, I. The weka data mining software: an update. *ACM SIGKDD Explorations Newsletter 11*, 1 (2009), 10–18.

[6] HU, Y., XIN, G., SONG, R., HU, G., SHI, S., CAO, Y., AND LI, H. Title extraction from bodies of html documents and its application to web page retrieval. In *SIGIR '05: Proceedings of the 28th annual international ACM SIGIR conference on Research and development in information retrieval* (New York, NY, USA, 2005), ACM, pp. 250–257.

[7] JACOBS, I., RAGGETT, D., AND HORS, A. L. HTML 4.01 specification. W3C recommendation, W3C, Dec. 1999. http://www.w3.org/TR/1999/REC-html401-19991224.

[8] LABER, E. S., DE SOUZA, C. P., JABOUR, I. V., DE AMORIM, E. C. F., CARDOSO, E. T., RENTERÍA, R. P., TINOCO, L. C., AND VALENTIM, C. D. A fast and simple method for extracting relevant content from news webpages. In *CIKM* (2009), D. W.-L. Cheung, I.-Y. Song, W. W. Chu, X. Hu, and J. J. Lin, Eds., ACM, pp. 1685–1688.

[9] LEVENSHTEIN, V. I. Binary codes with correction of deletions, insertions and substitution of symbols. *Doklady Akademii Nauk SSSR 163*, 4 (1965), 845–848.

[10] LUO, P., FAN, J., LIU, S., LIN, F., XIONG, Y., AND LIU, J. Web article extraction for web printing: a dom+visual based approach. In *Proceedings of the 9th ACM symposium on Document engineering* (2009), ACM, pp. 66–69.

[11] MAREK, M., PECINA, P., AND SPOUSTA, M. Web page cleaning with conditional random fields. *Cahiers du Cental 5* (2007), 1.

[12] MSHTML. http://msdn.microsoft.com/en-us/library/bb508516(v=VS.85).aspx. [Online; accessed 01-September-2010].

[13] NICOL, G., CHAMPION, M., HÉGARET, P. L., ROBIE, J., WOOD, L., HORS, A. L., AND BYRNE, S. Document object model (DOM) level 3 core specification. W3C recommendation, W3C, Apr. 2004. http://www.w3.org/TR/2004/REC-DOM-Level-3-Core-20040407.

[14] Opera Presto 2.1 – Web standards supported by Opera's core – Dev.Opera. http://dev.opera.com/articles/view/presto-2-1-web-standards-supported-by/. [Online; accessed 13-April-2011].

[15] PASTERNACK, J., AND ROTH, D. Extracting article text from the web with maximum subsequence segmentation. In *Proceedings of the 18th international conference on World wide web* (2009), ACM, pp. 971–980.

[16] SPENGLER, A., BORDES, A., AND GALLINARI, P. A comparison of discriminative classifiers for web news content extraction. In *Proceedings of RIAO 2010, 9th Int. Conf. on Adaptivity, Personalization and Fusion of Heterogeneous Information* (2010).

[17] SPENGLER, A., AND GALLINARI, P. Learning to Extract Content from News Webpages. In *Proceedings of the 2009 International Conference on Advanced Information Networking and Applications Workshops* (2009), IEEE Computer Society, pp. 709–714.

[18] SPENGLER, A., AND GALLINARI, P. Document structure meets page layout: loopy random fields for web news content extraction. In *Proceedings of the 10th ACM symposium on Document engineering* (2010), ACM, pp. 151–160.

[19] TAN, P.-N., STEINBACH, M., AND KUMAR, V. *Introduction to Data Mining*. Addison-Wesley, 2005.

[20] The WebKit Open Source Project – CSS (Cascading Style Sheets). http://www.webkit.org/projects/css/index.html. [Online; accessed 13-April-2011].

[21] VAN KESTEREN, A. HTML 5 differences from HTML 4. W3C working draft, W3C, Aug. 2009. http://www.w3.org/TR/2009/WD-html5-diff-20090825/.

[22] VIEIRA, K., DA SILVA, A. S., PINTO, N., DE MOURA, E. S., CAVALCANTI, J. M. B., AND FREIRE, J. A fast and robust method for web page template detection and removal. In *CIKM '06: Proceedings of the 15th ACM international conference on Information and knowledge management* (New York, NY, USA, 2006), ACM, pp. 258–267.

[23] WANG, J., CHEN, C., WANG, C., PEI, J., BU, J., GUAN, Z., AND ZHANG, W. V. Can we learn a template-independent wrapper for news article extraction from a single training site? In *Proceedings of the 15th ACM SIGKDD international conference on Knowledge discovery and data mining* (2009), ACM, pp. 1345–1354.

[24] WANG, J., HE, X., WANG, C., PEI, J., BU, J., CHEN, C., GUAN, Z., AND LU, G. News article extraction with template-independent wrapper. In *Proceedings of the 18th international conference on World wide web* (2009), ACM, pp. 1085–1086.

[25] WebKit. http://webkit.org/. [Online; accessed 01-September-2010].

[26] WIKIPEDIA. Comparison of layout engines (cascading style sheets) — Wikipedia, the free encyclopedia, 2010. [Online; accessed 22-September-2010].

[27] World Wide Web Consortium (W3C). http://www.w3c.org/. [Online; accessed 14-September-2010].

[28] XUE, Y., HU, Y., XIN, G., SONG, R., SHI, S., CAO, Y., LIN, C.-Y., AND LI, H. Web page title extraction and its application. *Information Processing and Management 43*, 5 (2007), 1332–1347.

A Versatile Model for Web Page Representation, Information Extraction and Content Re-Packaging[*]

Bernhard Krüpl-Sypien
Ruslan R. Fayzrakhmanov[†]

Wolfgang Holzinger
Mathias Panzenböck
Inst. of Information Systems,
DBAI Group
TU Wien, Austria
{kruepl, fayzrakh,
holzing, panzen}
@dbai.tuwien.ac.at

Robert Baumgartner
Lixto Software GmbH
Vienna, Austria
baumgartner@lixto.com

ABSTRACT

On today's Web, designers take huge efforts to create visually rich websites that boast a magnitude of interactive elements. Contrarily, most web information extraction (WIE) algorithms are still based on attributed tree methods which struggle to deal with this complexity. In this paper, we introduce a versatile model to represent web documents. The model is based on gestalt theory principles—trying to capture the most important aspects in a formally exact way. It (i) represents and unifies access to visual layout, content and functional aspects; (ii) is implemented with semantic web techniques that can be leveraged for i.e. automatic reasoning. Considering the visual appearance of a web page, we view it as a collection of gestalt figures—based on gestalt primitives—each representing a specific design pattern, be it navigation menus or news articles. Based on this model, we introduce our WIE methodology, a re-engineering process involving design patterns, statistical distributions and text content properties. The complete framework consists of the UOM model, which formalizes the mentioned components, and the MANM layer that hints on structure and serialization, providing document re-packaging foundations. Finally, we discuss how we have applied and evaluated our model in the area of web accessibility.

[*]This work is funded by the Austrian Forschungsförderungsgesellschaft FFG under grant 819563.

[†]Supported by the Erasmus Mundus External Cooperation Window Programme of the European Union.

Categories and Subject Descriptors

H.3.3 [**Information Systems**]: Information Search and Retrieval; H.5.2 [**Information interfaces and presentation**]: User Interfaces—*User-centered design*

General Terms

Design, Experimentation, Human Factors

Keywords

web page model, web information extraction, web page understanding, web adaptation, gestalt theory

1. INTRODUCTION

When typical users browse the Web, they experience two main artefacts: *pages,* which are independent units of information that can be addressed by an URI, and *visual objects,* which appear on these documents. We developed a methodology that creates a novel kind of model for such web pages by using *visual objects* (*VOs*) in its core. A VO is a distinguishable item—as it is drawn on screen by a browser—of arbitrary granularity. The main benefit of our model is that it abstracts away and is independent from actual implementation details, providing a more natural and workable description of pages and VOs than common DOM tree representations.

The motivation in the development of the versatile model has been our frustration about the current pre-dominance of tree-based approaches for web page analysis and information extraction. We believe that an adequate web page model has to include all content and functional aspects of web pages, but in particular also all layout related information, including information about the visual appearance that is only available after applying rendition standards, CSS and Javascript by a web browser. Such information is usually missing from tree-based models. Contrarily, our versatile model includes all these aspects and starts from atomic VOs (single words and images) that are directly retrieved from a web browser component.

Together with the VOs themselves, our versatile model

stores (1) parts of their DOM tree structure to provide additional hints for the web page understanding process; (2) any additional facts about VOs that have been found by detection algorithms; (3) relations between VOs and grouping information; (4) logical information, which links to generic, web and domain specific concepts. We implemented our model in the form of an ontology and for now simply refer to it as the *unified ontological model* (*UOM*).

As can be seen in Figure 1, our web page representation consists of layers numbered from 1–9. Starting with the source code at the bottom (1) and going upwards, data is transformed into increasingly workable and meaningful representations. Initially, we employ web browser components at layer (2) (prior variants were already disclosed in previous publications [15]) and extract *basic VOs* (*bVOs* or *blocks*) from client-side web pages that may consist of HTML, CSS and Javascript code. bVOs are words (later joined to text lines that share a common formatting), web page elements (such as image, web form element), or graphical objects, that correspond to some arbitrary area on the canvas. bVOs at layer (3) have positions, sizes and content; text bVOs also have font style properties, which correspond to the computed CSS attributes. In the next and last step of pre-processing at layer (4), spatial relationship properties such as distances, direction and alignment are computed and written back into the model. Also, the model is optimized by removing invisible and non-significant bVOs and joining adjacent bVOs that share common formatting and visual appearance.

The further processing is structured to mimic human perception of web pages. At layer (5), we transform bVOs into VOs by augmenting them with the additional information (saliency, similarity, etc.) and strive for performing those grouping steps that we believe to be innate to every sighted human being. We identify **gestalt figures**—a group of VOs—according to gestalt theory described in Section 3.1. At layer (6), generic document knowledge is employed to identify logical objects such as tables, lists, and grids. Layer (7) uses web specific knowledge to identify logical objects such as navigation menus or pagination elements sharing common web patterns. At layer (8), domain specific knowledge is employed to identify such logical objects as blog, news articles, comments, etc. Finally, at layer (9), we make use of all the information retrieved so far to re-order and re-package the identified concepts for different target groups such as blind or mobile users.

The upper layers (5–8) employ a number of different enrichment methods. Bottom-up approaches are used to identify low-level structures in the UOM; these approaches correspond to the scanning that sighted users apply when they look for specific information. Top-down approaches deal with web pages from a high-level perspective, utilizing statistical properties that can only be realized when considering a page as a whole. Common to all of these approaches is that they enrich the UOM of a particular page; thus, they all share a common interface and can be implemented using a wealth of different languages. We implemented *enrichers* in a number of different programming and query languages. Also, a new, dedicated pattern language (*SWPML*) has been developed in the course of our work that enables an easy formulation of enrichers.

Enrichers for all layers 5–8 have already been devised manually. In a case study, we also created a prototype implemen-

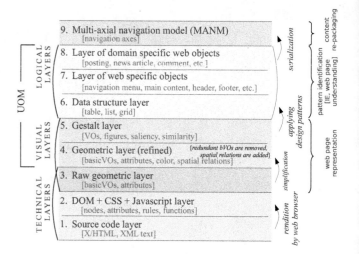

Figure 1: Layers of web page representation and understanding.

tation for all other parts of the model, data acquisition and transformation, and thus believe that we can demonstrate the benefit of our approach.

The remainder of the paper is organized as follows: in the next section we discuss related art. In Section 3 we first describe the theoretical foundation of our system, which is the gestalt theory, and continue by presenting details of our model. Section 4 shows how we make use of our model for information extraction. Section 5 explains how we exploit the model for content re-packaging with our multi-axial navigation method. Section 6 discusses how we evaluate the feasibility of our model by performing a case study in web accessibility. Finally, we conclude with Section 7.

2. RELATED WORK

While we are not aware of any related work that uses a gestalt model for web page understanding and web information extraction, a number of researchers have addressed web information extraction problems by following a visual approach.

In [18], Spengler et al. consider both tag tree and visual web page representations for automatic content extraction for the example of news articles. The problem is formulated as a classification task, where every region of web page layout is associated with the respective label from a predefined set. The authors apply loopy CRF and represent the web page model as a graph, which reflects a sequence of leaf nodes according to depth-first traversal over a tag-tree and relations between neighbouring regions that have similar visual features, such as size, colour, style, etc.

In [13], He et al. use the VIPS algorithm for web page segmentation and extraction of images along with their captions, building relations between images and linked web pages, and improving the precision of information retrieving. For the VIPS algorithm, relations such as the distance between neighbouring visual objects is utilized, where visual objects are DOM tree elements. Thereby, this approach is limited to the solution of some particular IE tasks.

In [14], Kovacevic et al. represent a web page as an adjacency multigraph, in which nodes correspond to simple HTML objects (text, image, web form element) while arcs represent spatial relations between neighbouring elements, such as "immediately before", "immediately after", "immediately left", "immediately right". Their multigraph is a labelled graph, and every arc has a distance value in pixels and boolean values, which express the existence of alignment relations between objects along the axes x and y. They use this representation for web page classification.

In contrast, our approach of web page representation is suitable for wider range of tasks in the fields of information extraction and document understanding and uses different layers of granularity for the visual appearance and perception of web pages. Furthermore, our model is also suitable to be used with a large variety of methods and algorithms.

Related work from the accessibility community comprise the following: In [17, 7, 6], the HearSay screen reader is introduced. The authors perform a structural and semantic segmentation of the DOM tree and represent web pages as a semantic partition tree. An XML-based audio dialog manager provides the navigation and speech output interface to the visually impaired user. HearSay also uses geometric clustering as one part of its segmentation algorithm. CSurf [16] is an extension of the HearSay screen reader with a context analysis model.

In contrast, our multi-axial navigation model (the MANM, used for content-repackaging), is created by analysing the visual appearance of web pages and represents element sequences according to their semantic and spatial relations. This model allows us to serialize documents in different ways, considering a variety of constraints such as a set of predefined object types that can appear on the web page, the maximal quantity of elements for serialization, etc. This model, along with the navigation methodology, has been described in one of our previous works [10] from an accessibility point of view.

3. GESTALT THEORY AND UNIFIED ONTOLOGICAL MODEL

3.1 Gestalt theory

In order to overcome ad-hoc approaches, we were looking for a psychological theory that would help us in getting a better understanding of the innate perceptive abilities of humans. We found such a theory in the gestalt theory, which was developed in the early 20th century [19]. The gestalt theory provides us with exactly those concepts that are needed to group VOs (i.e. gestalt primitives) into the large objects, gestalt figures, and to identify relationships between them, thereby improving processes of web page understanding and IE.

Gestalt theory was established by a group of psychologists in the beginning of the 20th century [19]. It is an attempt to explain how humans group single perceivable parts into a larger form or shape (*gestalt* in german). The resulting gestalt is of a special quality as it has to be considered as a wholeness, i.e. it is holistic and more than the sum of its parts.

The *praegnanz* principle says that humans strive for identifying shapes that look as good as possible, whereas good refers to a set of gestalt laws such as proximity, symmetry or closeness. Depending on the author, a varying number

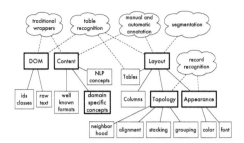

Figure 2: Parts of the UOM.

of gestalt laws were stated, which include: **(1)** *law of figure and ground*, which is about the distinction of foreground from background, based on perceptual saliency, where more important objects are perceived as being in front; **(2)** *law of proximity*, we perceive close objects as grouped together; **(3)** *law of similarity*, we perceive similar objects as grouped together; **(4)** *law of continuity*, oriented parts are integrated into a perceived whole if they are aligned with each other; **(5)** *law of closure*, objects are grouped together when enclose a shape; **(6)** *law of common fate*, objects are perceived as grouped if they share common temporal behaviour such as movement.

These laws allow us to add additional features that are based on perception to our web page model. These features provide us with the necessary information to create robust algorithms that mimick the way humans recognize elements on web pages. Our approach is bootstrapped from the structural rendition information and CSS model that is generated by and accessible from a web browser. In contrast to more general computer vision tasks, our atomic elements where analysis starts from are CSS boxes and not pixels; we do not have to deal with low-level information at all. We also can start from visual elements such as words that can be accessed via the web browser API and calculate gestalt features for them.

Gestalt theory together with its laws build adequate and sound foundation to represent web pages for information extraction and content re-packaging, as we will show in more detail now.

3.2 Unified Ontological Model

To make a web page understandable for the computer, we automatically generate its ontological model, taking into account its visual representation. We call this model the **unified ontological model** (UOM) [9]. It consists of two main layers: a *visual* and a *logical* layer, and it is encoded in RDF, with concept definitions spelled out in OWL.

The *visual layers* of the UOM are a conglomerate of several specialized ontologies that provide us with a rich set of visual characteristics. They are based on the *geometrical model* (GM) and the *gestalt model* (GsM) (layers 4–5, cf. Figure 1). The main object of the GM is a **basic visual object** (bVO or block), which is a rectangle wrapping some particular area on the web page canvas and defined by the coordinates of its top-left and bottom-right corners. The visual properties of a bVO are defined by a set of CSS boxes, a single CSS box, or by one of its components (e.g. its left border or a single word of its textual content). The main attributes of a bVO are its background features (colour or background image), border (width, colour, style), text content (text, text colour, font size and style), type of the wrapped information (text, image, web form element, etc.) as well as the painting order, which is used to identify the visibility of overlapped bVOs. The painting order is calculated according to the visual formatting model defined in the W3C CSS specification [4]. The GM is also capable to represent spatial relations between bVOs such as distance, direction, alignment ("left-aligned", "right-aligned", "top-aligned", "bottom-aligned", etc.) and topological relations (we use RCC8 [8]). Distance and direction can be expressed either quantitatively (using pixels and degrees) or qualitatively (using linguistic variables and membership functions that are defined for their values). Qualitative values have been defined experimentally: E.g., useful values for the distance relation are "very close", "close", "not far", "far", and "very far"; practical values for the direction relation are "north-of", "north-east-of", "east-of", etc.

The main objects of the GsM are *visual object* (*VO*) and *figure*. A **VO** is a visible object of a web page enriched with concepts such as font saliency, colour entropy, similarity, etc. and corresponds to an arbitrary set of bVOs. A **figure** is a visual object created by utilizing gestalt laws and corresponds to a set of VOs.

As explained in Section 3.1, we leverage gestalt laws to add additional visual features and relations to our visual objects. Practically, we defined RDF classes and attributes for these gestalt laws: For **(1)** the gestalt law of figure and ground, we add foreground and background RDF classes for VOs, which stack objects on top of each other based on their painting order. Also, the computed CSS attribute *background − image* is considered and the RDF background class will be set accordingly by our system. In contrast, the content of the CSS box will be considered as foreground. For **(2)** the law of proximity, we use distance relationships in the GM. These spatial relations are used for finding sets of neighbouring VOs and for hierarchical segmentations [12]. For **(3)** the law of similarity, we consider visual features such as color, font saliency, position, size, etc., which are utilized in our top-down approach (cf. Sec. 4.1). For **(4)** the law of continuity, the UOM contains alignment and direction relations between VOs, which are used in tasks such as list, grid, and table detection [11]. To support **(5)** the law of closure, the UOM provides information about position, size and spatial relations between VOs, as well as their shape. According to **(6)** the law of common fate, we store some of the dynamic properties of VOs by checking for the presence of possibly shared Javascript events such as mouseOver in the DOM tree.

The visual layers provide all necessary information that is appropriate for automatic reasoning and applying design patterns for semantic interpretation of VOs and spatial relationships defined between them (cf. Sec.4), which is important both for IE and web page understanding. This interpretation is done thanks to the logical layers.

The logical layers of the UOM are used to describe *logical objects* (LO) of a web page and to establish relationships between VOs and LOs (layers 6–8, cf. Figure 1): They describe the content and the functional aspects of a web page. There are three main levels of abstraction: (i) data structure representation model for describing structures such as table, grid, list; (ii) model of web specific web objects which allows us to represent general web specific concepts as navigation menu, main content, header, footer, etc; (iii) model of domain specific web objects, where concepts such as posting, news article, comment, etc. can be represented.

The process of linking VOs to LOs is a web patterns identification process which is done via applying enrichers. This process can be represented as a search of particular design patterns on the web page that correspond to the particular LOs along with relations defined between them. Enrichers can be encoded as naked Java code working directly on the RDF model, a useful method to describe patterns that involve complex computation. For simpler figures that do not involve complex computation, SPARQL queries are sufficient and our preferred way to recognize patterns in the model.

Listing 1 Example of a visual object in Notation 3 format (abridged)

```
@prefix  :    <http://www.dbai.at/apage#>.
@prefix g:    <http://www.dbai.at/gm#>.
@prefix gs:   <http://www.dbai.at/gestalt#>.
@prefix l:    <http://www.dbai.at/lm#>.

:c127_128_129
  a  gs:Foreground ,g:HtmlText ,g:HtmlElement ,
     l:NavMenuItem ;
  g:colorCode "rgb(0,_51,_153)" ;
  gs:colorEntropy "0.34792330342";
  g:coordinateX1 "706"^^xsd:int ;
  g:coordinateX2 "864"^^xsd:int ;
  g:coordinateY1 "355"^^xsd:int ;
  g:coordinateY2 "370"^^xsd:int ;
  g:eastOf :c450_451_452 , :c460_461_462 ;
  g:fontCode "arial;11;normal;400" ;
  gs:fontSaliency "68.7608215"^^xsd:double ;
  g:height "15.0"^^xsd:double ;
  g:htmlStringValue "Press_&_Events" ;
  g:leftAlignedWith :c120_121 , :c133_134 ;
  g:marginNorth "16.0"^^xsd:float ;
  g:marginSouth "0.5"^^xsd:float ;
  g:marginWest "41.0"^^xsd:float ;
  g:northOf :c133_134 ;
  g:southOf :c120_121_122 ;
  g:width "158.0"^^xsd:double ;
  l:partOf :c925_492 .
```

Listing 1 displays the part of the UOM that describes some geometric, gestalt and logical properties of a visual object, which is stored in RDF and shown in Notation 3. As can be seen, we introduced RDF classes for visual features such as color entropy or font saliency that we use to apply gestalt laws.

4. INFORMATION EXTRACTION

In this section, we show how we leverage our model for information extraction by querying it. Page analysis takes place from layer (5) of our model in Figure 1 and can be considered as a continued enrichment process, which writes additional, logical information into the model at layers (6–8).

We start with a top-down method that considers the page in its entireness in Sec. 4.1. The following subsections discuss bottom-up methods that make use of additional knowledge: Sec. 4.2 shows how we leverage the semantic web representation of our model to create SPARQL queries that can find web patterns, and Sec. 4.3 introduces our own, dedicated web pattern matching language.

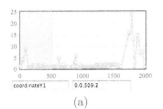 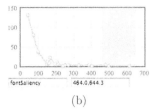

(a)　　　　　　　　　　　(b)

Figure 3: Visual object distribution graphs (VODGs) showing y1 coordinates and font saliency values.

4.1 VO Distribution Graphs

The gestalt layer (5) in Figure 1 provides access to a wealth of interesting information about a web page. Following gestalt principles, we start from the page as-a-whole and base our first method on statistical properties. Our method extracts common visual properties from VOs and strives for identifying gestalt figures based on these features. In particular, it employs the gestalt law of similarity, as it can be used to easily select various VOs anywhere on the page that share one or more visual properties, mimicking the way humans assess a page according to the similarity gestalt law. For each of these visual properties, statistical data is computed, and it is displayed how many VOs on the web page share the same value for a property, which can later be used for interactive or automatic analysis.

Figure 3 shows the distribution of two VO properties on a web page. Figure 3.a displays the vertical start positions of VOs, and Figure 3.b shows their font saliency[1] values. In both diagrams, the y axis denotes frequency (number of VOs that have a particular property value). These graphs are actually **VO property distribution graphs (VODGs)**. VODGs provide a view on a web page that is radically different from tree based representations, which commonly use XPath to select particular regions of a page. Contrarily, they provide a top-down overview of the page.

In fact, we can use VODGs for (a) *interactive selection of VOs* and (b) *page classification.*

Interactive selection of VOs. We developed an interactive tool, where users can easily select regions in VODGs to drill down to interesting objects in a fast way. Graphs can be generated for any VO property of interest. The example shows how to select all large text objects that appear in the top quarter of the page: In Figure 3.a, you can see a range that selects all VOs with *coordinateY1* positions from 0 to 509.2. In Figure 3.b, those VOs are selected that have a *fontSaliency* value which is greater than 464.

In our current prototype, we can display and use VO properties such as: *coordinateX1*, *fontSaliency*, *marginNorth*, *colorEntropy*. We can store range selections in three possible way: (1) absolute range values; (2) relative range values; (3) range values that are relative to statistical properties such as mean or median. Stored selections can be recalled on new web page instances. Also, our prototype system provides means to compose several VODG range selections in a boolean language. Users can use this method for the

intuitive and fast selection of sub-parts of a page, e.g. a particular column, all headings, the page footer etc.

Page classification. VODGs could also be used for automatic page classification and genre detection. Figure 3.a shows a typical bottom-heavy page with its main content towards the end of the page—the bulk of VOs starts just after 1500px down the page; Figure 3.b shows a typical font saliency value distribution of a newspaper web page—there is much text with small font saliency, and a decreasing number of text VOs with higher font saliency. Such an automatic classification algorithm has yet to be implemented by us though. We can also see how machine learning techniques could be applied for the automatic classification of pages starting from a learning set.

4.2 SPARQL

Whereas VODGs are an excellent tool for gestalt analysis and the top-down assessment of web pages, we can also do information extraction in a bottom-up manner. The bottom-up method is another way to employ gestalt laws and is particularly useful for the gestalt laws of proximity and continuity, e.g. when it takes into account properties describing distance and alignment.

Due to the UOM representation, we can describe our knowledge of common web patterns by specifying SPARQL queries that operate on the gestalt layer (5) of our model. These queries take into account VO properties such as neighbourhood and alignment, but also more advanced gestalt properties such as common fate, which is temporal behaviour that some VOs might share (e.g. a pop-up).

In Listing 2, we aim to identify web patterns that appear regularly on newspaper websites and which we call *article teasers*.

Article teasers are short summaries of articles that are shown on front pages in order to gain interest at the reader. They are characterized by a heading, which is text that is above of and larger than other text, and an image, often left of the teaser text. With our language and the help of SPARQL, we can describe exactly that: the query selects two text objects $t1, t2$ and an image *img*, where text $t1$ appears above $t2$ (*northOf*) and is in larger font than text $t2$ (has a larger *fontSaliency*, which is a measure for the perceived font size that has been defined by us). Finally, there has to be an image *img* which is positioned left of (*westOf*) the teaser text. We found that this description is (i) more intuitive for describing web patterns, and (ii) a quite general form to describe web patterns that works also on many unknown web pages. A thorough quantitative evaluation has yet to be undertaken.

SPARQL/Update is used to write the results back into the model by tagging individual VOs as part of an article instance. In the example of Listing 2, the heading of such article is marked with the *ArticleStart* type.

4.3 Simplified Web Pattern Matching Language

As discussed, the RDF graph representing the UOM can be queried using SPARQL. SPARQL is an ideal language to query rich ontologies. It enables you to retrieve node sets using complex conditions over the relations of these nodes. The node sets have a defined number of named members. Many visual patterns on the Web though contain a variable number of elements and thus need to be represented as mul-

---

[1]Font saliency is one of our gestalt properties that is a combined measure of font size, font weight, colour contrast and colour uniqueness. Texts with large font saliency are perceived as *standing out*.

Value returns in buyers market

"AFFORDABLE" IS the new buzzword in the Irish art market and Dolan's, the Galway-based art auction house, was keen to emphasise value ahead of its sale at the Rochestown Park Hotel in Cork last Sunday. Some 234 lots went under the hammer and auctioneer Niall Dolan said 85 per cent sold.

Figure 4: Example of an article teaser.

Listing 2 SPARQL query that finds article teasers – with a heading and a picture – in the UOM

```
PREFIX g: <http://www.dbai.at/Gestalt#>

INSERT { ?t1 rdf:type l:ArticleStart }

WHERE {
    ?t1  rdf:type g:HtmlText .
    ?t2  rdf:type g:HtmlText .
    ?img rdf:type g:HtmlImage .
    ?t1  g:northOf ?t2 .
    ?img g:westOf ?t2 .
    ?t1  g:fontSaliency ?f1 .
    ?t2  g:fontSaliency ?f2 . FILTER (?f1 > ?f2)
}
```

---

tiple SPARQL queries connected through logic outside of SPARQL.

A simple example for such patterns would be tables with a variable number of rows and columns, pagination elements or even words in a line and lines in a paragraph. The pattern in Listing 2 only looks at single words and not whole lines or paragraphs. This is because the whole text of the analysed document is tokenized so that the coordinates of each word can be determined. After all, a text node can span multiple lines. For some applications, finding single words is what you want, but for the article teaser pattern it would be better to find whole lines and paragraphs.

A line is basically a sequence of words; a paragraph is a sequence of lines. Using SPARQL, one can select the first word of a line and use consecutive queries to find the successive words of the line. After the first line of an paragraph is found, one can use further SPARQL queries to find successive lines. This rather simple task needs quite a lot of SPARQL queries to be performed and logic written in another language to glue the queries together.

This was the reason for the idea to develop the **Simplified Web Pattern Matching Language** (the **SWPML**), a domain specific language implemented as a Java API. This new API uses concepts loosely inspired by regular expressions. This means simple patterns can be repeated, marked optional and grouped to more complex patterns. The biggest difference to regular expressions lies in the way two patterns are connected. For regular expressions, the connection is defined through the one-dimensional direction of the text. With SWPML, the condition how two sub-patterns are connected has to be expressed explic-

itly, because this relation is not implicitly clear in the two-dimensional space of a web page.

Listing 3 Code snippet that finds article teasers in SWPML syntax

```
// Find an image and call it "img"
image().as("img")

// Pair the image with some other pattern:
.pair(

// The other pattern is a vertically
// aligned pair of text blocks.
// The top block is called "h" (header)
// and the bottom one "t" (text):
vpair(block().as("h"), block().as("t"))

// The font saliency of the header has
// to be bigger than the one of the text:
.where(
    "h.first.first.font_saliency > " +
    "t.first.first.font_saliency"))

// The image has to be left of the text:
.where("img.left_of(t)")
```

A pattern to find article teasers then looks like the Java code shown in Listing 3. This expression returns a pattern object which can be passed to a pattern matching engine that searches through a RDF graph. In order to do so, the engine translates the pattern to multiple SPARQL queries.

Listing 4 Helper methods

```
public static RepeatedPattern line() {
    return horizontal(word().repeat(1, -1))
        .uniform("$0.font_saliency");
}
public static RepeatedPattern block() {
    return vertical(line().repeat(1, -1))
        .uniform("$0.first.font_saliency")
        .uniform("$0.x1");
}
public static ConnectedPattern horizontal(
        ConnectedPattern pattern) {
    return pattern.by("$0.left_of($1)");
}
public static ConnectedPattern vertical(
        ConnectedPattern pattern) {
    return pattern.by("$0.top_of($1)");
}
public static PairPattern vpair(
        Pattern top, Pattern bottom) {
    return vertical(top.pair(bottom));
}
```

The pattern uses a number of static methods, like **word** or **image**, which build basic patterns that find HTML elements containing an arbitrary word or image. The **vpair** (vertical pair), **horizontal** and **vertical** methods are helper functions. These methods define the spatial relation between certain patterns. Their definition is shown in Listing 4. You can specify that a pattern is repeated by calling its **repeat** method. This method returns a pattern object providing a **by** method, which has to be used to declare the condition through which the found sub-matches are connected. This condition is expressed as a string, where $0 and $1 are references to a pair of matches of the sub-pattern. The **uniform** method is a convenience method. It can be used to define the uniformity of a certain property within a repeated pattern. E.g. `uniform("$0.x1")` translates into `by("$0.x1 == $1.x1")`.

In the `line` and `block` methods (also in Listing 4), the `repeat` method is used. In these functions, the increased expressiveness of SWPML can be observed. While the rest can be expressed purely in SPARQL, the repetition of sub-patterns needs more than SPARQL alone can provide.

The major benefit of SWPML is conciseness and thus readability and maintainability of the patterns. No logic outside of SWPML is needed in order to describe complex patterns with variable elements. Also, the API abstracts from the RDF representation of the UOM and will thus allow the SPARQL based backend to be replaced without the need to rewrite any pattern. This might be useful if it turns out that SPARQL is too slow and needs to be replaced by a domain specific query engine that heavily optimizes on geometric query operations.

5. CONTENT RE-PACKAGING

Content re-packaging, as we use the term within this paper, is the process of transforming the information from one representation—in our case the logical layers that we derived from a web page—into another output medium. Our approach is based on dynamic re-ordering of information for different needs and the serialization thereof. Our targets for content re-packaging are: (i) text–to–speech transformation for accessibility (ii) reformatting for mobile clients (iii) restructuring for automated information extraction. In order to perform these processes, it is not sufficient to consider the page as just a bag full of recognized figures: the grouping, nesting and ordering structures must be discovered and preserved in a way that survives the transformation process into a new target medium. Therefore, we devised another model, which we call **multi-axial navigation model** (the **MANM**) to provide the basic infrastructure that is needed for dynamic re-ordering and serialization. The MANM, along with the navigation methodology for blind users and its implementation, was presented in our earlier work (cf. [10, 5]). In this paper, we would like to present the MANM formally from the viewpoint of content re-packaging.

The *multi-axial navigation model* [5, 10] corresponds to the ninth layer of our layered representation of a web page (cf. Fig. 1). It provides us the means to re-package web page structure for different purposes, based on its logical structure. The MANM is built on top of the UOM and contains a set of generated axes that are applied to the visual representation of a web page with a glance both on its logical structure. Axes are the basis for serializing logical elements in different ways. An *axis* A is a sequence $(a_1, a_2, \ldots, a_n)$, where n is the length of the axis A and $n \geq 0$. An axis is a serialization of the logical objects of a web page. If $n = 0$ then the axis A is *empty*. An *axis type* T refers to the set of axes which share the same semantics or are generated using the same algorithm. The axis type represents an abstract rule for reading certain concepts of the underlying model. Both axes and axis types should have meaningful names that describe their purpose or name of the corresponding logical object.

There are several features that can be defined for an axis: *length, coverage, direction.*

Length of the axis A is a quantity of elements in the axis and is represented as $\|A\|$. Length can give an estimation of

the axis extent which can be expressed for instance as the time needed to read the axis. If an axis corresponds to the particular logical object, then length gives us an estimation of its size from the viewpoint of a serialization.

Coverage of the axis A is the ratio of the quantity of LOs of the web page that are located in axis A against the quantity of all LOs on the web page: $\|A\|/|\Theta|$. It gives us an estimation of the amount of information that we get following this axis relative to all information on the web page.

A *spatial direction* (or *direction* of an axis is defined whenever each two successive elements are related by the same spatial direction relation. If the direction relation is east–of or west–of, the spatial direction of the axis is horizontal; if the direction relation is north–of or south–of, the spatial direction is vertical. These two directions are the most important ones encountered in practice. More generally, we can define the direction as the angle between the horizontal screen axis x and the line interpolated through the geometric positions of the axis' elements. If this interpolation is not possible, the axis does not poses a spatial direction (undefined direction).

For every item of some axis A, concepts such as *spatial* μ_{sp}^A (1) and *semantic* μ_{sem}^A (2) *neighbourhoods* can be defined.

$$\mu_{\mathrm{sp}}^A(a_i, \varepsilon_{\mathrm{sp}}) \equiv_{def} \{\langle j, B, d \rangle \mid b_j \in B \wedge B \in \mathcal{A} \wedge B \neq A\}, \quad (1)$$

where $0 \leq d \leq \varepsilon_{\mathrm{sp}}$ is a distance between $a_i \in A$ and b_j which can be gotten from the GM; $\mathcal{A}$ is a set of all axes defined on the web page.

$$\mu_{\mathrm{sem}}^A(a_i, \varepsilon_{\mathrm{sem}}) \equiv_{def} \{\langle j, B, w \rangle | b_j \in B \wedge B \in \mathcal{A} \wedge B \neq A\}, \quad (2)$$

where $\varepsilon_{\mathrm{sem}} \leq w \leq 1$ is the strength of the semantic relation between LOs $a_i \in A$ and b_j which is available from the LM.

The *semantic relation* can be defined for the pair of axis types T_A and T_B as well. It is a triple $\langle T_A, T_B, w \rangle$, where $w \in [0, 1]$. *Semantic neighbourhood* μ_{ax}^A for the axis A on the web page can be defined as follows:

$$\mu_{\mathrm{ax}}(A, \varepsilon_{\mathrm{ax}}) \equiv_{def} \{\langle B, w \rangle \mid B \in \mathcal{A} \wedge B \neq A\},$$

where $\varepsilon_{\mathrm{ax}} \leq w \leq 1$ is the strength of the semantic relation between axis type T_A ($A \in T_A$) and T_B ($B \in T_B$). The semantic neighbourhood between LOs and types of axes can be additionally restricted by distance.

Axes in MANM have their role as a solid navigation structure for web pages, the basis for creating serialization of a page. The pairwise relations between axes allow us to choose those paths through pages and across axes that accommodate the requirements of the users. Consider the case of re-packaging a web page's structure based on the LOs *navigation menu*, *news article* and the first three comments that are related to it: At first, we retrieve all items from the navigation axis; after that we retrieve all items from the article axis which goes through its content and, considering spatial relations, we retrieve the first three items of the axis that traverses the comments (cf. [5]).

6. CASE STUDY IN WEB ACCESSIBILITY

In order to evaluate the utility of our model, we performed a case study in web accessibility. We implemented an end-to-end prototype application (ABBA) that goes all the way through the enrichment stack from layers 1–9, using and perusing our model. Our case study was to build an enhanced web screen reading environment for blind users on top of our model.

Screen readers are programs that read the content displayed on computer screens and convert it into audio (using text-to-speech) or tactile information (using special Braille output devices). ABBA's goal is not to replace existing screen reading software such as JAWS [1] or WindowEyes [2], but to inject additional navigation options for the blind user. In our case study, we implemented a web application that can be used as a proxy for a user's screen reading software. Fig. 5 shows ABBA's architecture.

In our prototype, we distinguish between *design phase* and *run-time phase*, and also between two user roles of *blind users* and *expert users*:

At the *design phase*, expert users define LOs and enrichers to extract visually interesting patterns and ways to navigate through them. At the fully automatic *run-time phase*, blind users make use of these definitions and get access to enhanced navigation options on web pages, i.e. access to logical objects.

We decided to pick up the domain of newspaper websites and implemented the full stack of our versatile web page model (cmp. with Figure 1) as follows:

Design phase. Expert users bootstrap the process by working on the logical layers 6–8 of our model—in fact and for the sake of our experiment, we took the role of expert users by ourselves. In a first step, we defined LOs for each of the layers 6–8. We assume that objects defined on layer (6) are document generic and can be recognized by the vast majority of literate, sighted people. We further assume that objects on layer (7) are known to most web users, and that special domain specific knowledge is only needed for objects on layer (8).

In our case study, we were quite simplistic about the data structure layer (6) of our model. Instead of creating sophisticated algorithms to extract objects such as lists and tables from visual information, we just relied on the limited support for identification of tables and lists that web screen readers already provide. This is commonly completely HTML tag based and thus prone to errors. To overcome this problem in particular in table recognition, we did our own extensive research that considers cases tables have to identified purely on visual appearance [15][11], and we plan on integrating these methods to our system. For now, we just defined a few LOs for lists and tables and took the shortcut.

We spent much more effort on layers 7 and 8. We identified common web patterns that appeared frequently in the newspaper domain. Some of these web patterns are: *article teasers, pagination blocks, headers* and *footers*. We defined LOs that reflect each pattern, and re-engineered web patterns, partly with the help of web pattern libraries [3]. For each of the patterns, query code was generated. E.g., Section 4 contains coded examples for the identification of article teasers.

There is no technical difference between our model layer 6–8, but they are useful because they draw the lines of reusability. When adding a new domain, definitions of layers 6 and 7 will be reused. The distinction of layers 6 and 7 will allow us to separate document generic methods from methods that take into account the dynamic, temporal and linking properties of web pages.

The system was implemented in Java, Jena, and Ruby,

additionally using modelling tools such as Protege. Currently, we do not use a dedicated triple store but simply use file system serializations. Queries are stored directly in the model.

Run-time phase. At run-time, screen reader users access our system by directing their own environment (text-to-speech or Braille line) to our web application proxy. Our ABBApp prototype—implemented in JRuby on Rails— forwards any HTTP request it receives to a Firefox 4 based web browser component that we created. The component turns Firefox into a headless rendition engine that works in fully autonomous mode, i.e. any user interaction and dialogs are suppressed. Moreover, it implements a TCP/IP server that accepts URIs as input and returns web pages in our own geometric representation (layer 3) as output. The representation is a set of bVOs in RDF Notation3 triple format, i.e. a more basic version of List. 1 that contains only coordinate, content and CSS formatting related triples. Our Firefox component also exposes an interface that informs about any run-time changes to a page (e.g., when new page content is added after a few seconds), which is not currently used in our ABBApp prototype. Using this component, source code from layer (1) is retrieved and resolved by the component itself into layer (2) format, and bVOs in layer (3) format a returned to and stored by the prototype.

At this stage, the simplification process is triggered that groups adjacent bVOs that share the same formatting together, creating full-fledged VOs that also contain spatial relations between them. The result is written back into the model, leaving us with a layer (4) type representation of the page. Now, gestalt related enrichers started that compute gestalt properties and add it to the model (cmp. Sect.3.1). At this point, the model is ready to be queried.

Pre-defined web pattern queries are run against the model to identify known web patterns. In our prototype, these queries can work top-down VODG style or bottom-up. We defined queries for navigation blocks, headers, footers, article teasers, newspaper articles, pagination and carrousel controls. Any positive results will be augmented into the model, establishing VO–LO relations. This rich model is then the input to the MANM layer (9), which generates navigation axes for the user. Finally, the axes are injected as additional jump lists into the original HTML of the page, which is returned to the user's screen reader.

Overall, it was rewarding to see that the model proved to be sufficient in handling the case study. We were able to represent concepts on input, intermediate and output levels of the different components, all joined together in a single, unified ontological model. The versatility of the model allowed us to create purely gestalt based queries, queries that mix different aspects, and to augment VOs at any time, when an enrichment process could contribute additional facts. The ABBApp prototype system will be made available on-line.

7. CONCLUSION AND FURTHER WORK

This article presented our versatile model for web page representation, information and content re-packaging. We introduced the 9-layer enrichment stack and our unified ontological model at the centre, where processes start from the bottom by transforming and augmenting visual objects into more and more workable and representation, finally extracting interesting web patterns and figures.

We introduced our gestalt model approach, which is soundly

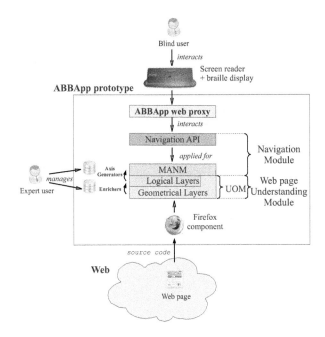

Figure 5: Architecture of the ABBApp prototype.

based on gestalt theory and lets us deal with web pages in a much more meaningful way than common tree-based methods.

We then showed how to leverage our model to identify web patterns, drilling down top-down using distribution graphs, and going bottom-up with the help of queries, also defined in a new, dedicated pattern matching language that we created to tackle that problem.

We did extensive research on content re-packaging and serialization and presented our navigation methodology that is based on the notion of axes.

Finally, we provided information about our case study in web accessibility, which shows that our model is useful in at least this particular application domain.

In summary, we believe that our approach is novel regarding the way how we consider and interact with web pages, both for information extraction and content re-packaging. Based on actual web page renditions, our versatile model provides unified access to a variety of page object properties.

In the future, we are going to continue work on our model and approach in these directions: We will be looking into automatic page classification based on gestalt properties; we will add more enrichers, providing a richer set of properties for the upper layers of our model; we will continue our work with gestalt laws and in particular will look into temporal behaviour of VOs; finally, we will continue look for even more natural descriptions of web patterns, as they appear in real life.

8. REFERENCES

[1] Freedom Scientific: JAWS for Windows Screen Reading Software (retrieved April 2011). http://www.freedomscientific.com/products/fs/jaws-product-page.asp.

[2] GW-Micro - Window-Eyes (retrieved April 2011). http://www.gwmicro.com/Window-Eyes/.

[3] Welie Design Pattern Library (retrieved April 2011). http://www.welie.com/patterns/.

[4] Cascading Style Sheets Level 2 Revision 1 (CSS 2.1) Specification, 2009.

[5] R. Baumgartner, R. R. Fayzrakhmanov, W. Holzinger, B. Krüpl, M. C. Göbel, D. Klein, and R. Gattringer. Web 2.0 vision for the blind. In *Proc. of Web Science Conference 2010 (WebSci10)*, page 8, Raleigh, USA, 2010.

[6] Y. Borodin, J. P. Bigham, A. Stent, and I. V. Ramakrishnan. Towards one world web with HearSay3. In *Proc. of the International Cross-Disciplinary Workshop on Web Accessibility (W4A' 08)*, pages 130–131, New York, USA, 2008. ACM Press.

[7] Y. Borodin, J. Mahmud, and I. Ramakrishnan. The HearSay Non-Visual Web Browser. (Vxml):128–129, 2007.

[8] A. G. Cohn. *Qualitative spatial representation and reasoning techniques*, volume 1303, pages 1–30. Springer Berlin, Berlin, Germany, May 1997.

[9] R. R. Fayzrakhmanov, M. C. Göbel, W. Holzinger, B. Krüpl, and R. Baumgartner. A Unified ontology-based web page model for improving accessibility. In *Proc. of the 19th international conference on World Wide Web (WWW'2010)*, pages 1087–1088, Raleigh, USA, 2010. ACM.

[10] R. R. Fayzrakhmanov, M. C. Göbel, W. Holzinger, B. Krüpl, A. Mager, and R. Baumgartner. Modelling Web navigation with the user in mind. In *Proc. of the International Cross Disciplinary Conference on Web Accessibility (W4A '2010)*, page 4, Raleigh, USA, 2010.

[11] W. Gatterbauer, P. Bohunsky, M. Herzog, B. Krüpl, and B. Pollak. Towards domain-independent information extraction from web tables. In *Proceedings of the 16th international conference on World Wide Web*, WWW '07, pages 71–80, New York, NY, USA, 2007. ACM.

[12] W. Gatterbauer, B. Krüpl, W. Holzinger, and M. Herzog. Web information extraction using eupeptic data in web tables. In *In Proceedings of the 1st International Workshop on Representation and Analysis of Web Space (RAWS 2005)*, pages 41—-48, Prague, Czech Republic, 2005. VSB - Technical University of Ostrava.

[13] X. He, D. Cai, J.-R. Wen, W.-Y. Ma, and H.-J. Zhang. Clustering and searching www images using link and page layout analysis. *ACM Trans. Multimedia Comput. Commun. Appl.*, 3, May 2007.

[14] M. Kovacevic, M. Diligenti, M. Gori, and V. Milutinovic. Visual adjacency multigraphs — a novel approach for a web page classification. In *Proc. of the Workshop on Statistical Approaches to Web Mining (SAWM'2004)*, pages 38–49, 2004.

[15] B. Krüpl and M. Herzog. Visually guided bottom-up table detection and segmentation in web documents. In *Proceedings of the 15th international conference on World Wide Web*, WWW '06, pages 933–934, New York, NY, USA, 2006. ACM.

[16] J. Mahmud, Y. Borodin, and I. Ramakrishnan. CSurf:

A Context-Driven Non-Visual Web-Browser. 1(c), 2007.

[17] I. V. Ramakrishnan, A. Stent, and G. Yang. Hearsay: enabling audio browsing on hypertext content. In *Proc. of the 13th International Conference on World Wide Web (WWW '04)*, pages 80–89, New York, NY, USA, 2004. ACM.

[18] A. Spengler and P. Gallinari. Document structure meets page layout: loopy random fields for web news content extraction. In *Proc. of the 10th ACM Symposium on Document Engineering (DocEng'10)*, pages 151–160, New York, USA, 2010. ACM.

[19] M. Wertheimer. Untersuchungen zur lehre von der gestalt. *Psychological Research*, 1:47–58, 1922. 10.1007/BF00410385.

Document Visual Similarity Measure For Document Search*

Ildus Ahmadullin, Jan P. Allebach
Purdue University
School of Electrical and Computer Engineering
West Lafayette, IN 47907, USA

Niranjan Damera-Venkata, Jian Fan,
Seungyon Lee, Qian Lin, Jerry Liu,
Eamonn O'Brien-Strain
Hewlett-Packard Laboratories
Palto Alto, CA 94304, USA

ABSTRACT

Being able to automatically compare document layouts and classify and search documents with respect to their visual appearance proves to be desirable in many applications. We propose a new algorithm that approximates a metric function between documents based on their visual similarity. The comparison is based only on the visual appearance of the document without taking into consideration its text content. We measure the similarity of single page documents with respect to distance functions between three document components: background, text, and saliency. Each document component is represented as a Gaussian mixture distribution; and distances between the components of different documents are calculated as an approximation of the Hellinger distance between corresponding distributions. Since this distance obeys the triangle inequality, it proves to be favorable in the task of nearest neighbor search in a document database.

Categories and Subject Descriptors

J.7 [**Computers in other systems**]: Publishing

General Terms

Algorithms

1. INTRODUCTION

Being able to effectively compare documents with respect to their visual similarity is an important task in many applications, such as document retrieval, document classification, and managing large document databases. Comparing visual similarity of documents is essential in searching for documents of a specific style in a large document database or just looking for the one most similar to the query. For example, a user may want to browse and search for a greeting card in a large database sorted not with respect to the cards' themes and content, but taking into consideration visual similarity.

The use of Gaussian Mixture Models has proven to be an effective tool in image similarity tasks. Goldberger et al. [6] proposed modeling position and color of the image regions by GMMs. The distance between GMMs was calculated by an approximation to Kullback-Leibler (KL) divergence. In [2], we used this KL divergence approximation to calculate a document similarity measure. It showed promising results and proved to be computationally inexpensive. However, due to the fact that it does not obey the triangle inequality, it fails to be practical in large database document search tasks.

Many efficient nearest neighbor search algorithms presented in the past rely on the fact that the similarity measure is a metric and obeys the triangle inequality. The nearest neighbor search algorithms in [3] make use of a set of reference points in the dataset. The best choice for the set of reference points is a complex issue that depends on the tradeoff between the cost of pre-computation to set up the database before any query item is presented, and the cost of finding the closest match to a particular query item, once the database is set up. For our application, the pre-computation can be performed off-line, and is not so constrained. This loosely suggests that the optimal reference set should uniformly sample the database. Once the reference set has been chosen, the search for an item in the database that is closest to a particular query proceeds sequentially by eliminating candidates that are either too close or too far from a particular reference item, relative to the distance between the query and the reference item. The decision threshold is the distance from the query to the closest item found thus far in the search. Thus, the threshold is continually updated, as the search proceeds. This method significantly reduces computational time.

In this paper, we present a modified version of the algorithm in [2] that uses an approximation to the Hellinger distance metric to calculate a document similarity measure. Even though we can only approximate the true Hellinger distance metric, our test results show that the approximation with the Unscented Transform [4] provides promising results.

2. DOCUMENT COMPONENTS

Prior to the component composition, the document is scaled to a single size that is used for all types of documents.

*This work was supported by Hewlett Packard company.

To meaningfully process both landscape and portrait orientations, we choose the scaled document shape to be square.

2.1 Document Background Component

The background component D^b of a document page is extracted by erasing all of the text present on the document page. Note that this component includes not only the background images that do not overlap with text regions, but also the ones that are underneath the text, if there are any. Prior to feature extraction, the component is converted from RGB to L*a*b* color space. Next, we extract a set of five-dimensional feature vectors (x, y, L, a, b) consisting of each pixel coordinate (x, y) and the corresponding L*a*b* color value at that pixel coordinate. Thus, we represent the document background as a feature set in five-dimensional space. We then search for dominant clusters in the feature set, assuming that they can be effectively represented as Gaussians. The Expectation Maximization algorithm is used to determine the maximum likelihood parameters of a mixture of Gaussians that best represents the feature set. The number of Gaussians in each mixture is determined according to the Minimum Description Length principle [1].

2.2 Document Text Box Component

The document text box component D^t is used to emphasize the location of text in a document. After enclosing all text in the document in bounding boxes; we define D^t as a binary image, where pixels corresponding to text bounding boxes are assigned a value of 1, while all remaining pixels have a value of 0. In order to be able to accomplish this successfully, we need a way to detect text in documents. This can be done with the help of open source PDF document text extraction software or optical character recognition, in case the document is given in an image format.

The feature set of the text component consists of two-dimensional vectors (x, y) of the coordinates of pixels that correspond to the text bounding boxes. After constructing a mixture of Gaussians using the EM and MDL algorithms, we come up with a probabilistic representation of the location of text in the document. Note that we do not consider the actual text content.

2.3 Document Saliency Component

The concept of saliency is used in document feature extraction to represent regions or objects that are visually most important. Salient regions or regions that stand out with respect to their neighborhood are the most distinctive; and similarity of such regions in different documents should play a substantial role in the visual comparison of these documents. These regions can be background images in the case of advertisement brochures or text in the case of bright, colorful headers in magazines.

Our computation of the saliency value for each pixel follows the approach in [7]. The proposed algorithm is very attractive with its outstanding performance and computational efficiency. It is based on exploitation of features of color and luminance, and detects not only boundaries of salient regions but also salient colors.

As in the case of the background component, we initially convert the document to the L*a*b* color space and then blur the document with a Gaussian filter with $\sigma = 3$ pixels to eliminate fine texture details. The saliency of each pixel of a document is calculated as the squared Euclidean distance between the L*a*b* value at that pixel location in the blurred document, and the average L*a*b* value of the entire document: $S(x, y) = (L(x, y) - L_{ave})^2 + (a(x, y) - a_{ave})^2 + (b(x, y) - b_{ave})^2$.

The document saliency component is represented as a grayscale image D^s, where each pixel is assigned the saliency value defined above. The feature set of the saliency component consists of the three-dimensional vectors $(x, y, S(x, y))$ of the coordinates of each pixel of the document along with its saliency value. Just as in the previous two cases, with the help of the EM and MDL algorithms, we obtain a corresponding probabilistic representation of saliency in the document in terms of a Gaussian mixture model.

3. HELLINGER DISTANCE

Once the features of a document are represented by Gaussian mixture distributions, it is natural to proceed with calculating similarity measures between the components of two documents by comparing their GMMs. Among these comparison functions, the most preferred one would be the one that obeys the triangle inequality - the condition that proves to be a very important property in nearest neighbor search algorithms.

In this paper, we present the Hellinger distance which is defined as a scaled L_2 norm between the difference of the square roots of two distributions:

$$H(f, g) = \frac{1}{\sqrt{2}} \left\| \sqrt{f} - \sqrt{g} \right\|_2 \qquad (1)$$

After simple mathematical operations, we see that the Hellinger distance takes the following form:

$$H(f, g) = \sqrt{\frac{1}{2} \int \left(\sqrt{f} - \sqrt{g} \right)^2} = \sqrt{\frac{1}{2} \int f - \int \sqrt{fg} + \frac{1}{2} \int g}$$

$$= \sqrt{1 - \int \sqrt{fg}} \qquad (2)$$

Here, we make use of the fact that the integral of a probability density function equals one. The integral of the product of the square roots of two distributions in the last expression is known as the Bhattacharyya coefficient $BC(f, g) = \int \sqrt{fg}$. It is referred to as a measurement of the overlap between two probability distributions, and is used to determine relative closeness of the distributions. Hence, we can rewrite expression (2) in the following form:

$$H(f, g) = \sqrt{1 - BC(f, g)} \qquad (3)$$

Note that the Bhattacharyya coefficient takes on values between 0 and 1; and therefore the Hellinger distance is bounded from above by 1. The Bhattacharyya coefficient can also be expressed as an expectation of a certain function with respect to one of the probability distributions:

$$BC(f, g) = \int \sqrt{fg} = \int \sqrt{\frac{g}{f}} f \qquad (4)$$

$$= \int \sqrt{\frac{f}{g}} g \qquad (5)$$

To calculate the expectation we use the Unscented Transform (UT) [4]. This transform is widely used to approximate the expectation of a nonlinear function with respect to some probability distribution. It is based on the intuition that it is easier to approximate the distribution rather than the nonlinear function. Let $X \backsim f(X) = N(\mu, \Sigma)$ be a d-dimensional Gaussian random variable and $h : R^d \rightarrow R$ be any non-linear function. Then, N sigma-points $x_i \in R^d$ are chosen so that their mean is μ and covariance is Σ. Then the values of a non-linear function h at each of the chosen points are used to approximate the expectation

$$\int h(x) f(x) dx \approx \frac{1}{N} \sum_{i=1}^{N} h(x_i) \qquad (6)$$

In the case of a Gaussian Mixture Model $f = \sum_{i=1}^{M} \alpha_i f_i$, where each of the Gaussians has a mean μ_i and a covariance Σ_i, we choose the sigma points in the following way

$$x_j^i = \mu_i + (\sqrt{d\Sigma_i})_j, \; j = 1, ..., d$$

$$x_{d+j}^i = \mu_i - (\sqrt{d\Sigma_i})_j, \; j = 1, ..., d$$

In this formula, $(\sqrt{d\Sigma_i})_j$ denotes the jth column of the matrix square root of the covariance matrix Σ_i multiplied by the number of dimensions. Let $h = \sqrt{\frac{g}{f}}$. We calculate the Bhattarcharyya coefficient as defined in (4)

$$BC(f,g) = \int \sqrt{\frac{g}{f}} f = \int h \sum_{i=1}^{N} \alpha_i f_i = \sum_{i=1}^{N} \alpha_i \int h f_i$$

$$\approx \frac{1}{2d} \sum_{i=1}^{M} \alpha_i \sum_{j=1}^{2d} h(x_j^i) \qquad (7)$$

Thus, we calculate the Hellinger distance between two GMMs by approximating (4) by (6) and substituting it into (3). Note that the approximation (6) can be applied to either (4) or (5). In our experiments we did not see any significant difference between either method of calculation.

4. DOCUMENT COMPONENT WEIGHT COMPUTATION

Once we can compare two documents separately with respect to their background, text, and saliency components, we set the overall document similarity measure to be a weighted sum of the distances between the corresponding GMMs. It is noteworthy to mention that the weights can also be regarded as the degree of relative importance of each of the components in the overall document visual similarity comparison. Hence we require each of the weights to be nonnegative.

One of the important and desirable applications of the document similarity measure is document style recognition. We wish to efficiently distinguish between documents of different styles, i.e. given two sets of documents from different publication types, we would like our similarity measure to automatically recognize the publication type and to tell us which document belongs to which type. Therefore, it is reasonable to train our weights with respect to a training set that consists of a collection of documents labeled according to their type.

For our training purposes we used a dataset of 410 documents. The dataset contained five different classes each containing from 45 to 175 documents, ranging from magazine pages to newsletters from various departments at Purdue University [8]-[12]. We tried to include in the collection a variety of documents with different complexity and visual appearance: black and white and colorful documents, documents consisting of mostly text, and documents with a lot of bright images on the background.

We train the weights in our document similarity measure stochastically using Neighborhood Component Analysis described in [6]. Every document is labeled with respect to its class. We calculate the distances between the components of every pair of documents, and combine them in distance matrices $[d_{i,j}^b]$, $[d_{i,j}^t]$, and $[d_{i,j}^s]$ that correspond to the distances between background, text, and saliency components for documents D_i and D_j. Let $\mathbf{d}_{i,j} = <d_{i,j}^b, d_{i,j}^t, d_{i,j}^s>$ and $\mathbf{w} = <w_b, w_t, w_s>$ be vector representations of the distances and corresponding weights. The final similarity measure between D_i and D_j is calculated as

$$m(D_i, D_j) = \mathbf{w^T} \mathbf{d}_{i,j} = w_b d_{i,j}^b + w_t d_{i,j}^t + w_s d_{i,j}^s$$

In searching for the optimal weights w_b, w_t, and w_s, we rely on the assumption that documents in the same class should be more similar than documents in different classes. Hence our objective is to select the weights in such a way that the same class documents will appear close to each other, whereas the documents in different classes will be further away from each other. We let the probability that the document D_i chooses a document D_j as its neighbor be defined as follows

$$p_{i,j} = \begin{cases} \frac{\exp(-\mathbf{w^T}\mathbf{d}_{i,j})}{\sum_{k \neq i} \exp(-\mathbf{w^T}\mathbf{d}_{i,k})}, & i \neq j \\ 0, & i = j \end{cases} \qquad (8)$$

Let $C_n = \{i | D_i \in \text{Class } n\}$ be the set of indices of documents D_i in Class n, where n varies from 1 to 5. Then the probability that the document D_i is correctly classified can be calculated as

$$p_i = \sum_{j \in C_i} p_{i,j} \qquad (9)$$

As the objective of our problem, we define the expected fraction of documents correctly classified.

$$f(\mathbf{w}) = \sum_i \sum_{j \in C_i} p_{i,j} = \sum_i p_i \qquad (10)$$

We obtain the optimal weights $\mathbf{w}$ by maximizing the objective function (10). Note that the probability function (8) is smooth and differentiable with respect to $\mathbf{w}$, therefore (10) can be maximized by gradient ascent or any other suitable optimization algorithm. However, (8) is not convex. In order to avoid local maximum solutions, one need to repeat the optimization algorithm a few times by randomly choosing the initial values of $\mathbf{w}$.

5. EXPERIMENTAL EVALUATION

The training and testing of the weights was performed using a 3-fold validation technique. The dataset was partitioned into three equal size subsets, such that each subset contained approximately an equal number of randomly selected documents from each class. The training was performed on the two subsets and tested on the remaining one. The procedure was then repeated three times so that each of the subsets participated in training and testing of the algorithm. We used the K nearest neighbor algorithm (KNN), where K = 7, with our document similarity measure $m(D_i, D_j)$ to assign class labels to the documents in the test subset. The accuracy of each of the three runs of the validation procedure is indicated in Table 1, where we show the percentage of correctly classified documents in each class of the test subset obtained in each of the validation tests. The final set of weights is calculated by averaging the results from each run.

	Class 1	Class 2	Class 3	Class 4	Class 5
Size	75	175	60	55	45
Run 1	100%	100%	56.25%	62.5%	95.83%
Run 2	100%	93.1%	88.23%	66.67%	100%
Run 3	73.3%	91.38%	70.59%	75%	95.83%

Table 1: Percentage of the correctly classified documents in each class of the test set. Each row shows the results from a different validation run; top row shows number of document samples in each class.

A desirable property of a document visual similarity algorithm is the ability to compare documents of arbitrary style. In order to check the applicability of our method to address this type of task, we conducted a different experiment. The dataset for this experiment consisted of documents from a number of different publications of newsroom magazines at Purdue University. We did not constrain the dataset to contain a certain number of document classes. In this experiment we instead concentrated on having a pool of documents that contained as many different types as possible. We tried to include not only documents of the same type, but also the ones that differ from each other in many different aspects, such as colorfulness of images, text structure, plain and colored background. The goal was to check how appropriate the visual similarity measure is, when comparing perceptually different documents. For each experiment, we chose a query document, sorted all of the documents with respect to their similarity to the query, and picked four approximately equally spaced samples from the list. The result of this comparison is shown in Figure 1.

6. CONCLUSION

We proposed a new method for calculating a document visual similarity measure that obeys the triangle inequality. This is especially important in searching similar documents in large databases since fast searching algorithms heavily rely on this property. Since documents of the same type share similar visual attributes, we conclude that this document similarity measure could also be effectively used in a document style recognition task.

7. REFERENCES

[1] Rissanen, J. 1983. A universal prior for integers and estimation by minimum description length. *The Annals of Statistics 11 (2)*, 417-431.

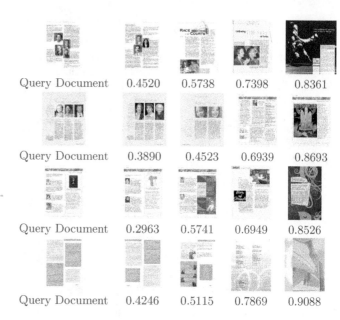

Figure 1: Documents, sorted with respect to their similarity to the query. Below each document we indicate the measure of its distance from the query.

[2] Ahmadullin, I., Fan, J., Damera-Venkata, N., Lim, S.H., Lin, Q., Liu, J., Liu, S., O'Brien-Strain, E., Allebach, J.P. 2011. Document similarity measires and document browsing. In *Proc. of SPIE-IS&T Electronic Imaging Conference*, San Francisco, CA, 7879.

[3] Shapiro, M. 1977. The choice of reference points in best-match file searching. *Communications of the ACM*, 20(5), 339-343.

[4] Julier, S. J., and Uhlmann, J. K. 2004. Unscented filtering and nonlinear estimation. In *Proceedings of the IEEE* 92(3), 401–422.

[5] Goldberger, J., Greenspan, H., Gordon, S. 2003. An efficient similarity measure based on approximations of KL-divergence between two Gaussian mixtures. *In Proc. of International Conference on Computer Vision*, Nice, France, October 2003, 1, 487-493.

[6] Goldberger, J., Roweis, S., Hinton, G., Salakhutdinov, R.. 2004. Neighborhood Component Analysis. In *Proceedings of Neural Information Processing Systems (NIPS)*, Vancouver, Canada, December 2004, 513-520.

[7] Achanta, R., Hemami, S., Estrada, F., Susstrunk, S. 2009. Frequency-tuned salient region detection. In *Proc. of IEEE International Conference on Computer Vision and Pattern Recognition*, Miami, FL, June 2009, 1597-1604.

[8] Purdue C. S. Annual Report 2003-2004. www.cs.purdue.edu/about_us/annual_reports/2003-2004

[9] Purdue C. S. Annual Report 2004-2005. www.cs.purdue.edu/about_us/annual_reports/2004-2005

[10] Purdue C. S. Annual Report 2006-2007. www.cs.purdue.edu/about_us/annual_reports/2006-2007

[11] Purdue College of Liberal Arts Magazine. www.cla.purdue.edu/news/magazine/documents/2004 Fall.pdf

[12] Purdue American Studies Newsletters 2003-2009. www.cla.purdue.edu/idis/american-studies/news_events/newsletters.html

Paginate Dynamic and Web Content

Fabio Giannetti

Hewlett-Packard Laboratories

1501 Page Mill Road – M/S 1161

Palo Alto, CA – 94304

+1 650 857 5085

fabio.giannetti@hp.com

ABSTRACT

Highly customized and content driven documents present substantial challenges in producing sophisticated layout. In fact, these are apps that usually look like well-designed documents. A concrete example is e-books. E-books have re-flowing requirements to allow the user to read them on a plethora of devices as wells as change the font size and font style. Meanwhile this increases the flexibility of the medium, it loses common features found in books like footnotes, marginalia (a.k.a. side notes), pull-quotes and, floats. This paper introduces an approach on extending the concept of galley to a generalized document design instrument. The proposed solution has the aim of providing an easy and flexible, yet powerful, way to express complex layout for highly dynamic and re-flowing content. To achieve this goal, not only it is important to express all the areas available within the page or page region, but also identify a mean to efficiently map content to them. To serve this purpose, a role based mapper has been introduced linking both flow and out-of-flow content.

Categories and Subject Descriptors

I.7 DOCUMENT AND TEXT PROCESSING - I.7.2 Document Preparation: Format and notation and markup. I.7.4 Electronic Publishing: Print publishing for variable data driven templates.

General Terms

Documentation, Design, Standardization, Languages, Theory.

Keywords

Web Printing, Dynamic Publishing, Page Layout, Pagination, Document Layout, Galley, Footnotes, Marginalia, Floats, Content Flow.

1. THE CHALLENGE

The production of highly sophisticated documents typically requires designers to manually combine the content characteristics and the selected layout. This approach is efficient when the page layout is hand crafted and the content is known in advance or can be post-edited.

In a content and web driven world, where content is published and consumed directly by the final user, controlling the pagination and layout is a challenge since it is neither hand crafted nor known in

advance [1]. Yet, readers expect the experience at the same level of typography and pagination coherency, available in more traditional publications like magazines or books.

On one hand, publishers are increasingly demanding the capability to produce content that can be easily re-purposed across several media [2], whilst maintaining the media advantages and characteristics. On the other hand, consumers want to consume content in any media of their choice without sacrificing the quality that could be experienced if the document was specifically designed for a particular media.

In order to address publishers' easiness to produce high quality documents across different media and consumers' wishes to have the best experience, we need to ensure that three aspects related to document layout are kept separate: page layout, pagination model, and content mapping.

A further aspect is that across media, pagination is represented differently. In web pages, for instance, is represented by the ability of linking pages that usually have a non-liner sequence. In linearly linked paginated content (e.g.: e-books, printed material, and so on), it is relevant to provide table of contents and internal references. Moreover, it is important that these are expressed in a language agnostic approach so, that they can be converted in different web, mobile, and print enabling formats, such as HTML5+CSS3 and XSL-FO.

In the next sections we will propose a generalized galley approach that addresses the page layout and content mapping. The pagination model is not addressed since there are models available in XSL-FO 1.1 and CSS3 that are directly applicable.

2. PAGE LAYOUT BASED ON GENERALIZED GALLEYS

In complex document design, the graphic artist is the responsible for allocating areas for the different content, e.g., footnotes, side notes, and float.

This is achievable when all the content is known in advance and can be laid out with high degree of precision. In the case of content driven publications, or even re-flowable content, this approach is not applicable. Yet, publishers would like to present their publications with the highest degree of complexity and accuracy providing the fundamental design principles that makes paginated media easy and pleasant to consume.

The idea introduced in this chapter allows extending the way pages are described using a generalized concept of galley.

A galley is defined as a content area that serves the placement of ancillary of associated content to the main flow(s) [3]. These areas are usually either represented as ad-hoc regions, or carved out from existing regions that accommodate content. Since we are discussing the use of these areas in the specific of re-flowable content we will explain the application of this concept only for the latter case.

In conventional design, galleys are implicitly linked to the corresponding content type, usually referred as "*out-of-flow*", and are related to a concept of anchor point. For instance, a footnote galley is anchored to the bottom of the region or page and accommodates footnotes content coming from its corresponding flow. If the content appearing in that region/page does not contain footnotes, there is no out-of-flow content and no anchor point. As a result, the galley will never be rendered. On one hand, this constitutes a great advantage compared to custom-built regions, since these will always be rendered, even if there is no out-of-flow content to accommodate. On the other hand, it is a big limitation since, differently from traditional regions, galleys' area cannot be directly addressed or referenced. A clear example of this limitation is the usage of column number and column gap adopted by several formats, e.g. CSS3 [4].

The CSS3 Template Layout Module [5] provides an alternative solution to galleys. The approach is based on a letter based matrix design. The idea is to have a matrix of letters to identify the position of each individual element and then express for each of these the relative position, or stacking and its size. Figure 1 illustrates a typical newspaper layout.

Figure 1: CSS3 Template Layout Module Newspaper Layout

Example 1 illustrates CSS3 notation for the newspaper layout.

```
<style type="text/css">
 body {
  height: 100%;
  @page :first {
  display: "A A A A A A A A" / 5cm
           ". . . . . . . . . ." / 0.25cm
           "B . C C C C C C" / *
           "B . C C C C C C" / *
           "B . C C C C C C" / *
```

```
           "B . C C C C C C" / *
           "B . C C C C C C" / *
           "B . D D D D D D" / *
           "B . D D D D D D" / *
           "B . E E E . F F F" / *
           "B . E E E . F F F" / *
           "B . E E E . F F F" / *
           * 3em * 3em * 3em * 3em *
    }
 h1 {position: a; border-bottom: thick; margin-bottom: 1.5em}
 #toc {position: b; margin-right: -1.5em; border-right: thin;
    padding-right: 1.5em}
 #leader {position: c; columns: 4; column-gap: 3em}
 #art1 {position: d; columns: 4; column-gap: 3em; border-top: thin}
 #art2 {position: e; columns: 2; column-gap: 3em}
 #art3 {position: f; columns: 2; column-gap: 3em}
</style>
```

Example 1: CSS3 Template Layout Module. Layout Definition and Mapping

CSS3 Layout Model showcases a grid layout where letters indicate content areas and dots spacing. When the letters are repeated through rows and/or columns, the associated content will be placed there. This is achieved using the position: qualifies for the styled element.

Unfortunately the CSS3 model is trying to address both the layout design and mapping at the same time and we believe these introduces three limitations:

1. lack of flexibility in the design; for instance, it is not possible to express changing stacking directions as this is only left to the content;

2. matrix design; to design a content area, letters must be repeated for all the rows or columns the area occupies and there is no support for conditional areas that may appear or not. These may live within the area allocated for the content using an implicit mechanism.

3. inability of addressing spacing areas; areas that are separating content (gaps) and/or represents decorations (represented in the layout design as dots) are not directly addressable. These could be represented as letters (e.g. making distinction between lowercase and uppercase) but it would make the design even more difficult to read.

The proposed approach addresses all these needs generalizing the galley concept to the entire region/page layout allowing the designer to express all the areas as galleys. Each galley is now explicitly added and it has a concept of extent. The extent can either have a fixed size or a range. Using this concept it is possible to achieve sophisticated galley control, like expressing footnote areas that are not rendered in case there is no footnote content, as well as a maximum size.

This introduces three benefits:

1. galley areas are now **controllable**. Size and stacking direction or relative location are two aspects that increase the ability of describing complex page layouts with simple instructions;

2. galleys are **addressable** and **identifiable**. Through the introduction of "*role-based-mapper*" it is possible to re-direct any of type of "out-of-flow" content to any galley;

3. **anchored** galleys can be **limited in number** to avoid crowded and awkward situations.

The design of a galley based region is based on a concept of nested elements, each one is cascading its properties providing the context for the inner elements. There are two main properties that allow generating complex and flexible design: extent and stacking. The extent has two properties modifiers the –min and –max, these modifiers allow the designer to express the size boundaries of the galley and its existence in case no content will flow into it. This is achieved simply by expressing an extent-min to zero.

The stacking properties provide control of how these galleys are placed respectively to each other and, for the outermost one, to the region or page container. There are five potential values: left-to-right (ltr), right-to-left (rtl), top-to-bottom (ttb), bottom-to-top (btt) and anchored.

The anchored case is particularly interesting since it refers to floating elements that are linked to a particular content and needs to be placed nearby that content. Usually this generates text runaround and the float size can be either fixed or left to the content. When the flowing content is truly dynamic, there is no possibility to predict where the floats will appear and how big they are. One approach is to dedicate a side or top area and have all of these stacking there, but, while easier to handle, it is definitely less appealing from a document design perspective.

The proposed solution allows the designer to express a galley placeholder for the anchored floats. This placeholder can have a size range or can be left entirely content driven. The immediate advantage of having such placeholder is that is possible to restrict the number of times that floats may appear in a page, region, or column, allowing a greater control over the final appearance of the document. The second advantage is that styling can be applied making the float content independent from alignment, borders, padding, and other area related stylistic needs.

Before we delve into the dynamic aspects and out-of-flow control, we want to represent the example in Figure 1 using the proposed notation to highlight the difference in capabilities and limitations expressed above.

```
<page name="CSS3LayoutTemplateModelExample">
  <region name="Newspaper" x="..." y="...">
    <galley name="main-wrapper" extent="auto" stacking="ttb">
      <galley name="A" extent="5cm"/>
      <galley name="gap1" extent="0.25cm"/>
      <galley name="sideandcontent" extent="auto" stacking="ltr">
        <galley name="B" extent="20%"/>
        <galley name="gap2" extent="3em"/>
        <galley name="content-CD" extent="80%" stacking="ttb">
          <galley name="C" extent="auto"/>
          <galley name="D" extent="auto"/>
        </galley>
        <galley name="content-EF" extent="20%" stacking="ltr">
          <galley name="E" extent="50%"/>
          <galley name="gap3" extent="3em"/>
          <galley name="F" extent="50%"/>
        </galley>
      </galley>
    </galley>
  </region>
</page>
```

Example 2: CSS3's Newspaper Example using the Galley Design Approach

Now, to better understand and illustrate the capabilities of the proposed approach, we are introducing few examples that illustrate design capabilities difficult to achieve with existing solutions:

- Figure 2: a different sized column document having two independent marginalia areas (one for each column) and a central footnote area (one for all columns);
- Figure 3: a four column document having a separate unified area for quotes and two footnote areas for columns 1-2 and 3-4;
- Figure 4: a two columns document with anchored floats that are controlled in position and number for the different columns.

In the following sub-sections we describe these examples in greater detail.

2.1 Marginalia and Central Footnote

In this first example, the document presents both side notes (and marginalia) and footnotes. In this case the out-of-flow content placed in the marginalia needs to be visually aligned with the related content in one of the columns meanwhile the out-of-flow content in the footnote can be merged. This optimized landscape layout clearly highlights the two parts of the document and easily navigates the reader through context dependent notes (side) and traditional footnotes.

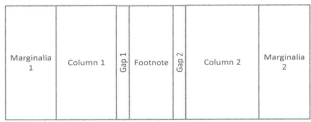

Figure 2: Two Columns Document with Marginalia and Central Footnote

The corresponding format to express the design is captured in Example 3. The page layout is simple to design and it has a main galley that defines the stacking order, as left to right, for all the other galleys. An interesting design feature is the possibility of giving to the footnote galley an extent-min of zero and an extent-max of ten percent. This allows the formatter to eliminate the footnote galley in case there is no content. The footnote wrapper also incorporates the gap1 galley so then, in case of no footnotes, the gap is not repeated.

```
<page name="Example1">
  <region name="MarginaliaAndCentralFootnote" x="..." y="...">
    <galley name="main-wrapper" extent="auto" stacking="ltr">
      <galley name="marginalia1" extent="15%"/>
      <galley name="column1" extent="24%"/>
      <galley name="ftn-wrapper" extent-min="0%" extent-max="10%">
        <galley name="gap1" extent="10%"/>
        <galley name="footnote" extent="90%"/>
      </galley>
      <galley name="gap2" extent="2%"/>
      <galley name="column2" extent-max="44%" extent-min="34%" />
      <galley name="marginalia2" extent="15%">
    </galley>
  </region>
</page>
```

Example 3: Two Columns, Two Marginalia Areas and One Footnotes Area

2.2 Quotes and Columns Related Footnotes

In this second example, the design gets more complex and separates the document into four columns that have their own sub areas for footnotes, whilst allowing flowing of content among them, and a unified area for quotes.

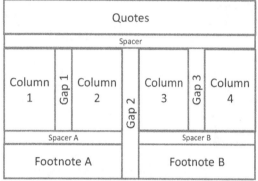

Figure 3: Four Columns Document with Quotes Area and Columns Related Footnotes

In the code Example 4, it is possible to identify three major design aspects: the first is the presence of a galley wrapper that defines the top to bottom representation of the layout; the second is that both footnote galleys and the quotes galley is collapsible, and the third is the combination of left-to-right and top-to-bottom stacking to provide the main gap galley separating the two bi-column blocks.

```
<page name="Example2">
 <region name=" VeryComplex" x="..." y="...">
  <galley name="main-wrapper" extent="auto" stacking="ttb">
   <galley name="quotes-wrapper" extent-min="0%" extent-max="10%">
    <galley name="quotes" extent="80%"/>
    <galley name="spacer" extent="20%"/>
   </galley>
   <galley name="cols-wrapper" extent="70%" stacking="ltr">
    <galley name="cols1-2-wrapper" extent="49%" stacking="ltr">
     <galley name="inner-wrapper" stacking="ttb">
      <galley name="cols1-2" stacking="ltr">
       <galley name="column1"/>
       <galley name="gap1" extent="1%"/>
       <galley name="column2"/>
      </galley>
      <galley name="ftnA" extent-min="0%" extent-max="15%">
       <galley name="spacerA" extent="10%"/>
       <galley name="footnoteA" extent="90%"/>
      </galley>
     </galley>
    </galley>
    <galley name="gap2" extent="2%"/>
    <galley name="col3-4-wrapper" extent="49%" stacking="ltr">
     <galley name="inner-wrapper" stacking="ttb">
      <galley name="cols3-4" stacking="ltr">
       <galley name="column3"/>
       <galley name="gap3" extent="1%"/>
       <galley name="column4"/>
      </galley>
      <galley name="ftnB" extent-min="0%" extent-max="15%">
       <galley name="spacerB" extent="10%"/>
       <galley name="footnoteB" extent="90%"/>
      </galley>
     </galley>
    </galley>
   </galley>
  </galley>
 </region>
</page>
```

Example 4: Four Columns, Two Footnotes Areas and One Quotes Area

2.3 Anchored Floats Management

The third and last example addresses another complex design feature: anchored floats. Anchored floats are out-of-flow elements that are relevant only to a particular part of the flow and must be placed nearby. The anchor represents the location where the float was generated. This is the reference point for alignment, positioning, and content run-around effects. When the content is neither pre-selected nor controlled, floats can happen everywhere. This is particularly true in news feeds or article extracts (e.g., daily digests). Many news headers and digest carry a picture float to illustrate or support the story. This can potentially lead to a proliferation of floats not only making the overall design unappealing, but also sensibly reducing the readability of the produced document. Using the proposed galley design it is possible to directly reference these elements in the page layout, consequently allowing to place them more efficiently and control the number appearing per page, region, or even column.

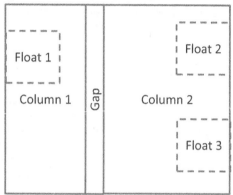

Figure 4: Two Columns Document with Anchored Floats

The page design in Example 5 illustrates the management of anchored floats. Each of the two column galley contains a nested galley definition that supports the anchored floats. The extent-min to zero and extent-max to "auto" it allows the float to be entirely content driven or fully collapsed; it also has the stacking property set to "anchor" to instruct the formatter to place the float relative to the content where it was generated. The last aspect is the introduction of max-occurrences attribute. This enables the designer to select the maximum number of instances that an anchored float may appear in that column. In this particular example, since the right column is larger, the designer preferred having more floats and aligning them towards the exterior of the page/spread.

```
<page name="Example3">
 <region name="AnchoredFloatsHandling" x="..." y="...">
  <galley name="main-wrapper" extent="auto" stacking="ltr">
   <galley name="column1" extent="30%">
    <galley name="col1-floats" display-align="start"
            extent-min="0%" extent-max="auto"
            stacking="anchor" max-occurrences="1"/>
   </galley>
   <galley name="gap" extent="2%" />
   <galley name="column2" extent="68%" >
    <galley name="col2-floats" display-align="end"
            extent-min="0%" extent-max="auto"
            stacking="anchor" max-occurrences="2"/>
   </galley>
  </galley>
 </region>
</page>
```

Example 5: Two Columns with Anchored Floats

This has been partially addressed in LaTeX using "counters" [7]. In LaTeX is possible to express the max number of the top and bottom floats and how much space they can occupy in the page. The proposed model goes a level further leveraging the galley model hierarchy and allowing specifying these constraints for all the galleys and, as a sum of all, the entire page. The overall space in the page can be limited indirectly limiting the float areas for each individual galley.

3. ROLE BASED MAPPER

Once the page has been designed and all the galleys defined, the designer expresses which flow and out-of-flow elements are going where. This is done by using a role based mapper. The role concept allows the user to express elements in the content that belongs to a particular class. There are already XML based formats that provide footnotes, floats, headers, and footers, but very few allowing marginalia, pull quotes, etc... It is not our intention to limit the designer's capability to refer to selected content by using special naming convention. It is clear, though, that for interoperability needs, a subset of convention should be defined across HTML5+CSS3 and XSL-FO, to support the W3C stack. XSL-FO already provides a simpler flow-map [6] concept that can be extended.

In order to illustrate our approach in representing the role based mapper, we will describe the instantiation of it using the previously introduced examples. Figure 5 depicts the flow and out-of-flow mappings in a graphical form.

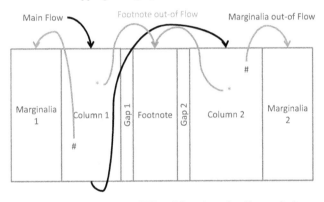

Figure 5: Flow and Out-of-Flow Mappings for Example 1

The role based map is comprised of a set of assignments. Each assignment has the specific role of mapping source elements to destination elements. For the flow source elements, the assignment is done towards a galley area. For out-of-flow elements, the assignment is done from a galley element to another galley element. Each assignment can have several source elements as well as destination elements. This syntax allows expressing the logic linkage of content flows into areas. In Example 6, the assignment with role content is allowing the main flow to flow from column 1 to column 2.

The other assignment with role marginalia is redirecting the columns' specific content to the corresponding marginalia galleys and lastly the assignment with footnotes role is combining the footnotes into a single galley.

```
<role-map name="Example1">
        <assignment role="content">
                <source href="URL"/>
                <dest ref="column1"/>
                <dest ref="column2"/>
        </assignment>
        <assignment role="footnotes">
                <source ref="column1"/>
                <source ref="column2"/>
```

```
                <dest ref="footnote"/>
        </assignment>
        <assignment role="marginalia">
                <source ref="column1"/>
                <dest ref="marginalia1"/>
        </assignment>
        <assignment role="marginalia">
                <source ref="column2"/>
                <dest ref="marginalia2"/>
        </assignment>
</role-map>
```

Example 6: Role Based Mapper for Example 1

The second example that we introduced in the previous section had a more complex layout. This reflects a slightly more complicated mapper, but as can be evinced from Example 3, the map description easily evokes the graphical representation in Figure 6.

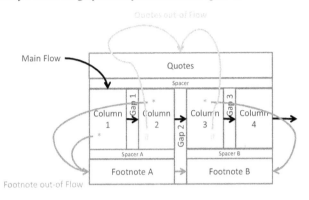

Figure 6: Flow and Out-of-Flow Mappings for Example 2

This close relationship between the graphical representation of a flow map and its XML representation provides a non-coding alternative to design complex dynamic document. Preserving some degree of WYSIWYG capabilities is crucial, since designers are accustomed with graphical tools and high degree of control over the final result; something that they have to trade with dynamic documents.

```
<role-map name="Example2">
        <assignment role="content">
                <source href="URL"/>
                <dest ref="column1"/>
                <dest ref="column2"/>
                <dest ref="column3"/>
                <dest ref="column4"/>
        </assignment>
        <assignment role="footnotes">
                <source ref="column1"/>
                <source ref="column2"/>
                <dest ref="footnoteA"/>
        </assignment>
        <assignment role="footnotes">
                <source ref="column3"/>
                <source ref="column4"/>
                <dest ref="footnoteB"/>
        </assignment>
        <assignment role="quotes">
                <source ref="column1"/>
                <source ref="column2"/>
                <source ref="column3"/>
                <source ref="column4"/>
                <dest ref="quotes"/>
        </assignment>
</role-map>
```

Example 7: Role Based Mapping for Example 2

The last of the examples is quite simple and leverages all the above mentioned mapping techniques. The only noticeable aspect is the capability of having assignments that address content in nested galleys, deciding where the out-of-flow content generated in the parent galley needs to be placed.

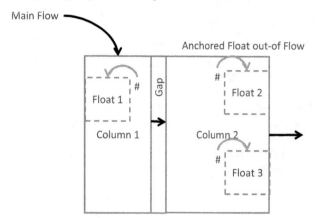

Figure 7: Flow and Out-of-Flow Mapping for Example 3

```
<role-map name="Example3">
        <assignment role="content">
                <source href="URL"/>
                <dest ref="column1"/>
                <dest ref="column2"/>
        </assignment>
        <assignment role="floats">
                <source ref="column1"/>
                <dest ref="col1-floats"/>
        </assignment>
        <assignment role="floats">
                <source ref="column2"/>
                <dest ref="col2-floats"/>
        </assignment>
</role-map>
```

Example 8: Role Based Mapping for Example 3

4. EXTENDING THE MAPPER FOR DECORATIVE CONTENT

A further usage of the Role Based Mapper is to extend it to support decorative content. In many publications when there are columns, footnotes, marginalia, and/or quotes, it is important to provide a separation space between the content areas. This is achieved in the above presented page model using empty galleys that act as spacers. This could be just fine, but it is common to have a more explicit separation element, e.g., a short line in footnotes, or a fancy decorative pattern between columns. The decorative content can be simply seen as content and mapped to the galley representing the separator or gap, as illustrated in Example 9.

```
<role-map name="Example1">
        <assignment role="content">
                <source href="URL"/>
                <dest ref="column1"/>
                <dest ref="column2"/>
        </assignment>
        <assignment role="content">
                <source href="http://content.svg#gap1"/>
                <dest ref="gap1"/>
        </assignment>
        <assignment role="content">
                <source href=" http://content.svg#gap2"/>
                <dest ref="gap2"/>
        </assignment>
        ...
</role-map>
```

Example 9: Role Based Mapping for Example 1 with Decorative Content for Gaps

The two available gaps can now be filled with the Scalable Vector Graphics (SVG) nodes containing, for instance, a repeated pattern. Meanwhile gap1 and gap2 content is mapped, only gap2 is rendered because gap1 is wrapped with the footnote galley and have an extent-min to zero. Only if there is some out-of-flow content that is linked to the footnote role, gap1 will be rendered showing the mapped SVG content. To detect the existence of out-of-flow content belonging to a particular role, it is necessary to have a way to, either implicitly, or explicitly, reference it at the content tags or attribute levels. The next section explores the currently used W3C standards to represent dynamic, and web content, it identifies techniques and elements that can support the role based mapping.

5. A NOTE ON PAGE SEQUENCING

Dynamic and content driven documents, when paginated will require more than one page. It is necessary then to express a mechanism to identify the sequence of pages and their repeatability. Most common designs have for instance a first page, left and right pages and a different last page. XSL-FO already addresses this problem using a "master" page sequence concept as illustrated in Example 10.

```
<fo:page-master-sequence master-name="even_odd">
    <fo:repeatable-page-master-alternatives>
        <fo:conditional-page-master-reference master-name=" odd-page"
                                             odd-or-even="odd"/>
            <fo:conditional-page-master-reference master-name=" even-
page"
                                             odd-or-even="even"/>
    </fo:repeatable-page-master-alternatives>
</fo:page-master-sequence>
```

Example 10: XSL-FO's Page Master Sequence Concept

The master page sequence specifies the order and/or conditions in which pages should be presented. This model can be used in conjunction with a role-map so that the role map can point where the flow and out-of-flow content will be directed. Example 11 illustrates the complete example in XSL-FO notation.

```
<fo:root xmlns:fo="http://www.w3.org/1999/XSL/Format">
    <fo:layout-master-set>
        <fo:page-master master-name="odd-page">
            <fo:region region-name="main-region">
                <fo:galley galley-name="main-wrapper" extent.optimum="auto"
                                             stacking-direction="ltr">
                    <fo:galley galley-name="left-wrapper" extent.optimum="29%"
                                             stacking-direction="ttb">
                    ...
                    </fo:galley>
                    <fo:galley galley-name="gap" extent.optimum="2%"/>
                    <fo:galley galley-name="right-wrapper" extent.optimum="69%"
                                             stacking-direction="ttb">
                    ...
                    </fo:galley>
                </fo:galley>
            </fo:region>
        </fo:page-master>
        <fo:page-master master-name="even-page">
            <fo:region region-name="main-region">
                <fo:galley galley-name="main-wrapper" extent.optimum="auto"
                                             stacking-direction="ltr">
                    <fo:galley galley-name="left-wrapper" extent.optimum="69%"
                                             stacking-direction="ttb">
                    ...
                    </fo:galley>
                    <fo:galley galley-name="gap" extent.optimum="2%"/>
                    <fo:galley galley-name="right-wrapper" extent.optimum="29%"
                                             stacking-direction="ttb">
                    ...
                    </fo:galley>
                </fo:galley>
            </fo:region>
        </fo:page-master>
```

```
<fo:page-master-sequence master-name="even_odd">
    <fo:repeatable-page-master-alternatives>
        <fo:conditional-page-master-reference
            master-name="odd-page" odd-or-even="odd"/>
        <fo:conditional-page-master-reference
            master-name="even-page" odd-or-even="even"/>
    </fo:repeatable-page-master-alternatives>
</fo:page-master-sequence>
    <fo:flow-map flow-map-name="test" master-reference="even_odd">
        ...
        <fo:out-of-flow-assignment role="xsl-footnotes">
            <fo:out-of-flow-source-list>
                <fo:galley-name-specifier anchor-galley-
reference="column1"/>    <fo:galley-name-specifier anchor-galley-
reference="column2"/></fo:out-of-flow-source-list>
                <fo:out-of-flow-target-list>
                    <fo:galley-name-specifier galley-name-reference="footnote"/>
                </fo:out-of-flow-target-list>
        </fo:out-of-flow-assignment>
    </fo:flow-map>
</fo:layout-master-set>
    <fo:page-sequence master-reference="even_odd">
        <fo:flow flow-name="">
            <fo:block/>
        </fo:flow>
    </fo:page-sequence>
</fo:root>
```

Example 11: Full Pagination Model (Page Master Sequence, Galleys and Role Based Mapper)

This code generates a connection loop between the pages in the sequence, internally between galleys (i.e. column1 with colunm2) and externally between pages (i.e. odd page column2 with even page column1) as illustrated in Figure 8.

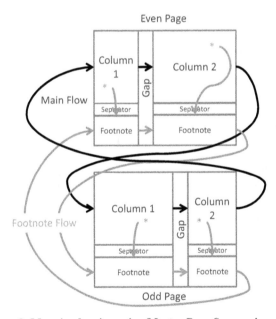

Figure 8: Mapping looping using Master Page Sequencing

6. PUTTING IT ALLTOGETHER: CONTENT TYPES AND ROLES

Once the page layout has been designed using the galleys and a role based mapping has been defined, the generation of the out-of-flow content is dependent on the source flow. In order to correctly associate this content to the corresponding role in the map, we can envisage similar techniques for various XML dialects. Moreover,

some already provide content "classes" that match one of previously identified roles (footnotes, floats and marginalia) and some can just be extended to do so.

In the following sub sections we will identify key aspects of the W3C standards that can serve our purpose. We will start with the most tailored to paginated media and which is used to paginate highly dynamic content: XSL-FO, then we will address HTML5+CSS3 which represents the standard for web pages.

6.1 XSL-FO

XSL-FO is the W3C's standard that specifically deal with paginated media. It already supports a concept of pagination, master pages and flow mapper. Extending XSL-FO to support the Role Base Mapper and the Galley Based Page Master is relatively straight forward and it is possible to adapt the generic representation specified in this article to the specific FO dialect. To exemplify how this could be achieved the full Example 1 Galley design and Mapper has been converted in FO notation, as illustrated in Example 12.

```
<fo:root xmlns:fo="http://www.w3.org/1999/XSL/Format">
    <fo:layout-master-set>
        <fo:page-master master-name="master-page">
            <fo:region region-name="main-region">
                <fo:galley galley-name="main-wrapper" extent.optimum="auto"
                    stacking-direction="ltr">
                    <fo:galley galley-name="marginalia1" extent.optimum="15%"/>
                    <fo:galley galley-name="column1" extent.optimum="24%"/>
                    <fo:galley galley-name="ftn-wrapper" extent.minimum="0%"
                        extent.maximum="10%">
                        <fo:galley galley-name="gap1" extent.optimum="10%"/>
                        <fo:galley galley-name="footnote" extent.optimum="90%"/>
                    </fo:galley>
                    <fo:galley galley-name="gap2" extent.optimum="2%"/>
                    <fo:galley galley-name="column2" extent.maximum="44%"
                        extent.minimum="34%" />
                    <fo:galley galley-name="marginalia2" extent.optimum="15%"/>
                </fo:galley>
            </fo:region>
        </fo:page-master>
        <fo:flow-map flow-map-name="test">
            <fo:flow-assignment role="xsl-flow">
                <fo:flow-source-list>
                    <fo:flow-name-specifier flow-name-reference="all-my-
content"/>
                </fo:flow-source-list>
                <fo:flow-target-list>
                    <fo:galley-name-specifier galley-name-reference="column1"/>
                    <fo:galley-name-specifier galley-name-reference="column2"/>
                </fo:flow-target-list>
            </fo:flow-assignment>
            <fo:out-of-flow-assignment role="xsl-footnotes">
                <fo:out-of-flow-source-list>
                    <fo:galley-name-specifier anchor-galley-
reference="column1"/>    <fo:galley-name-specifier anchor-galley-
reference="column2"/></fo:out-of-flow-source-list>
                <fo:out-of-flow-target-list>
                    <fo:galley-name-specifier galley-name-reference="footnote"/>
                </fo:out-of-flow-target-list>
            </fo:out-of-flow-assignment>
            <fo:out-of-flow-assignment role="xsl-marginalia">
                <fo:out-of-flow-source-list>
                    <fo:galley-name-specifier anchor-galley-
reference="column1"/>
                </fo:out-of-flow-source-list>
                <fo:out-of-flow-target-list>
                    <fo:galley-name-specifier galley-name-
reference="marginalia1"/>
                </fo:out-of-flow-target-list>
            </fo:out-of-flow-assignment>
            <fo:out-of-flow-assignment role="xsl-marginalia">
                <fo:out-of-flow-source-list>
                    <fo:galley-name-specifier anchor-galley-reference="column2"/>
```

```
            </fo:out-of-flow-source-list>
            <fo:out-of-flow-target-list>
                <fo:galley-name-specifier galley-name-
reference="marginalia2"/>
            </fo:out-of-flow-target-list>
        </fo:out-of-flow-assignment>
    </fo:flow-map>
</fo:layout-master-set>
<fo:page-sequence master-reference="">
    <fo:flow flow-name="">
            <fo:block/>

    </fo:flow>
</fo:page-sequence>
</fo:root>
```

Example 12: Galley Based Page and Role Based Mapper for XSL-FO

XSL-FO already has several built-in elements that are required to perform the role based mapping. In particular, the presence of:

- fo:footnote, fo:marginalia [7] and fo:float to identify the reference point of the out-of-flow content and may provide content to be placed in the main flow;

- fo:footnote-body, fo:marginalia-body and fo:float to specify the content that goes into the out-of-flow.

It is important to notice that the fo:float element represents both the reference point as well as the out-of-flow content. This is because the float is an anchored stacked element, so just its position within the main flow is the relevant information.

Beside these obvious elements, it would be possible to apply the role based mapping to virtually any fo: elements in the document. This can be achieved using the generic class attribute and assigning a name to the class that corresponds to the desired role in the mapper. The only restriction would be for parts that are already non-flowing like absolutely position elements.

6.2 HTML5+CSS3

HTML 5 does not support any specific construct to represent content pagination. However, CSS3, through the Paged Media Module [8], addresses the definition of pages and page sequences to enable printing. The main drawback of the Paged Media Module is the substantial lack of region definition and mapping. The regions that are defined are only ancillary regions to the main body and are referenced as top-left, top-center-left and so on, without specifying size and other characteristics. For what concern the content mapping, the semantics allowed are to connect page definitions with style types, so then different page designs can be used. This allows using alternative pages, e.g., portrait vs. landscape, as shown in Example 13, but does not allow to correctly place content in different page layout elements.

```
@page portrait { size: portrait }
@page landscape { size: landscape }
div { page: portrait }
table { page: landscape }
```

Example 13: CSS3 Paged Media Page Selection for Different Content

Also the web designer has to provide the pages definition and express all the potential/wanted combinations of content and page: selector to ensure the desired result.

There is a complete disconnection between the HTML elements and the areas in the page where these should be placed. Even when these are made through CSS3, they are associated using an implicit connection. In the CSS3 Generated Content for Paged Media

Module [9], the footnotes section introduces a way to express a footnote area in the page. The footnote area is not specifically designed, it is added as necessary and position relatively (or absolutely) to the parent container as illustrated in Example 14.

```
@page {
    @footnote {
            float: bottom page;
            width: 100%;
    }
}
```

Example 14: Footnotes Area Definition using CSS3

A similar capability is available in XSL-FO 1.x. A footnote area is generate as part of the body area and placed at the top, bottom or sides. Meanwhile this is fine for the very traditional footnotes; it does not work when the footnotes need to be directed from many content areas to separate footnote areas (e.g. one footnote area per column) or if there are several different footnote areas (e.g. a footnotes and comments/examples area).

In order to extend the HTML5+CSS3 capabilities in supporting both the galley design and mapper, we would like to introduce a new structural tag called pagination. The pagination tag contains all the page definitions to be used and the associated mapper. The pages and mapper can be defined by the user, as part of the HTML design and/or injected by another agent. Moreover, having the pagination content separated from the body and from the CSS style-sheet permits making it entirely optional. In case the HTML document is not supposed to be paginated nor printed, e.g. online video or music service, the designer can force the omission of the pagination tag. This could be interpreted as a specific wish to disable pagination and printing capabilities.

Example 15 illustrates how the HTML5 structure is expanded to accommodate the pagination tag and its content. The content structure can be used leveraging the notation introduced in the previous sections; the only change is the introduction of a CSS like selector named html: to support standardization over the various HTML out-of-flow features.

The role-mapper has in its html:content section three separate source entries. Each entry is pointing to a different URL. This is to demonstrate that it is possible, when the pagination is applied, to reference to other web pages and concatenate all of them to obtain a full document. For instance, this could be used for an article composed by several sections or a book composed by chapters or any other content that is better consumed on the screen in several chunks; but when printed, it can be concatenated.

We believe that for paginated models to be successful in a web centric approach, most of the definition needs to be automatically generated and inferred simply using conventions.

```
<html>
  <head>
    <title>HTML5 Representation</title>
  </head>
  <pagination>
    <page name="Example1">
      <region name="MarginaliaAndCentralFootnote">
        <galley name="main-wrapper" extent="auto" stacking="ltr">
          <galley name="marginalia1" extent="15%"/>
          <galley name="column1" extent="24%"/>
          <galley name="ftn-wrapper" extent-min="0%" extent-max="10%">
            <galley name="gap1" extent="10%"/>
            <galley name="footnote" extent="90%"/>
          </galley>
          <galley name="gap2" extent="2%"/>
          <galley name="column2" extent-max="44%" extent-min="34%" />
          <galley name="marginalia2" extent="15%">
        </galley>
```

```
      </region>
    </page>
    <role-map name="Example1">
      <assignment role="html:content">
          <source href="url-1"/>
          <source href="url-2"/>
          <source href="url-3"/>
          <dest ref="column1"/>
          <dest ref="column2"/>
      </assignment>
      <assignment role="html:footnotes">
          <source ref="column1"/>
          <source ref="column2"/>
          <dest ref="footnote"/>
      </assignment>
      <assignment role="html:marginalia">
          <source ref="column1"/>
          <dest ref="marginalia1"/>
      </assignment>
      <assignment role="html:marginalia">
          <source ref="column2"/>
          <dest ref="marginalia2"/>
      </assignment>
    </role-map>
  </pagination>
  <body>
    <h1>Enter your header here</h1>
    <p>Enter your body text here</p>
  </body>
</html>
```

Example 15: HTML 5 Representation for Pagination Model

By default, the web browser will be responsible for providing suitable pagination models based on its capability and/or combination of the operating system and connected devices. For instance, it is possible that the paginated model is provided by the printer manufacturer based on the printer model and its capabilities, as illustrated in Figure 9.

The user could have defined either the pagination model and/or the Role Based Mapper. In this case, the designer's defined portion is overriding any potential settings that the web browser has defined or obtained. If the designer had defined only a portion of the pagination element, the Web Browser or the Device Driver can complement adding the missing part. A common scenario is that the designer has created the role-map portion, so s/he can correctly map the various flow and out-of-flow elements in the page layout, but s/he did not design the page for a specific format and is delegating this to the web browser and device specific needs.

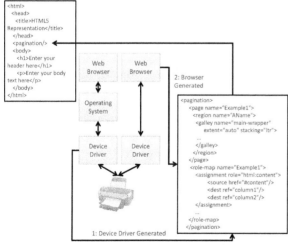

Figure 9: Web Browser or Device Driver Pagination Model Generation

HTML 5 does not have specific elements to address the footnotes, marginalia (a.k.a. side notes) and floats, but the standard presents various suggested methods to address the issue.

6.2.1 Footnotes

For instance, in order to express a footnote it is possible to either use the title attribute or define an empty anchor element that refers to some content placed anywhere within the document.

```
<p> <b>Customer</b>: Hello! I wish to register a complaint. Hello. Miss?
<p> <b>Shopkeeper</b>: <span title="Colloquial pronunciation of 'What do you'">Watcha</span> mean, miss?
```

Example 16: HTML 5 Footnotes using Title Attribute

In Example 16, the title attribute is used to express the additional content/comment to be displayed. Using the proposed approach, it will be just enough to add the class attribute, with the html: selector, to the span element indicating that this belongs to a footnote.

```
<p> <b>Customer</b>: Hello! I wish to register a complaint. Hello. Miss?
<p> <b>Shopkeeper</b>: <span class="html:footnote" title="Colloquial pronunciation of 'What do you'">Watcha</span> mean, miss?
```

Example 17: HTML 5 Title Based Footnote with Class Attribute

When the web browser identifies the html:footnote class, it places the span content in the flow galley and redirect the title attribute to the galley mapped to the html:footnote role.

The more generic approach to support footnotes in HTML5 is to use an interrelated anchor, as illustrated in Example 18.

```
<p> Norman: I don't. <sup><a href="#fn1" id="r1">[1]</a></sup>
<p> Interviewer: You told me you did!
...
<section>
 <p id="fn1"><a href="#r1">[1]</a> This is, naturally, a lie,
but paradoxically if it were true he could not say so without
contradicting the interviewer and thus making it false.</p>
</section>
```

Example 18: Advanced Methodology to Support Footnotes in HTML5

The a element acts as a footnote element that identifies where in the flow the footnote happens and provides the anchor point. It has a href pointing to the footnote body and an id to be cross-referenced. The p element inside the section has the footnote body content and a nested a element to point back to the footnote originating element in the flow. To support the Role Based Mapper it is sufficient to add the attribute class to the a element which is generating the footnotes; the web browser will simply follow the link indicated in the href to retrieve the footnote body and place it in the mapped galley. Example 19 illustrates the changes.

```
<p> Norman: I don't. <sup><a href="#fn1" id="r1" class="html:footnotes">[1]</a></sup>
<p> Interviewer: You told me you did!
...
<section>
 <p id="fn1"><a href="#r1">[1]</a> This is, naturally, a lie,
but paradoxically if it were true he could not say so without
contradicting the interviewer and thus making it false.</p>
</section>
```

Example 19: HTML5 Notation for Footnotes with Proposed Extension

6.2.2 Marginalia or Side Notes

Marginalia is supported through the usage of the aside element. The content wrapped by the aside element represent the marginalia and can be extended adding a class attribute that contain the html:marginalia selector.

```
<p> <span class="speaker">Customer</span>: I will not buy this record, it is
scratched.
<p> <span class="speaker">Shopkeeper</span>: I'm sorry?
<p> <span class="speaker">Customer</span>: I will not buy this record, it is
scratched.
<p> <span class="speaker">Shopkeeper</span>: No no no, this's'a
tobacconist's.
<aside class="html:marginalia">
 <p>In 1970, the British Empire lay in ruins, and foreign
nationalists frequented the streets — many of them Hungarians
(not the streets — the foreign nationals). Sadly, Alexander
Yalt has been publishing incompetently-written phrase books.
</aside>
```

Example 20: HTML5 with Extension to Support Marginalia

6.2.3 Floats

The identification of floats and positioning is defined using the CSS2.x float attribute [11]. This attribute identifies where the float is positioned considering its anchor, e.g., left or right. This implies (as before), that a convention must be applied and the CSS notation has to be used in conjunction with the html:float selector.

Clearly, the document designer can decide to ignore all these conventions and simply use the naming strategies that are the most useful, as well as semantically relevant, to the document. Ad-hoc strategies works only if the designer is providing the paginate model as well as the mapper with the document. On one hand, if the designer does not provide the paginate model and mapper the web browser will not be able to correctly place content in a visually pleasing layout for paginated content. On the other hand, if the designer follows the conventions s/he can ensure that the document will be correctly paginated no matter what the pagination device could be.

7. CONCLUSIONS AND NEXT STEPS

Personalization and customization are increasingly adding focus to the dynamic aspect of the content so that dynamic content in Web pages is becoming the rule. The presented approach enables web and print specific content to be better presented on paginated media such as printers, e-book readers, tablets, and other mobile devices where pagination is more effective than scrolling.

Pagination is currently addressed by several formats using different approaches, each one with its merits, but often applicable only to the specific targeted format. This paper attempts defining a more generic solution to the problem trying to abstract the layout and mapping into a high level description and then map it back into the specific format.

We understand that for this proposal to have success changes in the XSL-FO and HTML5+CSS3 specifications are required. For this reason, it is our intention to bring the solution to the attention of the various working groups and identify a path forward.

8. REFERENCES

[1] F.Giannetti, Mapping strategy for web-driven magazines with personalized advertisement and content, *Multimedia Tools and Applications*, Volume 43, Number 3, 327-343 (2009)

[2] R. Vetter, Print, Mobile, and Online, *Computer, IEEE Computer Society*, Volume 44, Number 1, 9-11 (2011)

[3] H. Kingston, The design and implementation of the Lout document formatting language. *Software— Practice and Experience* 23, 1001–1041 (1993)

[4] Cascading Style-Sheet 3 (CSS3) Multi-Column Layout Module: http://www.w3.org/TR/css3-multicol/

[5] Cascading Style-Sheet 3 (CSS3) Template Layout Module: http://www.w3.org/TR/css3-layout/

[6] eXtensible Style-Sheet – Formatting Objects (XSL-FO) 1.1 Specification, fo:flow-map: http://www.w3.org/TR/xsl/#fo_flow-map

[7] Controlling LaTeX Floats: http://mintaka.sdsu.edu/GF/bibliog/latex/floats.html

[8] Cascading Style-Sheet 3 (CSS3) Flexible Box Layout Module: http://www.w3.org/TR/css3-flexbox/

[9] Cascading Style-Sheet 3 (CSS3) Paged Media Module: http://www.w3.org/TR/css3-page/

[10] Cascading Style-Sheet 3 (CSS3) Generated Content for Paged Media Module: http://www.w3.org/TR/css3-gcpm/#footnotes

[11] eXtensible Style-Sheet – Formatting Objects (XSL-FO 2.0) Design Notes: http://www.w3.org/TR/xslfo20/

[12] Cascading Style-Sheet 2.1 (CSS2.1) Floats: http://www.w3.org/TR/2010/WD-CSS2-20101207/visuren.html#floats

A Novel Physics-Based Interaction Model for Free Document Layout

Ricardo Piccoli, Rodrigo Chamun, Nicole Cogo,
João Oliveira and Isabel Manssour

Centro de Pesquisa em Computação Aplicada
Faculdade de Informática — PUCRS
Porto Alegre — Brazil
{oliveira,manssour}@inf.pucrs.br

ABSTRACT

Marketing flyers, greeting cards, brochures and similar materials are expensive to produce, since these documents need to be personalized and typically require a graphic design professional to create. Either authoring tools are too complex to use or a predefined set of fixed templates is available, which can be restrictive and difficult to produce the desired results. Thus, simpler design tools are a compelling need for small businesses and consumers. This paper describes an interactive authoring method for creating free-form documents based on a force-directed approach, traditionally applied for graph layout problems. This is used for automatically distributing and manipulating images, text and decorative elements on a page, according to forces modeled after physical laws. Such approach can be used for enabling easy authoring of personalized brochures, photo albums, calendars, greeting cards and other free-form documents. A prototype has been developed for evaluation purposes, and is briefly described in this paper. Evaluation results are presented as well, showing that users enjoy the experience of designing a page by interacting with it, and that end results can be satisfactory.

Categories and Subject Descriptors

H.4 [**Information Systems Applications**]: Miscellaneous; I.3.6 [**Computer Graphics**]: Methodology and Techniques—*Interaction Techniques*; I.7.2 [**Document and Text Processing**]: Document Preparation—*Format and notation, Photocomposition/typesetting*

General Terms

Algorithms, Design, Experimentation

Keywords

automatic document layout, authoring, personalized documents, force-directed methods, physics-based simulation

1. INTRODUCTION

Authoring solutions for personalized material such as greeting cards, brochures, photo albums, and marketing collaterals are usually based on static templates produced by graphic designers, filled with content provided by a user. Good examples of such systems are Snapfish[1] and Shutterfly[2]. However, those templates can be expensive to create and maintain and are difficult to adapt to different purposes, media sizes or customers. Recent years have seen considerable advances on automatic document generation and adjustment, but providing non-specialists with a pleasant interface for producing simple documents has always been a challenge: editors have too many operations or are very complicated, turning the task of designing a simple document such as a greeting card into a laborious experience.

This paper describes a new interaction model for creating documents in a quick and intuitive way. A variety of content combinations (text, images, backgrounds and decorative items) and document form factors can be used to create different documents. The aim is to provide a better, easier experience at designing loosely-formatted documents. We do not intend to replace complex software systems to produce magazines, but assist small business owners in designing their own leaflets, and enable users with no experience in graphic design to produce their own personalized greeting cards, calendars, photo albums and others.

In our approach, items float freely on the page as new items are inserted or moved around, so that the user (designing a simple brochure for his shop, for example) always sees a page where items are well-spread, do not overlap and he may concentrate on placing a single item knowing that existing items will react to it and place themselves on better positions. Items float on the page and react to each other, trying to be as far away from each other as possible. This is performed through the use of a force-directed method, i. e., a physics model based on electrical repulsion. Items have different charges according to their sizes and the physics model was adapted for the sake of an easier manipulation of the objects. The use of force-directed methods seems to be quite new in the context of document layout. The approach used in this paper is briefly described in Section 3. Details of related approaches are discussed in Section 2.

The end result of our approach is an intuitive authoring method that is able to produce good-looking pages with

---

[1] http://www.snapfish.com
[2] http://www.shutterfly.com

small effort. We also provide extra facilities like anchoring an item to a specific position, or anchoring an item to another (so that a picture and a price tag may be moved together easily, for example). After the user is done, a printable document is generated as a PDF [3] file. Sections 4 and 5 describes the development of this interaction model and document generation in detail.

To assess the method described in this paper, a simple tool for authoring personalized calendars was developed. Additionally, a preliminary user evaluation was performed using this prototype, and the obtained results show that the interaction is in fact intuitive and produce results quickly. Some users reported that constructing a layout using the proposed method feels like playing a game, and they are usually less concerned about the design details (for instance, aligning exactly each item on the page on a grid). To show that the end results can be satisfactory, the prototype tool was used to generate sample results as well, which are presented in this paper. The user evaluation and sample results are presented and discussed in Section 6, and the final considerations and future works are discussed in Section 7.

2. RELATED WORKS

Several approaches for automatically constructing mixed-content documents have been proposed in the recent years [8]. Most approaches focused on producing mass-customized documents for Variable Data Printing (VDP) [23], or adapting some content for various media and sizes. In our approach, the creation of the document must be interactive, in order to allow the user to easily specify the desired layout. This is not the case with VDP, where most of document customization takes place in the latest stages of the printing process, thus still relying on templates and reducing the potential for personalization. Additionally, the generated documents must be free-form, as opposed to a grid-based document.

Typical document design systems such as Scribus [21] and InDesign[3] are very rigid in the sense that they require each item to be moved and aligned independently from others, since items cannot interact with each other. For example, items have to be spread on the page manually by the user, whereas in our approach they spread over the page automatically. The problem of automatically producing document layouts is best stated as a constrained optimization problem, where a number of geometric constraints guide an optimization method for finding a good placement and size for each document item [8]. In this sense, most works on document layout are based of constraint solving problems, which can be very expensive to solve. When the layout must be produced quickly, either the document layout model must be simplified or an approximate solution is used [8]. Our approach is based on force-directed methods [16], which have been extensively used for graph drawing and visualization [2, 11], but these approaches seem to never have been used for document layout.

TEX [20] is notable for its use of stretchable space (also known as *glue*) to join pieces of text in a page, which works similarly to a physical model of a spring. However, it only works on horizontal or vertical directions, at specific locations or explicitly defined between pairs of boxes [20]. This works well for low-level formatting, where boxes must be well-aligned (i. e. to form words, lines, paragraphs, and so

---

[3]http://www.adobe.com/products/indesign

forth), but in free-form documents items can be influenced by others in all possible directions. Moreover, in our scenario documents are assembled interactively, causing items to regularly find new positions in the page, and thus a different model is needed in our case.

Purvis et al. [18] proposed a genetic algorithm for finding a good placement of document items, using a combination of aesthetic measures [6] to guide the heuristic search. Unfortunately, this process is slow and aesthetics often interfere with each other, so the user may be required to tune aesthetic parameters and start over until a satisfactory layout is obtained. Jacobs et al. [10, 22] describe an approach for creating grid-based documents using templates that adapt for different sizes. However, as pointed out by the authors, producing an adaptive template is a complex task, which is not suitable for a non-expert user, and thus professional templates are still required. The approach by Lin [13] is similar, but appears to be slower and requires an initial document for adapting its layout to variable content. Oliveira [4] describes a method for producing newspaper-like layouts, based on repeated bisection. However, items on a page are not free-form as in brochure-like documents, but rather they are constrained to the hierarchical partition.

In all of these approaches, the end user has little or no involvement when producing the document. Either the task is fully automatic for mass-production or the user is forced to use only the previously-designed page templates.

The work from McCormack et al. [15] uses constraint solving to enable a user to create diagrams that are adjustable to different viewing conditions. It uses an interactive authoring tool as in our work, but as the authors pointed out, the use of constraint solving techniques is only viable when the geometric space on the page is discrete, as in a grid-based document. In our case, documents must be loosely-formatted.

A recent work from Ali et al. [1] appears to be similar to the work proposed in this paper in its use of physics. However, their system requires a template database and a constraint-solving system for initial layouts, which limits the use of physics only as a post-processing stage. Nevertheless, their model handles item repulsion similarly to our work (see Section 4.2), but requires a graph of adjacencies between items, obtained from the template. One interesting difference is the use of a gas pressure model for automatically resizing items during interaction, whereas in our case the resizing of items is performed by the user. Unfortunately, the paper lacks details regarding user interaction and experimental data, making it difficult to compare to the method from this paper.

3. PROPOSED APPROACH

The approach described in this paper is a semi-automatic method for document layout. In other words, the layout is constructed by a non-specialist user by means of an authoring tool, which automatically moves items on a page, to avoid item overlapping and bad distribution over the page [6].

User interaction involves selecting items from a set of pictures and textual content, dragging these to one ore more empty pages. Automatic distribution of content occurs in real-time, i. e., the page is reorganized as the user constructs it. For practical purposes, this work assumes some simplifications in the interaction model:

- The page is loosely-formatted, i. e., there are no columns or grids;

- Pictures and text are rectangle-shaped;

- Items cannot be rotated, but may be scaled or deleted;

- Collision detection is not performed [16], i. e., items may overlap with each other, if necessary (see Section 4.2 for details).

The user may drag images and text boxes to an empty page and those items float around: they repel each other using an electrical repulsion model and as a consequence of that repulsion content is automatically spread on the page (see Figure 1). As some item is resized or moved other items recede to provide space for it. Moving items makes other items react to that movement by moving as well, and a chain-effect of several editing software is avoided: to move one item from A to B one has to move another item from B to C to make room for it, and to move the second one has to move a third from C to D and so on. In our approach items distribute themselves and react to the movement of the first item.

After completing the desired layout, the final document must be produced by a rendering engine in order to obtain a printable PDF [3] file (see Section 5).

4. INTERACTION MODEL

While traditional physics simulation models usually attempt to replicate real world physics [16], we found that these lead to non-intuitive user cues and a poor user experience. Thus, the simulation engine used in our method has been built from scratch using a simplified model for the physical forces, aimed at reducing user surprise.

The user is initially presented with an empty document, and the final document is constructed interactively, as the user inserts or manipulates content. Any number of pages can be used to create a document, but each page is independent, and no attempts are made to perform automatic pagination [17] or flowing items between pages.

Items already placed in the page continuously interact with each other. At the same time, the user can still interact with the system, performing any of the following actions:

- Inserting new items into the page;

- Deleting items from the page;

- Moving items around;

- Resizing items.

Items are either text chunks, color blocks or pictures, which are represented by rectangles in this paper's approach. An item i is defined by its central position (x_i, y_i), and size (w_i, h_i). For simplicity, we assume that content is already available from some source or created by the user prior to the interaction with the layout system.

The entire system works according to the user events mentioned above, as well as the physical simulation itself, which is carried out in real-time. Each simulation step computes new positions for all items present in a page, if distribution is necessary. For efficiency, the simulation may execute several steps for each time the document frame is drawn on the screen. This will be described in more detail in Sections 4.1 and 4.3.

4.1 Physics model

The problem of finding new positions of items (for a better distribution on a page) can be modeled as an energy minimization approach, similar to graph drawing methods [2].

Although energy minimization can be achieved efficiently with general optimization heuristics such as simulated annealing [5], a force-based simulation is used in this paper. The forces are computed for each time step and the items are moved accordingly. Although slower, this allows the user to interact with the system and follow any changes while it works in finding a good placement. Despite being a standard approach for graph layout, its use is new in the general document layout context. Thus, several modifications were necessary to adapt the force-directed method for this context.

From a usability perspective, items should move smoothly on the page, so that the user is able to easily follow the items and thus feel more familiarized when interacting with the system. Therefore, a *velocity* parameter (vx_i, vy_i) was also added for each item i, allowing items to move by inertia. Section 4.3 discusses this in more detail.

Given this description, an equilibrium state of this system is defined as the state that minimizes the total kinetic energy ε, i. e.,

$$\varepsilon = \sum_i w_i h_i \left(vx_i^2 + vy_i^2 \right). \tag{1}$$

Item positions and velocities are computed using simple Euler integration [16]. This method updates positions and velocities explicitly in discrete time steps Δ_t, until the system reaches an energy minimum, i. e., $\varepsilon = 0$. To compute new velocities (and consequently, item placement), it is necessary to compute the forces generated between items first.

4.2 Repulsion forces

The basic working force is the electrical repulsion between items. Every item has an influence on every other item (i. e., visibility issues are not considered), based on the distance between each other and their sizes. These forces are computed pairwise. When a page is overfull with items, the repelling forces will not be enough to prevent items from intersecting. Although collision detection could be used to prevent intersection between items [16], it showed to cause some instability in earlier tests, because of the increased kinetic energy from repulsion. However, items will always try to be as far apart as possible from each other, so overlapping seldom occurs.

The repulsion force between two items uses the simplification from Di Battista et al. [2] for Coulomb's Law. Thus, the repulsion force vector $\overrightarrow{f_{u,v}}$ of item v over item u is defined as

$$\overrightarrow{f_{u,v}} = (fx(u,v), fy(u,v))$$
$$fx(u,v) = \frac{q_v \, \alpha}{\sigma(u,v)^2} \frac{\Delta_x}{\sigma(u,v)} \tag{2}$$
$$fy(u,v) = \frac{q_v \, \alpha}{\sigma(u,v)^2} \frac{\Delta_y}{\sigma(u,v)},$$

where $q_v = w_v h_v$ is the *charge* of item v, $\Delta_{x,y}$ is the displacement in x or y, α is the repulsion constant that can be tuned according to user preferences, and $\sigma(u,v)$ measures

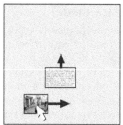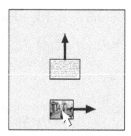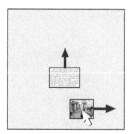

Figure 1: Moving items according to physics laws: the lower item moves from left to right and the upper item recedes as the lower one approaches.

the distance between u and v in the page. The distance between two items is usually measured between their central positions. Thus, the charge parameter q is set to be proportional to the item's area, since larger items need to be further away from others to avoid overlapping. Distance can also be measured in different ways, each achieving different results. Section 4.6 discusses two approaches for computing $\sigma(u, v)$.

Since multiple items will repel an item u, a resultant force must be computed so that a repelling direction and intensity is known in order to move item u. Thus, for every item u, the resultant force vector $\overrightarrow{F_u}$ is computed as

$$\overrightarrow{F_u} = \sum_{v \neq u} \overrightarrow{f_{u,v}} \quad (3)$$

From the resultant forces, the new velocities and positions can be computed for the next time step.

4.3 Making items move

After computing the resultant force vectors $\overrightarrow{F_i}$ for every item, they must be applied in order to move items for one time step. We developed a simple physical simulation consisting of force, velocity and position updates only, computed using the Euler integration method. Although Euler integration is usually unstable for real-time physical simulation [16, 7], we found it to be simple and stable enough for the purposes of this work. Moreover, it is efficient enough to be used in real-time in an interactive system, such as the one proposed in this paper.

To compute positions, assuming that an item i is positioned at $\left(x_i^t, y_i^t\right)$ at an instant t, we compute the positions for $t + 1$ as

$$\begin{aligned} x_i^{t+1} &= x_i^t + \Delta_t\, vx_i \\ y_i^{t+1} &= y_i^t + \Delta_t\, vy_i. \end{aligned} \quad (4)$$

For velocities, adding a damping (i. e., friction) coefficient is required, as the inertia could cause the system to move forever and not settle. Thus, velocity for item i is computed as

$$\begin{aligned} vx_i^{t+1} &= \left(vx_i^t + \Delta_t\, fx_i\right)(1 - \delta) \\ vy_i^{t+1} &= \left(vy_i^t + \Delta_t\, fy_i\right)(1 - \delta), \end{aligned} \quad (5)$$

where δ is a damping coefficient in the range $[0 \ldots 1]$, being $\delta = 0$ a frictionless system and $\delta = 1$ a static system.

4.4 Constraining items to the page

Repulsion forces are not enough to keep items distributed inside the page, since they can repel each other apart infinitely far. Therefore, it is necessary to incorporate more forces into our model, so that items are confined within the document page.

An approach that works well is to add virtual or "wall" forces in the page boundaries, so that escaping items are always pushed back inside.

Figure 2 shows a simple way to do this. In this case, forces are added for each side of the page (i. e., top, bottom, left and right), and follow the items in both axes, and using the same charge q as the item being pushed. This results in all items attempting to float to the center of the page. The dispute for this space results in items being spread evenly over the page.

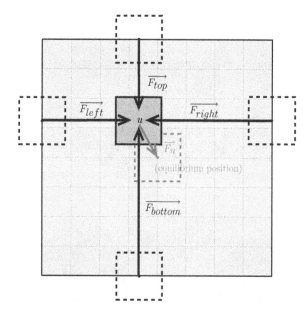

Figure 2: Constraining an item by adding virtual forces, forcing it to the center of the page.

4.5 Anchoring & grouping items

Sometimes, a user might wish to "glue" an item at a certain location in the page, and not allow it to move away from that location. This will be referred in this paper as *anchoring* an item from now on.

The force model can be adapted to support anchoring, by creating a new attribute (ax_i, ay_i) to store the anchoring location for item i, and a force to attract item i to the anchor point. This force is then added to the resultant force $\vec{F_i}$.

An intuitive force model that can be used for anchoring is a spring model. Hooke's law [2] can be used to keep (x_i, y_i) close to (ax_i, ay_i), while still allowing the item to move when under excessive pressure. The spring model was further simplified by assuming the spring rest length to be zero, since the desired position for an item is exactly over its anchor location. Thus, the spring force vector $\vec{g_i}$ of item i is described as

$$\begin{aligned} \vec{g_i} &= (gx\,(i)\,, gy\,(i)) \\ gx\,(i) &= \gamma\,\beta\,\Delta_x \\ gy\,(i) &= \gamma\,\beta\,\Delta_y, \end{aligned} \tag{6}$$

where β is the spring stiffness constant, γ is a user adjustable parameter in the range $[0\ldots1]$, and $\Delta_{x,y}$ are the displacements in x and y. The γ parameter was introduced for flexibility: it can be used to allow the user to fine tune the forces of individual items, by making the distance constraint between an item and its anchor point more or less rigid.

Although intuitive, this method can make the physical system more unstable due to the amount of movement produced when stretching the spring too far from its anchor point. For more stability, the force model can be dropped and the item's position can be directly manipulated after the items are moved at the end of the current time step.

The anchoring idea can be extended for a finer control of the layout as well. For instance, items can be anchored to each other, so they always end up near the same location. This idea was implemented in a prototype but was not included in the user evaluation from Section 6.3.

4.6 Measuring distance between items

A measure of distance must be defined in order to compute repulsion forces. The initial attempt was to measure the euclidean distance σ between items u and v as

$$\sigma\,(u, v) = \sqrt{(x_u - x_v)^2 + (y_u - y_v)^2} \tag{7}$$

between the center (x, y) of items u and v.

From early testing we found that, although this measure provided a realistic behavior physically, it caused the items to move too much on the page until they reached an equilibrium. In some circumstances this behavior may cause difficulties for a user to follow and interact with the system.

To circumvent this effect, an alternative distance measure was developed. It is based on the same euclidean distance from Equation (7). However, when two items u and v intersect in the x or y axes, the intersecting axis is cut off from the equation, i. e., only one direction component of the distance is used:

$$\sigma\,(u, v) = \begin{cases} |x_u - x_v| & y\text{-only intersection} \\ |y_u - y_v| & x\text{-only intersection} \\ \sqrt{(x_u - x_v)^2 + (y_u - y_v)^2} & \text{otherwise.} \end{cases} \tag{8}$$

From our tests, this approach seemed to be more stable than the one from Equation (7), because as items come

closer to each other, less movement is generated, causing the physical system to stabilize more quickly.

Figure 3 illustrates the behavior of each different distance measure when dragging item u against item v.

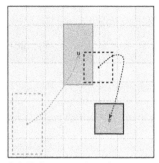 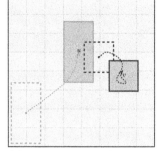

(a) Behavior when forces act on both axes.

(b) Behavior when forces act in a single axis, when colliding.

Figure 3: Repelling behavior using different approaches for measuring distance. Item u is being dragged by a user towards item v, which gets repelled in different ways as a consequence.

4.7 Stability

The final document can only be sent to a rendering engine when $\varepsilon = 0$, thus it is necessary to ensure that the physical system will be stable enough to converge. As shown in Section 4.3, a damping coefficient is necessary to ensure that the total sum of forces at each time step will eventually converge to zero. However, given the poor approximation obtained with the Euler equations when the time step is large, the force vectors may grow larger and cause a well-known phenomenon of *overshooting* [7, 16]. Moreover, the physical system may naturally end up in an oscillatory behavior.

For enhancing simulation stability, the time step Δ_t and damping δ parameters can be tuned for different values, depending on the number of items, page size, and others. Small time steps result in a more stable and precise simulation [16], but require more processing power.

To guarantee that the physical system will eventually stop without completely freezing it, i. e., $\delta = 1$, the repulsion constant α and spring stiffness β can be automatically damped over simulation steps by a small factor, e. g. 0.01, until the system stops or the user performs a new operation on the page, such as adding a new element. In the latter case, the constants are reset to their original values previous to damping. The damping factor also automatically increases as more items are added to the page, reducing instability.

5. GENERATING THE DOCUMENT

After deciding the final layout, the user can produce a PDF file through a rendering engine. The document description is given as a page rectangle (W, H) and a set of item rectangles (xc_i, yc_i, w_i, h_i) for each item i.

Content is placed over each rectangle independently by an external rendering engine. For this, LaTeX [12], Scribus [21] or the iText Java library [14] can be used to place pictures and textual content inside each rectangle. Other engines may be used, as long as they support some programming capabilities, such as font resizing in a text box. Both LaTeX and iText renderers were implemented in a prototype, achieving similar results.

To place pictures, the width and height of the picture are simply scaled to the rectangle size. We do not attempt to perform automatic cropping [24], thus the user interface is responsible for preserving the original picture's aspect ratio as the user resizes a picture.

For textual content, attributes such as font type and color may be defined for each different text. However, since we treat text as chunks that are completely contained in their enclosing rectangle, the font sizes are automatically computed so that the text completely fills the rectangle. Automatic justification and line breaking are also required. In this paper, a simple greedy line-break algorithm [9] was used along with left aligned text, but other methods may be used.

Example camera-ready documents created by the rendering process are presented in Section 6, produced by a prototype authoring tool that uses iText for rendering.

6. RESULTS AND EVALUATION

To produce results and evaluate the proposed method, a prototype was developed for authoring personalized calendars, which will be described briefly in this section. From this prototype, examples of different documents that can be produced are presented in Section 6.2.

A preliminary user evaluation was performed using this prototype tool as well, and will be presented in this section.

6.1 Case study: a personalized calendar authoring tool

To test the proposed interaction model, a simple authoring tool was created, intended for creating personalized calendars by a user. Calendars can be created, opened, saved and exported to a PDF file. Twelve day-month panels are already provided as pictures in the user interface, and the user can select pictures from his own collection and create text boxes as well. Other available items are colors boxes and page backgrounds. These decorative items are not affected by the physics engine, and can be freely placed and overlapped by the user. The main content is distributed automatically by the physics engine, but the possibility of anchoring (Section 4.5) was not included, for the purposes of the user evaluation (Section 6.3). In addition, the user has two different pages available for selection and editing. It is also possible to switch between portrait and landscape orientations. Figure 4 shows the main user interface of the authoring prototype.

After finishing the document, the user can produce a PDF file intended as a printing material, using a print option that renders the document into a PDF as described in Section 5.

The authoring tool has been implemented using the Processing API [19] in Java, and deployed as an applet in the Web[4].

It is important to note that the tool is still a prototype and has a limited set of capabilities regarding style: there is only one font style available for textual items, a limited range of colors, and a single page size (A4). Although it is possible to produce simple but aesthetically pleasing documents, the main purpose of the prototype is to evaluate the

---

[4]Our prototype can be tested at http://www.cpca.pucrs.br/BrochureLayout

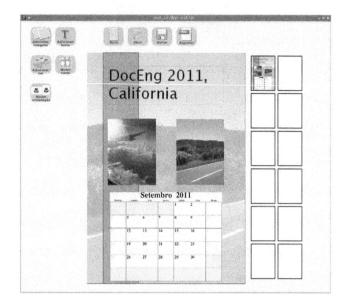

Figure 4: Screen shot of the prototype authoring tool.

actual interaction with the physics model, not to produce professional-looking documents.

6.2 Application examples

This section shows some example pages produced with our interactive approach. Because of the limitations of the prototype, the sample documents are not professional-looking, although they demonstrate what types of documents can be produced by the proposed interaction method. Moreover, the examples demonstrate how pictures, text and decorations can easily be combined into a document, and how the interaction model (Section 4) and the rendering process (Section 5) are integrated in the authoring tool. Figure 5 shows four examples of calendar pages, produced by the authoring tool described in Section 6. An example of a 5x8 personalized greeting card and photo album are shown in Figure 6 and were created with a more complete version of the tool that allows anchoring and different page sizes.

Figure 6: Examples of a photo album page and a 5x8 greeting card.

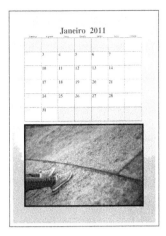

Figure 5: Sample personalized month calendars generated by the prototype.

6.3 User evaluation

To evaluate the proposed interaction method, a simple user study was performed. The experiment consisted in obtaining volunteers to test our prototype and produce a personalized calendar using pictures from two local soccer teams, according to the volunteers' personal tastes. There were no restrictions on the number of pictures or the time required to produce a document, but each volunteer could only produce at most two pages of a calendar. A minimum of instructions were given to the volunteers regarding the operation of the prototype tool and what were the goals of the evaluation. They were further instructed to stop and generate a final PDF of their documents whenever they felt the document looked as desired, or have given up using the tool. At the end of the process, the volunteers were asked to fill out a questionnaire, and had the option to send the produced calendar by e-mail.

In order to be as simple as possible, the version of the prototype used in the evaluation did not contain anchoring features. This enabled a better assessment of the interaction proposed in this paper, because the user might otherwise get distracted by the user interface features, and not pay attention to the interaction model. The physical parameters (described in Section 4) were tuned to the the following values:

- Electric repulsion was set to $\alpha = 150$;

- Velocity damping was set to $\delta = 0.01$;

- Time step was set to $\Delta_t = 0.005$;

- 500 simulation steps per frame;

- 33 frames per second (roughly);

- Repulsion forces were not damped.

For obtaining volunteers for this evaluation, a kiosk was assembled in the main hall of our university's computer science department, during rush hours. All volunteers were undergraduates from computer science, computer engineering and information systems courses.

In total, the evaluation spanned 3 week days, 2 hours per day, and a total of 27 people aged from 17 to 30 years old participated in this evaluation. These volunteers had no previous contact with the prototype.

After testing the authoring prototype, the volunteers were asked to answer a questionnaire consisting of five questions, regarding the interaction model implemented in the prototype:

1. Did the automatic distribution aid you for creating the page more quickly?

2. How did you feel about the automatic positioning of page items?

3. How much did you like the final result?

4. How did you feel about manipulating (moving, deleting, resizing) images on the page?

5. Would you use such a program?

The volunteers had the liberty of answering as many questions as they desired. Questions 1 and 5 were yes/no questions, while the answers for questions 3 and 4 were required to be given in a $1 - 5$ likert scale (i. e., very unsatisfied, unsatisfied, partially satisfied, satisfied and very satisfied). Question 2 had four levels, but encouraged the volunteer to answer the question subjectively. A space for comments was available for every question, allowing the volunteers to express their opinions better.

6.4 Final results and discussion

The evaluation results are presented in Figures 7, 8, 9, 10 and 11 and the comments from each volunteer are also discussed in this section.

As shown in Figure 7, the majority of the volunteers answered question 1 agreeing that the tool is responsive and quick to produce results.

The results from question 2 are shown in Figure 8. From the comments, most users found the interaction immediately intuitive or got used after a while. Few users however, reported feeling somewhat uneasy with the interaction, either because the automatic positioning interferes too much in the layout, or that a finer control is required to be more usable.

Regarding the final document layout and the generated PDF file (question 3), the volunteers were mostly partially

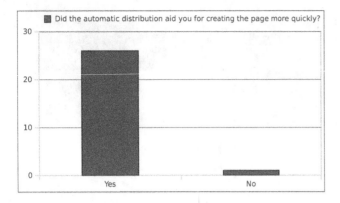

Figure 7: Evaluation results from question 1.

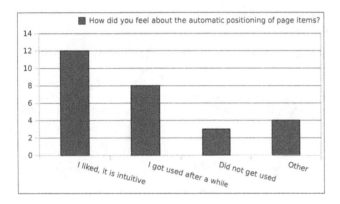

Figure 8: Evaluation results from question 2.

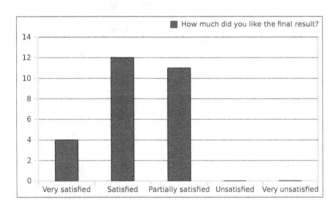

Figure 9: Evaluation results from question 3.

satisfied, as shown in Figure 9. Most likely this is due to the currently limited capabilities of the authoring tool.

As for the image manipulation (question 4), the final results are shown in Figure 10. Twelve users reported that the interaction is intuitive, provides reasonable results quickly and the available operations seem to be clear as well. Five users reported user interface issues unrelated to the interaction model, such as glitches in the prototype tool, the location for certain operations or lack of some facilities such as adding a large set of images to the page at once. Nine users did not get used to the interaction model, and reported a lack of control of the layout, but suggested the use of a fixing mechanism (such as the anchoring method described

in Section 4.5), or the splitting of the document page into areas for grouping items together.

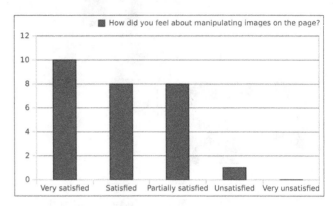

Figure 10: Evaluation results from question 4.

Finally, the results for question 5 are shown in Figure 11. Most of the volunteers approved the prototype tool and would use it for creating their personalized documents. The main reason reported by the volunteers was either because there are no immediately obvious alternatives available or that the tool is really pleasant to use and easy to interact. Few complained of a lack of more features to control positioning of items on the page and felt that, for their purposes, simpler tools such as Microsoft's PowerPoint or Paintbrush would be sufficient.

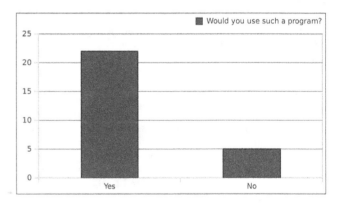

Figure 11: Evaluation results from question 5.

7. CONCLUSIONS AND FUTURE WORKS

This work presented a new interaction model for authoring personalized documents, such as photo albums, brochures, leaflets and others. It is based on a physics engine developed specifically for this purpose. A prototype authoring tool has been developed using the physics engine, and was deployed as a Web application[5] to create personalized photo albums, greeting cards and calendars. Although this tool is only a prototype and does not contain a wide range of artwork, shapes and design choices that would be needed in a whole product offering, the basic principles of the tool are functional, and usable to create small projects and experiment with the proposed model.

---

[5]http://www.cpca.pucrs.br/BrochureLayout

Marketing collaterals are usually expensive to create since it typically requires expert graphic design. Thus, we believe that such automated method will help lower the cost of generating layouts for small businesses (brochures, leaflets and others), allowing these documents to be individually customized. Moreover, a quick and intuitive method for creating documents will encourage the creation of personalized prints (photo albums, greeting cards and others) by non-designers, in an easier way than currently performed today by commercial applications, such as Snapfish[6] and Shutterfly[7].

For future works, there are several possible improvements to the current method. In particular, we are currently working on an extension to support arbitrarily shaped regions and cutouts (since these are common in brochures and greeting cards) and to support interactions among multiple polygonal items, since the current method can only handle rectangular items. A similar idea appears in the paper from Ali et al. [1], but few details are provided. Thus, we feel that further investigation on the use of polygons is still required, as there are implications for the physical model (e. g. performance and stability) and user interaction (e. g. designing and resizing arbitrary regions).

Other possibilities are the inclusion of grid capabilities on a page (for instance, dividing a page into rows or columns), and the addition of hierarchical groups of items for facilitating their placement together. It would also be interesting to experiment with more stable physics solvers, such as Verlet integration [16], and analyze speed/accuracy trade-offs.

Regarding the user study, we found that the obtained results – albeit preliminary – are encouraging, and most of the time the automatic distribution using a physical model is intuitive and pleasant to use. However, it might get in the way of the user at times, e. g. by occasionally moving items away from their previously set location. As the volunteers themselves suggested, an anchoring mechanism or the division of the layout into several areas would be able to solve this problem. Besides investigating the use of such features, we also intend to perform a comparative study with a commercial application for creating personalized documents (such as Snapfish or Shutterfly), to better understand how automatic placement helps users in assembling personalized documents.

8. ACKNOWLEDGMENTS

This paper was achieved in cooperation with Hewlett-Packard Brasil Ltda. using incentives of Brazilian Informatics Law (Law n°. 8.2.48 of 1991).

9. REFERENCES

[1] K. Ali, K. Hartmann, G. Fuchs, and H. Schumann. Adaptive layout for interactive documents. In *Proceedings of the 9th international symposium on Smart Graphics*, SG '08, pages 247–254, Berlin, Heidelberg, 2008. Springer-Verlag.

[2] G. D. Battista, P. Eades, R. Tamassia, and I. G. Tollis. *Graph Drawing: Algorithms for the Visualization of Graphs*. Prentice Hall PTR, Upper Saddle River, NJ, USA, 1998.

[3] R. Cohn. *Portable Document Format Reference Manual*. Addison-Wesley Longman Publishing Co., Inc., Boston, MA, USA, 1993.

[4] J. B. S. de Oliveira. Two algorithms for automatic page layout and possible applications. *Multimedia Tools Appl.*, 43(3):275–301, 2009.

[5] M. Fleischer. Simulated annealing: past, present, and future. In *WSC '95: Proceedings of the 27th conference on Winter simulation*, pages 155–161, Washington, DC, USA, 1995. IEEE Computer Society.

[6] S. J. Harrington, J. F. Naveda, R. P. Jones, P. Roetling, and N. Thakkar. Aesthetic measures for automated document layout. In *DocEng '04: Proceedings of the 2004 ACM symposium on Document engineering*, pages 109–111, New York, NY, USA, 2004. ACM Press.

[7] L. Hilde, P. Meseure, and C. Chaillou. A fast implicit integration method for solving dynamic equations of movement. In *Proceedings of the ACM symposium on Virtual reality software and technology*, VRST '01, pages 71–76, New York, NY, USA, 2001. ACM.

[8] N. Hurst, W. Li, and K. Marriott. Review of automatic document formatting. In *DocEng '09: Proceedings of the 9th ACM symposium on Document engineering*, pages 99–108, New York, NY, USA, 2009. ACM.

[9] N. Hurst, K. Marriott, and P. Moulder. Minimum sized text containment shapes. In *DocEng '06: Proceedings of the 2006 ACM symposium on Document engineering*, pages 3–12, New York, NY, USA, 2006. ACM.

[10] C. Jacobs, W. Li, E. Schrier, D. Bargeron, and D. Salesin. Adaptive grid-based document layout. In *SIGGRAPH '03: ACM SIGGRAPH 2003*, pages 838–847, New York, NY, USA, 2003. ACM Press.

[11] M. Kaufmann and D. Wagner. *Drawing Graphs, Methods and Models*, volume 2025 of *LNCS*. Springer, 2001.

[12] L. Lamport. *LaTeX: A Document Preparation System*. Addison-Wesley, Reading, Massachusetts, USA, Boston, MA, USA, 1986.

[13] X. Lin. Active document layout synthesis. *Proceedings. Eighth International Conference on Document Analysis and Recognition, 2005.*, pages 86–90 Vol. 1, 29 Aug.-1 Sept. 2005.

[14] B. Lowagie. *iText in Action*. Manning Publications Co., Greenwich, CT, USA, 2010.

[15] C. McCormack, K. Marriott, and B. Meyer. Authoring adaptive diagrams. In *Proceeding of the eighth ACM symposium on Document engineering*, DocEng '08, pages 154–163, New York, NY, USA, 2008. ACM.

[16] M. Müller, J. Stam, D. James, and N. Thürey. Real time physics: class notes. In *ACM SIGGRAPH 2008 classes*, SIGGRAPH '08, pages 88:1–88:90, New York, NY, USA, 2008. ACM.

[17] M. F. Plass. *Optimal pagination techniques for automatic typesetting systems*. PhD thesis, Stanford, CA, USA, 1981.

[18] L. Purvis, S. Harrington, B. O'Sullivan, and E. C. Freuder. Creating personalized documents: an optimization approach. In *DocEng '03: Proceedings of*

---

[6]http://www.snapfish.com
[7]http://www.shutterfly.com

the *2003 ACM symposium on Document engineering*, pages 68–77, New York, NY, USA, 2003. ACM.

[19] C. Reas and B. Fry. *Processing: A Programming Handbook for Visual Designers and Artists*. The MIT Press, Sept. 2007.

[20] D. Salomon. *The Advanced TEXbook*. Springer-Verlag, Berlin, 1995.

[21] C. Schäfer and G. Pittman. *Scribus: Open-Source Desktop Publishing*. FLES books Ltd., United Kingdom, 2009.

[22] E. Schrier, M. Dontcheva, C. Jacobs, G. Wade, and D. Salesin. Adaptive layout for dynamically aggregated documents. In *IUI '08: Proceedings of the 13th international conference on Intelligent user interfaces*, pages 99–108, New York, NY, USA, 2008. ACM.

[23] R. Sellman. Vdp templates with theme-driven layer variants. In *DocEng '07: Proceedings of the 2007 ACM symposium on Document engineering*, pages 53–55, New York, NY, USA, 2007. ACM.

[24] V. Setlur, S. Takagi, R. Raskar, M. Gleicher, and B. Gooch. Automatic image retargeting. In *MUM '05: Proceedings of the 4th international conference on Mobile and ubiquitous multimedia*, pages 59–68, New York, NY, USA, 2005. ACM.

Reflowable Documents Composed from Pre-rendered Atomic Components

Alexander J. Pinkney
Document Engineering Lab.
School of Computer Science
University of Nottingham
Nottingham, NG8 1BB, UK
azp@cs.nott.ac.uk

Steven R. Bagley
Document Engineering Lab.
School of Computer Science
University of Nottingham
Nottingham, NG8 1BB, UK
srb@cs.nott.ac.uk

David F. Brailsford
Document Engineering Lab.
School of Computer Science
University of Nottingham
Nottingham, NG8 1BB, UK
dfb@cs.nott.ac.uk

ABSTRACT

Mobile eBook readers are now commonplace in today's society, but their document layout algorithms remain basic, largely due to constraints imposed by short battery life. At present, with any eBook file format not based on PDF, the layout of the document, as it appears to the end user, is at the mercy of hidden reformatting and reflow algorithms interacting with the screen parameters of the device on which the document is rendered. Very little control is provided to the publisher or author, beyond some basic formatting options.

This paper describes a method of producing well-typeset, scalable, document layouts by embedding several pre-rendered versions of a document within one file, thus enabling many computationally expensive steps (e.g. hyphenation and line-breaking) to be carried out at document compilation time, rather than at 'view time'. This system has the advantage that end users are not constrained to a single, arbitrarily chosen view of the document, nor are they subjected to reading a poorly typeset version rendered on the fly. Instead, the device can choose a layout appropriate to its screen size and the end user's choice of zoom level, and the author and publisher can have fine-grained control over all layouts.

Categories and Subject Descriptors

I.7.2 [**Document and Text Processing**]: Document Preparation—*format and notation, markup languages*; I.7.4 [**Document and Text Processing**]: Electronic Publishing

General Terms

Algorithms, Documentation, Experimentation

Keywords

PDF, COGs, eBooks, Document layout

1. INTRODUCTION

In recent years, the consumption of documents on mobile devices, such as eBook readers, has increased dramatically. However,

the visual quality of a document on these devices is often lacking, when compared to other digital document systems (see figure 1). The result of an eBook reader's layout engine is often visually unappealing, with uneven spacing in consecutive lines of text, poor justification, and the lack of a sophisticated hyphenation system.

This is a far cry from the quality of typesetting available from PDF or PostScript documents. These vector-based, device-independent page description languages are able to create a digital version of the document that is identical to its printed counterpart. These page description languages, coupled with high-quality typesetting systems (such as TeX, troff or Adobe InDesign) have produced an expectation that digital documents will be of similar quality to that achievable through hand composition. TeX and Adobe InDesign, in particular, have excellent support for many of the subtle nuances used by hand compositors, which are often overlooked by more basic typesetting packages (e.g. automated support for kerning and ligatures). This quality does not come without a price: the algorithms used to calculate the layout are computationally expensive and so are run only once, to produce a PDF with a fixed layout targeted at a fixed page size.

eBook readers, it seems, have had to take a step backwards to simpler (and, therefore, less computationally expensive) algorithms to maximise the battery life of the device. The result is that the high-end hyphenation, kerning, and ligature support has had to be sacrificed and the on-screen result is reminiscent of the output of an HTML rendering engine or a very basic word processor.

This paper investigates an alternative approach to generating the display for an eBook reader. Here, the text is pre-rendered (using a high-quality typesetting algorithm) in several column widths, prior to display, when the document is created. At view time, the most appropriate column width is selected for display, the system balancing between excessive white space and multiple columns. Section 2 examines the problems posed by current eBook readers in further detail, while section 3 presents our initial prototype solution to some of these problems.

2. PROBLEMS WITH CURRENT EBOOK READERS

Three formats currently dominate the eBook market: EPUB and Mobipocket, which allow the document to be formatted to fit the device, and PDF, which does not. (Amazon's proprietary Kindle format is derived from Mobipocket; PDF and EPUB are open standards.) Both the EPUB and Mobipocket formats are largely based on XHTML. Whilst the use of XML-derived formats allow the semantic structure of documents to be very well defined, in general their presentation can only be specified in a very loose manner. The

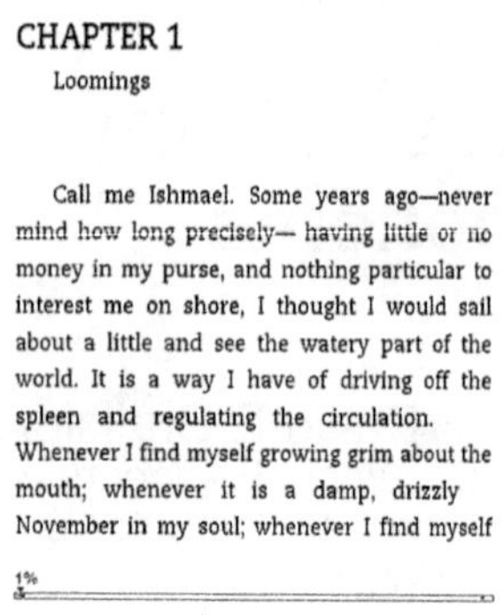

Call me Ishmael. Some years ago—never mind how long precisely— having little or no money in my purse, and nothing particular to interest me on shore, I thought I would sail about a little and see the watery part of the world. It is a way I have of driving off the spleen and regulating the circulation. Whenever I find myself growing grim about the mouth; whenever it is a damp, drizzly November in my soul; whenever I find myself

Figure 1: The Kindle 3 appears to primarily use justified text, falling back to ragged-right when inter-word spacing would become too large.

user is often presented with a choice of typefaces and point sizes, allowing the reader software to render the document in essentially any arbitrary way it chooses.

Conversely, PDF is entirely presentation-oriented, stemming from its origin of being compiled PostScript. PDF, therefore, will often include no information on the semantic structure of the document, and will consist simply of drawing operators which describe the document pages. There is no compulsion for these drawing operators to render the page in an order that might be considered sensible: for example, if a PDF generator program decided to render every character on a page in alphabetical order, or radially outwards from the centre, the resulting file would still be semantically valid, and the result might well be unnoticeable to the end user. This lack of imposed semantic structure can make it difficult to infer the best way to 'unpick' PDF files to allow their content to be reflowed into a new layout.

Since an XHTML-derived format has no fixed presentation associated with it, this must be calculated each time the document is displayed, in a similar manner to the way an interpreted programming language needs to be interpreted each time it runs. For an eBook reader to maximise its battery life (the human reader will be annoyed if the device dies just before the climax of a novel!), the 'interpretation' needs to be as simple as possible — i.e. the algorithm used must not be too complex, since the more CPU cycles spent executing it, the less time the CPU can spend idle, and hence the greater the drain on the battery. Furthermore, the longer that is spent formatting the output, the longer the delay between page turns on the device, and with the speed of CPUs used in these devices (< 500 MHz) it does not take too large an increase in computation for the page turn to become noticeable.

2.1 Hyphenation and Line-Breaking

eBook readers typically use a 'greedy' algorithm to lay out their text — that is, they place as many words as will fit onto the current line without exceeding it, then start a new line and continue. Although this algorithm is optimal in that it will always fit text onto the fewest possible lines, it often causes consecutive lines to have wildly varying lengths, accentuating either the 'ragged-right' effect of the text, or, in the case of justified text, the inter-word spacing. In general, eBook readers will only hyphenate in extreme cases — indeed the Kindle 3 seems not to do so at all. Knuth and Plass[7] developed a more advanced line-breaking algorithm (now used by TeX) which attempts to minimise large discrepancies between consecutive lines by considering each paragraph as a whole. TeX also uses the hyphenation algorithm designed by Liang[8], which has been ported to many other applications.

To AV V. Wa fi fl
To AV V. Wa fi fl

Figure 2: Examples of various letter-pairs and their kerned (left) or ligature (right) equivalents, as typeset by TeX.

2.2 Other Typographical Techniques

Other techniques employed during hand-typesetting and high-quality electronic typesetting include the use of kerning and of ligatures. Kerning involves altering the spacing between certain glyph pairs in order to produce more consistent letter spacing, whilst ligatures are single-glyph replacements for two or more single glyphs which may otherwise have clashing components. Some examples of these are shown in figure 2. Kerning requires a table of kern-pairs, specific to each font; values from this table must then be looked up for every pair of adjacent glyphs in the document. Ligatures may or may not need to be inserted: if the component characters of the ligature lie over a potential hyphenation point, it cannot be decided whether to replace them with the ligature until it is known whether the hyphenation point needs to be used.

3. A GALLEY-BASED APPROACH

Our proposed solution, of precomputing several text variants, revisits an approach to typesetting from before the advent of desktop publishing. In the days before DTP, newspaper articles were typeset into long columns known as *galleys*. Since all columns in the newspaper would be of uniform width, all articles could be typeset into galleys of the same measure, and then broken as necessary between lines, in order to slot into the final layout of the newspaper. Once the text has been set in this manner, with appropriate hyphenation and justification, the individual lines can be treated as atomic units that will never have to be re-typeset. In essence, each article is 'compiled' only once, but can be used anywhere in the final layout without penalty.

It is this behaviour we wish to emulate. So long as the atomic components of the document are tightly specified, and the reader software can obey the associated drawing instructions (essentially treating them as pre-typeset blocks), the resulting display of the document will be of as high typographic quality as that of the original galley, and the requirement for further computation will be vastly reduced. In order to permit aesthetic layout for a wider range of screen sizes, it seems sensible to create a document containing multiple renderings of the same content, and simply choosing the 'best fit' rendering when the document is displayed.

3.1 A Sample Implementation

Our sample implementation is built around our existing work in PDF and Component-Object Graphics (COGs)[1], but there is no reason why it could not be implemented in any other format capable of tightly specifying page imaging operations. It builds on existing software, principally *pdfdit*, in conjunction with *COG Manipulator*, as these tools are already capable of producing modular documents with tightly specified rendering.

3.1.1 The COG Model

The Component Object Graphic (COG) model was developed to enable the reuse of semantic components within PDF documents, by breaking the traditional graphically-monolithic PDF page into a series of distinct, encapsulated graphical blocks, termed COGs. In its original incarnation, the COG model did not account for any

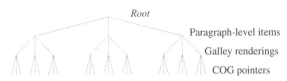

Figure 3: A simple document structure tree. The first level below the root represents all paragraph-level items: headings, paragraphs, figures etc. These items have one child for each galley rendering of the document. These in turn have one child for each COG comprising their content — in the case of a paragraph or heading: its lines; in the case of a figure: the figure itself and any associated caption.

relationship between individual COGs — it was simply designed as a method by which document components could be easily reused or reordered. The COGs it generates are largely at the granularity of a paragraph, and can still be imaged onto the page in any arbitrary order, independent of reading order.

In order to implement our galley-based design, it is necessary to decrease this granularity, such that each line of text is represented by a separate COG. However, it is also important that the semantic structure of the document is explicitly stored. This is principally so that the reading order of the COGs is maintained, and also so that the reader software can identify paragraphs, headings etc. to enable them to be laid out correctly.

The COG model takes advantage of the fact that the PDF specification allows the content of a page to be described by an array of streams of imaging operators, rather than the more commonly encountered single, monolithic stream. Unfortunately, this array can only be one-dimensional, meaning that while it can enforce the reading order, it cannot be used to, say, group lines into paragraphs. Since the PDF specification allows essentially arbitrary insertion of data structures into a document (PDF readers which do not recognise these will simply ignore them), this flexibility was used to embed a simple tree structure representing the paragraphs, in parallel to the COGs themselves (an example of which is shown in figure 3). At the level of its leaves, this tree simply contains pointers to the COGs which make up the content of the document. In the simplest case, where the document contains only one rendering (and thus the paragraph-level items have only one child) the COGs pointed at by the leaves can simply be rendered in order, adding vertical space as appropriate.

3.1.2 The Source Document

Since the majority of available tools for producing COGged PDFs rely on the typesetting package *ditroff*, it was decided to use this as the basis for the source document. Ditroff is particularly amenable to many of the features required here — it is quite happy to have its page length set to large numbers — one sample document used a page length of 2000 inches (approximately 50 metres) with no complaints from ditroff. The line length was set to a small value (approximately two inches) in order to produce a narrow column of text. Following this, the actual document content was inserted several times, and the line length incremented, producing one document effectively containing multiple galley renderings of the same content.

3.1.3 pdfdit

Having generated the source document, it was processed with ditroff to generate the intermediate code used to feed each typesetter post-processor. This output is very expressive, and, unlike

Figure 4: Sample renderings from the Acrobat plugin at page widths of 42, 48, and 54 em.

TEX's DVI, contains enough information that post-processors are easily able to locate the start and end of lines and paragraphs within the document. This meant that only minimal changes were needed to be made to the *pdfdit* package described in [1] to implement our design.

The first change necessary was to decrease the granularity of the output COGs, producing them at the line level, rather than at the paragraph level. Secondly, some method of generating the requisite tree representing the document structure was required. This was solved by simply using the point at which the original version of pdfdit would have started a new paragraph-level COG, and, instead, starting a new paragraph-level block entry in the document structure tree. Each subsequent line-level COG produced can then be added as a child of this block.

Once the entire output file has been parsed, the tree representations of the various width galleys are amalgamated per-paragraph, as indicated in figure 3, and finally the PDF file is serialised, replete with COGs and content tree.

3.1.4 Acrobat Plugin

The decision to use Acrobat as an eBook 'emulator' stemmed once again from the available existing COG-based tools, as well as the extensive API and developer support available for Acrobat. Moving a COG on a PDF page is as simple as deleting its associated spacer object from the content array of the page, creating a new spacer containing the COG's desired new position, and then adding that back to the content array.

Since, by this point, most of the computationally expensive typesetting has already been carried out, the algorithm used to lay out the lines of the galleys can be very simple. The plugin chooses the most appropriate galley width to lay out, based on the current page width, and according to some measure of aesthetics, and then simply lays the document out line by line, with appropriate vertical spacing, until no more lines will fit in the current column. Any subsequent columns which will fit on the same page are then laid out in the same manner.

3.1.5 Layout and Metrics

Since galleys of text lend themselves to being used in a columnar format, a method of fitting columns appropriately to the available page width must be devised. A sensible first approach is simply to calculate how many columns of each galley rendering will fit, by adding the galley width to a specified minimum inter-column

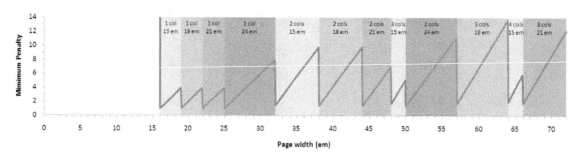

Figure 5: Graph showing the minimum penalty value of all galleys in a reflowable document, over a range of page widths. The particular document used contained four galleys; these were rendered at widths of 15, 18, 21 and 24 em, with a minimum gutter width of 1 em. Each vertical band highlights a range of page widths within which only the horizontal spacing of the page is altered. The boundaries between vertical bands represent a switch between galley renderings — the galley used and number of columns is as annotated on the graph.

spacing, and dividing the page width by this. The remainder of this division will then specify the total extra amount of horizontal whitespace required, which can then divided up and inserted between the columns. A simple measure of aesthetic quality here is to apply a linear penalty for any extra whitespace required, as we seek to keep page margins and column gutters to a minimum.

As the page width increases, so must the widths of the intercolumn gutters. In accordance with the extra-whitespace penalty, each galley rendering will produce penalties which vary in a sawtooth manner as the width of the page is increased. With a careful choice of galley widths, when these sawtooth penalties are overlaid, and the galley producing the minimum penalty chosen at each page width, a flatter and finer-toothed penalty graph emerges, as shown in figure 5.

In addition to penalising extra whitespace, wider columns should, in general, be favoured over narrower ones, i.e. for a given page width, fewer, wider columns are generally considered preferable to a greater number of narrower columns. By multiplying the existing penalty by a smaller-than-linear function of the number of columns (experiments have been carried out with both logarithms and roots) the penalty may be subtly increased for greater numbers of columns. The formula for the penalty used in figure 5 is $P = (C + W_{ex}) \cdot \sqrt{N_{cols}}$, where P is the penalty, W_{ex} is the extra whitespace required to be inserted, N_{cols} is the number of columns which are required to fill the width of the page, and C is a positive constant. The purpose of the constant is to prevent the penalty from ever evaluating to zero, which would have the effect of disregarding the weighting of the number of columns. Figure 5 uses $C = 1$.

4. CONCLUSIONS AND FUTURE WORK

This paper outlines our initial exploration of the idea of using pre-rendered galleys for eBooks. So far, our initial implementation has generated multicolumn layouts that look acceptable, and we believe there is mileage in continuing to investigate this method. However, there is still a lot of work to be done. Firstly, a very simple formula is used to determine which column width variant to select, and we are investigating the suitability of other methods of determining aesthetically pleasing layouts (such as those outlined in [2, 3, 4, 5, 6, 9]). Also, our system does not currently allow the font size to be changed (since it is fixed when the galleys are created). One approach to allow the font size to be changed would be to scale smaller column width variants up to larger columns. For example, if the 15 em wide variant is scaled up to 18 em, then text would be scaled up by 20% — the equivalent of converting 10 pt text to 12 pt.

It should also be noted that optimal placement of floating blocks cannot be 'compiled out' in the same manner that hyphenation and line breaking can; these will still need to be positioned into the relevant places as the document is displayed. If the simple approach is taken that floats should be placed at the top of a column or after another float, a document layout somewhat reminiscent of this one will emerge, although the floats will inevitably tend to drift towards the end of the document, away from their desired position.

Finally, to confirm that this method has validity it needs to be implemented in an actual eBook system, rather than simulated in Acrobat. There, it will be possible to compare the performance of our system with both a normal eBook renderer, and one that has been enhanced to use a sophisticated hyphenation and justification algorithm.

5. REFERENCES

[1] S. R. Bagley, D. F. Brailsford, and M. R. B. Hardy. Creating reusable well-structured PDF as a sequence of component object graphic (COG) elements. In *Proceedings of the 2003 ACM Symposium on Document Engineering*, pages 58–67. ACM Press, 2003.

[2] H. Y. Balinsky, J. R. Howes, and A. J. Wiley. Aesthetically-driven layout engine. In *Proceedings of the 2009 ACM Symposium on Document Engineering*, pages 119–122, 2009.

[3] R. Bringhurst. *The Elements of Typographic Style (v 3.2)*. Hartley & Marks, 2008.

[4] E. Goldenberg. Automatic layout of variable-content print data. Master's thesis, University of Sussex, 2002.

[5] S. J. Harrington, J. F. Naveda, R. P. Jones, P. Roetling, and N. Thakkar. Aesthetic measures for automated document layout. In *Proceedings of the 2004 ACM Symposium on Document Engineering*, pages 109–111. ACM Press, 2004.

[6] R. Johari, J. Marks, A. Partovi, and S. Shieber. Automatic yellow-pages pagination and layout. Technical report, Mitsubish Electric Research Laboratories, 1996.

[7] D. E. Knuth and M. F. Plass. Breaking paragraphs into lines. *Software — Practice and Experience*, 11:1119–1184, 1981.

[8] F. M. Liang. *Word Hy-phen-a-tion by a Com-put-er*. PhD thesis, Stanford University, 1983.

[9] L. Purvis, S. Harrington, B. O'Sullivan, and E. C. Freuder. Creating personalized documents: an optimization approach. In *Proceedings of the 2003 ACM Symposium on Document Engineering*, pages 68–77. ACM Press, 2003.

Introduction of a Dynamic Assistance to the Creative Process of Adding Dimensions to Multistructured Documents [*]

Pierre-Édouard Portier
Université de Lyon, CNRS
INSA-Lyon, LIRIS, UMR5205
F-69621, France
pierre-edouard.portier@insa-lyon.fr

Sylvie Calabretto
Université de Lyon, CNRS
INSA-Lyon, LIRIS, UMR5205
F-69621, France
sylvie.calabretto@insa-lyon.fr

ABSTRACT

We consider documents as the results of dynamic processes of documentary fragments' associations. We have experienced that once a substantial number of associations exist, users need some synoptic views. One possible way of providing such views relies in the organization of associations into relevant subsets that we call "dimensions". Thus, dimensions offer orders along which a documentary archive can be traversed. Many works have proposed efficient ways of presenting combinations of dimensions through graphical user interfaces. Moreover, there are studies on the structural properties of dimensional hypertexts. However, the problem of the origins and evolution of dimensions has not yet received a similar attention. Thus, we propose a mechanism based on a simple structural constraint for helping users in the construction of dimensions: if a cycle appears *within* a dimension while a user is creating a new dimension by the aggregation of existing ones, he will be encouraged (and assisted in his task) to restructure the dimensions in order to cut the cycle. This is a first step towards a rational control of the emergence and evolution of dimensions.

Categories and Subject Descriptors

H.3.7 [**Information Storage And Retrieval**]: Digital Libraries—*User issues*

General Terms

Human factors, Design

Keywords

dimensions, hyperstructures, multistructured documents

[*]This work was co-funded by the National Research Agency under the AOC project: ANR-08-CORD-009.

1. INTRODUCTION

The ideas we would like to introduce appeared to us while working with members of the Jean-Toussaint Desanti Institute[1] (ENS-Lyon[2]). Philosophers from the Institute are building a digital edition of the handwritten archives of French philosopher Jean-Toussaint Desanti (1914-2002).

Digital editing covers the whole editorial, scientific and critical process that leads, eventually, to the publication of an electronic resource. In case of manuscripts, the editing process mainly consists in the transcription and critical analysis of digital facsimiles. In terms of computations, such an analysis comprises at least two aspects. The first one is the association of an annotated textual document with the images of a handwritten manuscript. The second one is the formalization of the uncovering of interesting associations between fragments of the archive. For example, a set of pages is identified as a first version of a well-known published work.

In the context of our main use case, users, with a strong philosophical background, are studying a large collection (90170 handwritten pages organized hierarchically in 1288 collections) of handwritten documents. Their main concern is in finding meaningful orderings for the documents of the archive. Without this preliminary work, the archive could hardly reach the many potentially interested readers. In order to fulfill this ordering task, the users have to find associations between heterogeneous documentary fragments (images of manuscript pages, transcriptions, intervals of text, polygonal zones extracted from the images, etc.).

Therefore, the system we developed (DINAH) let the users create ternary relations for representing associations between documentary fragments (see Figure 1).

In this work, we will be interested in finding an effective way for the users to manage the growing complexity of the associations they create between documentary fragments.

2. A NEED FOR SYNOPSIS

As far as our experiments with the users of the J.T. Desanti's Institute can tell us, when a substantial number of associations have been created, the naive graph representation implied by the SVO (Subject-Verb-Object) metaphor fails at providing the synoptic views the users need.

One possible way for providing such views relies in the

[1]http://institutdesanti.ens-lyon.fr/
[2]http://www.ens-lyon.eu/

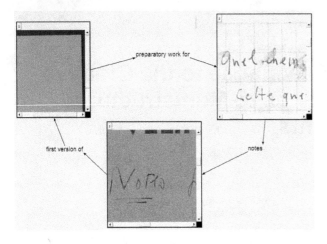

Figure 1: Screenshot of the graph-oriented module for the creation and the visualization of associations

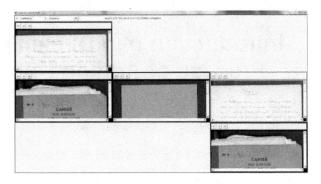

Figure 2: Screenshot of the dimension-based visualization module of DINAH (X: *d.anteriority* ; Y: *d.analysis*)

organization of the associations into relevant subsets we call "dimensions". The abstract function of a dimension is to group similar ways of associating documentary fragments. For example, to a dimension named "anteriority" could belong all the instances of the associations labeled "is a draft for", "is a preprint of", "is a first version of", "is a preparatory work for", etc.

How do dimensions appear? What governs the process of their construction? First of all, as far as a specific domain is concerned, one could often find *a priori* and pertinent dimensions. For example, in case of textual documents, the TEI (Text Encoding Initiative) [5] can certainly bring some interesting insights into the set-up of dimensions. However, we are mainly interested by situations with insufficient *a priori* knowledge for managing the complexity of the relations. Thus, while users are creating new associations we would like to help them define meaningful dimensions. To that purpose, we will introduce a new semi-automatic methodology for the construction of dimensions.

3. RELATED WORK ON DIMENSIONS

The dimensions we now introduce are binary relations in a mathematical sense (i.e. sets of pairs). Therefore, the association between two documentary fragments (e.g. A is_a_preparatory_work_for B) is represented by the membership of the pair of documentry fragments to a dimension representing this kind of association (e.g. $(A, B) \in$ d.preparatory_work).

Two dimensional models will differ by the structural constraints the pairs belonging to a dimension have to meet. Thus, by combining a few structural constraints (invertibility, partial functions, cyclic relations, ...), we now introduce well-known dimensional models.

3.1 Hyperorder

Hyperorders [1] are based on binary relations. A hyperorder is defined by the pair: $< F, \{D_1, D_2, \ldots, D_n\} >$ where the second member of the pair is a set of binary relations called *dimensions*.

3.2 Zzstructure

A zzstructure [4] is a hyperorder with two additional restrictions:

- The dimension are invertible: for each dimension D_m, there is a dimension D_m^{-1}.

- The dimensions are partial functions.

Since the dimensions are partial functions, a cell can only be the subject of at most one association along a specific dimension. Moreover, since the dimensions are invertible, and the inverse dimension must also form a partial function, a cell can only be the object of at most one association along any specific dimension. Thus, a zzstructure offers a linear way of navigating along dimensions (without the need of any hyperlinking engine).

The users can be offered presentations like the one of Figure 2. The computer screen is mapped to a Cartesian space. The dimension "d.anteriority" has been affected to the horizontal axis and grows positively to the right. Since each dimension is invertible, the dimension "inv(d.anteriority)" is affected to the horizontal axis of the screen and grows positively to the left. Similarly, the "d.analysis" dimension has been affected to the vertical axis of the screen.

Finally, in order to interpret unambiguously the representation of a zzstructure, we need to specify the meaning of the fragments on rows and columns others than the ones crossing at the cursor. Two interpretations have been proposed [4]. An illustration of the first one called "H view" is given by the Figure 3. Given this representation one could deduce, for example, that the pairs $(F4, F2)$ and $(F3, F5)$ are members of the dimension $D1$ while the pair of documentary fragments $(F4, F3)$ is a member of the dimension $D0$. However one shouldn't deduce that the pair $(F2, F5)$ is a member of the $D1$ dimension. In other words, in case of the "H view", apart from the two main axes crossing at the cursor, only the vertical juxtapositions of documentary fragments are meaningful. Similarly, in case of the so-called "I view", apart from the two main axes, only the horizontal juxtapositions of fragments are meaningful.

3.3 Edge-colored graph

It has been proved [3] that a zzstructure is theoretically equivalent to an edge-colored graph. The edges with a same color don't have to form a partial function on the set of nodes. Thus, for example, a documentary fragment can be linked to more than one author by a relation named "author". A zzstructure can't directly model such a situation. However by adding to a zzstructure a "d.clone" dimension

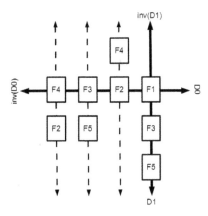

Figure 3: Illustration of the "H view" of zzstructures

along which cells can be cloned, it becomes possible to model multi-pointing links [1].

However, by using edge-colored graph, we are losing some of the good navigational properties of zzstructures: a dimension doesn't have to be a strictly linear structure but branching can occur.

3.4 Semantic Web

Finally, "semantic webs" [2] are edge-colored graph with an additional restriction: the edges with a same color cannot make cycles.

3.5 synthesis

We want to explore how a dimensional data modeling and presentation framework could, at least partially, answer the need for synoptic views. However, none of the dimensional models introduced above are taking into account the problem of the creation of dimensions. Therefore, we introduce a mechanism for helping the users in the process of building meaningful dimensions that will provide powerful synoptic views on the documentary archive.

4. METHODOLOGY FOR THE CONSTRUCTION OF DIMENSIONS

Our dimensional model add to the zzstructure an acyclicity constraint. We have chosen this constraint since its violation is often meaningful. We shall now explain this point in more details.

4.1 The discriminating power of cycles

We are assuming that when two relationships between two given nodes are making a cycle and are not each other inverse, then they don't semantically belong to the same dimension. In other words, it is assumed that the "unity of meaning" of a dimension would be lost if such cycles could occur.

However, the zzstructure model allows dimensions to have a ring structure. Indeed, since a node can only be the subject (resp. object) of at most one association, the only kind of possible intra-dimensional cycle is a ring. But it can be observed that each time such a ring structure is mentioned in the context of zzstructures, it is in a navigational context only. As an example, T.H. Nelson while describing "The main mechanisms of ZigZag" references "wheels" as[4]:

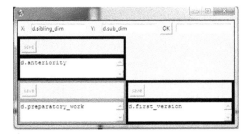

Figure 4: *d.anteriority* **is defined as an aggregation of the dimensions** *d.first_version* **and** *d.preparatory_work*

ringrank of cells that effectively turns, operationally, as one or more pointers step around it. Ringranks with stepping pointers are used for a number of repetitive operations or data: (1) "next dimension" (2) "next view" ...

If we extend the previous argument to cycles of n nodes with $n > 2$, we can encounter two situations:

- $n - 1$ edges belong to a dimension D while 1 edge belongs to D^{-1}, the inverse of D (this is a form of transitivity).

- At least two different dimensions, which are not each other inverse, are involved in the cycle.

We didn't take into account the case of an identity relation. For example, if "x is-a y" and "y is-a x", it may seem like we have a cycle inside a dimension...But really, x is obviously equals to y.

Therefore, from this analysis, it appears clearly that[2]:

where cyclicity exists, it is always asymmetric

So, it is at least reasonable to emphasize our acyclicity constraint in the context of the zzstructure dimensional model.

By design, the dimensional interface of Figure 2 will not allow the creation of a cycle inside a dimension. But then, how can we claim that this additional constraint will help the users for the creation of meaningful dimensions? In order to answer this, we have to introduce one last component of our dimensional model: a way of building new dimensions from the composition of old ones.

4.2 The composition of dimensions

A new dimension can be created by aggregation of other dimensions. This ability combined with dimensional representations (e.g. Figure 2) truly reveals the synopsis power of dimensions. For example, a dimension *d.anteriority* can be defined as an aggregation of the dimensions *d.first_version* and *d.preparatory_work*. We introduced two new dimensions (*d.sub_dim* and *d.sibling_dim*) for the users to manipulate this aggregation structure (see Figure 4).

4.3 An interactive use of the acyclicity constraint

We let the users create associations between documentary fragments of the archive with the dimensional framework introduced above. For example, they may have used the dimensions *d.first_version* and *d.preparatory_work* as in Figure 5.

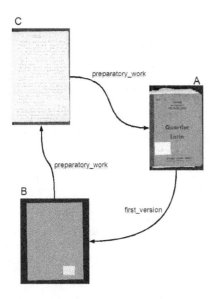

Figure 5: An example use of the dimensions *d.first_version* **and** *d.preparatory_work*

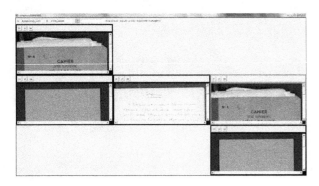

Figure 6: When a cycle occurs, the users are offered a dimensional view centered on the conflicting situation. (X: *d.preparatory_work* **; Y:** *d.first_version*)

For building synoptic views, they can group dimensions with the aggregation mechanism introduced above. However, this can make a cycle appear within the newly created dimension... This would happen on our previous example (Figure 5) if it was decided to aggregate *d.first_version* and *d.preparatory_work* into a *d.anteriority* dimension.

When such a cycle is detected, the aggregation process is suspended and the user is offered a dimensional view centered on the conflict (see Figure 6). We have been able to see that each time such a restructuring opportunity is offered to the users, interesting and meaningful information is formalized!

5. CONCLUSION

Our lightweight methodology offers a simple mechanism for dynamically promoting a rational structuring of the dimensions. From the reduction of intra-dimensional cycles, knowledge is gained either about the inverse of dimensions or about the structuring of the dimensions. This work is a first step towards a generic mechanism for assisting the users in creating multistructured documents in the context of dimensional hypertext systems.

6. ACKNOWLEDGMENTS

We would like to thank the team of researchers from the Jean-Toussaint Desanti Institute for their collaboration during the development of DINAH.

7. REFERENCES

[1] J. Goulding, T. Brailsford, and H. Ashman. Hyperorders and transclusion: understanding dimensional hypertext. In *Proceedings of the 21st ACM conference on Hypertext and hypermedia*, pages 201–210. ACM, 2010.

[2] C. Joslyn. Semantic webs: a cyberspatial representational form for cybernetics. *Cybernetics And Systems Research*, pages 905–910, 1996.

[3] M. McGuffin et al. A comparison of hyperstructures: Zzstructures, mspaces, and polyarchies. In *Proceedings of the fifteenth ACM conference on Hypertext and hypermedia*, pages 153–162. ACM, 2004.

[4] T. Nelson. A cosmology for a different computer universe: data model, mechanisms, virtual machine and visualization infrastructure. *Journal of Digital Information*, 5(1), 2006.

[5] C. Sperberg-McQueen, L. Burnard, et al. *Guidelines for electronic text encoding and interchange (TEI P5)*. http://www.tei-c.org/Guidelines/P5/, 2011.

Interoperable Metadata Semantics with Meta-Metadata: A Use Case Integrating Search Engines

Yin Qu, Andruid Kerne, Andrew M. Webb and Aaron Herstein
Interface Ecology Lab, Texas A&M University
College Station, Texas
{yin, andruid, andrew, aaron}@ecologylab.net

ABSTRACT

A use case involving integrating results from search engines illustrates how the meta-metadata language facilitates interoperable metadata semantics. Formal semantics can be hard to obtain directly. For example, search engines may only present results through web pages; even if they do provide web services, they don't provide them according to a mutually interoperable standard.

We show how to use the open source meta-metadata language to define a common base class for search results, and how to extend the base class to create polymorphic variants that include engine-specific fields. We develop wrappers to extract data from HTML search results from engines including Google, Bing, Delicious, and Slashdot. We write a short meta-search program for integrating the search results, reranking them, and providing formatted HTML output. This provides an extensible formal and functional semantics for search. Meta-metadata also directly enables representing the same integrated search results as XML or JSON. This research can profoundly transform the derivation and representation of interoperable metadata semantics from a multitude of heterogeneous wild web sources.

Categories and Subject Descriptors

H.1.m [**Information Systems: Models and Principles**]

General Terms

Design, Documentation, Human Factors, Management

1. INTRODUCTION

We use the meta-metadata language and architecture to integrate results from search engines. *Metadata* can be employed to describe the numerous documents published online by businesses, organizations, and individuals. A *metadata schema* describes the vocabulary and structure used to represent and communicate a type of metadata. *Meta-metadata*

is a language for describing schemas, wrappers that map information sources to schemas, and more. We define *metadata semantics* as the conjunction between a document and its metadata, including content, descriptions, citations, and afforded operations. *Interoperable* metadata semantics are obtained from multiple sources, translated into consistent representations, and then processed independently of the initial sources. Interoperability can be impaired by the fact that publishers often use their own metadata schemas and representations. For example, a user may find it difficult to search across multiple engines since they use different structures and interfaces to present results.

We build on *meta-metadata* [10] as a structural basis for obtaining and using interoperable metadata semantics in applications across system boundaries. Meta-metadata is an open source, cross-platform formal language, architecture, and wrapper repository addressing metadata structure, extraction, translation, bridging, and presentation [8]. The heart of meta-metadata is a type system for *information sources*, each published with a consistent DOM structure and addressing scheme. With meta-metadata, developers author wrappers to represent metadata vocabularies and structures, information source selectors to automatically select wrappers, rules to extract metadata into program objects, and semantic actions to connect metadata with business logic. Meta-metadata does not impose particular schemas on metadata-consuming applications, allowing for flexibility. It comes with a repository of inheritable wrappers, encouraging re-use across contexts.

This paper develops a use case that derives interoperable metadata semantics to integrate results from multiple search engines. We show how meta-metadata helps developers invoke and re-use interoperable metadata semantics.

2. PRIOR WORK

Standard metadata schemas, such as Dublin Core [14], were developed to capture common needs for interoperable metadata semantics across systems. Metadata semantics that use the same standard schema are directly interoperable, since no structural or vocabular ambiguities prevent unified processing. Sometimes data represented with one schema can be translated to a more standard one by associating synonymous fields across schemas, achieving interoperability. For example, the Library of Congress developed a crosswalk between Dublin Core and MARC [11].

However, as the ecosystem of needs and tasks grows in complexity, the pure standards approach to metadata is inadequate. Diverse metadata is a fact of life. It is impossible for standards to keep up or attain sufficient flexibility. The cost of creating translations is high. In contrast, meta-metadata does not rely on external standard metadata schemas. Instead, it allows developers to extend existing schemas, or start anew, according to their needs.

The Semantic Web [4] approaches interoperable metadata semantics with *ontology alignment* [1] and practice guidelines like Linked Data [3]. An ontology is a set of related metadata schemas (i.e. classes) and instances (i.e. resources) described by *formal semantics*, typically represented as a triple store (e.g. RDF or N3). A triple consists of a subject, predicate, and object. Any relationship between any pair of objects can be described. Ontology alignments map classes, fields and resources from one ontology to equivalents in another, resolving ambiguities. Linked Data identifies resources by URIs. Related resources in the same or other systems are bi-directionally linked using formal semantics.

Semantic Web triples constitute a general solution for resolving ambiguities. However, the situated and evolving nature of knowledge makes formal semantics costly to create and maintain [12]. In practice, they can involve extensive development of source specific "mediators" and optimizations [5]. As a result, The Semantic Web's availability and scalability are limited. Again, meta-metadata does not rely on rigid formal semantics from publishers. It allows developers to author schemas based on their own situated contexts, and use wrappers to automatically extract metadata semantics from informal representations into schematized structures.

Like meta-metadata, Piggy Bank [6] addresses scraping metadata into structured forms. It has been used to support Exhibit [7], a structured data presentation framework. The resulting system facilitates document publishing. One important difference between Piggy Bank / Exhibit and meta-metadata is that meta-metadata provides an object-oriented type system for wrappers. This helps reduce the effort of authoring domain models and extraction rules. Wrappers authored by independent developers can be re-used with inheritance and polymorphism, propagating interoperability across contexts.

3. INTEGRATING SEARCH ENGINES

Developers and power users author **meta-metadata wrappers** to specify metadata structures, extraction rules, operations, and presentation rules. The following wrapper (abbreviated for space) specifies a metadata schema **search** with a collection of **search_result**s inside, as well as semantic actions (which we will talk about soon):

```
<meta_metadata name="search" extends="compound_document">
  <collection name="search_results" child_type="search_result"/>
  <semantic_actions>
    <get_field name="search_results" />
    <for_each collection="search_results" as="search_result">
      <get_field object="search_result" name="link" />
      <parse_document now="true">
        <arg name="location" value="link" />
      </parse_document>
    </for_each>
  </semantic_actions>
</meta_metadata>
```

while **search_result** is defined in another wrapper, consisting of properties (or fields) commonly seen in search results:

```
<meta_metadata name="search_result" extends="metadata">
  <scalar name="heading" scalar_type="String"
          navigates_to="link" layer="10.0" />
  <scalar name="snippet" scalar_type="String" />
  <scalar name="link" scalar_type="ParsedURL" />
  ...
</meta_metadata>
```

In the above samples, attribute **extends** indicates inheritance, as the reserved word "extends" does in Java. Wrappers **compound_document** and **metadata** are primitive structures defined by the system. Wrapper **search** will inherit fields and attributes from **compound_document**. Wrapper inheritance represents an *is-a* relationship, which means support for polymorphism. For example, in a semantic action where **search** is expected, using a **google_search** is permissible. By inheritance and polymorphism, different metadata schemas can be translated into a consistent representation containing common fields, e.g. **google_search** and **slashdot_search** (which we will define later) being used as **search**, reconciling schematic differences. Meta-metadata maintains implicit mappings between fields initially defined in the base wrapper and inherited in derived wrappers for polymorphism and translation of schemas.

Attributes **navigates_to** and **layer** define *presentation rules*. The former makes a field navigable, using another field as the underlying target (**link** in the above sample); the latter indicates visual priority, to sort fields for presentation. Other supported presentation rules include hiding a field, or emphasizing it with a style.

The example specifies a group of *semantic actions* in wrapper **search**. These are inherited by derived wrappers that do not override and define their own. Semantic actions can include variable definitions, control structures, and bridge functions that connect metadata with applications. The semantic actions here take the collection of **search_result**s and iterate over each, forming and parsing **document** objects with **parse_document**, with the potential to derive further semantics. Semantic actions form a high-level abstraction of afforded operations, re-usable across applications and platforms. Developers extend or re-define bridge functions by simply overriding corresponding methods.

To attach *extraction rules*, we derive a new wrapper from **search** for Google Search:

```
<meta_metadata name="google_search"
               type="search" parser="xpath">
  <selector url_stripped="http://www.google.com/search" />
  <collection name="search_results"
              xpath="//div[@id='res']//div//ol//li[@*]" />
    <scalar name="heading" xpath=".//h3/a" />
    <scalar name="snippet" xpath=".//div[@class='s']" />
    <scalar name="link" xpath=".//h3/a/@href" />
  </collection>
</meta_metadata>
```

We use **type** to re-use an existing wrapper without defining new fields. We refer to fields defined in the base wrapper by name, e.g. **search_results** and **heading**, to attach extraction rules. Currently supported extraction rules include XPath and regular expressions, and direct binding of XML or JSON to objects through the S.IM.PL (Support for Information Mapping in Programming Languages) de/serialization engine [13]. Relative XPaths are supported inside nested structures, as shown for **heading**, **snippet** and **link**. Selectors specify the URL pattern or MIME type for a particular source, which the runtime uses to find an appropriate wrapper given a location and mime type.

We define similar wrappers for other publishers' search engines, including Yahoo Buzz, Bing, Slashdot, Delicious, and Tumblr (some wrappers and XPaths are omitted for space):

```
<meta_metadata name="yahoo_buzz_search"
               type="search" parser="xpath">
 <selector url_path_tree="http://buzz.yahoo.com/search"/>
 <collection name="search_results" xpath="//div...//dl">
   <scalar name="heading" xpath="./dt/a" />
   <scalar name="snippet" xpath="./dd[...]/a" />
   <scalar name="link" xpath="./dt/a/@href" />
 </collection>
</meta_metadata>
```

```
<meta_metadata name="bing_search_xpath"
               type="search" parser="xpath">
 <selector url_stripped="http://www.bing.com/search" />
 <collection name="search_results" xpath="//div.../ul/li[*]">
   <scalar name="heading" xpath=".//h3/a" />
   <scalar name="snippet" xpath=".//div[@class='sa_cc']/p" />
   <scalar name="link" xpath=".//h3/a/@href" />
 </collection>
</meta_metadata>
```

```
<meta_metadata name="slashdot_search"
               type="search" parser="xpath">
 <selector url_stripped="http://slashdot.org/index2.pl" />
 <collection name="search_results" xpath="//div...">
   <scalar name="heading" xpath=".//a" />
   <scalar name="link" xpath=".//a/@href" />
   <scalar name="snippet" xpath="..." />
   <scalar name="author" xpath="..." />
   <collection name="tags" xpath="...">
     <scalar name="tag_name" xpath="." />
     <scalar name="link" xpath="./@href" />
   </collection>
 </collection>
</meta_metadata>
```

The *meta-metadata compiler* automatically translates wrapper specifications into **metadata classes**, in the form of Abstract Data Types in target programming languages, including Java, C#, and Objective C. A metadata class serves as a mapping between an information source and its internal representation in program. The resulting metadata classes in Java for wrapper `search` and `search_result` are:

When search requests are processed at runtime, the correct `Search` subclass and associated meta-metadata will be selected. Wrapper `google_search` and others that specify `type="search"` will be mapped to the data structure of Java class `Search`. Wrappers extending `search` will be mapped to the appropriate metadata subclasses extending Java class `Search`. Instances will be constructed and populated. Semantic actions will be invoked.

S.IM.PL annotations generated by the meta-metadata compiler (see above) enable de/serialization of metadata objects from / to XML or JSON for storage and communication. For example, here results from a Google Search are serialized:

Using these wrappers, we wrote a Java program in less than 50 statements to mix search results from the engines with meta-metadata and present the results (Figure 1). It takes a query as input, uses meta-metadata to make requests to the search engines. All resulting metadata objects are of Java class `Search`, allowing the program to simply iterate over the collection field `search_results` to mix search results. The program then uses the meta-metadata runtime's

```
@simpl_inherit public class Search extends CompoundDocument
{
  @simpl_collection("search_result")
  private ArrayList<SearchResult> searchResults;
  ...
}

@simpl_inherit public class SearchResult extends Metadata
{
  @simpl_scalar private MetadataString    heading;
  @simpl_scalar private MetadataString    snippet;
  @simpl_scalar private MetadataParsedURL link;
  ...
}
```

```
<search mm_name="google_search"
location="http://www.google.com/search?q=japan+earthquake">
<search_result heading="BBC News - Japan earthquake"
  snippet="Japan quake relief budget passed ..."
  link="http://www.bbc.co.uk/news/world-asia-pacific-12711226">
</search_result>

<search_result heading="Japan Quake Map"
  snippet="Time-lapse visualisation of the March 11, 2011 ..."
  link="http://www.japanquakemap.com/">
</search_result>

<search_result heading="Powerful Quake and Tsunami ..."
  snippet="Mar 11, 2011 ... Japan was filled with ..."
  link="http://www.nytimes.com/2011/03/12/world/asia/...">
</search_result>

<search_result heading="Japan Earthquake: New Explosion ..."
  snippet="Mar 13, 2011 ... A hydrogen explosion reportedly ..."
  link="http://abcnews.go.com/International/japan-earthqua...">
</search_result>

<search_result heading="Widespread destruction from Japan ..."
  snippet="Mar 11, 2011 ... The morning after Japan was ..."
  link="http://articles.cnn.com/2011-03-11/world/japan.qua...">
</search_result>
...
</search>
```

built-in DHTML rendering capabilities to present the re-ranked integrated results to the user. The support for polymorphism preserves specialized fields such as "tags" and "author" from Slashdot in the integrated results. The program and results, along with other examples, are avaiable online [8].

4. CONCLUSION AND FUTURE WORK

Meta-metadata functions as a basis for interoperable metadata semantics for developing applications connecting published information sources, metadata schemas, and users. Meta-metadata wrappers integrate metadata structure, extraction rules, semantic actions, and presentation rules. The meta-metadata runtime automates the process of structurally extracting, translating, and connecting metadata semantics to applications, supporting software development. Inheritance and polymorphism encourage re-use of the repository and sharing of wrappers, transferring interoperability across time and space.

Meta-metadata has been used to develop a variety of applications using metadata semantics. There are presently wrappers for digital libraries, products, restaurant reviews, and social media, including Google Books, the ACM Portal, Amazon, Urbanspoon, Wikipedia, and Flickr. One student uses it to integrate RSS feeds in a novel interface. Another uses it to scrape artwork data with semantics from a museum web site for an exhibition planning tool. Our creativ-

Figure 1: Meta-search for "japan earthquake", from Google, Bing, Yahoo Buzz and Slashdot. Note that specialized fields such as "tags" and "author" from Slashdot are preserved.

ity support tool, *combinFormation* [9], uses meta-metadata to operate on and present rich semantics to users amidst a holistic visual information presentation.

We envision meta-metadata as a foundation for building applications operating on metadata semantics from one or many sources. As it builds on common mechanisms like HTTP and XPath, and does not impose rigid external semantics, any template-based web site or collection can be wrapped and turned into a situated source of metadata semantics, making semantics immediately available and usable without waiting for publishers to adopt semantic web standards. Thus, meta-metadata translates the wild web into an ecosystem of interoperable semantic information. The repository allows sharing and re-use of metadata schemas / wrappers for each source. The runtime hides the complexity of handling network connections, extracting metadata semantics, and connecting to applications. As a result, applications that need metadata semantics buried in XML or HTML documents on the web, especially those that integrate multiple sources, can benefit from meta-metadata. These include browsing and searching applications [15] that use metadata fields as facets, bibliography management tools that support citation chaining and berry picking [2], planning assistants that depend on rich metadata semantics (e.g. weather, schedules, locations, etc.), and visualizations that use metadata semantics to convey stories.

In the future, we will work on connecting meta-metadata with other standards and systems such as the Semantic Web, Linked Data, and Exhibit, to extend its applicability. We will continuously extend the expressiveness of the meta-metadata language based on emerging needs and use cases. We seek collaborations with developers, curators, and publishers to enhance future derivation of metadata semantics to make good use of the world's information resources.

5. REFERENCES

[1] G. Antoniou and F. van Harmelen. *A Semantic Web Primer*. The MIT Press, 2004.

[2] M. Bates. The design of browsing and berrypicking techniques for the online search interface. *Online Information Review*, 13:407–424, 1993.

[3] T. Berners-Lee. Linked data. *International Journal on Semantic Web and Information Systems*, 4, 2006.

[4] T. Berners-Lee et al. The semantic web. *Scientific American*, 284:34–43, 2001.

[5] H. Glaser et al. CS AKTive Space: building a semantic web application. In *Proc. of ESWS*, pages 417–432. Springer Verlag, 2004.

[6] D. Huynh et al. Piggy bank: Experience the semantic web inside your web browser. *Proc. of ISWC*, 2005.

[7] D. Huynh et al. Exhibit: lightweight structured data publishing. In *Proc. of WWW*, 2007.

[8] Interface Ecology Lab. Meta-Metadata Guide. http://ecologylab.net/research/simplGuide/metaMetadata/index.html.

[9] A. Kerne et al. combinformation: Mixed-initiative composition of image and text surrogates promotes information discovery. *ACM TOIS*, 27:5:1–5:45, 2008.

[10] A. Kerne et al. Meta-metadata: a metadata semantics language for collection representation applications. In *Proc. of CIKM*, 2010.

[11] Library of Congress. Dublin Core/MARC/GILS crosswalk. http://www.loc.gov/marc/dccross.html, 2008.

[12] C. C. Marshall and F. M. Shipman. Which semantic web? In *Proc. of HYPERTEXT*, 2003.

[13] N. Shahzad. S.IM.PL serialization: Translation scopes encapsulate cross-platform, multi-format information binding. Master's thesis, Texas A&M University, 2011.

[14] S. Weibel et al. RFC 2413: Dublin core metadata for resource discovery. *RFC*, 1998.

[15] K.-P. Yee et al. Faceted metadata for image search and browsing. In *Proc. of SIGCHI*, 2003.

Automatic Text Summarization and Small-World Networks

Helen Balinsky
Hewlett-Packard Laboratories
Long Down Avenue
Bristol BS34 8QZ, UK
Helen.Balinsky@hp.com

Alexander Balinsky
Cardiff School of Mathematics
Cardiff University
Cardiff CF24 4AG, UK
BalinskyA@cardiff.ac.uk

Steven J. Simske
Hewlett-Packard Laboratories
3404 E. Harmony Rd. MS 36
Fort Collins, CO USA 80528
Steven.Simske@hp.com

ABSTRACT

Automatic text summarization is an important and challenging problem. Over the years, the amount of text available electronically has grown exponentially. This growth has created a huge demand for automatic methods and tools for text summarization. We can think of automatic summarization as a type of information compression. To achieve such compression, better modelling and understanding of document structures and internal relations is required. In this article, we develop a novel approach to extractive text summarization by modelling texts and documents as small-world networks. Based on our recent work on the detection of unusual behavior in text, we model a document as a one-parameter family of graphs with its sentences or paragraphs defining the vertex set and with edges defined by Helmholtz's principle. We demonstrate that for some range of the parameters, the resulting graph becomes a small-world network. Such a remarkable structure opens the possibility of applying many measures and tools from social network theory to the problem of extracting the most important sentences and structures from text documents. We hope that documents will be also a new and rich source of examples of complex networks.

Categories and Subject Descriptors

I.5 [**Pattern Recognition**]: Applications—*text processing*; I.7 [**Document and Text Processing**]: Miscellaneous; I.2.7 [**Natural Language Processing**]: Text analysis

General Terms

Algorithms, Theory

Keywords

Text Summarization, small world network, unusual behavior detection, Helmholtz principle

1. INTRODUCTION

Text summarization is an important area in Natural Language Processing (NLP). Since manual summarization of large documents is a very difficult and time-consuming task, there is high demand for effective, fast and reliable automatic text summarization tools and models. This becomes especially important with an exponential growth in the number of electronically available documents on the Internet and enterprise intranets.

In the "Introduction to the Special Issue on Summarization" Radev et al. [17] define a *summary* "as a text that is produced from one or more texts, that conveys important information in the original text(s), and that is no longer than half of the original text(s) and usually significantly less than that. Text here is used rather loosely and can refer to speech, multimedia documents, hypertext, etc." From this simple and intuitive definition, we can see that summaries should be very similar to lossy data compression; i.e., we should preserve important information and make summaries short. Since not every data can be compressed, we need to develop a model for texts that capture possible redundancies in textual data.

There are two types of summarizations, an *abstractive summarization* and an *extractive summarization*. In the abstractive summarization the goal is to represent main concepts and ideas of a document by paraphrasing of the source document in clear natural language. Automatic abstractive summarization is a very difficult task and is still in its infancy. In extractive summarization, the goal is to extract the most meaningful parts of documents (sentences, paragraphs, etc.) to represent main concepts of the document. Automatic extractive summarization is a much more developed area with many different approaches and tools [5]. In this article, we address only extractive summarization, with the main goal being the extraction of meaningful sentences or paragraphs.

To extract the most meaningful sentences or paragraphs from a text document, we need to define a *ranking function* for these text elements. The standard approach to define a ranking function is the so called *graph theoretic approach*. Modelling texts with graphs has long history in NLP research and has been used under different names and in different setups: a paragraph or sentence *relationship map* in [18], LexRank in [4], etc. In the graph theoretic approach, documents are modelled by graphs (networks) with sentences (or paragraphs) as the nodes. The edges between the nodes are used to represent the relationships between pairs of nodes and defining such relationships is a crucial task. After that, a ranking function is defined as a measure of the importance of the nodes in such a graph. In all previous approaches, edges have been defined by thresholding some measures of similarity from information retrieval. With these approaches, there is no control over the types of resulting graphs, and very often the resulting graphs are "noisy" without reliable ranking functions. The range of ranking functions can be very narrow, or only a small number of nodes will have small values of ranking functions. For example, if we use the degree function for ranking and we have 40% of sentences with their degrees in the range [100, 102] and 30% of sentences with their degrees in the range [98, 99], it is difficult to believe that sentences in the first range are more important that in the second range. It would be much better to have ranking functions with a large range of values, with only a small number of top values and a long tail of small values, as in the power-law distributions.

During last ten years, we have witnessed increased research activities in different social, biological and man-made networks. Research on the theory of social networks has demonstrated that most observed graphs have many centrality measures with the large range of values and heavy-tail type distributions [8, 16]. Such measures include *betweenness, centrality, clustering, degree, eigenvector centrality*, etc. It has been shown that social graphs preserve their crucial properties even after pruning a large number of (unimportant) nodes. Another example is electrical grids or road networks, where a failure of some nodes does not affect network properties significantly. This is precisely what we want in the automatic summarization: we can think about the extracting of a small number of important sentences and paragraphs as removing a large number of unimportant sentences, whilst preserving the structure of a document.

The main goal of this article is to define a relationship between sentences and paragraphs that produce graphs with the same properties as social networks. There are two provable approaches to construct graphs with such desired properties: the theory of *Affiliation Networks* [9] and *Kronecker Graphs* [10]. In this article we shall construct our graphs as affiliation networks with affiliations defined by the Helmholtz Principle for topics and an unusual behavior detection in textual data that was developed in [1, 2].

The paper is organized as follows. In Section 2 we describe precisely the problem we are going to solve. In Section 3 we present our solution and define our main objects – a one parameter family of graphs – and briefly explain what it means for a graph to be a small world. In Section 4, numerical results for several documents are presented. Finally, conclusions and future work are discussed in the Section 5.

2. PROBLEM STATEMENT

In this article we model documents by graphs (networks) $G = (V, E)$ with sentences (or paragraphs) as the vertex set V. In network theory, a vertex is very often called a node. The edge between two nodes is used to represent the relationships between a pair of entities.

In the past several years, the study of many real world networks has greatly expanded: the Internet, social networks, power grids, connections in brain, etc. People have observed many important characteristics of real networks such as a high degree of clustering, a small number of edges and the small average distance, fast searchability, densification, scale free behavior, etc. [8, 16]. Such properties allow us to explain and to calculate many remarkable characteristics of networks, i.e. hubs, ranking, stability under connection failures etc.

The main task of this article is to define relationships between sentences and paragraphs in text documents that result in graphs with similar properties and topology. More precisely, we would like to define relations between sentences and paragraphs which produce graphs with a small world topology [19, 14, 6, 7]. Since text documents are made by humans and for humans, it is natural to expect that during writing, an author will present his main ideas and concepts in a manner similar to biological networks.

Why are we interested in graphs with a small world topology? The main reason is when a graph topology becomes a small world, we can reliably define the most important nodes and edges of such a graph by measuring their contributions to the graph being a small world. This will give us a mechanism to define the most important sentences and paragraphs. After that, we can use these important sentences and paragraphs to create an automatic text summary. We hope that such an approach will bring a better understanding of complex logical structures and flows in text documents.

Several text data mining researchers have already used ideas from social networking and the concept of a small world for keyword extraction in documents [13]. They constructed graphs by selecting words as nodes. They also introduced edges between two words if and only if these two words appear in the same sentence: these are termed co-occurrence graphs.

Another application of a small world topology can be found in [12], wherein three different methods to model the semantic similarity between documents are presented. In [12], it is shown that document similarity based on a small world topology outperforms the standard cosine similarity measure in the case of the document representation as a bag of words.

3. OUR SOLUTION

The theory of *Affiliation Networks* [9] is probably the most powerful tool for creating and modelling of social networks. The underlying idea behind affiliation networks is that in social networks there are two types of players: actors and societies. They are related by the affiliation of the actors to the societies. In an affiliation network, the actors become the nodes and we say that two actors are related if they both belong to at least one common society. As it was shown in [9], there are precise rules for creation societies and actors that result in affiliation networks with the desired properties. In our case, the sentences will be the actors. The society should be responsible for grouping sentences in topics, and topics should be modelled by unusual activities in documents. The necessary tools for topic and unusual behavior detection in textual data was developed in [1, 2].

Nearest sentences (or paragraphs) in a document are often logically connected to create a flow of information. So, we always create an edge between a pair of nearest sentences (or paragraphs). We shall call such edges *local* or *sequential* connections.

However, a document has a more complicated structure. Different parts of a document–either proximate or distant–can be logically connected. We would like to connect sentences (or paragraphs) if they have something important in common and are talking about something similar. For example, to clarify something, an author can recall or reference in one location words that also appear in another location. Such references will create *distant* relations inside the document.

Our approach to the problem of defining relations between sentences is as follows.

1. We construct a *one-parameter family* of meaningful words *MeaningfulSet*(ϵ) for the document. These sets will be our societies for constructing corresponding affiliation networks.

2. We connect two sentences (or two paragraphs) by the edge if and only if they have at least one word from the *MeaningfulSet*(ϵ) in common or if they are a pair of consecutive sentences.

3. We demonstrate that by selecting the *MeaningfulSet*(ϵ) using Helmholtz's principle such graphs become a small world for some range of the parameter ϵ. In fact, for most of documents the range around $\epsilon = 2$ will produce the desired properties.

The simplified diagram is shown in Figure 1.

If a set of meaningful words *MeaningfulSet*(ϵ) is too small, then only the local relations will be present and the graph will look like a regular graph. If, however, we select too many meaningful words, then the graph will look like a large random graph with too many edges.

To be more specific, let us consider the size of "societies" for some text documents. We present in Fig.2 the size of the *MeaningfulSet*(ϵ) for three different documents as a function of ϵ. For our examples, we chose the following three documents: the State of the Union address 2000 given by President Bill Clinton, the State of the Union address 2011 given by President Barack Obama and the Book of Genesis from the Natural Language Toolkit corpus. In all three documents, the rapid drop of the size of the *MeaningfulSet*(ϵ) can be observed within some vicinity of $\epsilon = 0$. We performed many experiments, which demonstrated that this type of behavior is typical for many real-world text documents with at least thirty sentences.

Such a rapid drop in the size of *MeaningfulSet*(ϵ) always happens for some positive ϵ and can be easily detected automatically by looking into the highest value of the derivative of the curve.

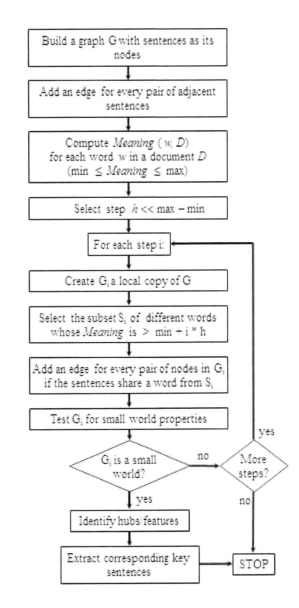

Figure 1: Simplified diagram for determining document small world structure

After finding the range of the parameter ϵ corresponding to a small number of edges, a small mean distance and high clustering (see below), we can define an extractive summary as follows:

1. Select a measure of centrality for small world networks.

2. Check that for the corresponding range of the parameter ϵ this measure of centrality has a wide range of values and the heavy-tail distribution.

3. Select sentences and paragraphs with the highest ranking as a summary. We can also select the highest ranking paths in the graph if some coherence in the summary is desired.

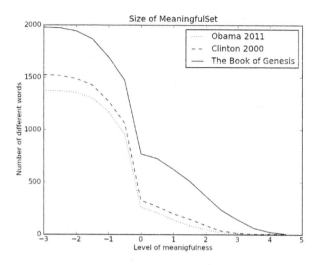

Figure 2: Size of *MeaningfulSet*(ϵ) **as function of** ϵ

For the reader's convenience we briefly describe the Helmholtz principles for calculating the measure of an unusual behavior in text documents. We also present some characteristics and feature of networks with small world topology.

MeaningfulSet(ϵ) and the Helmholtz Principle

Let D denote a text document and P denote a part of this document. P could be a paragraph of the document D if the document is divided into paragraphs, or several consecutive sentences. In documents without explicit paragraph structure we use four or five consecutive sentences as P.

Based on the Helmholtz Principle from the Gestalt Theory of human perception, defined in [1, 2], we can determine a *measure of meaningfulness* of a word w from D inside P. If the word w appears m times in P and K times in the whole document D, then we define the *number of false alarms* $NFA(w, P, D)$ by the following expression

$$\binom{K}{m} \cdot \frac{1}{N^{m-1}}, \tag{1}$$

where

$$\binom{K}{m} = \frac{K!}{m!\,(K-m)!}$$

is a binomial coefficient. In (1) the number N is $[L/B]$ where L is the length of the document D, B is the length of P in words and $[L/B]$ denotes the integer part of L/B.

As a measure of meaningfulness of the word w in P we are using

$$Meaning(w, P, D) := -\frac{1}{m} \log NFA(w, P, D) \tag{2}$$

The justification for using $Meaning(w, P, D)$ was given in [1] based on arguments from statistical physics and asymptotic behavior of binomial distributions.

Following [1, 2], we define a set of meaningful words in P as words with $Meaning(w, P, D) > 0$ and larger positive values of $Meaning(w, P, D)$ give larger levels of meaningfulness.

Given a document subdivided into paragraphs, we define the *MeaningfulSet*(ϵ) as a set of all words with

$$Meaning(w, D) > \epsilon,$$

where $Meaning(w, D)$ is the maximum of $Meaning(w, P, D)$ over all paragraphs P. In general, we do not require paragraphs to be disjoint. If our document does not have a natural subdivision into paragraphs, then we will use as a paragraph several consecutive sentences (typically four or five consecutive sentences).

For a sufficiently large positive ϵ, the set *MeaningfulSet*(ϵ) is empty. For a small enough negative $\epsilon << 0$ the set *MeaningfulSet*(ϵ) contains all the words from D. For real world documents, we observe that the size of *MeaningfulSet*(ϵ) has a sharp drop from the total number of words to zero around some critical value $\epsilon_0 > 0$ as shown in Fig.2.

Since *MeaningfulSet*(ϵ) with a nonnegative ϵ will play the major role in our approach to automatic text summarization, we would like to check that *MeaningfulSet*(0) can indeed be used for representing text in a natural language.

Zipf's well-known law for natural languages states that, given some corpus of documents, the frequency of any word is inversely proportional to some power γ of its rank in the frequency table, i.e. frequency(rank)≈const/rank$^\gamma$. Zipf's law is mostly easily observed by plotting the data on a *log-log* graph, with the axes being log(rank order) and log(frequency). The data conform to Zipf's law to the extent that the plot is linear. Usually Zipf's law is valid for the upper portion of the log-log curve and not valid for the tail.

In [3] it was shown that Zipf's law is the only possible outcome of an evolving communicative system under a tension between two communicative agents. The speaker's economy tries to reduce the size of the dictionary, whereas the listener's economy tries to increase the size of the dictionary. Without going into technical details, this means that the *MeaningfulSet*(0) should also obey Zipf's law in order to properly represent topics and text.

For all of the words in all of the Presidential State of the Union Addresses, we plot rank of a word and the total number of the word's occurrences in log-log coordinates, as shown in Figure 3.

Using Zipf's law for only the meaningful words of this corpus ($\epsilon = 0$), we use the individual State of the Union Addresses as paragraphs in the calculation of our measure of meaningfulness (2). We plot the rank of a meaningful word and the total number of the word's occurrences in log-log coordinates, as shown in Figure 4. We still can observe Zipf's law, although the curve becomes smoother and the power γ becomes smaller.

If we increase the level of meaningfulness (i.e. increase the ϵ), then the curve becomes even smoother and conforms to Zipf's law with smaller and smaller γ. This is exactly what

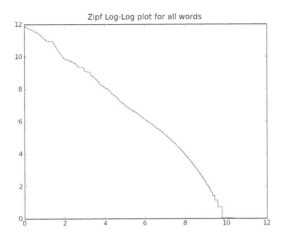

Figure 3: Zipf's law for all of the Presidential State of the Union Addresses

we should expect for good feature extraction and dimensionality reduction: to decrease the number of features and to decorrelate data.

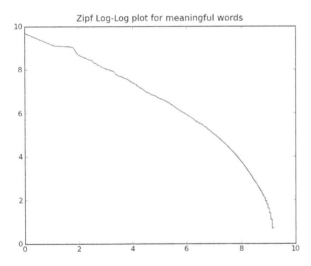

Figure 4: Zipf's law for only the meaningful words in the all of the Presidential State of the Union Addresses

Very similar results can be observed for many different documents and collections. We also demonstrate in [1] that $MeaningfulSet(\epsilon)$ are extremely powerful for document classifications.

Let us now define our main object – the one parameter family of graphs $Gr(D, \epsilon)$ for a document D.

We first pre-process the document D by splitting the words by non-alphabetic characters and down-case all words. After that we apply stemming (lemmatization).

Let us denote by

$$S_1, S_2, \ldots, S_n$$

the sequence of consecutive sentences in the document D.

We can also use a sequence of consecutive paragraphs or other parts of the document. But for clarity let us concentrate on sentences.

The graph $Gr(D, \epsilon)$ has sentences $S_1, S_2, \ldots, S_n$ as its vertex set. Since the order of sentences is important in documents, and since the nearest sentences are usually related, we will add an edge for every pair of consecutive sentences (S_i, S_{i+1}). This will also guarantee connectivity of our graph to avoid unnecessary complications with several connected components.

Finally, if two sentences share at least one word from the set $MeaningfulSet(\epsilon)$ we will connect them by the edge. This will define our family of graphs $Gr(D, \epsilon)$.

For a sufficiently large positive number ϵ, we have

$$MeaningfulSet(\epsilon) = \emptyset,$$

and thus $Gr(D, \epsilon)$ degenerates into the path graph Fig.5.

Figure 5: $Gr(D, \epsilon)$ for large enough ϵ

If a text document has its own set of keywords, title words, etc., then it is always wise to extent the set $MeaningfulSet(\epsilon)$ by adding such additional words to the set $MeaningfulSet(\epsilon)$.

Characteristics of a Small World

To formalize the notion of a small world, Watts and Strogatz defined in [19] the *clustering coefficient* and the *characteristic path length* of a network.

Let $G = (V, E)$ be a simple, undirected and connected graph with the set of nodes $V = \{v_1, \ldots, v_n\}$ and the set of edges E. Let us denote by l_{ij} the geodesic distance between two different nodes v_i and v_j. The geodesic distance is the length of the shortest path counted in the number of edges in the path. Watts and Strogatz define in [19] the characteristic path length (or *the mean inter-node distance*), L, as the average of l_{ij} over all pairs of different nodes (i, j):

$$L = \frac{1}{n(n-1)} \sum_{i \neq j} l_{ij}.$$

The diameter L_D is the maximum of all inter-node distances l_{ij}.

Our graph $Gr(D, \epsilon)$ depends on the parameter ϵ, so the characteristic path length becomes the function $L(\epsilon)$ of the parameter ϵ. Obviously, $L(\epsilon)$ is also a non-decreasing function of ϵ.

Clustering is a description of the interconnectedness of the nearest neighbors of a node in a graph. Clustering is a non-local characteristic of a node and goes one step further than the degree. Clustering plays an important role in study of many social networks [16]. There are two widely-used measures of clustering: *clustering coefficient* and *transitivity*.

The clustering coefficient $C(v_i)$ of a node v_i is the probability that two nearest neighbors of v_i are themselves nearest neighbors. In other words,

$$C(v_i) = \frac{number\ of\ pairs\ of\ neighbors\ of\ v_i\ that\ are\ connected}{number\ of\ pairs\ of\ neighbors\ of\ v_i},$$

i.e.,

$$C(v_i) = \frac{t_i}{q_i(q_i-1)/2},$$

where q_i is a number of the nearest neighbors of v_i (degree of the vertex) with t_i connections between them. $C(v_i)$ is always between 0 and 1. When all the nearest neighbors of a node v_i are interconnected, $C(v_i) = 1$, and when there are no connections between the nearest neighbors, as in trees, $C(v_i) = 0$. Importantly, most real-world networks exhibit strong clustering. In [19], Watts and Strogatz proposed to calculate a *clustering coefficient* (or *mean clustering*) for an entire network as the mean of local clustering coefficients of all nodes:

$$C_{WS} = \frac{1}{n} \sum_{v_i \in V} C(v_i),$$

where n is the number of vertices in the network. In [19] the several examples of C_{WS} for real-world networks were presented: for the collaboration graph of actors $C_{WS} = 0.79$, for the electrical power grid of the western United State $C_{WS} = 0.08$ and for the neural network of the nematode worm *C.elegans* $C_{WS} = 0.28$.

Historically, C_{WS} was the first measure of clustering in the study of networks and is a widely used and very popular characteristic. There is also another measure of clustering, which has become increasingly popular – that of *transitivity*. It is important to mention that the clustering coefficient and the transitivity are not equivalent and that they can produce substantially different numbers for a given network. Many researchers consider the transitivity to be a more reliable characteristic of a small world than the clustering coefficient (see [16] for more detailed explanations).

Transitivity is a very important and natural concept in social networks. In mathematics, a relation R is said to be transitive if aRb and bRc together imply aRc. In networks, there are many different relationships between pairs of nodes. The simplest relation is "connected by an edge". If the "connected by an edge" relation was transitive it would mean that if a node u is connected to a node v, and v is connected to w, then u is also connected to w. For social networks this would mean that "the friend of my friend is also my friend".

Perfect transitivity can only occurs in networks where each connected component is a complete graph, i.e. all nodes are connected to all others. In general, the friend of my friend is not necessarily my friend. But intuitively, we can expect a high level of transitivity between people.

In the case of our graphs $Gr(D, \epsilon)$, the transitivity would mean that if a sentence S_i describes something similar to a sentence S_j, and S_j is also similar to a sentence S_k, then S_i and S_k probably should also have something in common. So,

it is natural to expect a high level of transitivity in $Gr(D, \epsilon)$ for some range of parameter ϵ.

We can quantify the level of transitivity in graphs as follows. If u is connected to v and v is connected to w, then we have a path uvw of two edges in the graph. If u is also connected to w, we say that the path is a triangle. Let us define the transitivity of a network as the fraction of paths of length two in the network that are triangle:

$$C = \frac{(number\ of\ triangles) \times 3}{(number\ of\ connected\ triples)},$$

where a "connected triple" means three nodes u, v and w with edges (u, v) and (v, w). The factor of three in the numerator arises because each triangle will be counted three times during counting all connected triples in the network.

Let us present some typical values of transitivity for social networks. The network of film actor collaborations has been found to have $C = 0.20$ [15]; a network of collaborations between biologists has $C = 0.09$; a network of who sends an email to whom in a large university has $C = 0.16$.

4. EXPERIMENTAL RESULTS

For our numerical experiments, as before we chose three different text documents: the State of the Union address 2000 given by President Bill Clinton, the State of the Union address 2011 given by President Barack Obama, and the Book of Genesis. As mentioned above, we first pre-process the documents by splitting the words by non-alphabetic characters and down-case all words. After that we apply stemming (lemmatization). We use natural paragraphs for the first two documents. For the Book of Genesis, we define a paragraph as any four nearest sentences. For the three documents, the numbers of sentences, paragraphs, words and different words are presented in Table 1.

Table 1: Document statistics

Document	Senten.	Paragr.	Words	Dict.
Obama, 2011	435	95	7083	1372
Clinton, 2000	533	133	8861	1522
The Book of Genesis	2343	N/A	35250	1975

To better understand the properties of networks $Gr(D, \epsilon)$, we should examine the different measures and metrics (see Chapter II in the book [16] for a comprehensive list of such characteristics).

First of all, let us plot the number of edges in $Gr(D, \epsilon)$ for all three documents as a function of ϵ, Fig.6.

As we can see, there is a dramatic drop in the number of edges in $Gr(D, \epsilon)$ for some ranges of positive values of ϵ. These are areas where we expect to observe some remarkable properties of the graphs $Gr(D, \epsilon)$.

The plot of the characteristic path lengths for the three selected text documents is shown in Fig.7. The example values of the characteristic path length $L(\epsilon)$ are shown in Table 2.

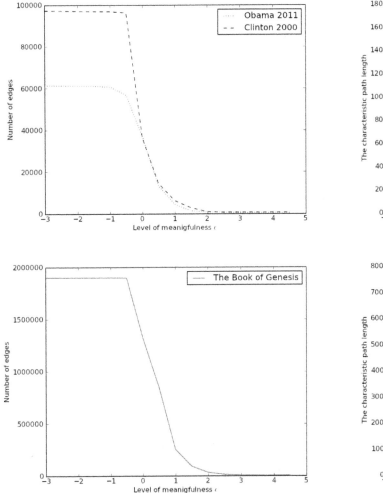

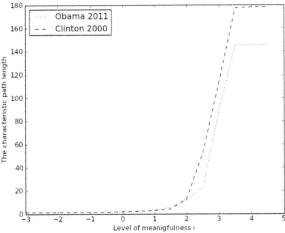

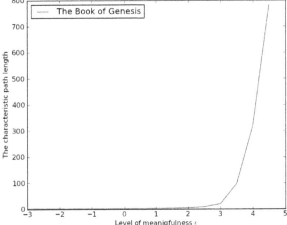

Figure 6: Number of edges in $Gr(D, \epsilon)$ as function of ϵ

Figure 7: The characteristic path length of $Gr(D, \epsilon)$ as a function of ϵ

Results of calculating the clustering coefficient for our three one-parameter family of graphs are presented in Fig.8.

By looking at Fig.6, Fig.7 and Fig.8 we can clearly see that in the range $\epsilon \in [1.0, 2.5]$ the network $Gr(D, \epsilon)$ is a small world in the case of State of the Union address 2000 given by President Bill Clinton and in the case of State of the Union address 2011 given by President Barack Obama. Both documents have a *small degree of separation*,high mean clustering C_{WS} and a relatively small number of edges. For the Book of Genesis, the range $\epsilon \in [2, 3]$ also produces a small world with even more striking values of the mean clustering C_{WS}.

Results of calculation of the transitivity for the three one-parameter family of graphs are presented in Fig.9. By looking at Fig.9 we can clearly see that in the range $\epsilon \in [1.0, 2.5]$ the network $Gr(D, \epsilon)$ has high transitivity in the case of State of the Union address 2000 given by President Bill Clinton and in the case of State of the Union address 2011 given by President Barack Obama. For the Book of Genesis in the range $\epsilon \in [2, 3]$ the transitivity is also quite high (>0.6).

From Table 1, we can see that $Gr(D, \epsilon)$ has 435 nodes in the Obama 2011 address, 533 nodes in the Clinton 2000 address, and 2343 nodes in the case of the Book of Genesis. So, it is not easy to represent such graphs graphically. A much nicer picture can be produced for the graphs with paragraphs as the node set. We will connect paragraphs by the same rule: two paragraphs are connected if they have meaningful words in common. In the case of the State of the Union address 2011 given by President Barack Obama we have 95 paragraphs, and for the value $\epsilon = 2$, the corresponding graph is presented in Fig.10. On this picture we can see how several highly-connected nodes make a small world topology.

If we want to rank nodes according to some ranking function, it is important that this function gives us a wide range of values. Let us look into ranking of sentences according to their degree. We sort all nodes in $Gr(D, \epsilon)$ in decreasing order of degree and get a degree sequence $d(\epsilon) = \{d_1(\epsilon), \dots, d_n(\epsilon)\}$, where $d_1(\epsilon) \geq d_2(\epsilon) \geq \dots \geq d_n(\epsilon)$. Consider, for example, the first fifty values of d_i in the case of the President Obama speech. To have a reliable selection of five, ten or

Table 2: Some values of $L(\epsilon)$ for the three documents

ϵ	Obama	Clinton	The Book of Genesis
-1.0	1.358748	1.319542	1.309066
0.0	1.622702	1.773237	1.527610
1.0	2.937931	2.861523	2.079833
1.5	5.514275	3.945697	2.580943
2.0	12.274517	12.715485	3.727103
2.5	22.471095	52.442205	7.280936
3.0	89.049007	113.237971	18.874327
3.5	144.854071	177.272814	96.873744
4.0	145.333333	178.000000	317.638370
4.5	145.333333	178.000000	779.802265

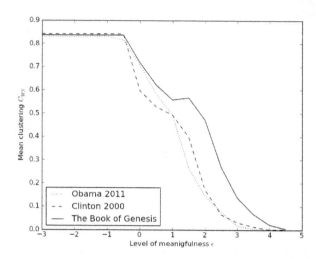

Figure 8: The clustering coefficient C_{WS} of documents

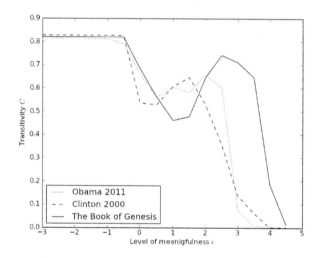

Figure 9: The transitivity C of documents

more highest-ranked sentences, we need a wide range of values of the degree function. Let us plot $d(\epsilon)$ for several values of ϵ and the first fifty elements, Fig.11. We scale the value such that the largest one, $d_1(\epsilon)$, is equal to one. We can clear see that the values $\epsilon = 1.0$ and $\epsilon = 2.0$ have the best dynamic range and this is where our graphs have a small world topology.

The most connected sentence in Obama address (for $\epsilon = 2$) is

"The plan that has made all of this possible, from the tax cuts to the jobs, is the Recovery Act."

with degree=29.

Let us also present the two most important paragraphs in Obama address (for $\epsilon = 2$) extracted from a graph that uses paragraphs as nodes and the degree as measure of centrality:

The first paragraphs is:

"The plan that has made all of this possible, from the tax cuts to the jobs, is the Recovery Act. That's right, the Recovery Act, also known as the stimulus bill. Economists on the left and the right say this bill has helped save jobs and avert disaster. But you don't have to take their word for it. Talk to the small business in Phoenix that will triple its workforce because of the Recovery Act. Talk to the window manufacturer in Philadelphia who said he used to be skeptical about the Recovery Act, until he had to add two more work shifts just because of the business it created. Talk to the single teacher raising two kids who was told by her principal in the last week of school that because of the Recovery Act, she wouldn't be laid off after all. "

And the second most important paragraph is

"Now, the price of college tuition is just one of the burdens

facing the middle class. That's why last year, I asked Vice President Biden to chair a task force on middle class families. That's why we're nearly doubling the child care tax credit and making it easier to save for retirement by giving access to every worker a retirement account and expanding the tax credit for those who start a nest egg. That's why we're working to lift the value of a family's single largest investment, their home. The steps we took last year to shore up the housing market have allowed millions of Americans to take out new loans and save an average of $1,500 on mortgage payments. This year, we will step up refinancing so that homeowners can move into more affordable mortgages."

A crucial issue with automatic text summarization is its *evaluation*. Unfortunately, there is no universally accepted strategy and toolset for evaluating summaries. The problem is that humans produce summaries with a wide variance and there is no agreement on what should be a "good" summary.

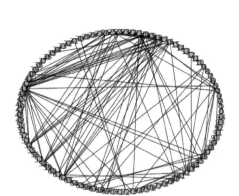

Graph of paragraphs

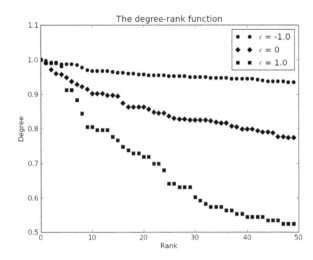

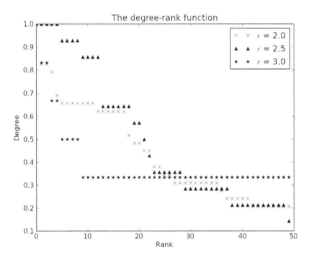

Figure 11: The degree-rank function for different values of ϵ.

Figure 10: The Graph $Gr(D, \epsilon)$ with paragraphs as nodes, $\epsilon = 2$, Obama 2011

We manually created 10 sentences summaries on a large set of documents and they look very good from the point of view of people we interviewed. As the next step we are planning to investigate in more details one of the metric used in NIST Annual Document Understanding Conferences. This is so called the ROUGE metric (Recall-Oriented Understudy for Gisting Evaluation). This metric calculate n-gram overlap between automatically created summaries and written by the human summary.

We should also point out that our approach is suitable for large documents, where complicated network structures can be observed. For short texts and news our approach is not expected to produce the desired results.

5. CONCLUSIONS

The main goal of this article is to model large textual documents as graphs with small world topology. Such models should allow better automatic text summarization by using different measures of centralities of sentences and paragraphs in documents. More precisely, we model a text document by *one parameter graphs* with sentences (or paragraphs) as the set of its nodes. We define edges by a carefully selected (using the Helmholtz principle form [1, 2]) family of meaningful words.

When the parameter goes from a negative to a large positive value, the graphs transform from large random graphs to regular graphs. We demonstrated that a remarkable transition into a small world topology happens in between. The transition has a high derivative value on the curves illustrated herein, indicating it can be used as a single descriptive feature for the degree of connectedness of a document.

We think that observed transitions into a small world topology can have many applications far beyond automatic summarization. The transition into a small world topography can be quantified based on the peak derivative and used as a

single metric for defining the style and/or characterizing the "interconnectedness" of the content evaluated. Among next steps, we will investigate the extent to which these and other transition parameters depend on authors' style and topics. Based on these findings, we can then see how these features can be used for identifying authors (e.g. to find potential plagiarism in parts of a document), comparing authors for style (e.g. to suggest further reading), and for rating the style of different authors.

Having a small world topology is also important in many ranking problems since in such graphs different nodes have different contributions to a graph being a small world.

As the next step, we are going to investigate different measures and metrics for complex networks, such as the eigenvector centrality, Katz centrality, hubs and authorities, betweenness centrality, power law and scale-free networks, etc. We expect that such measures will play an important role in document summarization.

It is also very important to investigate if the small-world graphs introduced in this paper have Kleinberg's type of distance decay. This would be an indication of how easily and quickly we can search and navigate in text documents.

Another direction for the future work is in investigating the natural orientation of edges in our graphs, i.e. the directed graph structure. For two connected sentences, we can say which one is the first and which one is the second, according to their position in a document. This will make such a graph look like a small WWW-type network and PageRank-type algorithms may produce interesting rankings of nodes.

In a seminal paper [11], Lescovec et al. made two remarkable empirical observations about many known social networks. Namely, they demonstrated that real-world networks become *denser* over time, and their diameters effectively become *smaller* over time. We can also introduce a *time* parameter t in our model by looking into the first t sentences of a document. It would be very interesting to investigate the time evolution of our graphs and to check if we have similar effects.

We manually check that summaries created from such small world graphs are very good for a large collection of different documents. Unfortunately, there is no generally-accepted "gold" standard for the evaluation of summaries. Perhaps the most popular tool currently used is the ROUGE (Recall-Oriented Understudy for Gisting Evaluation) metric. Our next step will be evaluating our algorithms with this measure.

Many early automatic text summarization methods used several heuristics like the cue method, title method and location method. It would be very interesting to see if our model can automatically generate such methods.

We hope that tools and algorithms from the theory of complex networks will be useful in text data mining–and that documents will be also a source of examples of complex networks.

6. REFERENCES

[1] A. Balinsky, H. Balinsky, and S. Simske. On the Helmholtz principle for data mining. In *Proceedings of 2011 International Conference on Knowledge Discovery, Chengdu, China , April 2011*.

[2] A. Balinsky, H. Balinsky, and S. Simske. On Helmholtz principle for documents processing. In *Proc. of the 10th ACM symposium on Document engineering*, 2010.

[3] B. Corominas-Murtra, J. Fortuny, and R. V. Solé. Emergence of zipf's law in the evolution of communication. *Phys. Rev. E*, 83(3):036115, Mar 2011.

[4] G. Erkan and D. R. Radev. LexRank: graph-based lexical centrality as salience in text summarization. *J. Artif. Int. Res.*, 22:457–479, December 2004.

[5] K. S. Jones. Automatic summarising: The state of the art. *Information Processing and Management*, 43:1449–1481, 2007.

[6] J. Kleinberg. Navigation in a small world. *Nature*, 406:845, 2000.

[7] J. Kleinberg. The small-world phenomenon: An algorithmic perspective. In *Proc. 37th Annual ACM Symposium on Theory of Computing*, pages 163–170, 2000.

[8] J. Kleinberg and D. Easley. *Networks, Crowds, and Markets: Reasoning About a Highly Connected World*. Cambridge University Press, 2010.

[9] S. Lattanzi and D. Sivakumar. Affiliation networks. In *STOC '09 Proceedings of the 41st annual ACM symposium on Theory of computing*, 2009.

[10] J. Leskovec, D. Chakrabarti, J. Kleinberg, C. Faloutsos, and Z. Ghahramani. Kronecker graphs: An approach to modeling networks. *Journal of Machine Learning Research*, 11:985–1042, February 2010.

[11] J. Leskovec, J. Kleinberg, and C. Faloutsos. Graphs over time: Densification laws, shrinking diameters and possible explanations. In *ACM SIGKDD International Conference on Knowledge Discovery and Data Mining(KDD)*, 2005.

[12] J. Leskovec and J. Shawe-Taylor. Semantic text features from small world graphs. In *Subspace, Latent Structure and Feature Selection techniques: Statistical and Optimization perspectives Workshop, Slovenia*, 2005.

[13] Y. Matsuo, Y. Ohsawa, and M. Ishizuka. A document as a small world. *Lecture Notes in Computer Science*, 2253:444–448, 2001.

[14] M. Newman. Models of the small world. *J. Stat. Phys.*, 101:819, 2000.

[15] M. Newman. The structure and function of complex networks. *SIAM Rev.*, 45:167–256, 2003.

[16] M. Newman. *Networks: An Introduction*. Oxford University Press, 2010.

[17] D. R. Radev, E. Hovy, and K. McKeown. Introduction to the special issue on summarization. *Computational Linguistics*, 28(4):399–408, December 2002.

[18] G. Salton, A. Singhal, M. Mitra, and C. Buckley. Automatic text structuring and summarization. *Inf. Process. Manage.*, 33:193–207, March 1997.

[19] D. Watts and S. Strogatz. Collective dynamics of small-world networks. *Nature*, 393:440, 1998.

Efficient Keyword Extraction
for Meaningful Document Perception

Thomas Bohne   Sebastian Rönnau   Uwe M. Borghoff

Universität der Bundeswehr München
Werner Heisenberg 39
München, Germany
{thomas.bohne,sebastian.roennau,uwe.borghoff}@unibw.de

ABSTRACT

Keyword extraction is a common technique in the domain of information retrieval. Keywords serve as a minimalistic summary for single documents or document collections, enabling the reader to quickly perceive the main contents of a text. However, they are often not readily available for the documents of interest.

Common keyword extraction techniques demand either a large data collection, a learning process, or access to extensive amounts of reference data. By relying on additional linguistic features (e.g. stop word removal), most approaches are language-restricted. Moreover, the extracted keywords usually pertain to the entire document, rather than only to the portion that is of interest to the reader.

In this paper, we present an efficient and flexible approach to summarize selections of text within a document. Our solution is based on a keyword extraction algorithm that is applicable to a variety of documents, regardless of language or context. This algorithm relies on the Helmholtz principle and extends a recently presented approach. Our extension covers the features of a weighting algorithm while providing a self-regulation capability to allow for more meaningful results. Furthermore, our approach takes into account the document structure in order to enhance pure statistic summarizations. We evaluate the efficiency of our approach and present results with meaningful examples. In addition, we outline further applications of our approach that allow for enhanced document perception as well as for meaningful document indexing and retrieval.

Categories and Subject Descriptors

H.3.3 [**Information Storage and Retrieval**]: Information Search and Retrieval; I.7.3 [**Document and Text Processing**]: Index Generation

General Terms

Algorithms, Performance

Keywords

information retrieval, keyword extraction, text data mining, single document, heuristic algorithm

1. INTRODUCTION

The explosive expansion of document repositories in the last decades has driven the development of various methods that support the user to find and perceive documents quickly. Nowadays, search engines and meta-information systems help to select documents that might fit the users' profile or match a search query. Early information retrieval systems usually relied on user-defined boolean search requests. Recent approaches often assign a numeric score to documents or terms that might be of interest to the user in order to classify the sheer size of possibly matching documents [22]. The contribution of this paper is an algorithm that extracts meaningful keywords from a single document to enable the reader to quickly ascertain the main idea of a portion of text.

In the following we would like to clarify notations and classify our approach. A *summary* is a condensed version of the original document, preserving its content [26]. Automatic summarization methods extract the most important content from a text and present it to the user in less space. Automatic summarization is a task in the information retrieval and natural language processing (NLP) domain. In general, NLP requires a deep knowledge of the specific language to process. This relates not only to the language itself but also to the domain of the corresponding document and is especially important to avoid an incorrect interpretation.

Single document summarizations can be classified either *generic* or *query-relevant* [26]. A generic document summary is generated upon the document itself. For each document, the resulting summary is deterministic. Query-relevant summarizations, however, are performed towards a given query. Here, the summarization is created in a way that matches the keywords given by the query in order to receive more specific results. In this context, the user requires previous knowledge about the content of the document to formulate a meaningful query. Furthermore, summaries can be classified *abstraction-based* which use natural language generation technologies to paraphrase sections of the source document. *Extraction-based* summaries depend directly on the lexical content of the document. The proposed approach focuses on generic summarization based on keyword extraction.

Single keywords may not reflect the exact meaning of the underlying text, however a list of keywords may provide an interpretation of the content to the user [24]. Keywords act as a simple method to enable the user to quickly grasp the content of a document. They are also used for index generation, similarity measure, and for browsing or clustering document collections [24]. We use keywords as a high level description of a document because they can be interpreted individually and independently from each other. Authors usually do not assign keywords manually unless they are instructed to do so because it is annoying and time consuming. Additionally, an author will probably assign keywords which are most likely to draw attention to the document [24]. Therefore, it is uncertain whether author-assigned keywords correspond precisely to the content of the document. Automatic keyword extraction has to ensure that the most meaningful keywords are extracted. One challenge is to find a good balance between the quantity of extracted keywords and the actual content. Another one is to prevent common words like "and", "she", etc. (also called *stop words*) to be listed as keywords. In general, both challenges are met by corresponding pre-processing procedures and algorithms that rely on training data.

Training data and pre-processing steps require that the language and the domain of the document is known. Especially for training data, large databases may be needed to ensure meaningful results. This is not an issue in the domain of search engines. Problems may arise, however, if the domain of the document is not well known (e.g. in historic books). In some cases, a connection to a training database may not be possible or acceptable. Note that most keyword extraction approaches are designed to index whole repositories, not single documents.

In this paper, we address these issues by proposing a novel keyword extraction approach that is language-independent and self-sufficient. Our central contribution is a keyword extraction algorithm that relies on the Helmholtz approach, which has previously been applied to the domain of keyword extraction by Balinsky et al. [5]. Its main advantage is its flexibility and efficiency as well as its independence from training data.

The flexibility of the keyword extraction algorithm allows the analysis of documents as a whole, as well as only portions of it. This enables the user to scale its focus. Especially for larger documents, the difference between the content of single sections may differ significantly from the content of the whole document.

As already stated before, the quantity of the extracted keywords may become an issue. We do not only extract keywords but weight them within the set of keywords with respect to a corresponding probability model. We use tag clouds to visualize this weighting [13]. They allow users to quickly gain the sense of the document or the actual section. We consider tag clouds to be a simple method for human reception of documents. Extensively used, they require the underlying keyword extraction algorithm to act efficiently to ensure the usability of the approach. The proposed algorithm runs quickly and allows continuous use.

The remainder of this paper is organized as follows. First, we discuss related work in Section 2. Afterwards, we propose the heuristic model used in our approach in Section 3, before proposing its application to single documents in Section 4. In Section 5, we evaluate our approach in terms of speed

and discuss the meaningfulness of the results. Finally, we propose different applications of our approach in Section 6 and discuss the approach itself in Section 7.

2. RELATED WORK

In this section, we focus on three major topics. We introduce document summarization approaches with a focus on techniques applicable to single document summarization. A large variety of keyword extraction approaches exists in information retrieval. We aim to address the most relevant for our approach. We briefly present tag clouds as a common visualisation techniqe for document summarization via keywords.

2.1 Document Summarization

Only a few single document summarization approaches are available. Teng et al. [23] proposed a single document summarization by calculating sentence similarity in a document to define topics. The sentences of each topic are being rated and sentence position is taken into account. The best scoring sentences result in a summary. Features such as sentence and paragraph position, length or commendation weight are taken into account by other approaches, e.g. by computing lexical chains in a text to identify significant sentences [6]. Most approaches create the summaries by identifying the most meaningful sentences and utilize them as extracts to create a summary [23].

The DARPA Topic Detection and Tracking initiative addresses event detection in document streams [1]. Documents with a temporal order (e.g. news stories) can be considered as streams that are applicable to event detection. The developed methods aim to detect an indication for a new event. This event detection is more a theme detection approach than a classification task [30]. A number of automatic summarization techniques involve NLP [10]. NLP may help to reference context in a collection but requires extensive knowledge of the environment. A number of techniques propose that computational linguistics improve text analysis but with limited success [22].

2.2 Keyword Extraction Algorithms

Automatic keyword extraction directly depends on the quality and speed of the corresponding keyword extraction algorithm. The algorithm usually generates weights that are assigned to terms in documents. One of the most popular and widespread heuristic ranking functions has been proposed by Jones [14]. The definition of term specificity later became well known as IDF (Inverse Document Frequency) and in combination with TF (Term Frequency) it has proven to be extraordinarily robust. The Bayesian Decision Theory is a fundamental statistical approach for classification - terms are classified as keywords or not. In contrast to the naive Bayes model, the Hidden Markov Model does not assume statistical term independence. Conroy et al. [7] have trained a Hidden Markov Model to assign a summary likelihood to each sentence.

A graph-based document representation can be built upon lexical similarities or semantic chains in a document. The Google PageRank algorithm [20] and Kleinberg's HITS algorithm [15] are link-based search algorithms that take into account graph information. Litvak et al. [19] apply HITS to document graphs to determine the top ranked nodes in the graph in order to identify the most significant keywords.

Turney [24] is one of the first who has defined keyword extraction as a supervised learning task. He combines the genetic training algorithm Genitor with the heuristic Extractor to create GenEx. GenEx has twelve parameters to maximize performance of the algorithm [24]. Tuning such a number of parameters is a sophisticated, domain specific process.

The KEA algorithm replaces the genetic learning algorithm with a naive Bayes machine learning technique [10]. Documents with already assigned keywords create a model for keyword extraction [28]. More pre-processing steps must be performed until the first keywords can be extracted and the TF-IDF weight is used to distinguish keyword candidates. For the speed of learning algorithms it is essential that documents belong to the same domain.

A number of methodologies use linguistic features in combination with filtering and lexical analysis to identify relevant terms in a document. Kumar et al. [17] use an n-gram filtration to extract potential keyphrases. The LAKE algorithm couples linguistic analysis with a learning algorithm that uses features such as TF-IDF for scoring [4]. In general, simple probabilistic models show best performance [11], non-parametric algorithms do not require the use of a training set of documents.

2.3 Tag Clouds

There exist several techniques of keyword visualizations such as ordered lists. To increase the speed of perception we instantly visualize keywords via tag clouds. Tag Clouds became popular on Web 2.0 websites such as *Delicious* and *Flickr* and rapidly enjoyed popularity in other branches. Xexeo et al. [29] propose the use of clouds as the conceptual representation of a text. They define a *summary tag cloud* as a unique tag cloud which summarizes the content of a document set. Furthermore, they introduce the notation *differential tag cloud* which in contrast "highlights the particular features of each document with regard to the summary tag cloud".

Tag Clouds have been used to visualize search results and help the users to refine searches in database searches [16] and web queries [18]. CloudMine is a concept of document summarization with a tag cloud presented by Watters [27]. He defines the interpretation of the essential meaning of a document via tag clouds as semantic representation.

3. HEURISTIC MODEL

Principally, each word is assigned a weight which corresponds to the expected importance of that word. Keywords correspond the words with the highest weight.

In this section, we introduce our weighting model, which builds the heart of our keyword extraction approach. Afterwards, we briefly present an improved version of the algorithm presented by Balinsky et al. [5], which is basically an application of Helmholtz' Gestalt theory to the domain of textual documents. If a keyword occurs frequently in small space, a burst occurs, probably indicating a topic change. We discuss the issue of burst detection in the following, before presenting a normalization technique that is required to compare the results computed for different text lengths. Finally, we sketch the complete algorithm.

3.1 Weighting Model

Information retrieval processes usually take a document collection, preprocess the contained documents and perform the retrieval process on a single document with respect to the whole collection. We aim to extract meaningful information from a single document and therefore consider a single document as a collection of topics. We treat a single document as if it was a document collection. Therefore we divide it into windows with each window behaving as a document. In Section 4 we explain the document segmentation process in detail.

Generally, a *document* is a structured segment of text that appears in different shapes such as traditional book, paper, article, more recently E-Mail, Web page or source code. A *term* can be a single word, a word pair, a phrase within a document - we define single words as terms. The *term weight* measures the importance of a term with respect to an assigned text unit. The relevance of a term coincides with the relevance of the term for the summary. A weighting algorithm assigns each extracted keyword a term weight. Finding an adequate weighting algorithm is the main challenge in keyword extraction.

We utilize a term weighting model that applies weights to terms diverging from a predefined random process. The performance of a heuristic retrieval algorithm is directly related to how well it satisfies desirable constraints [8]. Here, a document is considered a bag of words, in other words it represents an unordered collection of terms. In Section 4.1, we will discuss the influence of structure in a document on the quality of the keyword extraction process. Similar to the commonly used weighting function TF-IDF, the effect of *term frequency* or *exhaustivity* should be taken into account. Exhaustivity encompasses the various topics covered by the extracted keywords. In addition, popular terms that distribute equally over the whole document should be penalized [14]. Therefore we define the *specificity* as the number of windows that contain the term.

Our weighting model combines the Helmholtz approach with the algorithm proposed by Amati and Rijsbergen [3] in order to extract keywords from a single document and to meet the constraints outlined before.

We consider a document D that is divided into N *windows* $w_1, ..w_N$. Our model incorporates the exhaustivity of a term t and also the distribution of t in the whole document, based on the count of the number of windows that contain the term t, according to Amati and Rijsbergen [3]. The *weight* of each term t is a decreasing function of the two probabilities $Prob_1$ and $Prob_2$ which will be discussed further in the following:

$$weight = (1 - Prob_2) \cdot (-log Prob_1) \qquad (1)$$

We identify the terms with $weight > 0$ as keywords in the corresponding window. The terms with the highest probability $Prob_1$ contain less information compared to the terms with low $Prob_1$. This conforms to the definition of entropy in information theory. Where $Prob_1$ refers to the whole document, $Prob_2$ is only related to the elite set of windows. The *elite set* of a term t consists of the windows w_t that contain the term t [3]. If the probability $Prob_2$ of a term in a window is low with respect to its elite set, the weight formula accounts a high amount of information to that term.

The first probability $Prob_1$ is derived from a random process that is based on term independence and the frequency of a term w in the document. Whereas $Prob_2$ appends the information gain obtained by the conditional probability of encountering a further appearance of t in the elite set. The nonparametric combination of $Prob_1$ and $Prob_2$ meets the constraints of a well-performing retrieval function as stated above.

We calculate $Prob_1$ using the Helmholtz approach presented in the next section. $Prob_2$ can be tuned with respect to the burstiness, which we will discuss in Section 3.3. As mentioned before, the term occurrence and therewith the term frequency of a term t in a window w depends directly on the length of the window w. We propose a normalization of the term occurrences in Section 3.4.

3.2 Helmholtz Approach

A recently published approach proposes to model the unusual activity based on the Gestalt theory in Computer Vision: Meaningful events appear as large deviations from randomness. Balinsky et al. have applied the Helmholtz Principle in the context of automatic keyword extraction [5]. The Helmholtz principle is defined as the statement that meaningful words appear as large deviations from randomness. In compliance with this model we presume the terms are equally distributed by pure chance. Furthermore we assume the term appearances in a document are independent, respectively there should be no common cause that one appearance benefits a second appearance of a term.

Initially we define the windows $w_1, ..w_N$ to have the same length and the term t to occur in some of them. K is defined as the sum of occurrences of t in the document D. $NFA_t(m, K, N)$ (2) is defined as *number of false alarms* of an m-tuple of a term t [5]:

$$NFA = \binom{K}{m} \cdot \frac{1}{N^{m-1}} \qquad (2)$$

Here, the number of false alarms measures the meaningfulness of a keyword. If t appears m times in some document and $NFA < 1$ then this is an unexpected event. If $NFA_t < \epsilon < 1$ it is called ϵ-*meaningful*. The Helmholtz approach exhibits good performance on various types of documents. It is a parameter-free method that delivers fast and flexible meaningful keywords without the use of training documents. It may be noticed that this algorithm does not consider the window frequency which takes into account the number of windows containing the term t.

The automatic removal of stop words is a valuable criteria for using the Helmholtz algorithm. Not only is manual stop word removal language specific and time consuming, it also excludes words that might be of higher informational value in its context. The name of british rock band "The Who" would never be identified as a keyword in a text where stop words have been removed before keyword extraction.

The main advantage of the Helmholtz approach lies in its speed and simplicity. Balinsky et al. [5] have applied the algorithm to containers of documents of different lengths as it is common in information retrieval. If the input stream does not consist of documents of equal size, the algorithm identifies only a few meaningful words in small documents and stop words are identified in documents that exceed the average size of a single file in the collection by far. Applied to a single document, that means a that constant window length is the major prerequisite for reasonable results. The suggested techniques for the cases "moving window technique" and "subdivision into equal size blocks" did not show significant improvement to the results. We propose a normalization technique in Section 3.4 to overcome that issue.

3.3 Burstiness in the Elite Set

A phenomenon that recently raised attention in the field of text data mining is the appearance of *bursts*. A sudden unexpected rise of term frequency in a short period of time is defined as burst [11]. Considering a document as a stream of content, the appearance of bursts can indicate significant topic changes. We incorporate the aspect of burstiness in our model with the $Prob_2$ as stated in Equation (1).

If the search for a term t in a text has been unsuccessful for a long time and suddenly t appears once, the expectation to find more terms rises. This effect of rising expectation is called the *aftereffect* of future sampling and similar to the notion of *burstiness* [9]. The probability of an observed term t contributing to the discrimination of a window is assumed to correlate to the probability of another appearance of t [3]. This probability is obtained by the conditional probability $p(m + 1|m, w)$ of the term t by the aftereffect model. This probability is only related to the *elite* set of windows w_t that contain the term t.

With regard to the assumption of term independence we assume that $m+1$ appearances of term t have been observed and apply Laplace's law of succession. The probability of $m + 2$ appearances is $\frac{m+1}{m+2}$. Laplace's law of succession models the aftereffect of the appearance a term in the elite set of windows. Assuming that $m - 1$ occurrences have been observed, the additional appearance is approximated with the following equation [3].

$$Prob_2(m) = \frac{m}{m + 1} \qquad (3)$$

The performance of several aftereffect models has been studied before [3]. The aftereffect model of Laplace's law of succession performs similar to other normalization processes e.g. the normalization with Bernoulli processes. Due to its simplicity and linear complexity it is considered to be the method of choice. Equation (1) and (3) lead to the following equation:

$$weight(m, w) = \frac{1}{m + 1}(1 - Prob_2) \cdot (-logProb_1) \qquad (4)$$

The weight function is a function of the term occurrence m, the number of windows N, and the number of windows in the elite set n.

$$weight = weight(m, N, n) \qquad (5)$$

The number of occurrences of t depends on the window length. In Section 4 we clarify reasons for varying window lengths. The term frequency m has to be normalized in order to compare windows of different lengths.

3.4 Normalization

Normalization of document length is a recurring topic in information retrieval (Section 3.1). Applied to our document model we re-scale the windows to the average window size in order to normalize the occurrences of t:

$$m_i' = m_i \cdot \frac{l(w_i)}{\text{avg\_l}} \qquad (6)$$

where avg_l depicts the average length of a window. Here, we count the number of terms in a window to determine the actual size, we present further techniques in Section 4.1. Let the average window size avg_l be 2000 and the terms in window $w_1 = 8000$ and in window $w_2 = 200$. If a term t appears 8 times we re-scale m_1 in w_1 to $m'_1 = 2$ and in w_2 to $m'_2 = 80$. The proposed normalization reduces the extracted keywords of above-average length windows satisfactorily but performs poorly on short windows.

The presented normalization formula postulates, that an increase of window length corresponds to a linear increase of the number of occurrences of a term t in the window. Recalling the model of term independence, that is, linear correlation is highly improbable. It is rather probable that new terms appear in the window. We assume the term frequency density to be a decreasing function on the window length [3]:

$$m'_i = m_i \cdot log \left(1 + \frac{\text{avg\_l}}{l(w_i)} \right) \tag{7}$$

We apply the normalization of term occurrences to the proposed weighting algorithm and replace m in Equation (1) with m'. The weight for each keyword can be calculated using the following algorithm. Each term t assigned a *weight* > 0 is considered a keyword.

3.5 Algorithm

Algorithm 1 shows a simplified version of the complete algorithm. Even if the Helmholtz approach does not require extensive pre-processing, we have introduced some pre-processing steps. They ensure that neither the casing of the words nor their position within a sentence do affect the result. Special characters and numbers are also excluded as we suppose them to be not very useful for keyword extraction in general. We will discuss the positioning of windows in the following section.

Algorithm 1 The whole keyword extraction algorithm requires only few pre-processing steps and ensures meaningful results by normalizing the occurrences.

{Pre-processing}
remove punctuation symbols
remove non-alphabetic characters
down-case terms
{Keyword extraction}
count occurrences of all terms t
partition documents into windows
for all windows w in D **do**
   count each term t in w
end for
{Normalization}
normalize window length
for all normalized windows w_{norm} in D **do**
   calculate *weight* of each term t in w_{norm}
end for

---

4. SINGLE DOCUMENT PROCESSING

In contrast to most information retrieval processes, we focus on keyword extraction for single documents. In this section we discuss different approaches to document segmentation, which is needed to compare the current segment with the whole document. Finally, we present our segmentation technique which is based on a self-adjusting windowing technique.

4.1 Document Segmentation

The (few) preprocessing steps stated in Algorithm 1 must be performed before document segmentation. As stated before, we divide the single document into windows. The windows $w_1, ..w_N$ in the document D consist of single terms, are non-overlapping and contain all terms in D. We propose three different criteria to determine the optimal window composition:

- Number of terms that compose a window
- Logical document structure
- Number of extracted keywords (Section 4.2)

The number of extracted keywords strongly depends on the size of the windows that compose the underlying document, which we will empirically show in Section 5.2. With a varying window size, the number of terms in a window changes. Consequently the risk of content-overlapping windows or content-splitting windows rises with increasing window size. We propose the use of document structure information to reduce that risk. Furthermore, we presume that there is no external corpus of documents available that could provide useful domain-specific information for the summarization process.

Document segmentation aims to identify topic boundaries within documents in order to subdivide into subtopic segments [12]. Assuming that each window w contains a subtopic of the document, we claim that our algorithm identifies more distinct keywords. Accordingly, a well performed segmentation process leads to better keywords in the outcome.

In principle, every syntactically correct text contains sentences. We take advantage of the sentence punctuation and mark sentences as single units of the document. Instead of varying the window size by single terms we increase or decrease the windows by single sentences to extract the claimed number of keywords (see Section 4.2). A similar segmentation approach creates windows that contain paragraph-sized units. Obviously these approaches have to be performed before preprocessing. So far the results with sentence and paragraph segmentation approaches showed a fair amount of success but were not the main scope of this paper.

Despite their usefulness, *linguistic features* are language-specific and domain-dependent. A scientific article contains an abstract at the beginning and a conclusion at the end. On the other hand a news story or Web blog might not conform with that structure.

A more reliable domain- and language-independent source of additional information is the document itself. Documents contain many types of structured data. The structures of a document can be used to obtain a maximum of information for the keyword extraction process. We illustrate the usefulness of document structure with the following example.

The popularity of XML files rose significantly in the last years as they also contribute to the ODF document format.

XML documents can contain meta-data, style-sheets and they also include content relevant information. XML content data is stored in leaves in the XML tree. Rönnau and Borghoff [21] have shown that the lowest two levels of the XML tree contain over 80% of all nodes. Since text nodes are often just leaf nodes of the parent nodes we propose a windowing technique that takes into account the tree structure of the XML document.

4.2 Self-Regulation

The number of keywords correlates to the window size. We claim that the window size is crucial for the success of the keyword extraction algorithm. A too small window size may split topics and generate not enough keywords whereas a too big window size may not satisfy the intended focus and lead to content-overlapping. In order to obtain a fixed number of keywords k in the window w we introduce a self-regulating algorithm for an adaptable window size.

Algorithm 2 Bisection method for window adaption

1: c = claimed number of keywords
2: left = 0
3: right = maximum window size
4: **while** left ≤ right **or** not found **do**
5:    size = (left + right) / 2
6:    k = extracted keywords with window size *size*
7:    **if** k > c **then**
8:       right = size - 1
9:    **else if** k < c **then**
10:      left = size + 1
11:    **else**
12:      found = true
13:    **end if**
14: **end while**

Algorithm 2 repeatedly bisects the interval of number of items that compose a window. The items can be single terms as well as structural units. In case of single terms we set the maximum window size for the initialization at the algorithm to half the number of terms in the document: $right = \frac{|D|}{2}$.

5. EXPERIMENTAL RESULTS

In order to evaluate our results we have to consider properties that make the use of keywords as a summary beneficial. First we measure the compression ratio, which depicts the savings of space that we obtain with the extracted keywords in contrast to the original text [25].

$$compression\ ratio = \frac{length\ of\ Summary}{length\ of\ Full\ text} \quad (8)$$

The second fundamental property measures the quality of information that is retained. Here, relevant examples show the meaningfulness instead of applying the properties *precision*, *recall* and *F-measure*. This is because the evaluation of the extracted keywords is a subjective task and can not be performed on single keywords but rather on the extracted collection as a whole. Nevertheless, we plan an extensive user study to prove the meaningfulness of our approach.

In this section, we evaluate the meaningfulness of the extracted keywords. Afterwards, we analyze the compression ratio in dependence of the window size. Finally, we evaluate the runtime efficiency of our implementation.

5.1 Meaningfulness

To depict the meaningfulness of the extracted keywords, we choose documents of different domains and different sizes with an easily deducible content. In general, extensive user studies have to be performed in order to prove the meaningfulness of a summarization algorithm. To date, we have not conducted a rigorous user study, but plan to do so in near future.

We show the meaningfulness using two examples. First, we have performed a keyword extraction on all abstracts of all ACM DocEng submissions. We have compared the extracted keywords to the author-defined keywords to estimate the meaningfulness. In the second example, we have extracted keywords from President Obama's State of the Union Address in January 2011. Here, our goal was to bring out the benefits of inner-document keyword extraction. Both examples are briefly explained in the following.

For the first example, we have accumulated all published ACM DocEng abstracts into one document. The submissions comprise 387 abstracts between the years 2001 to 2010. Each single abstract corresponds to a window, so that each window contains a subtopic of the document. Apparently the window sizes differ, which is handled by our normalization approach presented in Section 3.4. The next step, we apply Algorithm 1 on the document and extract keywords.

Table 1 shows the computed keywords in comparison to the author-assigned keywords for six representative abstracts. In our opinion, the extracted keywords match the corresponding paper well. However, some results are interesting. Stop word filters would have removed the extracted keyword "other" in row six, though in this context it refers to document relations which are an essential idea of the referred paper. There is no doubt that author defined keywords may outperform automatically extracted keywords but the results in row one and three show that the results of our algorithm come close. In fact, the extracted keywords in row two reveal that the paper focuses on "Chinese" language and the submission in row five uses "technical" documentation. These examples show the importance of a corresponding visualization technique for keywords, which we will discuss in Section 6.1.

The State of the Union Address of US President Barack Obama of January 25th 2011 is a single document that covers several topics that were current at that time. We have split up the text into 26 consecutive windows of equal size.

Table 2 shows the top keywords of some sample windows. They clearly illustrate the variety of topics within this single speech. Additionally, this example shows the importance of keyword extraction for small portions of a text to help the reader to depict the actual semantic context. We will propose a corresponding application in Section 6.1.

Title (Year)	Author-assigned keywords	Extracted keywords
Vector Graphics: From Postscript and Flash to SVG. (2001)	svg, flash, swf, pdf, postscript	svg, vector, graphics
Fast Structural Query with Application to Chinese Treebank Sentence Retrieval. (2004)	treebank, structural query, xml	pcrf, query, chinese, structural, flexible, corpus, search
Towards XML Version Control of Office Documents (2005)	version control, office applications, xml diffing	office, openoffice, diff, version, control, state-of-the-art, binary, xml, versioning, documents
A Document Engineering Environment for Clinical Guidelines. (2007)	clinical guidelines, xml, deontic operators, gem	computerization, mark-up, recommendations, guidelines, clinical, operators, medical
Logic-based Verification of Technical Documentation. (2009)	model checking, document verification	checker, specification, technical, documentation
Semantics-based change impact analysis for heterogeneous collections of documents. (2010)	document collections, document management, change impact analysis, semantics, graph rewriting	changing, documents, other, different, collections

Table 1: In this example, we have compared the author-assigned keywords with the automatically extracted keywords using all submissions of ACM DocEng.

Window	Top Keywords
5	how, innovation, future, what, change
12	rebuilding, infrastructure, high-speed, internet
16	freeze, spending, decade, chamber
21	afghan, qaeda, troops, taliban, thanks

Table 2: Extracted keywords for different parts of President Obama's State of the Union Address show significant content changes within the speech.

5.2 Compression Ratio

In order to visualize the compression ratio, we have accumulated the number of extracted keywords of all windows of a single document. We have performed the keyword extraction on more than 200 State of the Union Addresses of the Presidents of the USA[1] individually. The size of the documents varies between 8 KB and approximately 200 KB with more than 27 000 words, which is fairly large. Furthermore the use of language and style of writing varies for each document.

We have performed the weighting on each single term of the documents with a varying window size. The window sizes range from 0.25% up to 50% of the document length. The windows are none-overlapping and of equal size. Here, we do not consider document structure. Figure 1 shows the results. For a window size smaller than 1% of the document length, the increase of extracted keywords is significant, especially for larger documents. Obviously, a window size below 1% may not lead to meaningful keywords.

A window size of about 1% of the document length offers a good balance between meaningful results and a high number of keywords. A window size of more than 30% of the document size leads to topic-overlapping windows. Topic-overlapping windows reduce the weight of terms, so that the number of extracted keywords stagnates with a window size of more than 10% and even decreases abruptly in larger

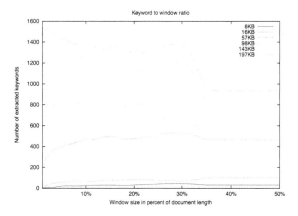

Figure 1: The compression ratio directly relates to the window size. The results of a small document and a large document are fairly comparable. A window size of over 30% leads to non-meaningful results.

documents with a window size of more than 30% of the document length.

To summarize, the window size is crucial for the extraction of the right number of meaningful keywords and should stay between 1% and 30% of the document length.

5.3 Performance

In this section, we aim to measure the speed of the algorithm and compare it to the well known TF-IDF weighting algorithm. We have implemented both algorithms in Java, using the current version Java SE 1.6.

Both keyword extraction algorithms were performed on text files of different length. We have measured the complete process of parsing the text, processing the words, segmentation of the document, and performing the weight calculation for each term. It is important to measure the whole process because document parsing and initialization of the

---

[1] http://stateoftheunion.onetwothree.net/

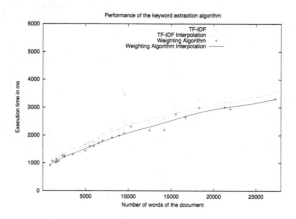

Figure 2: Performance of TF-IDF and our proposed algorithm applied to documents of different lengths

algorithms take time, depending on the document length. Pre-processing steps are sometimes excluded in literature. As we aim to give a hint on real-world scenarios, the pre-processing is part of our runtime evaluation.

The experiments have been set up on a commercial-of-the-shelf computer with an Intel Core 2 Duo processor with 2.66 GHz and 4 GB RAM, running Mac OS X 10.6.7.

To decrease the influence of the operating system and to minimize the effects of the JRE loading procedure, we have performed each test run four times. The total execution time is displayed in Figure 2. The runtime scales almost linearly for TF-IDF and our approach as well. A linear growth of the execution time is a crucial feature that enables applications to ensure responsiveness. Our algorithm performs slightly faster than TF-IDF. Here, we recall that in contrast to TF-IDF, our algorithm does not rely on stop word lists or other training data, whose time for creation is not listed here.

In conclusion, our approach appears to be fast for real-world applications, especially when run in a thread-based environment. Here, the keywords are extracted in parallel to the actual editing or viewing application.

6. APPLICATIONS

In this section, we propose three use-cases for our approach. First, we sketch an application that instantly presents a tag cloud consisting of the keywords which are actually displayed within a text. Afterwards, we discuss the importance of keywords in order to support less-skilled readers in catching the main contents of a document. Finally, we propose the use of our approach for e-mail filtering.

6.1 Cloud-based Perception Technique

Highlighting or underlining of portions of text facilitates skimming and enables the reader to quickly determine the content of a document [24]. Whereas highlighting is limited to a portion of text that is visible to the reader, tag clouds may comprise large portions of a document. Furthermore, they allow for fast adjustment with respect to the selected content and enable a weighted display of keywords to provide differential emphasis of each towards the central topic. The compact appearance of a tag cloud enables the reader to gain the sense of a topic at a single glance. Though they are inferior to standard alphabetical listing, user stu-

Look to Iraq, where nearly 100,000 of our brave men and women have left with their heads held high. American combat patrols have ended, violence is down, and a new government has been formed. This year, our civilians will forge a lasting partnership with the Iraqi people, while we finish the job of bringing our troops out of Iraq. America's commitment has been kept. The Iraq war is coming to an end.

Of course, as we speak, al Qaeda and their affiliates continue to plan attacks against us. Thanks to our intelligence and law enforcement professionals, we're disrupting plots and securing our cities and skies. And as extremists try to inspire acts of violence within our borders, we are responding with the strength of our communities, with respect for the rule of law, and with the conviction that American Muslims are a part of our American family.

We've also taken the fight to al Qaeda and their allies abroad. In Afghanistan, our troops have taken Taliban strongholds and trained Afghan security forces. Our purpose is clear: By preventing the Taliban from reestablishing a stranglehold over the Afghan people, we will deny al Qaeda the safe haven that served as a launching pad for 9/11.

Thanks to our heroic troops and civilians, fewer Afghans are under the control of the insurgency. There will be tough fighting ahead, and the Afghan government will need to deliver better

Figure 3: The text cloud displays the top keywords of the visible portion.

dies have showed that they were perceived useful for showing dynamic changing information. Not to mention their popularity, the acceptance rate for clients and customers is considerably high [13].

Figure 6.1 displays a tag cloud that comprises the displayed text. In order to grasp the information of a portion of text, we propose to display a tag cloud next to the text of interest as it is shown in Figure 6.1.

Referring the definition of summary tag clouds and differential tag clouds by Xexeo et al. [29], our tag cloud matches the definition of a differential tag cloud because it highlights the particular features of the selected portion of text with regard to the whole document. The performance of our algorithm enables rapid keyword extraction and therewith quick recalculation as the assigned text might change. Quick response time guarantees user interaction as tag clouds evoke human activity [13]. The self-regulation feature affects the denomination of the underlying windows and might be a tool for the user to adjust the granularity of the algorithm. Independent of a collection of training documents, this application would be executable on a standalone mobile device without Web access.

6.2 Website Comprehension

The basic skill of reading is a crucial condition for the participation in modern life. Although developed countries have a high literacy rate, there remains a large amount of people with rudimentary literacy skills. Rudimentary literacy skills include the ability to read basic texts and deduce the content of a document[2]. Recent approaches have been addressing this issue by providing text simplification programs in order to facilitate poor literacy readers to access documents [2]. We claim that our approach could be a applied to Web sites in order to outline the main aspect of the page to the visitor. The reader may select portions of text and perform analysis on the selected text in order to differentiate the content. Our approach is flexible and language independent what makes it accessible for people in less developed and non-English speaking countries.

---

[2]OECD: *Literacy in the Information Age: Final Report of the International Adult Literacy Survey*. Paris, 2000

6.3 E-Mail Labeling

We all know the issue of receiving large amounts of e-mails daily, insofar as they are not sorted into our spam folder we have to label and categorize them. In that case our approach serves two purposes. To automatically categorize incoming e-mails, the text of arriving e-mails is compared to the extracted keywords of the whole e-mail collection, whereas the windows consist of categorized e-mails. Furthermore, labels of categories in our e-mail archive comprise of the highest scored keywords in the category. Hence the labels of the category might vary over time as the topic of the category diverges. Considering the fact that communication is an ongoing development, we suggest an instant and flexible categorization.

7. DISCUSSION AND FUTURE WORK

In this section we interpret our results, reflect the main objectives of this paper and suggest future development.

In Section 4.1, we propose several techniques to improve our results by document structure. So far our experiments did not show significant improvement but further investigations might reveal potential for enhancements. We aim to perform further evaluations on scanned documents that contain very little structural information and take headings and subheadings into account that mark topic structure.

Heuristics seem to be effective for keyword extraction in a single domain but do not extend well to new domains [24]. For instance, the characteristics for keywords in scientific articles differ from the characteristics of keywords in news articles. Our algorithm is performed on single documents and therefore applied to single domains only. The benefit of a training a collection would be questionable since most training collections do not fit exactly the specific domain of the underlying text.

We have proven that the extracted keywords enable the reader to quickly ascertain the content of a portion of text. We do not aim to to provide a precise summary that can be generated with automatic summarization approaches based on semantics and NLP (see Section 2).

It is remarkable that no significant pre-processing has to be performed on text that is contained in the document because a large number of summarization approaches perform exhaustive pre-processing on their documents. Pre-processing as stop word extraction is generally language specific and time consuming.

The presented algorithm extracts keywords heuristically from a relatively small number of terms. Since heuristics only perform well on big numbers, the document size is crucial. The examples shown in Section 5 range from relatively small texts to book-size format. We argue that the need for a summary requires a certain amount of text, hence it does not make sense to apply our algorithm to very short documents.

The extraction performs well on large-sized documents, though the number of extracted keywords depends on the number of windows. This implies a reasonable window size. Section 5.2 suggests a correlation between window size and document length. The window size should not be chosen to be more than 10% of the document length because the degree of differentiation decreases considerably. We suggest an adjustable window size that can be actively varied in order to individually specify the granularity of the algorithm.

Implementing a varying window size is not an issue as the presented method is fast and applicable for environments with constant input and constant changing parameters.

As depicted in Section 6, we propose to implement applications that take advantage of the properties of our approach. We would like to perform extensive user studies with users of various cultural backgrounds in different languages to assess the benefits of such a tool.

8. CONCLUSIONS

In this paper, we have presented an efficient and flexible keyword extraction approach for arbitrary text documents. Its key features are independence from language, structure, and content. Unlike most other approaches, we do not rely on a training phase or extensive pre-processing steps.

Our algorithm satisfies the constraints of a well performing heuristic retrieval algorithm as it combines the Helmholtz approach with information related to the burstiness of terms. Furthermore we have extended the Helmholtz approach and have applied it to single documents in order to extract meaningful keywords from portions of texts. We have introduced self-regulating windows to achieve more meaningful results.

We have exemplified the meaningfulness of the results and compared them to human-assigned keywords. Additionally, we have evaluated the compression ratio as well as the efficiency of our approach. The runtime scales linearly to document size and performs better than the well-known TF-IDF approach.

Finally, we have proposed different applications of our approach. They incorporate the use of tag clouds to offer an enhanced visualization experience. This way, a fast perception of the content of a larger portion of text becomes possible.

Currently, we plan to conduct further investigations on various sorts of text that help us to develop a tag-cloud application on a mobile platform, based on the applications sketched in this paper. We plan to conduct an extensive user study regarding this application.

References

[1] J. Allan, J. Carbonell, G. Doddington, J. Yamron, Y. Yang, B. Archibald, and M. Scudder. Topic Detection and Tracking Pilot Study Final Report. In *Proc. of the DARPA Broadcast News Transcription and Understanding Workshop*, 1998.

[2] S. M. Aluísio, L. Specia, T. A. Pardo, E. G. Maziero, and R. P. Fortes. Towards Brazilian Portuguese automatic text simplification systems. In *Proc. of the 8th ACM sym. on Document engineering - DocEng '08*, page 240, New York, NY, USA, Sept. 2008. ACM.

[3] G. Amati and C. J. Van Rijsbergen. Probabilistic models of information retrieval based on measuring the divergence from randomness. *ACM Trans. Inf. Syst.*, 20:357–389, October 2002.

[4] E. D. Avanzo, B. Magnini, and A. Vallin. Keyphrase Extraction for Summarization Purposes: The LAKE System at DUC-2004. In *Document Understanding Conferences*, Boston, USA, 2004.

[5] A. A. Balinsky, H. Y. Balinsky, and S. J. Simske. On helmholtz's principle for documents processing. In

Proc. of the 10th ACM symp. on Document engineering, DocEng '10, page 283, New York, NY, USA, 2010. ACM.

[6] R. Barzilay and M. Elhadad. Using Lexical Chains for Text Summarization, 1997.

[7] J. M. Conroy and D. P. O'leary. Text summarization via hidden Markov models. In *Proc. of the 24th ann. int. ACM SIGIR conf. on Research and development in information retrieval - SIGIR '01*, pages 406–407, New York, NY, USA, Sept. 2001. ACM.

[8] H. Fang, T. Tao, and C. Zhai. A formal study of information retrieval heuristics. In *Proc. of the 27th ann. int. conf. on Research and development in information retrieval - SIGIR '04*, page 49, New York, NY, USA, July 2004. ACM.

[9] W. Feller. *An Introduction to Probability Theory and Its Applications, Vol. 2*. Wiley, 1967.

[10] E. Frank, G. W. Paynter, I. H. Witten, C. Gutwin, and C. G. Nevill-Manning. Domain-Specific Keyphrase Extraction. In *Proceedings of the Sixteenth International Joint Conference on Artificial Intelligence*, IJCAI '99, pages 668–673. Morgan Kaufmann Publishers Inc., July 1999.

[11] Q. He, K. Chang, E.-P. Lim, and A. Banerjee. Keep It Simple with Time: A Reexamination of Probabilistic Topic Detection Models. *Pattern Analysis and Machine Intelligence, IEEE Transactions on*, 32(10):1795–1808, 2010.

[12] M. A. Hearst. TextTiling: Segmenting Text into Multi-paragraph Subtopic Passages. *Computational Linguistics*, 1997.

[13] M. A. Hearst and D. Rosner. Tag Clouds: Data Analysis Tool or Social Signaller? In *Proc. of the 41st Hawaii Int. Conf. on System Sciences*, pages 1–10, 2008.

[14] S. Jones. A statistical interpretation of term specificity and its application in retrieval. *Journal of Documentation*, 28(1):11–21, 1972.

[15] J. M. Kleinberg. Authoritative sources in a hyperlinked environment. *Journal of the ACM*, 46(5):604–632, Sept. 1999.

[16] G. Koutrika, Z. M. Zadeh, and H. Garcia-Molina. CourseCloud. In *Proc. of the 12th Int. Conf. on Extending Database Technology Advances in Database Technology - EDBT '09*, EDBT '09, page 1132, New York, NY, USA, 2009. ACM.

[17] N. Kumar and K. Srinathan. Automatic keyphrase extraction from scientific documents using N-gram filtration technique. In *Proc. of the 8th ACM symp. on Document engineering - DocEng '08*, page 199, New York, NY, USA, 2008. ACM.

[18] B. Y.-L. Kuo, T. Hentrich, B. M. . Good, and M. D. Wilkinson. Tag clouds for summarizing web search results. In *Proc. of the 16th int. conf. on World Wide Web*, WWW '07, pages 1203–1204, New York, NY, USA, 2007. ACM.

[19] M. Litvak and M. Last. Graph-based keyword extraction for single-document summarization. In *MMIES '08 Proc. of the Workshop on Multi-source Multilingual Information Extraction and Summarization*, pages 17–24, Aug. 2008.

[20] T. Pagerank, C. Ranking, and B. Order. The PageRank Citation Ranking: Bringing Order to the Web. *World Wide Web Internet And Web Information Systems*, pages 1–17, 1998.

[21] S. Rönnau and U. Borghoff. Xcc: change control of xml documents. *Computer Science - Research and Development*, pages 1–17, 2010. 10.1007/s00450-010-0140-2.

[22] A. Singhal. Modern Information Retrieval : A Brief Overview. *IEEE Data Engineering Bulletin 24*, pages 35–43, 2001.

[23] Z. Teng, Y. Liu, F. Ren, S. Tsuchiya, and F. Ren. Single document summarization based on local topic identification and word frequency. In *Artificial Intelligence, 2008. MICAI '08. Seventh Mexican International Conference on*, pages 37 –41, oct. 2008.

[24] P. D. Turney. Learning algorithms for keyphrase extraction. *Inf. Retr.*, 2:303–336, May 2000.

[25] V. R. Uzêda, T. A. S. Pardo, and M. D. G. V. Nunes. Evaluation of Automatic Text Summarization Methods Based on Rhetorical Structure Theory. In *8th Int. Conf. on Intelligent Systems Design and Applications*, pages 389–394. Ieee, Nov. 2008.

[26] X. Wan and J. Xiao. Exploiting neighborhood knowledge for single document summarization and keyphrase extraction. *ACM Transactions on Information Systems*, 28(2):1–34, May 2010.

[27] D. Watters. Meaninful Clouds: Towards a Novel Interface for Document Visualization, 2009.

[28] I. H. Witten, G. W. Paynter, E. Frank, C. Gutwin, and C. G. Nevill-Manning. KEA: practical automatic keyphrase extraction. In *Proc. of the 4th ACM conf. on Digital libraries - DL '99*, pages 254–255, New York, NY, USA, Aug. 1999. ACM.

[29] G. Xexeo, F. Morgado, and P. Fiuza. Differential Tag Clouds: Highlighting Particular Features in Documents. In *2009 IEEE/WIC/ACM Int. Joint Conf. on Web Intelligence and Intelligent Agent Technology*, pages 129–132. IEEE, Sept. 2009.

[30] C. Yang. Discovering Event Evolution Graphs From News Corpora. *IEEE Transactions on Systems, Man, and Cybernetics - Part A: Systems and Humans*, 39(4):850–863, July 2009.

Building a Topic Hierarchy Using the Bag-of-Related-Words Representation

Rafael Geraldeli Rossi
Institute of Mathematics and Computer Science -
University of São Paulo
P.O. Box 668
São Carlos, SP - Brazil
ragero@icmc.usp.br

Solange Oliveira Rezende
Institute of Mathematics and Computer Science -
University of São Paulo
P.O. Box 668
São Carlos, SP - Brazil
solange@icmc.usp.br

ABSTRACT

A simple and intuitive way to organize a huge document collection is by a topic hierarchy. Generally two steps are carried out to build a topic hierarchy automatically: 1) hierarchical document clustering and 2) cluster labeling. For both steps, a good textual document representation is essential. The bag-of-words is the common way to represent text collections. In this representation, each document is represented by a vector where each word in the document collection represents a dimension (feature). This approach has well known problems as the high dimensionality and sparsity of data. Besides, most of the concepts are composed by more than one word, as "document engineering" or "text mining". In this paper an approach called bag-of-related-words is proposed to generate features compounded by a set of related words with a dimensionality smaller than the bag-of-words. The features are extracted from each textual document of a collection using association rules. Different ways to map the document into transactions in order to allow the extraction of association rules and interest measures to prune the number of features are analyzed. To evaluate how much the proposed approach can aid the topic hierarchy building, we carried out an objective evaluation for the clustering structure, and a subjective evaluation for topic hierarchies. All the results were compared with the bag-of-words. The obtained results demonstrated that the proposed representation is better than the bag-of-words for the topic hierarchy building.

Categories and Subject Descriptors

H.3.3 [**Information Search and Retrieval**]: Clustering

General Terms

Algorithm, Experimentation

Keywords

Document Representation, Text Mining, Topic Hierarchy

1. INTRODUCTION

Part of the data in the digital universe is in the textual format, like e-mails, reports, papers, and web-pages contents. To manage, search and browse in large text collections, topic hierarchies are very useful. Topic hierarchies are structural organizations where the documents are separated in a tree in which the nodes close to the root are general topics and the nodes far from the root are more specific topics. This type of structure allows an exploratory search. The user can browse in the textual document collection interactively through the topics of the nodes. Examples of topic hierarchies are the online directories of Yahoo[1] and Open Directory Project[2].

The manual building of a topic hierarchy for large text collections requires a huge human effort. Therefore methods to perform this task automatically have received great attention in the literature [13, 15, 9]. Generally hierarchical clustering methods are used to create the hierarchical structure organization automatically, and then a process of cluster labeling is performed. For both steps, the textual document collection has to be represented in an appropriate way.

The Vector Space Model (VSM) [14] is generally used to represent textual documents. In this model, each document is represented by a vector, and each position of this vector is a dimension (feature) of the document collection. A common approach based on the VSM is the bag-of-words. In this representation, each word in a textual document collection becomes a dimension in the vector space. However, this approach presents some problems such as the high dimensionality, the high sparsity data, and the word order is not kept.

Another problem presented by the bag-of-words is that concepts are usually represented by more than one word, such as "document engineering", "topic hierarchy", and "text mining". For instance, if we look at the features "machine learning" or "data mining", there is no doubt about the topic of the document. On the other hand, features compounded by single words as "learning" can represent documents about machine learning or about teaching, and "mining" can represent documents about data mining or mineral extraction.

Trying to obtain features compounded by more than one

---

[1]http://dir.yahoo.com/
[2]http://www.dmoz.org/

word, approaches were developed based on n-grams [5, 17, 4, 12], and based on set of words [10, 22, 19, 6, 21]. These approaches usually add the features to the bag-of-words, increasing the already high dimensionality. Some approaches analyze the entire collection to generate the features. This can have a high computational cost due the dimensionality, and generate features without meaning. Furthermore, most of the approaches perform a supervised feature selection, requiring labeled collections, which is not common in real world textual document collections.

This paper proposes an approach that generates related words and use them as features. The proposed approach, named *bag-of-related-words*, generates features from each document of a collection through association rules. The association rules [1] are used to discover relations among items in a dataset. Besides obtaining the relations of the words (items), the intention of using association rules is to reduce the dimensionality of the bag-of-words. This is possible because not all the words of a document will be used as features, but only the words that occur or cooccur with a frequency above a certain threshold.

In order to extract association rules from each document, a mapping of the document into transactions is necessary. This can be done considering the sentences, paragraphs, or a sliding window as transactions. Since the words are related in specific contexts of the documents, the mapping can produce more understandable features to the user. An analysis will be carried out on which way the mapping of the document into transactions produces better results.

Interest measures can be applied to prune the association rules. In this paper the Confidence, Lift, Yules'Q, Linear Correlation Coefficient, Mutual Information, Gini Index, J-Measure, and Kappa measures [8, 7, 3, 18] were used. An evaluation about which measure produces better results for the textual document clustering was carried out.

A representation is created based on the VSM with the features generated for each textual document. An hierarchical clustering is built and a process of cluster labeling is performed. Experiments were carried out to analyze the feasibility of the bag-of-related-words to build a hierarchical cluster and its contribution to the cluster labeling. When the bag-of-related-words were used to represent the documents, the clustering quality and the topics of the hierarchy were better than the bag-of-words.

This paper is organized as follows. Section 2 presents the related works about textual document representation using features compounded by more than one word. Section 3 presents the details of the proposed approach. Section 4 shows the results of the proposed approach and the bag-of-words for the textual document clustering and topic taxonomy labeling. Finally, Section 5 presents the conclusions and future works.

2. TEXTUAL DOCUMENT REPRESENTATION

The textual document representation is fundamental to build a topic hierarchy automatically or for some other pattern extraction from textual documents. Besides the bag-of-words representation, there are several works that generate features compounded by more than one word. Usually these features can be generated using n-grams (or statistical phrases) or using set of words.

N-grams are sequences of n words that appear in the text collection. Some works using n-grams are [5, 17, 4, 12]. A general characteristic of these approaches is that the use of all the sequences of n words produces a dimension much higher than the bag-of-words. Some approaches try to reduce the number of generated word sequences. For instance, in [17] the bigrams are generated if at least one of the components is frequent in a document.

Another way to represent textual documents is through features compounded by set of words. The words that compound this type of feature do not necessarily occur together or close to each other in a textual document. This representation can be considered an extension of the n-grams. Some works using set of words as features are [10, 22, 19, 6, 21]. Most of the works generally execute an cooccurrence analysis in a bag-of-words to generate features compounded by sets of words. Some approaches try to avoid this type of analysis. For instance, in the proposed approach by Lie. et al. [22], the "loose n-grams" were defined. The loose n-grams are set of words that cooccur in a limited space, as sentences, or a sequence of words (window). The relative position of the words are not considered. Only the words that appear in a minimum number of documents are used. The words that have a good discriminative power are separated. Then, sets of words compounded by words that cooccur in the collection are obtained, and these sets of words are filtered using the χ^2 method. The results demonstrated a little improvement in relation to the bag-of-words.

Most of the related researches about features compounded by more than one word generate a number of features much higher than the bag-of-words and most of the time a process of feature selection is necessary. This is computationally expensive due the high dimensionality of the collection and the need of labeled collections, which is not common in real textual document collections.

Moreover, some approaches that analyze the entire collection may obtain features without meaning. For instance, if we consider a collection of scientific articles, the feature "*introduction_conclusion*" can be generated. Besides, collections with few documents of some vocabulary may not have their features generated. For instance, considering 5000 documents about chemistry and 20 about computer science. Supposing that some words of the documents about computer science cooccur only in 5 documents, the minimum cooccurrence threshold must be of 0.001. On the other hand, a huge number of features about chemistry could be generated.

3. BAG-OF-RELATED-WORDS

The proposed approach, called bag-of-related-words, generates features compounded by a set of related words from textual documents. The main goal of this approach is to use related words that repeat over the document in limited spaces as features.

The related words are obtained through association rules. An association rule is a rule of the type $A \Rightarrow B$, in which A and B are groups of items, called itemsets, and $A \cap B = \emptyset$ [1]. The association rules discover relations among the itemsets in a dataset, in which $A \Rightarrow B$ means that when A occurs, B also tends to occur. Two classical measures to generate association rules are Support and Confidence. Support measures the joint probability of an itemset in a database, that is, $sup(A \Rightarrow B) = q(A \cup B)/Q$, in which $q(A \cup B)$ is the number of transactions that A and B occur together, and Q is the total number of transactions. Confidence indicates the probability of A and B occur together given that A occurred, that is, $conf(A \Rightarrow B) = q(A \cup B)/q(A)$. Usually minimum values of support and confidence are defined to generate the rules.

The four main steps to generate the features from the textual documents with the proposed approach are:

1. Mapping the textual document into transactions;

2. Extracting association rules from the transactions;

3. Using the itemsets of the rules to compound the features;

4. Using the features to construct the *document-term* matrix;

The mapping of the textual document into transactions allows the extraction of association rules. In this step, Text Mining preprocessing tasks as stopwords removal, and stemming can be applied. After the preprocessing, the transactions can be obtained considering the sentences, paragraphs, or a sliding window.

When sliding windows are used to map the transactions, the first transaction contains only the first word, the second transaction contains the two first words, and so on, until the window contains the number of words equal the defined size (l). After this, the sliding window slides one word and considers the next l words of the document. At the end of the document, the last (l) transaction contains the last l words, the last $l - 1$ transaction contains the $l - 1$ words, and so on, until the window contains only one word. This is done so that all the words are contained the same number of times in a sliding window.

The second step consists of extracting the association rules from the transactions which were mapped from the text. To illustrate the two first steps, consider the content of the Table 1. There is a text about Data Mining taken from Wikipedia[3], that we called "example text". Applying the stopword removal and stemming, and considering the sentences as transactions, it was possible to map the text into 10 transactions. Considering these 10 transactions, a minimum support threshold of 30, 0% and a minimum confidence threshold of 75%, 15 association rules were extracted. The values on the right of the rules are the support and the confidence respectively. It can be noticed that the itemsets of the rules on Table 1 are really related to the example text, as "data" and "mine", "larger data" and "sampl".

The proposed approach also considers features compounded by single words. The items of the rules with empty set like $\emptyset \Rightarrow data$ will be used as features compounded by single words. In this case the feature "data" would be generated.

The different ways to map the document into transactions can modify the number of transactions, the frequency of the words and their cooccurrences. The frequency of the words is modified due the number of repetitions of the same word in the transaction. For instance, the word "data" occurs 15 times in the example text from Table 1. If we consider the paragraphs as transactions, the word "data" occurs 11 times in the first transaction, and is counted only once. Also,

the mapping of paragraphs made the word "data" occurs in all the transactions (2 transactions), that is, the word "data" has a support of 100%. If we consider the sentences as transactions, the word "data" occurs at most three times in a transaction and occurs in 7 of 10 transactions. Thus its frequency is equal to 7 and its support is equal to 70%.

The cooccurrences of the words are also modified by the type of transaction. For instance, the mapping considering the paragraphs make the words cooccur more with other words than the mapping of sentences. Different values of minimum support are required due all the modifications caused by the different types of mapping. To illustrate the practical impact of the mapping, we present in Table 2 the extracted frequent itemsets considering the example text from Table 1. We used the following types of transactions and their respective minimum supports: sentences 30.0%, paragraphs 100.0%, size 5 sliding window 10.0%, size 10 sliding window 21.0%, size 20 sliding window 47.0%, and size 30 sliding window 58.0%. These minimum support values were chosen to generate a similar number of frequent itemsets.

It can be noticed that some itemsets, that are useful to identify the example text from Table 1, appear in all the types of mapping as *"process"*, *"mine"*, *"pattern"*, *"data"*, *"mine_pattern"*, *"data_mine"*, *"data_pattern"*, and *"data_mine_pattern"* (highlighted with italic font). Some other useful itemsets do not appear in all the types of mapping as *"mine_sampl"* (only in sentences, sliding windows of size 5, 10, and 20), *"pattern_sampl"* (it does not appear only in paragraphs), *"data_mine_sampl"* (only in sentences, and sliding windows of size 10 and 20), *"data_mine_process"* (only in paragraphs and size 30 sliding window), *"mine_pattern_process"* (only in paragraphs and size 30 sliding window), and *"data_mine_pattern_process"* (only in the size 30 sliding window). Given the differences of each type of mapping, this paper will analyze which mapping produces better results for the textual document clustering.

The third step consists of using the itemsets of the association rules obtained in the previous step to compound the features of a document. For instance, the rule *"data $\Rightarrow$ mine"* will generate the feature *"data_mine"*. In order to avoid that rules with the same itemsets generate different features, the items of the rule are sorted lexicographically or according to the order they occur in the textual document.

In this step, interest measures besides support and confidence can be used to obtain different relations among the itemsets, to rank and prune the association rules. The intention to use interest measures is to obtain more understandable features and reduce even more the dimensionality. The Confidence, Lift, Yules'Q, Linear Correlation Coefficient, Mutual Information, Gini Index, Kappa, and J-Measure measures [8, 7, 3, 18] were used.

The ranking and consequently the pruning of the rules using these measures can be different due their characteristic. To illustrate this, on Table 3 are presented the rankings of features considering a text about clustering validation measures [20] with the Confidence, Mutual Information and Kappa measures. Important features to describe the textual document appear in the rankings of these 3 measures as *"data set"*, *"measur valid"*, and *"cluster valid"*. Other important features appear only in the rankings of the Confidence and Mutual Information measures as *"contig matrix"*, *"data sample"*, and *"extern valid"*. Other features appear only in the ranking of one measure, as *"cluster data"* that

[3]http://en.wikipedia.org/wiki/Data_mining (May 13, 2010).

Table 1: Process of association rule extraction from a textual document.

Example Text
Data mining is the process of extracting patterns from data. Data mining is becoming an increasingly important tool to transform this data into information. It is commonly used in a wide range of profiling practices, such as marketing, surveillance, fraud detection and scientific discovery.
Data mining can be used to uncover patterns in data but is often carried out only on samples of data. The mining process will be ineffective if the samples are not a good representation of the larger body of data. Data mining cannot discover patterns that may be present in the larger body of data if those patterns are not present in the sample being "mined". Inability to find patterns may become a cause for some disputes between customers and service providers. Therefore data mining is not foolproof but may be useful if sufficiently representative data samples are collected. The discovery of a particular pattern in a particular set of data does not necessarily mean that a pattern is found elsewhere in the larger data from which that sample was drawn. An important part of the process is the verification and validation of patterns on other samples of data.

Transactions considering the sentences of the preprocessed example text.
1. data mine process extract pattern data
2. data mine increasingli import tool transform data inform
3. commonli wide rang profil practic market surveil fraud detect scientif discoveri
4. data mine uncov pattern data carri sampl data
5. mine process ineffect sampl good represent larger bodi data
6. data mine discov pattern present larger bodi data pattern present sampl mine
7. inabl find pattern disput custom servic provid
8. data mine foolproof suffici repres data sampl collect
9. discoveri pattern set data necessarili pattern found larger data sampl drawn
10. import part process verif valid pattern sampl data

Rules obtained considering the sentences as transactions.		
data ⇒ ∅ (80.0, 80.0)	data ⇒ mine (60.0, 75.0)	mine pattern ⇒ data (30.0, 100.0)
process ⇒ data (30.0, 100.0)	data ⇒ sampl (60.0, 75.0)	larger data ⇒ sampl (30.0, 100.0)
larger ⇒ sampl (30.0, 100.0)	sampl ⇒ data (60.0, 100.0)	pattern data ⇒ sampl (40.0, 80.0)
larger ⇒ data (30.0, 100.0)	mine ⇒ data (60.0, 100.0)	mine sampl ⇒ data (40.0, 100.0)
pattern ⇒ data (50.0, 83.3)	larger sampl ⇒ data (30.0, 100.0)	pattern sampl ⇒ data (40.0, 100.0)

Table 2: Frequent itemsets obtained through different ways of mapping the text from Table 1 into transactions.

Sentence		Paragraph		Size 5 Window	
Itemsets	**Sup.**	**Itemsets**	**Sup.**	**Itemsets**	**Sup.**
larger	30,0	*data*	100,0	bodi	10,9
process	30,0	mine	100,0	discoveri	10,9
mine	60,0	pattern	100,0	import	10,9
pattern	60,0	*process*	100,0	present	10,9
sampl	60,0	*data mine*	100,0	larger	16,3
data	80,0	*data pattern*	100,0	process	16,3
data process	30,0	data process	100,0	sampl	32,6
data larger	30,0	*mine pattern*	100,0	*mine*	38,0
larger sampl	30,0	mine process	100,0	*pattern*	42,4
mine pattern	30,0	pattern process	100,0	*data*	69,6
mine sampl	40,0	*data mine pattern*	100,0	mine sampl	9,8
pattern sampl	40,0	data mine process	100,0	pattern present	9,8
data pattern	50,0	data pattern process	100,0	data larger	12,0
data mine	60,0	mine pattern process	100,0	pattern sampl	14,1
data sampl	60,0	data mine pattern process	100,0	*mine pattern*	15,2
data larger sampl	30,0	-	-	data sampl	21,7
data mine pattern	30,0	-	-	*data mine*	30,4
data mine sampl	40,0	-	-	*data pattern*	30,4
data pattern sampl	40,0	-	-	data pattern sampl	9,8
-	-	-	-	*data mine pattern*	10,9

Size 10 Window		Size 20 Window		Size 30 Window	
Itemsets	**Sup.**	**Itemsets**	**Sup.**	**Itemsets**	**Sup.**
larger	28,9	*process*	56,1	*sampl*	72,6
process	30,9	*sampl*	70,1	*process*	76,9
sampl	54,6	*mine*	75,7	*mine*	77,8
mine	63,9	*pattern*	90,7	*pattern*	94,9
pattern	67,0	*data*	100,0	*data*	100,0
data	94,8	mine sampl	47,7	mine process	59,8
larger pattern	21,6	pattern process	54,2	pattern sampl	71,8
larger sampl	21,6	data process	56,1	data sampl	72,6
pattern process	21,6	pattern sampl	69,2	*mine pattern*	75,2
data larger	28,9	*mine pattern*	69,2	pattern process	75,2
data process	30,9	data sampl	70,1	data process	76,9
mine sampl	30,9	*data mine*	75,7	*data mine*	77,8
pattern sampl	42,3	*data pattern*	90,7	*data pattern*	94,9
mine pattern	44,3	data mine sampl	47,7	mine pattern process	58,1
data sampl	52,6	mine pattern sampl	47,7	data mine process	59,8
data mine	61,9	data pattern process	54,2	data pattern sampl	71,8
data pattern	64,9	data pattern sampl	69,2	*data mine pattern*	75,2
data larger pattern	21,6	*data mine pattern*	69,2	data pattern process	75,2
data larger sampl	21,6	data mine pattern sampl	47,7	data mine pattern process	58,1
data pattern process	21,6	-	-	-	-
data mine sampl	28,9	-	-	-	-
data pattern sampl	40,2	-	-	-	-
data mine pattern	42,3	-	-	-	-

Table 3: Ranking of the features for the Confidence, Mutual Information, and Kappa measures.

	Confidence			Mutual Information			Kappa	
Rank	Features	Score	Rank	Features	Score	Rank	Features	Score
1°	conting matrix	0,766	1°	data set	0,148	1°	measur valid	0,427
2°	effect uniform	0,663	2°	conting matrix	0,100	2°	measur normal	0,384
3°	effect mean uniform	0,600	3°	effect uniform	0,083	3°	cluster measur	0,361
4°	bound upper	0,562	4°	bound upper	0,073	4°	measur properti	0,351
5°	data set	0,552	5°	inform mutual	0,055	5°	extern measur	0,339
6°	cluster measur valid	0,336	6°	data sampl	0,049	6°	entropi measur	0,333
7°	inform mutual	0,331	7°	measur valid	0,047	7°	evalu measur	0,330
8°	rand statist	0,271	8°	rand statist	0,039	8°	mean measur	0,329
9°	measur valid	0,237	9°	data imbalanc	0,033	9°	measur select	0,328
10°	data sampl set	0,226	10°	extern valid	0,032	10°	cluster valid	0,328
11°	cluster valid	0,207	11°	data sampl set	0,031	11°	consist measur	0,327
12°	cluster mean measur	0,203	12°	effect mean	0,030	12°	measur section	0,327
13°	data simul	0,199	13°	data simul	0,030	13°	measur puriti	0,326
14°	effect mean	0,198	14°	sampl set	0,029	14°	defect measur	0,324
15°	sampl set	0,189	15°	effect mean uniform	0,028	15°	equival measur	0,323
16°	extern measur valid	0,184	16°	mean uniform	0,025	16°	measur result	0,321
17°	mean uniform	0,179	17°	cluster result	0,024	17°	inform measur	0,319
18°	class distribut	0,179	18°	class size	0,023	18°	cluster mean	0,310
19°	cluster mean	0,175	19°	cluster mean	0,022	19°	cluster result	0,303
20°	class data	0,172	20°	class distribut	0,019	20°	class cluster	0,283
21°	cluster measur	0,172	21°	measur properti	0,017	21°	cluster evalu	0,262
22°	extern valid	0,171	22°	cluster valid	0,016	22°	cluster data	0,261
23°	result valid	0,166	23°	measur normal	0,013	23°	cluster number	0,259
24°	measur normal	0,162	24°	defect measur	0,012	24°	data set	0,254
25°	class set	0,161	25°	cluster evalu	0,012	25°	cluster set	0,254

appears only in the ranking of the Kappa measure. The different interest measures used in this paper will be compared to verify which of them produce better results.

The fourth step consists of using the generated features to build a representation in the vector space model.

The bag-of-related-words avoids several problems mentioned in Section 2. For instance, the proposed approach does not analyze the whole dimensionality of the textual document collection. Instead, each document is analyzed individually. The number of words in each document can be orders of magnitude smaller than the number of words in the text document collection.

An interesting characteristic of the bag-of-related-words is that the setting of the number of words in the feature is not required. The rules will be generated until the values of support and some interest measure are higher than the thresholds informed by the user.

To take away the responsibility of the user of setting the minimum support value, and to allow the setting of this value according to each document, we proposed the following formula:

$$AutSup = \frac{GeneralMeanFreq}{NumberTrans} \quad (1)$$

where $GeneralMeanFreq$ means the sum of the frequency of all the words divided by the number of different words, and $NumberTrans$ means the number of transactions.

The definition of the thresholds of other interest measures can be also set automatically. In this paper we used a threshold based on mean ($\bar{x}$) of the interest measure values of the extracted rules in a document. According to the thresholds chosen by the user during the process, the dimensionality can be much lower than the bag-of-words, and the application of feature selection methods is not necessary. However, as the features are represented in the vector space model, these methods can be applied as an additional task.

Moreover, the proposed approach is independent on i) the

natural language processing, which is normally computationally expensive ii) the interference of domain specialists, and iii) knowledge base.

4. EXPERIMENTS AND RESULTS

Two experiments were carried out to analyze the feasibility to aid the topic hierarchy building using the bag-of-related-words and the bag-of-words. The first experiment was an objective evaluation of the quality of the hierarchical clustering. The second experiment was a subjective evaluation about the topics of the hierarchy.

We created 8 collections of proceedings from the ACM digital library for the experiments. Each collection has 5 classes and approximately 90 documents per class[4].

First we evaluated which way to map the document into transactions produced better results to the clustering task. We used only the frequent itemsets in this step of the evaluation so that the objective measures do not interfere in the results. We considerer six possibilities to map the textual document into transactions: sentences, paragraphs, and sliding windows of sizes 5, 10, 20, and 30. A set of minimum support thresholds was used for each way of mapping. Table 4 presents the minimum support thresholds used in the experiments and Figure 1 presents the number of features obtained. The values of the minimum supports were defined based on previous experiments. It can be noticed that most of the used minimum support made the approach obtain a lower number of features than the *bag-of-words* approach for most of the collections.

After evaluating which way of mapping produced better results, we focused on the better way of mapping and apply the interest measures mentioned in Section 3 to prune the number of features. For each interest measure, we also used a set of thresholds based on the number of generated features. We also considered the threshold based on mean ($\bar{x}$).

---

[4]http://sites.labic.icmc.usp.br:8088/ragero/Acm-Collection/

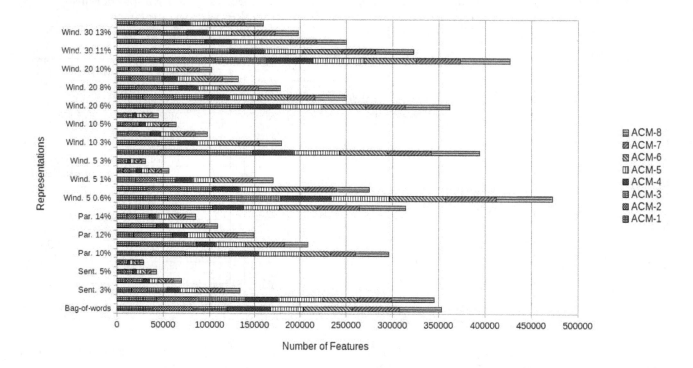

Figure 1: Number of features obtained by the different types of mapping and their thresholds.

Table 4: Minimum support values used in the experiments.

Type of Trans.	Min. Sup. Values
Sentence	2%, 3%, 4%, 5%, 6%
Paragraph	9%, 10%, 11%, 12%, 13%
Sliding Wind.: size 5	0.6%, 0.8%, 1%, 2%, 3%
Sliding Wind.: size 10	2%, 3%, 4%, 5%, 6%
Sliding Wind.: size 20	6%, 7%, 8%, 9%, 10%
Sliding Wind.: size 30	10%, 11%, 12%, 13%, 14%

Table 5 presents the used thresholds and Figure 2 shows the number of obtained features. As the measures Lift range from 0 to ∞, the standardization proposed in [11] was used.

Table 5: Thresholds of the interest measures used in the experiments.

Interest Measure	ACM	Reuters
Confidence	0,25; 0,50; $\overline{x}$	0,25; 0,50; $\overline{x}$
Lift	0,10; 0,20; $\overline{x}$	0,10; 0,20; $\overline{x}$
Yule's Q	0,50; 0,75; $\overline{x}$	0,50; 0,75; $\overline{x}$
Correlation	0,25; 0,50; $\overline{x}$	0,25; 0,50; $\overline{x}$
Mutual Information	0,005; 0,01; $\overline{x}$	0,01; 0,05; $\overline{x}$
Gini Index	0,005; 0,01; $\overline{x}$	0,01; 0,04; $\overline{x}$
Kappa	0,15; 0,20; $\overline{x}$	0,30; 0,50; $\overline{x}$
J-Measure	0,01; 0,02; $\overline{x}$	0,02; 0,05; $\overline{x}$

The variations of the bag-of-related-words will be represented between brackets. The type of mapping, the minimum support values, the interest measure, and the threshold

of the interest measure will be presented inside the brackets. For instance, the representation that used the mapping of sentences with a minimum support value of 2%, and the Confidence measure with a threshold of 70%, will be represented by [Sent. 2% Confidence 70%].

The Pretext tool [16] was used to generate the bag-of-words representation.

4.1 Textual Document Clustering Evaluation

In the first experiment, the feasibility of the bag-of-related-words and the bag-of-words to build a hierarchical clustering was analyzed. The UPGMA (Unweighted Pair Group Method with Arithmetic Mean) algorithm was used to build an hierarchical clustering. This is an agglomerative hierarchical clustering algorithm of the type *average-link*. The UPGMA algorithm obtains results as good as the bisecting-k-means for the textual document hierarchical clustering. We decided to use the UPGMA because the results of the bisecting-k-means depends on the choices of the initial seeds to partition the clusters [23].

The metric used to compare the hierarchical clusters was the *FScore* measure. This measure analyzes the hierarchy group and compares it with the document categories. The *FScore* measure uses the following concepts [23]:

- L_r: represents a class r from the collection;
- n_r: represents the number of documents of a class r;
- S_i: represents a subgroup i of the obtained clustering;
- n_i: represents the number of documents from group i;
- n_{ri}: represents the number of documents from group Si that belong to the class L_r.

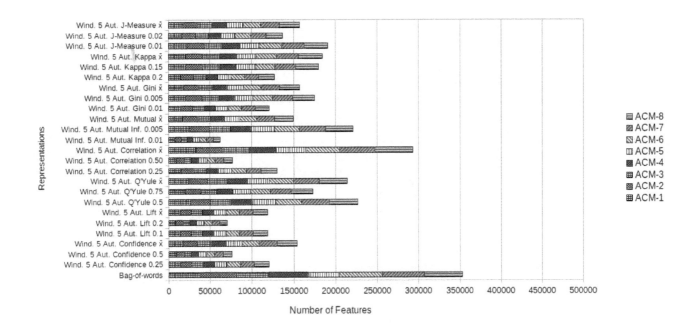

Figure 2: Number of features obtained by the representations using interest measures and their thresholds.

The *FScore* is an harmonic mean between precision (π) and recall (ρ) [2]. The precision and recall for a class L_r and a group S_i are the following:

$$\pi(L_r, S_i) = \frac{n_{ri}}{n_i} \qquad (2)$$

$$\rho(L_r, S_i) = \frac{n_{ri}}{n_r} \qquad (3)$$

After the *FScore* for each class and their respective groups were obtained, the *FScore* value for a class L_r is the biggest *FScore* value of that class for any group of the hierarchy T, that is:

$$F(L_r) = max_{S_i \in T} F(L_r, S_i) \qquad (4)$$

The overall FScore value for the entire clustering solution is the sum of the *FScore* values for all the classes (c) weighted by the class size. Then, the *FScore* value for the clustering is:

$$FScore = \sum_{r=1}^{c} \frac{n_r}{n} F(L_r) \qquad (5)$$

This measure ranges from $[0, 1]$, where 1 represents the maximal quality for the clustering.

Table 6 presents the representations that obtained the best *FScore* values for some ACM collection using only the frequent itemsets[5]. These results aim to evaluate which way to map the textual document into transactions can produce the better results. Using the frequent itemsets as features, the highest values of the *FScore* measure was obtained by

---

[5]The complete results are available in http://sites.labic. icmc.usp.br:8088/ragero/DocEng2011/docint.pdf

the bag-of-related-words for most of the collections (7 of 8 collections). There were considerable differences in comparison with the bag-of-words for some collections, such as ACM-2, with a difference of 0.111, and ACM-7, with a difference of 0.130.

For some collections, as ACM-1 and ACM-3, the best results were obtained by the representations based on the bag-of-related-words [Wind. 5 3%] and [Wind. 10 5%]. The reduction in the number of feature was approximately 91% and 80% respectively in relation to the bag-of-words.

In general, representations based on the mapping of size 5 sliding window obtained the best results. The automatic minimum support setting was used in the mapping of size 5 sliding window, since it obtained the best results. Table 7 presents a comparison of the results obtained by the automatic support and the other used thresholds. It can be noticed that the use of automatic minimum support setting presented results as good as the manually set minimum support.

The Friedman test was applied considering the 8 ACM collections to verify how the variation of the minimum support values affected the results for the different types of mapping analyzed. There was significant statistical difference for the results obtained by the sentence mapping, where the lowest minimum support thresholds obtained better results than the highest minimum support values.

Then the best values of minimum support for each type of mapping were chosen to compare the several types of mapping and the bag-of-words representation. The representations chosen were: [Sent. 2%], [Par. 10%], [Wind. 5 2%], [Wind. 10 5%], [Wind. 20 8%], and [Wind. 30 11%]. None of the representations presented statistical significant differences. It must be highlighted that the representation [Wind. 5 2%] presented better results than the bag-of-words and reduced the number of features more than 80%.

Table 6: Results of the representations using frequent itemsets that obtained the highest value of the FScore measure for some ACM collection.

Representação	ACM-1	ACM-2	ACM-3	ACM-4	ACM-5	ACM-6	ACM-7	ACM-8
Bag-of-words	0,779	0,818	0,821	0,854	0,842	0,892	0,800	*0,867
Wind. 5 0,06%	0,771	0,820	0,863	0,917	0,868	*0,910	0,812	0,792
Wind. 5 0,08%	0,788	0,845	0,891	0,900	*0,881	0,893	0,803	0,813
Wind. 5 2%	0,806	0,912	0,843	0,909	0,829	0,880	*0,930	0,795
Wind. 5 3%	*0,813	0,915	0,804	0,886	0,822	0,841	0,897	0,749
Wind. 10 5%	0,761	0,924	*0,904	0,891	0,816	0,868	0,904	0,817
Wind. 20 7%	0,765	0,874	0,877	*0,921	0,833	0,849	0,895	0,768
Wind. 30 11%	0,752	*0,929	0,893	0,889	0,814	0,872	0,919	0,764

Table 7: Results of the representations using the mapping of size 5 sliding window.

Representation	ACM-1	ACM-2	ACM-3	ACM-4	ACM-5	ACM-6	ACM-7	ACM-8
Bag-of-words	0.779	0.818	0.821	0.854	0.842	0.892	0.800	*0.867
Wind. 5 Aut.	0.802	0.886	0.889	*0.917	0.868	0.891	0.823	0.821
Wind. 5 0.06%	0.771	0.820	0.863	*0.917	0.868	*0.910	0.812	0.792
Wind. 5 0.08%	0.788	0.845	0.891	0.900	*0.881	0.893	0.803	0.813
Wind. 5 1%	0.784	0.886	*0.900	0.913	0.848	0.895	0.850	0.820
Wind. 5 2%	0.806	0.912	0.843	0.909	0.829	0.880	*0.930	0.795
Wind. 5 3%	*0.813	*0.915	0.804	0.886	0.822	0.841	0.897	0.749

Table 8: Results of the representations using interest measures that obtained the highest value of the FScore measure for some ACM collection.

Representation	ACM-1	ACM-2	ACM-3	ACM-4	ACM-5	ACM-6	ACM-7	ACM-8
Bag-of-words	0.779	0.818	0.821	0.854	0.842	0.892	0.800	*0.867
Wind. 5 Aut. Conf. 0.5	*0.855	0.882	0.886	0.922	0.911	0.884	0.799	0.744
Wind. 5 Aut. Lift 0.1	0.799	*0.889	0.892	0.918	0.901	0.913	0.786	0.814
Wind. 5 Aut. Lift Aut	0.799	*0.889	0.892	0.918	0.901	0.913	0.786	0.814
Wind. 5 Aut. Q'Yule 0.75	0.796	0.872	0.904	*0.926	0.902	0.910	0.763	0.824
Wind. 5 Aut. Corr. 0.25	0.789	0.854	0.896	0.922	*0.915	0.883	0.783	0.801
Wind. 5 Aut. Corr. 0.50	0.831	0.877	0.883	0.922	0.909	*0.925	0.805	0.778
Wind. 5 Aut. M.I. 0.01	0.853	0.883	*0.908	0.920	0.904	0.923	0.814	0.765
Wind. 5 Aut. Kappa $\bar{x}$	0.778	0.875	0.872	0.918	0.905	0.897	*0.815	0.803

Interest measures were applied considering the mapping of size 5 sliding window, since it obtained the best results. We also decide to use the automatic minimum support setting since they obtained results as good as the other minimum support values. Table 8 presents the results of the representations that obtained the best results for some ACM collection.

The representations that used interest measures obtained better results than the bag-of-words for 7 of 8 ACM collections (ACM-1, ACM-2, ACM-3, ACM-4, ACM-5, ACM-6, ACM-7). The interest measures reduced the number of features and obtained better results than the base representation ([Wind. 5 Aut.]) for the collections ACM-1, ACM-2, ACM-3, ACM-4, ACM-5, ACM-6, and ACM-8. All the representations on Table 8 always reduced the number of features in the minimum of 26%. In some situation, 82% of the number of features were reduced.

The Friedman test was applied again to verify if the thresholds of the used interest measures affected the values of the FScore. There was only statistical significant difference for the Kappa measure, in which the threshold based on mean presented better results than the threshold 0.15. Thus, the threshold variations of the interest measures reduced the number of features and kept the quality of the results. Also there was no evidence that the thresholds defined manually

were better than the thresholds based on mean. Then, we decided to compare the representations with interest measures and thresholds based on mean with the bag-of-words representation. Again, the Friedman test was applied and there were not statistical significant differences among the representations.

The obtained results demonstrated that the bag-of-related-words is appropriate for the textual document clustering. The highest FScore values was obtained by the bag-of-related-words for most of the ACM collections. It also can be highlighted that in most of the situations, the best results were obtained with a reduced number of features.

4.2 Topic Hierarchy Evaluation

In the second experiment, topic hierarchies were built and submitted to computer science domain experts. Three ACM collections for the subjective evaluation were used: ACM-3, ACM-4, and ACM-8. Table 9 presents the topics of these collections. The representation [Wind. 5 Aut. Correlation $\bar{x}$] was chosen to be compared with the bag-of-words and the Torch[6] tool was used to build the topic hierarchies.

Two hierarchies of each collection were given to 6 computer science domain experts and the representation used

---

[6]http://sites.labic.icmc.usp.br/marcacini/ihtc/

Table 9: Topics of the collections used for the subjective evaluation of the topic hierarchy.

Col.	Topic
ACM-3	Computer Architecture Education
	Architecture for Networking And Commu. Systems
	Privacy in the Electronic Society
	Software and Performance
	Web Information and Data Management
ACM-4	Embedded Networked Sensor Systems
	Research and Development in Info. Retrieval
	Parallel Algorithms and Architectures
	Volume Visualization
	Cross-Disciplinary Conf. on Web Accessibility
ACM-8	Mobile Ad Hoc Networking & Computing
	Knowledge Discovery and Data Mining
	Langu., Comp. and Tool Sup. for Embedded Systems
	Hypertext and Hypermedia
	Microarchitecture

to build the topic hierarchy was not specified. The following questions were evaluated:

1) Is it possible to identify clearly the topics of the collections based on the respective categories?

2) Does the browsing through the topic hierarchy conduced the user to a desired document set?

For both evaluations the used grades were:

0 - Nothing;

1 - Little;

2 - Fair;

3 - Good;

4 - Excellent.

Table 10 presents the means of the grades given to the domain experts for all the evaluated topic hierarchies. It can be noticed that in all the collections, the bag-of-related-words obtained better results than the bag-of-words.

Table 10: Mean of the results obtained in the subjective evaluation of the topic hierarchies for the ACM-3, ACM-4, and ACM-8 collection.

	ACM-3		ACM-4		ACM-8	
	BOW	BORW	BOW	BORW	BOW	BORW
Question 1	2.00	2.83	1.83	2.50	2.50	2.83
Question 2	2.00	2.66	2.00	2.50	2.16	2.66

To illustrate the topic hierarchies built using the bag-of-related-words and bag-of-words, Figures 3 and 4 presents the first level of the topic hierarchy for the ACM-3 and ACM-4 collections.

5. CONCLUSION AND FUTURE WORKS

Topic hierarchies are useful for managing, searching, and browsing in large text collections. To allow the automatic topic hierarchy building, the textual document collection must be represented in an appropriate way. The proposed approach for textual document representation, bag-of-related-words, demonstrated to be useful for the topic hierarchy building.

The textual document clustering, used to create the hierarchical structure of the document collection, presented better results when performed using the bag-of-related-words representation. Also the topics of the obtained hierarchy using the bag-of-related-words were more understandable for the users than the topics obtained by the bag-of-words. It must be highlighted that the initial dimensionality of the proposed approach, even containing features composed by

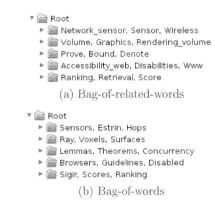

(a) Bag-of-related-words

(b) Bag-of-words

Figure 3: Example of the first level of the topic hierarchy using the bag-of-related-words and bag-of-words representations for the ACM-3 collection.

(a) Bag-of-related-words

(b) Bag-of-words

Figure 4: Example of the first level of the topic hierarchy using the bag-of-related-words and bag-of-words representations for the ACM-4 collection.

set of words, in most cases was much smaller than the bag-of-words with better results.

The mapping of a textual document into transactions using a size 5 sliding window presented the best results. The thresholds of interest measures used in the experiments in most cases do not present significant statistical difference. However, this setting is not trivial and can be time consuming for the user. An automatic threshold setting was used in this paper. The results obtained with the automatically set thresholds were equivalent or better than the manual threshold setting.

The analysis of each document individually, besides avoiding the entire dimensionality of the collection, allows the proposed approach be used in dynamic contexts, in which the analysis of the entire document collection to extract the features is unfeasible for each new document.

For future research, we are going to evaluate the pruning of redundant rules and redundant features to reduce even more the number of features without the need of a supervised feature selection method.

Acknowledgements

We would like to thank CNPq to the financial support.

6. REFERENCES

[1] R. Agrawal and R. Srikant. Fast algorithms for mining association rules in large databases. In *VLDB'94: International Conference on Very Large Data Bases*, pages 487–499. Morgan Kaufmann Publishers Inc., 1994.

[2] R. A. Baeza-Yates and B. A. Ribeiro-Neto. *Modern Information Retrieval*. ACM Press / Addison-Wesley, 1999.

[3] J. Blanchard, F. Guillet, R. Gras, and H. Briand. Using information-theoretic measures to assess association rule interestingness. In *ICDM'05: Internation Conference on Data Mining*, pages 66–73, 2005.

[4] M. F. Caropreso, S. Matwin, and F. Sebastiani. A learner-independent evaluation of the usefulness of statistical phrases for automated text categorization. *Text databases & document management: theory & practice*, pages 78–102, 2001.

[5] A. L. C. Carvalho, E. S. Moura, and P. Calado. Using statistical features to find phrasal terms in text collections. *Journal of Information and Data Management*, 1(3):583–597, 2010.

[6] A. Doucet and H. Ahonen-Myka. Non-contiguous word sequences for information retrieval. In *MWE'04: Workshop on Multiword Expressions: Integrating Processing*, MWE'04, pages 88–95. Association for Computational Linguistics, 2004.

[7] L. Geng and H. J. Hamilton. Interestingness measures for data mining: A survey. *ACM Computing Surveys*, 38(3):9, 2006.

[8] F. Guillet and H. J. Hamilton, editors. *Quality Measures in Data Mining*, volume 43 of *Studies in Computational Intelligence*. Springer, 2007.

[9] V. Kashyap, C. Ramakrishnan, C. Thomas, and A. P. Sheth. Taxaminer: an experimentation framework for automated taxonomy bootstrapping. *International Journal of Web and Grid Services*, 1(2):240–266, 2005.

[10] Y. Lie, H. T. Loh, and W. G. Lu. Deriving taxonomy from documents at sentence level. In A. H. do Prado and E. Ferneda, editors, *Emerging Technologies of Text Mining: Techniques and Applications*, chapter 5, pages 99–119. Information Science Reference, 1 edition, 2007.

[11] P. D. McNicholas, T. B. Murphy, and M. O'Regan. Standardising the lift of an association rule. *Computational Statistics & Data Analysis*, 52(10):4712–4721, 2008.

[12] D. Mladenic and M. Grobelnik. Word sequences as features in text-learning. In *ERK'98: Electrotechnical and Computer Science Conference*, pages 145–148, 1998.

[13] M. F. Moura and S. O. Rezende. A simple method for labeling hierarchical document clusters. In *IASTED'10: International Conference on Artificial Intelligence and Applications (IAI 2010)*, pages 363–371, 2010.

[14] G. Salton. *Automatic text processing: the transformation, analysis, and retrieval of information by computer*. Addison-Wesley Longman Publishing Co., Inc., 1989.

[15] F. F. Santos, V. O. de Carvalho, and S. O. Rezende. Selecting candidate labels for hierarchical document clusters using association rules. In Springer-Verlag, editor, *MICAI'10: Mexican International Conference on Artificial Intelligence*, 2010.

[16] M. V. B. Soares, R. C. Prati, and M. C. Monard. PRETEXT II: Descrição da reestruturação da ferramenta de pré-processamento de textos. Technical Report 333, ICMC-USP, 2008.

[17] C.-M. Tan, Y.-F. Wang, and C.-D. Lee. The use of bigrams to enhance text categorization. *Information Processing and Management*, 38(4):529–546, 2002.

[18] P.-N. Tan, V. Kumar, and J. Srivastava. Selecting the right interestingness measure for association patterns. In *ACM SIGKDD'2002: International Conferenceon Knowledge Discovery and Data Mining*, pages 32–41. ACM, 2002.

[19] R. Tesar, V. Strnad, K. Jezek, and M. Poesio. Extending the single words-based document model: a comparison of bigrams and 2-itemsets. In *DocEng'06: ACM Symposium on Document Engineering*, pages 138–146, 2006.

[20] J. Wu, H. Xiong, and J. Chen. Adapting the right measures for k-means clustering. In *SIGKDD'09: Proceeding of the International Conference on Knowledge Discovery and Data Mining*, pages 877–886. ACM, 2009.

[21] Z. Yang, L. Zhang, J. Yan, and Z. Li. Using association features to enhance the performance of naïve bayes text classifier. In *ICCIMA '03: International Conference on Computational Intelligence and Multimedia Applications*, page 336. IEEE Computer Society, 2003.

[22] X. Zhang and X. Zhu. A new type of feature - loose n-gram feature in text categorization. In *IbPRIA'07: Iberian Conference on Pattern Recognition and Image Analysis*, pages 378–385. Springer, 2007.

[23] Y. Zhao and G. Karypis. Evaluation of hierarchical clustering algorithms for document datasets. In *CIKM '02: International Conference on Information and Knowledge Management*, pages 515–524. ACM Press, 2002.

Local Metric Learning for Tag Recommendation in Social Networks

Boris Chidlovskii and Aymen Benzarti
Xerox Research Centre Europe
F–38240 Meylan, France
chidlovskii@xrce.xerox.com, aymen.benzarti@gmail.com

ABSTRACT

We address the problem of tag recommendation for media objects, like images and videos, in social media sharing systems. We propose a framework that 1) extracts both object features and the social context and 2) uses them to learn recommendation rules. The social context is described by different types of information, such as a user's personal objects, the objects of a user's social contacts, the importance of the user in the social network, etc. Both object features and the social context are used to guide the k-nearest neighbour method for the tag recommendation. We then enhance the kNN by the local topology adjustment on how the nearest neighbours are selected. We learn a local transformation of the feature space surrounding a given object which pushes together objects with the same tags and puts apart objects with different tags. We show how to learn the Mahalanobis distance metric on multi-tag objects and adopt it to the tag recommendation problem.

Categories and Subject Descriptors

H.3.3 [**Information Systems**]: Information Search and Retrieval—*Information filtering*; I.2.6 [**Artificial Intelligence**]: Learning; G.1.6 [**Mathematics of Computing**]: Optimization—*Convex programming*

General Terms

Algorithms, Experimentation, Performance

Keywords

social tagging, tag recommendation, metric learning

1. INTRODUCTION

Social media sharing sites like Flickr and YouTube contain billions of image and videos which are uploaded and annotated by millions of users. Tagging the media objects is proven to be a powerful mechanism that can improve search options for shared images and videos [4]. Tags play the role of metadata, however they are often provided in a free form reflecting the individual user's choice.

Despite this free individual choice, some common usage themes can emerge when people agree on the semantic description of a given media object or a group of objects.

The wealth of annotated and tagged objects on the media sharing sites can form a solid basis for tag recommendation [3, 10]. The recommender systems can be particularly useful in the bootstrap, querying and search modes. In the *bootstrap* mode, the recommender system suggests the most relevant tags for newly uploaded objects by observing the characteristics of the objects as well as their social context. In the *query* mode, a user annotating an image is recommended tags that can extend the existing image tags. Both modes can ease the annotation task for the user and help expand the coverage of the tags annotating the images. In *search* mode, the role of the recommender system is to provide search recommendations. This can be achieved by automated query expansion or through an interactive process of adding query terms.

When building a tag suggestion system one can target the social network as a source of collective knowledge. Recommender systems based on collective social knowledge have been proven to provide relevant suggestions [3, 9, 10, 11]. Some of these systems aggregate the annotations used in a large collection of objects independently of the users that annotate them [11]. Alternatively, the recommendations can be personalized by using the annotations for the images of a single user [3]. Both approaches come with their advantages and drawbacks. When the recommendations are based on collective knowledge the system can make good recommendations on a broad range of topics, but is likely to miss some recommendations that are relevant in a personal context [9, 10].

Methods for tagging objects in social media systems that take onto account both object similarity and the social context appears to be a winning strategy. Below we propose a tag recommendation method for both the bootstrap and querying recommendation mode, where the objects are images and the social context is defined by a number of important relations, like ownership or friendship. The main contributions of this work are the following:

1. We extend the standard k-nearest neighbours (kNN) method, by using both image features and the social content to guide the neighbour candidate selection. The candidate set is defined as the images owned by the user and his/her contacts.

2. We replace the kNN method for measuring the object similarity with a local distance metric. Similarly to other domains like face recognition, the local metric can be adjusted in each specific context.

3. We adopt the Mahalanobis metric, previously used in the binary classification and extend it to the multi-tag cases.

We run series of experiments on the Flickr image dataset and show that kNN with local Mahalanobis distances improves considerably the prediction quality over the baseline models learned with the identical features sets [1].

Figure 1: Flickr web interface with a flower image example, tagged with `colour`, `beautiful`, `flower` **and 17 more tags.**

1.1 Dataset and Preprocessing

To process the object features and social context, we have downloaded from Flickr media sharing site (see Figure 1) a large collection of users with their contacts, as well as images with their descriptions, comments and tags (using the Flickr API). Before deploying the tag recommendation techniques, we preprocess the images, their descriptions and the social network formed by users; we obtain three groups of features:

Visual features of images: they form d-dimensional vectors describing the visual image characteristics; we used the XRCE Generic visual toolbox [8] for extracting the bag of d=1024 visual words.

Textual features of images: they are given by the TFIDF representation extracted from image descriptions and comments.

Social features of any user: they describe her position in the social network, they include the hub and authority scores, six centrality measures including degree, closeness, betweenness ones, the clustering coefficient, PageRank value, four different clique numbers, etc. (see [5] for more detail on the social features).

All tags provided in the free text form are equally normalized. Although we consider information about images, their owners and owners' contacts, other kinds of additional information can be included in the consideration and downloaded from Flickr; possible choices are groups of interest, users' favourite images, image pools, etc. All the methods discussed below can be extended to the additional information accordingly.

2. KNN FOR TAG RECOMMENDATION

The k-nearest neighbours (kNN) is one of the most common classification methods. Nevertheless, it often yields competitive results, and in certain domains, when combined with prior knowledge, it has significantly advanced the state-the-art. The kNN rule classifies each unlabelled object by the majority label(s) among its k-nearest neighbours in the training set. The main idea of using this algorithm is to predict tags for a given image I by observing the k most similar neighbour images. Since the exhaustive similarity check is computationally prohibitive for all image pairs, we constrain the way how the image neighbours can be found. Let C denote the *neighbour candidate set* the most similar images are selected from. The candidate set C should be defined in a way that ensures a nearly optimal size-similarity trade-off that does not compromise changes to find the truly similar images. In the context of social networks, two instantiations of C are most plausible:

Personal space: C is the set of images owned by the user, $C = C_o$;

Social context: We take into account the social context of the owner and consider all images from the owner's contacts, $C = C_c$. Moreover, we consider a version when the candidate set is limited to the top L closest contacts where "closest" defines a similarity by jointly used tags. Once the candidate set C is determined for an unlabelled image I, the k most similar images are selected from C according to their similarity to image I. Once these k images are selected, their most m_k frequent tags are taken to annotate image I where m_k is the average number of tags in the selected images.

2.1 Metric learning

Although defining a good candidate set C is an important issue, the performance of kNN methods depends more on the distance metric used to identify nearest neighbours in set C. In the absence of prior knowledge, the standard Euclidean distance measures the dissimilarities between objects represented as feature vectors. The Euclidean distance, however, does not capitalize on any statistical regularity in the data that might be estimated from a large training set of examples. Ideally, the distance metric for kNN classification can be adapted to the particular problem being solved. As alternative to the Euclidean distance, we propose to learn a *Mahalanobis distance* metric between the feature vectors describing the images. This metric can be trained inside the kNN routine under the local regularity assumption. According to the assumption, the k nearest neighbours often have the same tags while images with different tags are separated by a large margin [7].

The squared Euclidean distance between two images described with d-vectors $\mathbf{x}_i$ and $\mathbf{x}_j$, defined as $dE(\mathbf{x}_i, \mathbf{x}_j) = ||\mathbf{x}_i - \mathbf{x}_j||^2 = (\mathbf{x}_i - \mathbf{x}_j)^T(\mathbf{x}_i - \mathbf{x}_j)$. The class of Mahalanobis distances generalizes the Euclidean one by adding a $d \times d$ positive semi-defined matrix A. The Mahalanobis distance $dA(\mathbf{x}_i, \mathbf{x}_j) = (\mathbf{x}_i - \mathbf{x}_j)^T A(\mathbf{x}_i - \mathbf{x}_j)$ can be seen as the (squared) Euclidean distance after applying a linear transformation L to all vectors $\mathbf{x}_i$ where $A = L^T L$ [6]. Finding the semi-positive matrix A should satisfy a set R of relative distance constraints on triples $(\mathbf{x}_i, \mathbf{x}_j, \mathbf{x}_k)$ from the training set S indicating that $\mathbf{x}_i$ and $\mathbf{x}_j$ are neighbours of the same tag while $\mathbf{x}_i$ and $\mathbf{x}_k$ have different tags.

The problem is defined as an instance of *semi-definite programming* (SDP) on a set of pairs $(\mathbf{x}_i, \mathbf{x}_j)$ such that $\mathbf{x}_i$ and $\mathbf{x}_j$ are neighbours sharing the same tag t. The objective function tries to minimize the sum of distances of pairs $(\mathbf{x}_i, \mathbf{x}_j) \in S$ and to satisfy the relative distance constraints. Formally it is defined as the constrained minimization problem:

$$min_A \sum_{(\mathbf{x}_i, \mathbf{x}_j) \in S} d_A(\mathbf{x}_i, \mathbf{x}_j), \quad \text{subject to}$$
$$d_A(\mathbf{x}_i, \mathbf{x}_k) - d_A(\mathbf{x}_i, \mathbf{x}_j) \geq 1, \forall(\mathbf{x}_i, \mathbf{x}_j, \mathbf{x}_k) \in R, A \geq 0. \quad (1)$$

The semi-definite programs are convex which guarantees that the global minimum can be efficiently achieved. By analogy with the SVM when not all constraints can be satisfied, slack variables ξ_{ijk} can be introduced to minimize the violation of certain constraints:

$$min_A \sum_{(\mathbf{x}_i, \mathbf{x}_j) \in S} d_A(\mathbf{x}_i, \mathbf{x}_j) + \sum_{(\mathbf{x}_i, \mathbf{x}_j, \mathbf{x}_k) \in S} \xi_{ijk}$$
$$s.t. \, d_A(\mathbf{x}_i, \mathbf{x}_k) - d_A(\mathbf{x}_i, \mathbf{x}_j) \geq 1 - \xi_{ijk}, \forall(\mathbf{x}_i, \mathbf{x}_j, \mathbf{x}_k) \in R,$$
$$A \geq 0, \xi_{ijk} \geq 0. \quad (2)$$

The distance can be learned globally, yielding one matrix A for all instances in the dataset. We will show however that learning the distance metrics locally gives a much larger performance gain. Once candidate set C is constructed for a test image I, we train the local metric dA_i with all images in set C. The k images nearest to image I are then determined according either to global metric d_A or local one d_{A_i}.

2.2 Learning distance metrics for multiple tags

We deploy a specialized CVX solver [2] in order to solve semi-define programs in (2) on the Flickr dataset. Note that all existing SDP solvers work in the binary and multi-class mode where an object is labelled with one tag t only. Since images on social media sharing sites can be annotated with multiple tags, we accommodate the Mahalanobis metric learning to the multi-tag case, in one of the following ways:

1. One obvious way is to learn the metrics in the per-tag mode. At the training step, for each tag t we can form the training set S_t with values 1 for images tagged with t and 0 otherwise. At the test step, the matrix A_t for tag t is applied to find the dA_t-nearest images and the decision is then made whether to tag I with t.

2. The per-tag metric learning schema is prohibitively expensive. The full tag set T may include thousands of tags, thus requiring to run the learning routine and to store matrix A_t for each tag t in the local setting. Our alternative is to learn one metric for all tags together. We consider two images with feature vectors $\mathbf{x}_i$ and $\mathbf{x}_j$ as (dis)similar as the function of tags they have in common. Two images are (fully) dissimilar if they have no one common tag. Two images are considered (fully) similar if they have *all* tags the same. When defining the set R of relative distance constraints on triples, only fully similar and fully dissimilar image pairs are considered.

3. Note that the above approach excludes all image pairs having some common tags. Moreover, relative distance constraints are generated for images I and J having the same tag sets which is not frequent in datasets like Flickr. To remedy this, we consider an option which accepts the partial similarity among images. Let T_i and T_j denote the tag sets of I and J, respectively. We introduce the tag overlap fraction τ_{ij} defined as $\tau_{ij} = 2|T_i \cap T_j|/(|T_i \cup T_j|)$ which gives the common fraction of tags for both images. Then we define the set R of relative distance constraints on triples $(\mathbf{x}_i, \mathbf{x}_j, \mathbf{x}_k)$ such that $\mathbf{x}_i$ and $\mathbf{x}_j$ are neighbours with at least one common tag while neither $\mathbf{x}_i$ nor $\mathbf{x}_j$ have a common tag with $\mathbf{x}_k$. The semi-definite program (2) is then rewritten as follows to integrate the partial tag overlaps:

$$\min_A \sum_{(\mathbf{x}_i, \mathbf{x}_j) \in S} d_A(\mathbf{x}_i, \mathbf{x}_j) + \sum_{(\mathbf{x}_i, \mathbf{x}_j, \mathbf{x}_k) \in R} \xi_{ijk},$$
$$s.t.$$
$$d_A(\mathbf{x}_i, \mathbf{x}_k) - d_A(\mathbf{x}_i, \mathbf{x}_j) \geq \tau_{ij} - \xi_{ijk}, \quad (3)$$
$$\forall (\mathbf{x}_i, \mathbf{x}_j, \mathbf{x}_k) \in R, A \geq 0, \xi_{ijk} \geq 0.$$

2.3 KNN with local metric learning

Algorithm 1 is the pseudo code of the training and testing steps for the kNN algorithm with the local Mahalanobis metric learning. To measure the performance of different methods, we use the standard measures of precision and recall extended to the multi-tag case. For each test image I, we compare the set of predicted tags to the true tag set. The precision Pr is defined as to (the number of tags in common between true tags and predicted tags)/(Number of predicted tags). The recall Re is defined as (the number of tags in common between true tags and predicted tags) / (number of true tags). The $F1$-score for image I is to $2\frac{Pr*Re}{Pr+Re}$. Then, all these measures are averaged over all images in the test set.

3. EVALUATION

To validate the idea of local metric learning we run series of evaluation tests on the Flickr data. In all cases, we address the tag recommendation task on images with multiple tags, where images are described with visual, textual and social features. Three series

---

Algorithm 1 kNN algorithm with integrated local Mahalanobis metric learning

Input: A training set S_l and test set S_u of images
Input: the rule to construct the candidate set C
Input: the neighbourhood size k
Output: Tag predictions

1: **for all** image I in the test set S_u **do**
2:     Extract the candidate set C_i for I from the training set S_l
3:     Learn Mahalanobis metric d_{A_i} from images in C_i
4:     **for all** image J in set C_i **do**
5:         Calculate weights $w_{ij} = d_{A_i}(I, J)$
6:     **end for**
7:     Select k images with the smallest w_{ij}
8:     Select all tags of the selected images
9:     Calculate the average size m_i of tag sets of selected images
10:    Sort the tags according to their frequency in selected images
11:    Return the top m_i tags as predicted tags for image I
12: **end for**

---

of tests have been run on the Flickr dataset. First, we learn SVM linear kernel models with combinations of visual, textual and social features. Second, we learn both global and local Mahalanobis metrics. Finally, we combine kNN with Euclidean and local Mahalanobis distance metrics for different candidate rules.

3.1 SVM baseline

As the baseline, we train Support Vector Machine linear models for multi-tag classification by combining different features to describing the images. We combine visual, textual and social features and construct three different sets. The first set contains visual image features only. In the second set we extend the visual features with the social features of their users. In the third set, we add the textual features of images. We run SVMLIB software (http://www.csie.ntu.edu.tw/cjlin/libsvm/) in the multi-label mode where we retain only images with 2, 5, 10 and 20 most frequent tags. Table 1 shows the number of images for each case and the precision of SVM models run with 5 folds.

Number of tags	2	5	10	20
Number of images	42,842	77,414	110,238	147,942
Feature set	Precision (%)			
Visual	75.20	50.79	29.49	19.07
Visual+social	78.04	57.56	38.60	21.70
Visual+textual+social	78.12	57.59	29.45	17.51

Table 1: Baseline evaluation with SVM models.

The accuracy drops quickly with the increasing tag numbers in all cases, due to the growing number of images and the higher complexity of the classification task. Adding social user features to the image features improves the accuracy in all cases, while adding the textual features can hurt the classification performance; this is caused by the high noise and sparseness of textual descriptions.

3.2 Metric learning

In the second series of tests we probe different options for the Mahalanobis metric learning on the Flickr dataset.

Different features sets: We first test different combinations of available feature sets (visual, textual and social). We apply the principal component analysis (PCA) to reduce dimensionality of feature vectors and speed up the training. In all cases, we reduced the

d-vectors $\mathbf{x}_i$ to 50 principal components before learning the Mahalanobis distance metric and comparing it to the Euclidean distance. Similarly to SVM models, the combination of visual and social features outperforms all other combinations.

Metric learning for multi-tag cases: Three versions of Mahalanobis metric learning have been implemented and tested, namely PerTag, AllFull and AllPartial (see Section 2.2). In each of 5 runs, 80% of images are randomly selected for training and remaining 20% is used for testing.

Global versus local metric: The global Mahalanobis metric was learnt from 1,000 images randomly sampled from the training set before (results are averaged over 5 runs). When applying the local metric learning, for each image I in the test set we extract the personal context C_0 in the training set.

Figure 2 shows the prediction performance for the global and local metric learning when only images with top 20 tags are retained. For both cases, the figure reports the average F1 values of the multi-tag learning and the Euclidean metric when k varies from 1 to 5.

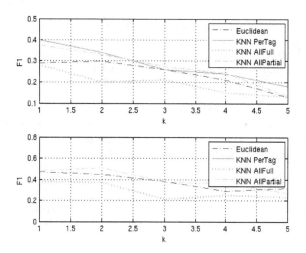

Figure 2: Global vs local multi-tag metrics learning for 20 tags.

The local metrics show up a smaller recommendation error as compared to the global metric. Among three methods for the multitag learning, the PerTag method performs generally the best, but the time needed to train all metric matrices for the Flickr data are prohibitive and the method is not reported for the local metrics case. Between AllFull an AllPartial methods, taking into account partial tag overlaps helps reduce the recommendation error. The price for this is a longer learning time due to a larger size of the constraint set R. The smallest errors are achieved for small k values which yield 5-12% gain over the Euclidean distance.

3.3 KNN with local metric learning

Finally we incorporate the local metric learning into the kNN and compare its performance to the SVM models learnt with exactly the same sets of images. Tests are run in 5 folds; the number of neighbours k is 3, the dataset is reduced to the top $T=5,10,20,50$ and 100 most frequent tags. Two versions of the kNN use the Euclidean distance: one uses the personal space C_0 where the candidate set is limited to owner's images, another uses the personal and social context where the candidate set C_c includes also the owner contacts' images (number of contacts is limited to $L=10$). The version

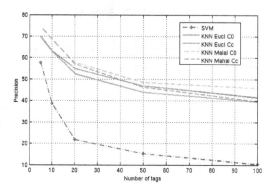

Figure 3: Comparing kNN with Euclidean and Mahalanobis metrics to SVM baseline.

of kNN with local metrics is based on personal context and the AllPartial method for multi-tag metric learning. Figure 3 reports the precision values for all the tests.

4. CONCLUSION

We propose a framework for tag recommendation in social media sharing systems. It uses both object features and the social context to guide the kNN-based classifiers which are better adapted for the large numbers of tags than SVM models. We then enhance the method by learning the Mahalanobis distance metric on multi-tag objects. Experiments of read data sets show that incorporating the local metric learning yields a comfortable performance gain.

5. REFERENCES

[1] T. Bogers. Movie recommendation using random walks over the contextual graph. In *Proc. 2nd Workshop on Context-Aware Recommender Systems (CARS)*, 2010.

[2] CVX. Matlab software for disciplined convex programming. In *http://cvxr.com/cvx/download*.

[3] N. Garg and I. Weber. Personalized, interactive tag recommendation for flickr. In *Proc. ACM RecSys'08*, pages 67–74, 2008.

[4] M. Gupta, R. Li, Zh.Yin, and J. Han. Survey on social tagging techniques. *ACM SIGKDD Explorations Newsletter*, 12:58–72, November 2010.

[5] M. Hovelynck and B. Chidlovskii. Multi-modality in one-class classification. In *WWW '10*, pages 441–450, 2010.

[6] P. Jain, B. Kulis, J. V. Davis, and I. S. Dhillon. Metric and kernel learning using a linear transformation. In *arXiv:0910.5932v1*, 2009.

[7] K. Q. Weinberger and L. K. Saul. Distance metric learning for large margin nearest neighbor classification. In *Proc. NIPS*, 2005.

[8] F. Perronnin and Ch. R. Dance. Fisher kernels on visual vocabularies for image categorization. In *Proc. CVPR*, 2007.

[9] S. Peters, L. Denoyer, and P. Gallinari. Iterative annotation of multi-relational social networks. In *Proc. ASONAM*, pages 96–103, 2010.

[10] A. Rae, B. Sigurbjörnsson, and R. van Zwol. Improving tag recommendation using social networks. In *Proc. RIAO'10*, pages 92–99, Paris, France, 2010.

[11] B. Sigurbjörnsson and R. van Zwol. Flickr tag recommendation based on collective knowledge. In *Proc. WWW '08*, pages 327–336, 2008.

Expressing Conditions in Tailored Brochures for Public Administration

Nathalie Colineau and Cécile Paris
CSIRO – ICT Centre
cnr. Vimiera and Pembroke Roads
Marsfield, NSW 2122, Australia
(+61) 2 9372 4222

nathalie.colineau@csiro.au,
cecile.paris@csiro.au

Keith Vander Linden
Department of Computer Science
Calvin College
Grand Rapids, MI, 49546, USA
(+1) 616 526 7111

kvlinden@calvin.edu

ABSTRACT

Citizen-focused documents in Public Administration devote considerable effort to the expression of conditions. These conditions are commonly expressed as statements of eligibility requirements for the programs being described, but they manifest themselves in other places as well, such as in feedback to readers in tailored informational brochures and as input fields on program application forms. This paper discusses how administrative conditions can be represented in a manner that supports both the eligibility reasoning required for the generation of citizen-tailored documents and also the automated generation of condition expressions in a variety of forms. The paper pays particular attention to the question of how a generation mechanism can allow authors to override the default forms of automated expression when necessary. The discussion is based on a prototype tailored delivery application whose knowledge base is implemented in OWL DL and whose output is constructed using Myriad, a platform for tailored document planning and formatting.

Categories and Subject Descriptors

H.5 [**Information Interfaces and Presentation**]: General; I.2.7; [**Natural Language Processing**]: Language Generation.

General Terms

Documentation, Human Factors.

Keywords

Tailored Information Delivery, Natural Language Generation, Public Administration.

1. INTRODUCTION

We are exploring the feasibility and usefulness of deploying tailored information delivery in the domain of Public Administration (PA). Achieving this goal requires that we develop a principled treatment of eligibility conditions. The representation, use and expression of these conditions are fundamental to both tailoring and to PA.

The work reported here is based on a collaboration with the Student Communications Team at Centrelink, the Australian Government's service delivery agency. This team is responsible for the creation and distribution of informational brochures for

entitlement programs for Australian students. For example, Figure 1(a) shows a three-page excerpt from Centrelink's paper brochure for ABSTUDY, an entitlement program that helps Indigenous Australians with the cost of studying (Centrelink, 2011). This excerpt provides a one-page overview of the ABSTUDY program and then a two-page description of the eligibility conditions for the program. This distribution of the volume of what might be called "raw" content (i.e., the first page) to the volume of conditions (i.e., two pages) illustrates the importance of conditions in this domain. An informal accounting of the three program brochures for Centrelink student payments shows that nearly half of the document content expresses eligibility conditions or the definitions that go with them.[1]

The goal of this collaborative project is to help Centrelink communicate more effectively with its constituency by producing tailored brochures. Centrelink currently writes its brochures for generic audiences. For example, the excerpt shown in Figure 1(a) makes very few assumptions about its readers, choosing, instead, to include careful descriptions of the eligibility conditions required for the program. This allows readers to determine for themselves whether or not they are eligible for the program. By contrast, a tailored brochure is personalized for a particular reader. For example, Figure 1(b) shows the one-page excerpt of an automatically produced brochure tailored for a reader who is known to be an Indigenous Australian citizen pursuing an approved course of study. This tailored excerpt covers the same information content as the generic version, but its assumptions about the reader allow it to replace the two pages of detailed descriptions of eligibility conditions with a short bulleted list summarizing the characteristics known to be true of the individual reader (i.e., "You've told us that you meet the following criteria: … so you should be eligible for this program.").

This example includes two forms in which conditions are expressed in citizen-focused, tailored PA documents: the obligative mode form (e.g., "You must be an Aboriginal …" see Figure 1(a)) and the simple declarative form (e.g., "You are an Indigenous Australian …" Figure 1(b)). We focus here on a mechanism that supports these and other related forms.

This paper describes an application that automatically produces tailored brochures for the Centrelink domain. It reviews related work, details the role conditions play in the domain and describes a prototype application that implements automatic tailoring. The paper focuses in particular on the representation, use and expression of conditions.

---

[1] Condition expressions and related definitions comprise 49.6% (6790/13687 words) of the ABSTUDY, Austudy & Youth Allowance brochures (Centrelink, 2011).

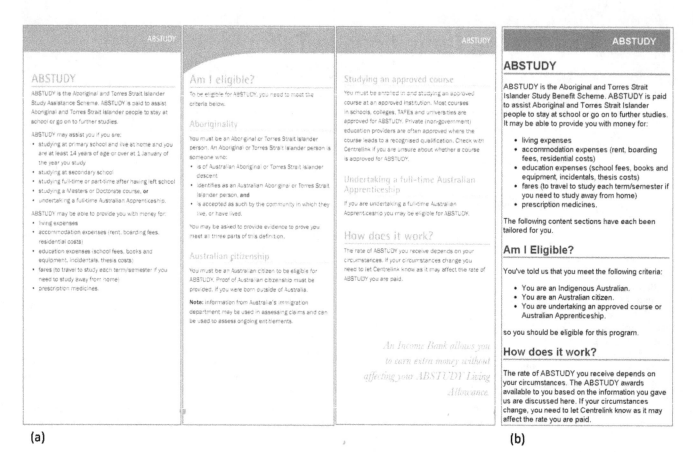

Figure 1. (a) a three-page excerpt from the current brochure; (b) a corresponding one-page excerpt from the tailored brochure.

2. RELATED WORK

The work described in this paper provides a treatment of tailored, citizen-focused documents in PA. It attempts to strike an effective balance between theoretical approaches to Natural Language Generation and applied approaches to Document Automation.

2.1 Natural Language Generation

This work is based on the theory and techniques developed in Natural Language Generation (NLG) for tailored delivery (Reiter & Dale, 2000). Bateman and Paris's work on the expression of logical conditions is a good example of a theoretically-oriented NLG approach to tailored expressions of logical conditions (Bateman & Paris, 1989). Their system produced a variety of literate output forms tailored for different users, but it required considerable effort from skilled analysts to configure the linguistic resources required to support the tailoring. In a typical PA organization like Centrelink, the volume of material that must be configured and presented is too great and the authors' training in technical analysis too limited for this approach to be practical.

More engineering-oriented approaches from the NLG community addressing the eHealth domain include STOP, which produced tailored smoking cessation letters (Reiter, Robertson, & Osman, 2003), and HealthDoc, which generated tailored health-education documents (DiMarco, et al., 2008). Though in a different domain, these systems are similar in scope and orientation to the current work. The work reported here tends to manage a greater volume of more coarsely-grained text units and is thus more limited in its ability to address detailed issues in sentence-level generation, e.g., referring expressions (DiMarco, et al., 2008), aggregation (Dalianis, 1999), and variation in expressional form (Bateman &

Paris, 1989)(Hovy, 1997). The work reported here follows more directly from applications built using the Myriad platform (Paris, Colineau, Lampert, & Vander Linden, 2010).

While the domain of eHealth is a useful starting point for this work, PA domains tend to place a heavy focus on eligibility conditions. Note that there is work on automatically recognizing conditional guidelines in existing medical texts (Georg, Hernault, Cavazza, Prendinger, & Ishizuka, 2009), but while such mechanisms have the advantage of starting from existing texts, they are not generally capable of extracting operational conditions of the form required for the tailoring being described here.

2.2 Document Automation

As a more practical alternative to NLG, Document Automation (DA) tools support tailoring by providing mail-merge features extended with conditional inclusion and exclusion of coarse-grained text units generally on the order of sections, paragraphs or perhaps sentences. There are, perhaps, a dozen commercial Document Automation (DA) systems (e.g., HotDocs, Exari, and Arbortext) that have been used in the legal profession to automate the production of custom-built legal documents (e.g., deeds of sale, standardized agreements, etc) and in the technical documentation field to produce model-specific product documentation. These tools support the inclusion or exclusion of text units as appropriate for given readers, but they are not generally powerful enough to support detailed forms of tailoring. For example, the feedback text shown in Figure 1(b). ("You've told us that you meet …") would be difficult to support in the general case with traditional DA tools because they do not support sufficiently fine-grained sentence construction mechanisms.

As was the case with NLG, PA's focus on conditions distinguishes it from the common DA domains of legal and technical documents.

2.3 Ontology Verbalization

Work in ontology verbalization attempts to take logical expressions written in an ontology language and generate textual descriptions that are both fluent and accurate. Recent work in this area has focused on representations using the Web Ontology Language (OWL) (Power & Third, 2010). This work tends to produce structural mappings from all possible ontological expressions to simple textual forms that are grammatical and correct, but may not be as fluent as might be desired.

The work described here implements ontology verbalization for the subset of OWL expressions most commonly used in citizen-focused PA documents, but it also provides a mechanism for authors to override the automated expression when logical expressions become too complex to be generated automatically with any fluency or when the texts must be more carefully crafted.

3. CONDITIONS

Conditions play a key role in both PA and in tailoring. In addition, the conditions used by PA organizations to make eligibility determinations for citizens are the same conditions that tailoring engines use to produce tailored documents for readers.

3.1 The Use of Conditions in PA

Conditions are fundamental to the citizen-focused aspects of PA. Centrelink determines who is eligible for a given program by checking their personal characteristics against the conditions associated with the program, known as *qualification* conditions. For example, the qualification conditions for the ABSTUDY program are expressed on pages 2 and 3 of the excerpt from the current brochure shown in Figure 1(a). They include aboriginality, citizenship, studying an approved course and undertaking an apprenticeship. Similar lists of qualification conditions head every program brochure that Centrelink produces.

Centrelink determines the appropriate level of payment for qualifying students by checking personal characteristics against the conditions associated with each element of the program, known as *payability* conditions. For example, Centrelink uses payability conditions to assign ABSTUDY recipients appropriate award categories, benefits, means tests and payment rates.

As an example of the importance of conditions, consider the ABSTUDY program, which supports 29 payment rates. Each rate is designed for students with different payability characteristics. There are a sufficient number of these rates that the current ABSTUDY brochure doesn't take the time to list them individually, electing, instead, to group them into sub-sets. For example, it includes the following discussion of the "rate" for students living at home with their parents, which is actually a gloss for the set of 9 rates designed for dependent children:

The 'at home' rate

The 'at home' rate of ABSTUDY Living Allowance applies if you:
- *are living at home*
- *do not qualify for the away, independent, or pensioner rates, or*
- *are in state care, have reached the minimum school leaving age for your state/territory, and your carer receives a Foster Care Allowance for you.*

Similar sections address other sub-sets of rates (i.e., "away from home" and "independent" rates). [2]

It is interesting to note that this excerpt only discusses conditions; this section of the ABSTUDY brochure contains no information about the rates themselves. In a sense, the expression of the conditions takes precedence in this case over the expression of "raw" content. This is typical in citizen-focused PA documents.

By contrast, a tailored output for the rates of ABSTUDY can leverage what it knows about the reader to reduce the number of rates that must be expressed and to use the saved space to include the actual rate the reader would receive. For example, the following output is automatically tailored for an 18-year-old Indigenous student living at home with her parents:

Rates of ABSTUDY

The rates listed here are fortnightly maximum payment rates and serve as a guide only. They are effective from 1 January 2011.

Because you are living at home, 18-20 years of age: $255.80

Here, the text expresses the conditions known about the reader as feedback, and then includes the actual payment rate. It shows how tailored content can be matched to the individual reader, which entails removing irrelevant material from the current brochure and potentially adding new material not in the current brochure.

3.2 The Use of Conditions in Tailoring

Conditions are also fundamental to tailoring. Tailoring mechanisms match the appropriate information units with each reader based on the unit's conditions and the reader's characteristics.

For example, the ABSTUDY program has 7 award categories, each having different payability conditions based on the reader's age, school status, living arrangements, etc. Usually a reader is eligible for only one category. A tailored delivery application can take advantage of this situation by choosing to express only the award for which the reader is eligible. In the case of ABSTUDY, this can reduce the six pages of information on awards down to a single page.

3.3 The Nature and Expression of Conditions

Conditions come in a variety of forms in citizen-focused PA documents, most of which can be represented by expressions built from simple property-value pairs. For example, the first qualification condition for ABSTUDY shown in Figure 1(a) specifies that the reader has a race property value of Indigenous Australian (i.e., *reader hasRace indigenousAustralian*). The potential values for the hasRace property would include Caucasian, African, Asian, etc.

Individual property-value pairs can then be combined using conjunction, disjunction and negation. For example, the current ABSTUDY brochure specifies four qualification conditions in parallel sections; readers are expected to discern the actual logical relationship: aboriginality AND citizenship AND (approved-course OR apprenticeship).

---

[2] The individual rates are listed on Centrelink's website, see http://www.centrelink.gov.au/internet/internet.nsf/payments/abstudy.htm.

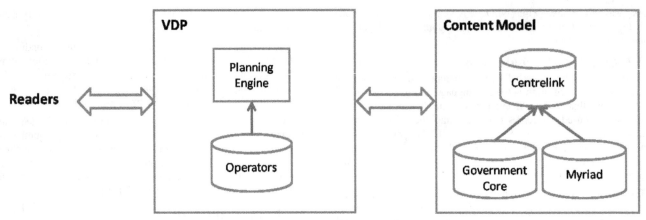

Figure 2. The Myriad platform architecture

There are complexities that arise in the nature of the conditions, but though they are often driven by political motivations, the conditions tend to be founded by practicality. Conditions that cannot be expressed using some formulation of feature-value pairs with conjunction and disjunction are rare.

In citizen-focused PA documents, conditions can take the following forms:

- **Declarative forms** – Informational brochures generally use one of the following forms:
  - Obligatory modality – *"You must be* an Indigenous Australian." – These expressions are the common way to express required eligibility expressions (e.g., see the "You must be …" expressions in Figure 1(a)).
  - Simple declarative – *"You are* an Indigenous Australian." – These declarations can be used to express tailored feedback to the reader (e.g., see "You've told us that you meet the following criteria: …" in Figure 1(b)).

  Traditional brochures use the first form; tailored brochures can include the second.
- **Query forms** – Application forms and query wizards express eligibility conditions in the form of queries:
  - Boolean query – *"Are you* an Indigenous Australian? Yes/No"
  - Multiple-choice query – *"What is your* race? Indigenous/White/Asian/African/…"

The prototype application described below produces these forms of expression as appropriate for the context.

4. GENERATION PLATFORM

The prototype application is built using Myriad, a platform for tailored information delivery (Paris, Colineau, Lampert, & Vander Linden, 2010). It acquires information on the current reader's characteristics and uses this information to plan and generate a brochure that is coherent in expression and tailored for the reader. This section discusses its two main architectural components, shown in Figure 2.

4.1 Document Planning

The document planning is done by the Virtual Document Planner (VDP), shown in the center of Figure 2. The VDP uses a text planning engine and declarative plan operators (Moore & Paris, 1993) to produce a discourse tree structured with rhetorical relations coded in Rhetorical Structure Theory (RST) (Mann & Thompson, 1988) and populated with information content. These

are well-known technologies in the Natural Language Generation (NLG) community (Reiter & Dale, 2000) and the specializations provided by the Myriad platform are documented elsewhere (Paris, Colineau, Lampert, & Vander Linden, 2010); cf. the use of RST in recognition (Georg, Hernault, Cavazza, Prendinger, & Ishizuka, 2009).

The Centrelink prototype application exhibits one key departure from traditional approaches to building NLG applications in that it works with coarse-grained information units. Traditional NLG applications have tended to include tactical generation components that work with units at the lexical and sentence levels. While the Myriad platform supports this traditional mode, its applications have tended to work with units at the level of paragraphs and sections.

The main disadvantage of coarse-grained NLG is the stylistic infelicities that can occur when structuring the units in different orders. The relative scarcity of these infelicities in citizen-focused PA documents has been detailed elsewhere (Colineau, Paris, & Vander Linden, 2011), but when they do occur, a repair mechanism is required. The repair mechanisms required for condition expression are discussed below.

The advantages of the coarse-grained approach to generation are that: (1) it is more efficient, both in terms of computation and representation; (2) it is more easily used by authors with limited technical training; and (3) it allows the application to rely on text units carefully written by professional authors to satisfy the legal and political requirements placed on the content.

4.2 Content Modeling

The information content is stored in the Content Model, shown on the far right of Figure 2. This model is implemented as an ontology using OWL-DL (Smith, Welty, & McGuinness, 2004). The conditions are also represented in the content model, and the conditional reasoning is implemented using HermiT (Motik, Shearer, & Horrocks, 2009). The conditions and conditional reasoning are discussed in detail in the next section.

OWL ontologies are made up of OWL *entities*, where an entity is either: a *class* (aka concept), a *property* (aka relationship) or an *individual* (aka instance). To support reuse of the various entities in this model, it is divided into the following three separate but interrelated ontologies:

- The **Government Core** ontology comprises the general administrative entities, e.g., classes like Person and Condition, properties like hasCondition and hasRace, and

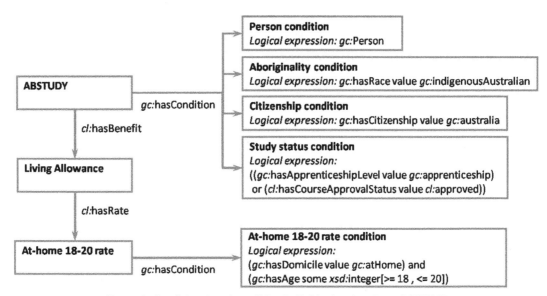

Figure 3. Conditional and condition individuals related to ABSTUDY

individuals like the legal values for hasRace (e.g., indigenousAustralian, Asian, ...). Cf. the more structural ontologies discussed elsewhere in the eGovernment literature (Peristeras, Tarabanis, & Goudos, 2009).

- The **Myriad** ontology comprises the entities required by the VDP for the purpose of document tailoring and generation, e.g., classes like Domain and User, and annotations like hasName and hasLongDescription. The Myriad ontology has no real content of its own and can, thus, be better seen as a "harness" that is used to categorize content from the other ontologies in a manner expected by the operators in the VDP's plan library. For example, the VDP expects all individuals representing expressible information units to have the following annotation values:

  o **Name** – The one-word or one-phrase title of an entity;

  o **Short description** – A one-sentence or one-paragraph-sized summary of the entity;

  o **Long description** – A one-paragraph or one-section-sized description of the entity

These annotations represent the coarse-grained information units from which the VDP builds tailored documents. For example, the name annotation for the individual representing the ABSTUDY program is "ABSTUDY", which is used as a title and header in Figure 1(b). Its short description annotation is a one-sentence summary of the program: "ABSTUDY is the Aboriginal and Torres Strait Islander ... go on to further studies" (see Figure 4 below, under the "ABSTUDY" list item), and its long description annotation is the paragraph and bulleted list shown in Figure 1(b) under the "ABSTUDY" heading (i.e., "ABSTUDY is the ... prescription medicines"). The VDP determines which of the description annotations to use in each context.

- The **Centrelink** ontology imports the entities from the Government Core and Myriad ontologies and integrates them within a full ontology for the Centrelink domain. It includes Centrelink-specific classes like Program and Award, properties like hasRate and hasTest, and individuals like ABSTUDY and Austudy.

As a matter of notation, the reminder of this paper prefixes all entities in the content model by one of the following codes: *gc:* for the government core ontology; *myriad:* for the Myriad ontology; and *cl:* for the Centrelink ontology.

5. REPRESENTING CONDITIONS

The Content Model allows conditions to be associated with any individual in the ontology for use in eligibility reasoning and in tailoring. Conditions are represented as reified condition individuals associated with the relevant domain individual and annotated with logical expressions written in textual form.

For example, Figure 3 shows a sub-set of the individuals related to ABSTUDY along with their conditions and logical expressions. On the left, there are three domain individuals representing: the ABSTUDY program; Living Allowance, one of ABSTUDY's key benefits; and the Living Allowance rate for students between 18 and 20 years of age living at home. These individuals are connected using properties defined for the Centrelink domain: programs have, amongst other things, benefits linked with the *cl:*hasBenefit property and benefits have, amongst other things, payment rates linked with the *cl:*hasRate property.

Figure 3 also includes condition individuals, which are associated with the domain individuals using the *gc:*hasCondition object property. The ABSTUDY individual has four conditions shown on the top right: personhood, aboriginality, citizenship and study status. The at-home rate individual has a single condition, shown on the bottom right. The logical expressions associated with each condition individual are written in textual form using Manchester OWL syntax.[3] For example, the at-home rate condition on the

---
[3] See http://www.co-ode.org/resources/reference/manchester_syntax/.

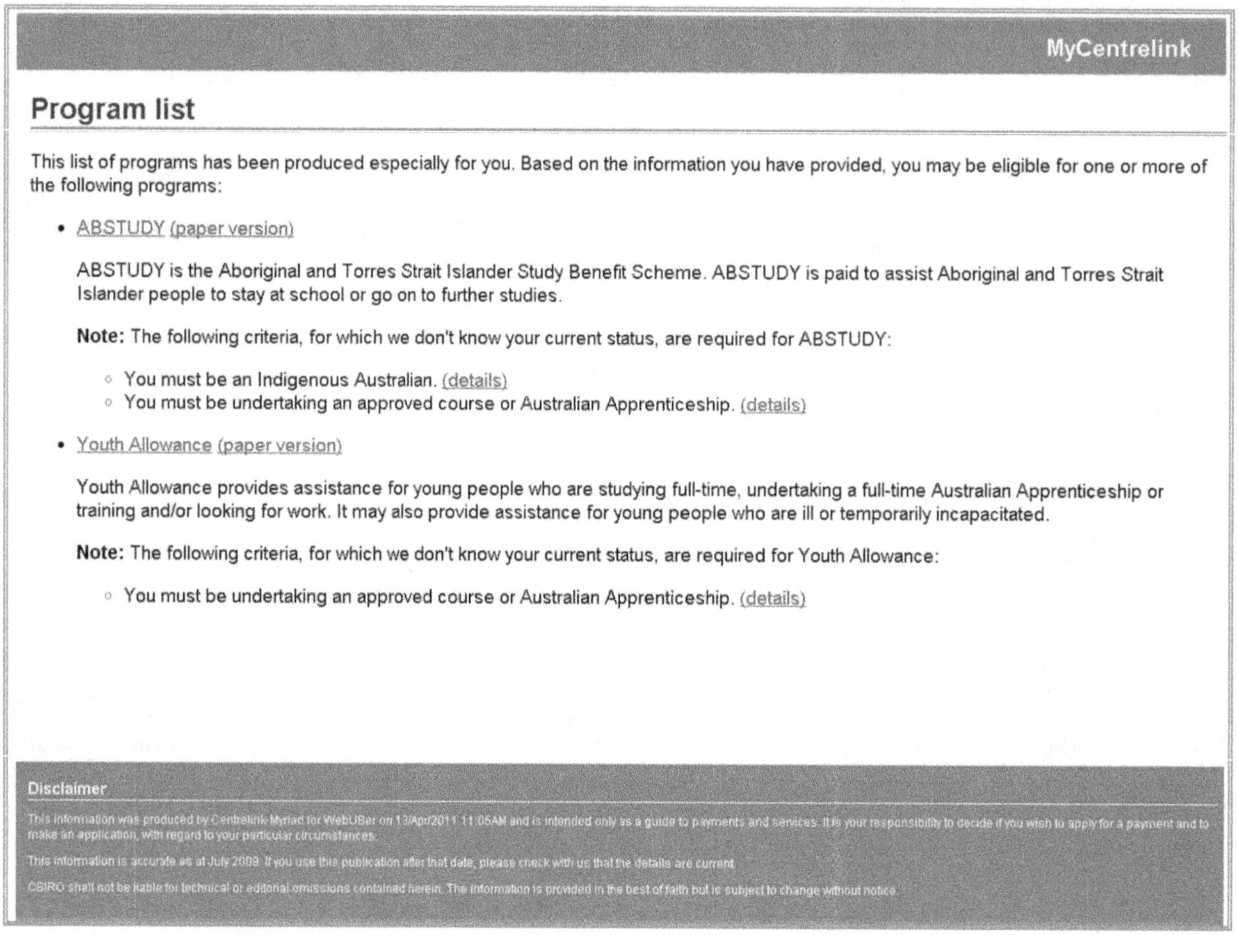

Figure 4. A program list for an 18-year-old Australian citizen living at home

bottom right indicates that the reader must be living at home and have an age between 18 and 20 inclusive. Note that the plan operators assume that: (1) the logical expressions are written with respect to the current reader, so the reader is not explicitly mentioned; and (2) readers being considered for sub-elements of a program must be eligible for the program itself, so the program's qualification conditions are not repeated in the rate's payability conditions.

Conditions are reified as separate individuals so that Myriad annotations can be attached to them separately from the annotations attached to the domain individual they modify. Also, authors can choose whether to associate multiple condition individuals with simple logical expressions, as is done for the ABSTUDY individual, or a single condition individual with a more complex logical expression, as is done for the at-home rate individual. The reasons for these decisions relate to expression and are discussed in the next section.

When the VDP constructs tailored presentations, it is configured to determine the eligibility of the current reader for the various elements of the content model and to make discourse planning decisions based on this determination. For example, Figure 4 shows a list of educational entitlement programs tailored for an 18-year-old Australian citizen. The list includes two of the three student programs currently represented in the content model: ABSTUDY, a program or Indigenous Australians, and Youth

Allowance, a program for all younger Australian students. The third program, Austudy, is for older students and is thus not listed here.

To tailor this list, the VDP's operators ask the content model to compute the reader's eligibility for each of the programs in its ontology. The content model recognizes three levels of eligibility:

- **Eligible** readers have characteristics that provably satisfy the logical conditions associated with the domain individual.
- **Ineligible** readers have characteristics that provably fail to satisfy the logical conditions.
- **Unknown** readers have characteristics that neither satisfy nor fail to satisfy the logical conditions.

The VDP's plan operators for the current prototype are configured to list those programs for which the reader is potentially eligible, that is, for which the reader's eligibility is determined to be either eligible or unknown. The reader is assumed to be uninterested in the programs for which he or she is not eligible.

The content model uses categorical reasoning to compute the reader's eligibility for each program by constructing an unnamed OWL class expression that conjoins all the condition individuals associated with the program. For ABSTUDY, this expression is as follows:

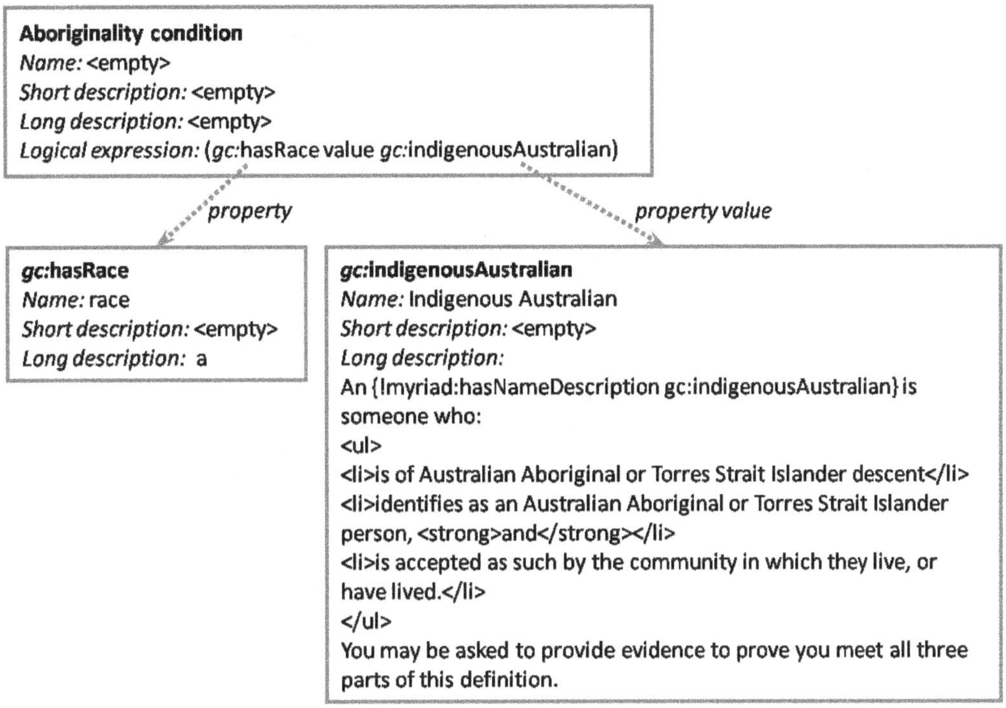

Figure 5. The individuals involved in the expression of the Aboriginality condition

*gc:*Person **and**

(*gc:*hasRace value *gc:*indigenousAustralian) **and**

(*gc:*hasCitizenship value *gc:*australia) **and**

((*gc:*hasApprenticeshipLevel value *gc:*apprenticeship) **or**
(*cl:*hasCourseApprovalStatus value *cl:*approved))

This expression conjoins the logical expressions for ABSTUDY's four conditions shown in Figure 3. The content model contains the following assertions about a reader who is an 18-year-old Australian living at home:

*gc:*Person **and**

(*gc:*hasCitizenship value *gc:*australia) **and**

(*gc:*hasAge value 18) **and**

(*gc:*hasDomicile value *gc:*atHome)

Given that the object properties used in these assertions are functional properties whose domains are specified by value partitions of mutually exclusive individuals, OWL classification reasoning is unable to determine whether the reader is an individual of the unnamed class. Neither is it able to determine that the reader is ineligible. It would need to know the reader's race and study status to make a determination. Thus, the reader's status is determined to be unknown. This leads the VDP to include the ABSTUDY program as one of the programs for which the reader is potentially eligible, see Figure 4. Note that the VDP's operators also provide feedback listing the conditions for which the reader's condition is unknown. It assumes that this will help the reader decide if the program warrants further investigation.

The program list in Figure 4 provides hyperlinks to web and paper versions of the two programs as well as to further details on each of the unknown conditions. If the reader chooses to create a paper brochure for ABSTUDY, the VDP will create a tailored version of the ABSTUDY brochure that includes, among other things, expressions of the rates for which the reader is potentially eligible. Given that the reader is 18 and living at home, there will be one

rate – the rate whose representation is shown at the bottom left of Figure 3 and whose expression was discussed at the end of Section 3.1. The content model selects this particular rate using the same reasoning process described above for selecting potentially eligible programs.

The prototype implements conditions using OWL has-value property expressions for data and object properties or existential conditions for numeric properties. It also supports conjunction and disjunction. These forms have been sufficient for the current application, but other forms, e.g., universal quantification and cardinality constraints, could be used as needed.

6. EXPRESSING CONDITIONS

The prototype provides support for expressing the conditions described in the previous section. It supports a default mechanism that automatically generates condition expressions, and it allows authors to override the automatic generation by manually specifying their own expression.

6.1 Automated Expression

The automatic expression mechanism supports OWL has-value conditions for object and data properties, existential conditions for numeric properties, and class-instance conditions. It also provides recursive support for conjunctions and disjunctions.

Figure 5 shows the individuals involved in the automated expression of the Aboriginality condition. The Aboriginality condition, at the top of the figure, gives a more detailed view of the same individual shown on the top right of Figure 3, including the Myriad annotations discussed in Section 4.2. The name and the short and long descriptions for this condition are empty. This figure shows the two entities referenced in the simple has-value logical expression: the *gc:*hasRace property and the *gc:*indigenousAustralian property value individual, along with their respective names and descriptions. When the condition individual provides no descriptions, as is the case here, plan

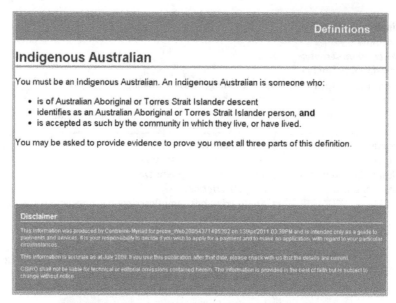

Figure 6. Full definition for Indigenous Australian

operators implementing one of the following patterns are used to automatically generate a condition expression.

Has-Object-Value conditions. To express a simple has-value expression for an object property, the VDP plan library provides plan operators that construct expressions as shown here.

The constructed name is the name of the property in the condition. In the example, this is the name of the hasRace property – "race". If the property has no name annotation, then the name of the property itself is used.

The constructed short description follows this pattern:

$$\left\{ \begin{array}{l} \text{"You are "} \\ \text{"You must be "} \end{array} \right\} + propertyLongDesc + valueName + \text{"."}$$

Here, the opening is chosen to match the desired expression type, discussed in Section 3.3, and this is followed by the long description of the property and then the name of the property value. For an obligatory mode declarative form of the example, the result is "You must be " + "a" + "Indigenous Australian" + "." You can see this constructed expression in the first bullet under ABSTUDY in Figure 4. Note that the incorrect indefinite article here is repaired as discussed below.

Note also that the long description of the gc:hasRace is not actually a description of the property at all; properties are never described individually. To support more flexible expression, the long description specifies a text designed to be used in the pattern just described and need not always be a simple indefinite article. For example, the gc:hasCourseApprovalStatus property has the long description "taking a", which leads to expressions like "You are taking an approved course.".

The constructed long description for a condition has the same structure as its short description but with the addition of the value's long description at the very end. The constructed long description for the example is shown in Figure 6. The definition page shown here uses the constructed name as the title and the condition's constructed long description as the text body. Note that the body starts with the short description and then includes the value's long description. See Figure 5 for the raw texts used here.

Data-Value conditions. To express a condition based on a numeric data property, using either in a has-value and an existential condition, the current prototype makes the assumption that the condition is an age condition and uses the following pattern:

$$\left\{ \begin{array}{l} \text{"You are "} \\ \text{"You must be "} \end{array} \right\} + \left\{ \begin{array}{l} X + \text{" years of age."} \\ X + \text{" years of age or older."} \\ X + \text{" years of age or younger."} \\ \text{" between "} + X + \text{" and "} + Y + \text{" years of age."} \end{array} \right\}$$

This pattern supports the four forms of age conditions most commonly used in Centrelink brochures. Support for other numeric conditions or other data-value properties could be added as needed.

Class-Instance conditions. To express class-instance conditions, the operators use the following pattern:

$$\left\{ \begin{array}{l} \text{"You are a "} \\ \text{"You must be a "} \end{array} \right\} + className + \text{"."}$$

For example, the prototype generates "You must be a person." for the person condition shown at the very top of Figure 3. As mentioned above, this particular example is never used by the prototype.

Conjoined/Disjoined conditions. To express sets of simple condition expressions combined recursively using conjunction and disjunction, the operators combine the individual condition expressions using "and" and "or" as appropriate.

For example, the automatically generated expression of the study-status condition shown in the middle of Figure 3 is:

> *You must be undertaking an Australian Apprenticeship, or you must be studying an approved course.*

This expression is serviceable but it is not the one used in the outputs shown in Figure 1(a) and Figure 3. Those expressions are hand-authored as described in the next section to make them somewhat shorter and more fluent. The longer and more complicated the logical expression is, the more imperative it becomes to hand-author a more readable form.

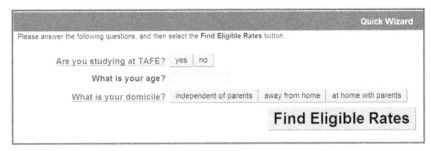

Figure 7. An automatically generated query wizard

Repair. When concatenating hand-authored strings in the manner described above, certain grammatical and stylistic infelicities occur, including incorrect indefinite article matching (e.g., "a Indigenous Australian"), improper capitalization in titles and sentences, and incorrect punctuation in conjoined and disjoined clauses. After the plan operators construct an expression, they repair these infelicities using simple text manipulation routines. Indefinite articles are set appropriately, titles are capitalized, non-titles are not capitalized, and commas are put in place of periods in conjoined or disjoined sentences.

These repair mechanisms are considerably weaker than proper NLG mechanisms, but they have proven to be adequate for the current prototype and their use greatly simplifies the task of constructing the content model and planning resources.

Query forms. The plan library also includes operators that construct query expression forms from condition representations. These query forms are suitable for use in the query wizard that the system uses to collect characteristics from the reader to be used in tailoring and could be extended to produce paper application forms.

For example, the query wizard shown in Figure 7 is constructed for the rates of ABSTUDY Living Allowance. It includes queries to collect all the reader characteristics needed to determine the appropriate rate of Living Allowance that the reader should receive. Though this doesn't guarantee in the general case that only one rate will be chosen, if the conditions on the individual rates are constructed in such a way that they are mutually exclusive and exhaustive and the reader answers all the questions, then only one rate will be chosen.

Note that the possible query forms include all the forms discussed in Section 3.3:

- The Boolean form, e.g., "Are you studying at TAFE?", is chosen when only one value of the given property is relevant. In this case, there are several possible values for the cl:hasStudyLevel property, but only one of those values appears in the conditions for the rates for ABSTUDY Living Allowance. Thus, a Boolean query is sufficient to determine which rate to choose. The construction pattern used here is:

$$\text{"Are you " } + valueName + \text{ "?"}$$

- The Multiple-choice form, e.g., "What is your domicile?", is chosen when several values of the given property appear in the relevant conditions. Here, there are three possible values for gc:hasDomicile that factor into the conditions on the rates. Thus, a Boolean query

form is insufficient to collect the necessary characteristics. The construction pattern used here is:

$$\text{"What is your " } + propertyName + \text{ "?"}$$
$$+ conditionButtons$$

One condition button is created for each value and the value's name is used as a label for that button.

- The numeric query form, e.g., "What is your age?", is chosen for numeric data properties. The pattern used here is:

$$\text{"What is your " } + propertyName + \text{ "?"}$$
$$+ numericInputField$$

The prototype can generate a query wizard such as this for any class or individual-property pair. For classes (e.g., cl:Program), it creates a wizard to distinguish the individuals declared to be instances of that class. For individual-property pairs (e.g., the Living Allowance-rate pair shown in Figure 7), it creates, a wizard to distinguish the individuals related to the given individual through the given property.

6.2 Manual Expression

One key departure of the current work from the more theoretical work on ontology verbalization is that the prototype is designed to allow the author to intervene and specify their own information content for any condition being expressed in the tailored document. This is important because the nature of the conditions supported by the expression techniques discussed in the previous section is limited. Some conditions don't easily fit into a "*You are a/n X.*" pattern. For these cases or when the conditions are complex enough to render the automatically generated expressions infelicitous for any reason, the prototype provides a manual-override mechanism.

Coarse-grained, tailored generation. Before discussing the manual override of condition expressions in particular, it is important to note that the entire generation mechanism is designed to construct tailored documents by piecing together the coarse-grained information units included as annotations in the content model. These units must all be hand-authored because the prototype provides no automated support for generating them automatically.

Pure NLG techniques would attempt to represent the key elements of these information units and generate them from first principles, but this paper argues that this is rarely practical in PA and is sometimes dangerous. For example, while it might be possible to represent the elements of the definition of Aboriginality shown in Figure 6, it would be difficult for authors untrained in content modeling to do this effectively. Further, while this statement of the race property is relatively simple logically speaking, it is complex politically speaking. The definition of Aboriginality given in the current brochure has been the subject of considerable attention by the Australian government and should, thus, be stated exactly as specified in the legislation.[4] Citizen-focused documents for PA have a legal status that makes any form of paraphrasing or approximation dangerous. This makes it difficult to deploy pure NLG in the PA context.

Hand-authored condition expressions. For reasons discussed above, the prototype provides automated support for generating condition expressions from their logical expressions, but it also provides a manual-override mechanism. This mechanism is based on the name and description annotations set for the condition individuals in the content model. In cases where authors would like to override the automated generation of condition expressions, they can compose an appropriate name and descriptions for the condition individual.

For example, to override the automated expression of the Aboriginality condition represented in Figure 5, the author must compose a name and two descriptions for the Aboriginality condition individual itself. This overrides the construction of the condition expression from the names and descriptions of the constituents of the logical condition, as described in the previous section.

Repair. The repair of infelicities of expression is a tricky issue with hand-authored texts. Generally, the plan operators accept hand-authored texts as given, but when alternating between simple and obligatory declaration forms, hand authored texts must be repaired. Again, the prototype does this using simple text manipulation routines, but performing these manipulations on hand-authored texts is more difficult than those done for automatically authored conditions. The current prototype replaces instances of "You must be" and "You are" in hand-authored texts to match the chosen form of expression and relies on the authors to ensure that the resulting text is grammatically and stylistically correct.

7. CONCLUSIONS

This paper has presented a prototype application that integrates automated condition expression with hand-authored conditions. The mechanisms that support this integration are detailed.

The paper argues that using coarse-grained generation with auto-generated conditions that can be overridden by authors provides a practical solution to tailored delivery in the PA domain. The mechanisms supporting this integration are demonstrated and shown to be feasible.

The next steps for this work are to build an authoring tool that makes these features conveniently and practically available to authors in PA. The current prototype provides a partial implementation of a frame-based authoring tool that can be used as the basis for this authoring tool, but it is likely that a higher-level interface metaphor will be needed to build a tool that authors can use to effectively drive the mechanism described in this paper. In addition, studies assessing the effectiveness of tailoring in PA generally and of the generated conditions in particular will be necessary.

8. REFERENCES

Bateman, J. A., & Paris, C. (1989). Phrasing text in terms the user can understand. *Proceedings of the 11th International Joint Conference on Artificial Intelligence*, (pp. 1511-1517).

Centrelink. (2011). Retrieved from http://www.centrelink.gov.au/

Colineau, N., Paris, C., & Vander Linden, K. (2011). Automatically Generating Citizen-Focused Brochures for Public Administration. *Proceedings of the 12th Annual International Conference on Digital Government Research*. College Park, Maryland, USA.

Dalianis, H. (1999). Aggregation in Natural Language Generation. *Journal of Computational Intelligence , 15* (4), 384-414.

DiMarco, C., Bray, P., Covvey, H. D., Cowan, D., DiCuccio, V., Hovy, E., et al. (2008). Authoring and Generation of Individualised Patient Education Materials. *Journal on Information Technology in Healthcare , 6* (1), 63-71.

Georg, G., Hernault, H., Cavazza, M., Prendinger, H., & Ishizuka, M. (2009). Fron Rhetorical Structures to Document Structure: Shallow Pragmatic Analysis for Document Engineering. *Proceedings of the ACM Symposium on Document Engineering* (pp. 185-192). Munich: ACM Press.

Hovy, E. (1997). Pragmatics and Natural Language Generation. *Artificial Intelligence , 43*, 153-197.

Mann, W. C., & Thompson, S. A. (1988). Rhetorical structure theory: Toward a functional theory of text organization. *Text , 8* (3), 243-281.

Moore, J. D., & Paris, C. L. (1993). Planning text for advisory dialogues: Capturing intentional and rhetorical information. *Computational Linguistics , 19* (4), 651-694.

Motik, B., Shearer, R., & Horrocks, I. (2009). Hypertableau Reasoning for Description Logics. *Journal of Artificial Intelligence Research , 36*, 165-228.

Paris, C. L., Colineau, N., Lampert, A., & Vander Linden, K. (2010). Discourse Planning for Information Composition and Delivery: A Reusable Platform. *Natural Language Engineering , 16* (1), 61-98.

Peristeras, V., Tarabanis, K., & Goudos, S. K. (2009). Model-driven eGovernment interoperability: A review of the state of the art. *Computer Standards & Interfaces , 31* (4), 613-628.

Power, R., & Third, A. (2010). Expressing OWL Axioms by English Sentences: Dubious in Theory, Feasible in Practice. *Proceedings of the 23rd International Conference on Computational Linguistics*. Beijing.

Reiter, E., & Dale, R. (2000). *Building Natural Language Generation Systems*. Cambridge University Press.

Reiter, E., Robertson, R., & Osman, L. M. (2003). Lessons from a Failure: Generating Tailored Smoking Cessation Letters. *Artificial Intelligence , 144* (1), 41-58.

Smith, M. K., Welty, C., & McGuinness, D. L. (2004). *OWL Web Ontology Language Guide*. Retrieved from http://www.w3.org/TR/owl-guide/

---

[4] See http://www.alrc.gov.au/publications/36-kinship-and-identity/legal-definitions-aboriginality.

Adaptive Layout Template for Effective Web Content Presentation in Large-Screen Contexts

Michael Nebeling, Fabrice Matulic, Lucas Streit, and Moira C. Norrie
{nebeling|matulic|norrie}@inf.ethz.ch
Institute of Information Systems, ETH Zurich
CH-8092 Zurich, Switzerland

ABSTRACT

Despite the fact that average screen size and resolution have dramatically increased, many of today's web sites still do not scale well in larger viewing contexts. The upcoming HTML5 and CSS3 standards propose features that can be used to build more flexible web page layouts, but their potential to accommodate a wider range of display environments is currently relatively unexplored. We examine the proposed standards to identify the most promising features and report on experiments with a number of adaptive layout mechanisms that support the required forms of adaptation to take advantage of greater screen real estates, such as automated scaling of text and media. Special attention is given to the effective use of multi-column layout, a brand new feature for web design that contributes to optimising the space occupied by text, but at the same time still poses problems in predominantly continuous vertical-scrolling browsing behaviours. The proposed solutions were integrated in a flexible layout template that was then applied to an existing news web site and tested on users to identify the adaptive features that best support reading comfort and efficiency.

Keywords

Adaptive layout, large display settings, new web standards

Categories and Subject Descriptors

H.5.2 [**Information Interfaces and Presentation**]: User Interfaces—*Screen design*

General Terms

Design, Human Factors

1. INTRODUCTION

Since the early days of the web and the realisation that a static presentation of web content was overly limiting, researchers, web developers and designers alike have sought to come up with methods to adapt content to a variety of constraints and limitations imposed by a number of factors, in particular the viewing context. Considering the wide variety of browsing devices, types of content to be delivered and consuming conditions, the task of providing a solution that tackles all these different concerns at the same time is formidable. This complexity has led software engineers and researchers to develop comprehensive frameworks with the flexibility to distribute content in a broad range of contexts as well as concentrate on specific subsets of these issues in order to solve them more effectively.

In the past decade, the booming emergence of small-form factor devices such as PDAs, smartphones and tablets have caused much research effort to be directed towards addressing the particular problems associated with this category of appliances, i.e. small screen size, reduced resources and particular interaction models. However, recent years have also seen an increase at the other end of the spectrum, namely that of large-size viewing terminals, which are also becoming more widespread[1]. Despite this clear trend towards higher resolutions and screen sizes, this segment has surprisingly received little attention from the web engineering community.

Studies show that large displays bring productivity benefits if the interfaces are suitably adapted to make use of the available screen space [6, 18]. For web sites, pages that present a greater degree of adaptability to larger screen dimensions can increase the user's experience and overall comfort when engaging with web content, as we will demonstrate in this paper. Yet, an investigation on news web sites that we conducted recently [17] shows that the vast majority of them fail to fully utilise available screen real estate on large displays, as the layouts of web pages prove to be mostly fixed and static. The result is the appearance of wide margins surrounding the main content as well as smaller text and multimedia elements. To some extent, the latter can be remedied using zoom features of web browsers, but this does not always yield satisfactory results and, more importantly, it does not solve the layout problem. This failure to adapt affects the browsing experience and to a greater or lesser degree detracts from the web site's general usability.

The reasons for using static layout have mostly to do with design issues (it is easier to design for fixed dimensions) and the perceived added complexity and costs that a system supporting highly adaptive and flexible page structuring would incur. The increasing diversity of the web-browsing device landscape is real, however, and so instead of devel-

---

[1]http://www.w3schools.com/browsers/browsers_display.asp

oping adaptive presentation engines for their content, publishers have opted to produce several versions of their web sites dedicated to specific device categories, mainly desktops and smartphones. There is currently no sign that content providers will address the needs of consumers with large and very large displays and so we feel that work needs to be done in that area. As a starting point, we believe that if the problem of web page layout adaptation could be posed purely in terms of web standards and dealt with in a relatively straightforward manner using native web technologies, it would provide an incentive to designers and publishers to consider more flexible layout strategies.

Based on the above philosophy, we show in this paper how layout adaptation can be achieved using standard web technologies and especially new features of HTML5 and CSS3 that lend themselves well to tackling this problem. By adopting this approach that is likely to be attractive to web developers and designers, we differentiate ourselves from prior work, which is mostly based on complex layout-generating systems implemented in proprietary browser software or requiring expensive server-side computations. The main technical contribution of this paper is an adaptive layout template that can accommodate a range of viewing situations and, because it is based on only native web technologies, can easily be applied to existing web sites (see Figure 3 at the end of the paper). In the latter part of the paper we report on the results of a preliminary user study conducted on such adapted news articles to evaluate the perception and impact of the different layouts that we create in this manner.

Our paper is structured as follows: we begin by discussing related work and relevant features of HTML5 and CSS3. This is then followed by an analysis of the required adaptation techniques and an implementation of these in the form of an adaptive layout template. Finally, we present and discuss the results of our user evaluation and conclude with an assessment of the offered functionality and its potential to improve the user experience.

2. RELATED WORK

As stated above, the adaptation of web sites to meet the constraints imposed by a heterogeneous array of consuming devices has been the subject of extensive research. In this area, one can generally distinguish two different categories of adaptation schemes: generational and transformational methods. The first type of techniques seek to build or generate dynamic, adaptive web pages usually using a dedicated server-based framework to retrieve raw content, often from a database, and then, based on a number of rules and settings, create a web page accordingly. Further adaptations can then also be performed with the help of client-side scripts. AMACONT [7] is an example of such a framework that explicitly addresses presentation and adaptive formatting of content based on the Hera web design methodology to generate reusable adaptive web documents. In the broader context of adaptive content delivery, there have also been a number of systems that address different aspects of context-specific content consumption such as multimedia content [15], web services [11] and user interfaces in general [3]. At the heart of these approaches, is the aim to provide device-independent access to content, often with a desire to separate the different architectural concerns, i.e. data management, application logic and presentation (see [4] and [8] for examples).

The second type of adaptation techniques take existing, already structured web pages and reformat them to fit a set of new constraints, typically those associated with a mobile device with smaller screen size and processing resources. Here also the literature abounds with research efforts which tackle this problem. Most of the more advanced techniques attempt one way or another to infer the semantic structure of the page using segmentation or parsing algorithms and alter it or build a new document from extracted page components so that the returned web page meets the constraints of the target client [5, 10]. Methods that analyse the content of web pages including multimedia elements such as images in order to remove extraneous items or portions thereof have also been proposed [14].

Looking at the layout-producing problem itself, a popular approach consists of determining layout requirements for a document (web or otherwise) to be dynamically generated as a set of constraint-based relationships or rules between the different components [12, 13, 19]. These constraints are typically defined in one or more templates that, when fed with input values representing target viewing conditions, generate with the help of an appropriate solver a suitable layout for the given content. One of the main difficulties here lies in the specification and characterisation of those design templates that have to abide by a number of aesthetic and content-driven rules while also allowing for a great degree of flexibility in order to adapt to a wide range of rendering contexts.

The problem with most of the techniques developed so far is that they either target a very specific category of terminals, namely small-form factor devices, or they are relatively complex to implement with pure web standards and technologies, which are usually favoured by web designers and developers. As a matter of fact, some authors of adaptive layout algorithms have even suggested extensions to CSS to enable constraint-based layout generation in a more web-compliant manner [1]. While these propositions may not have made it yet to the W3C consortium, the standards continue to evolve and expand with new specifications that address a number of problems including that of web page layout and adaptation. Specifically, the new HTML5 and CSS3 specifications bring features such as multi-column layout and advanced media queries to the table, which can be used to afford web sites a certain degree of presentation adaptivity and flexibility. We examine these new features in more detail in the next section.

3. REVIEW OF WEB STANDARDS

The HTML5 standard is currently still in working draft state[2], but many parts of the specification are considered stable and already implemented in some web browsers. At the syntactical level, HTML5 introduces two kinds of new elements. First, elements such as <nav> for a navigation block, <header> for the top and <aside> for side bar elements of a web page aim at replacing the generic block elements <div> and <span> with semantically more meaningful ones. Second, elements such as the <video> or the <svg> tag provide a standardised way of embedding multimedia elements such as videos or vector-based graphics directly in a web page.

With respect to adaptive layout techniques, the new semantic elements of HTML5 are useful for annotating page

---

[2]http://www.w3.org/TR/2010/WD-html5-20101019/

elements, which may open up new and simpler ways for the automatic adaptation of web page layouts based on document analysis. However, due to the poor backwards compatibility, it will probably take some time before web developers adopt these new standards for this purpose and refrain from using <div> or <span> elements with custom id or class attributes. Furthermore, the <video> tag introduces a new possibility to embed videos as regular DOM elements and thus allow them to be styled with CSS and manipulated using JavaScript. For example, the dimensions of a video could be changed when viewed on a large screen and, additionally, its content could be replaced with a high-definition version—this only using web standards.

The development of CSS3 is modularised and consists of various separate recommendations which are developed independently. An important module is the CSS3 Values and Units module[3] which defines a number of new length units that can be used to specify the dimensions of page elements now also relative to the browser viewport. These new units, most notably vw and vh referring to the width and height of the viewport, respectively, seem in theory very useful for constructing flexible layouts that dynamically scale with the size of the viewport, but unfortunately current browser support is minimal to non-existent. The most practical unit for creating adaptive layouts therefore remains em which always refers to the font size of the current element. If the dimensions of an element are specified using this unit, the element keeps its proportions to the text when the font size is changed. This is desirable if users are allowed to adjust the font size via the web interface or the browser. For pixel-based layouts, it is still a problem to match these to the corresponding physical units, as the browser would need to know the pixel density of the display, typically measured in dots per inch (DPI), in order to convert them. As it turns out, our experiments have shown that all major browsers currently work with the system DPI value, e.g. 96 on Windows systems, which can vastly differ from the actual physical value. As a consequence, absolute length units are not suitable in the screen context unless corrected using additional methods discussed later in the paper.

Since CSS2, the scope of a style sheet can be limited with respect to certain media types. For instance, different layouts can be specified for on-screen reading or printing. CSS3 expands this concept by providing access, not only to the media type, but also to media-specific features[4]. By querying these properties, certain styles can be included or excluded depending on various factors such as the size of the browser viewport or the whole screen, the orientation of the device, the aspect ratio, or the resolution of the output device, i.e. the pixel density mentioned before. Equally important is the fact that media queries are re-evaluated whenever a relevant variable is changed at runtime. This means that CSS media queries are sensitive to client-side events, but do not require event handling via JavaScript. As a result, media queries can be used to automatically switch the layout of a web page, for example, when the device orientation changes or the size of the browser window exceeds a certain value. The Media Queries module is currently in the Candidate Recommendation stage and so all major browsers in their newest versions support most parts of the specification.

Finally, the CSS3 Multi-column Layout module[5] introduces new CSS properties to lay out the content of web page elements in multiple columns. Elements inside a multi-column container are thereby treated the same way as content in regular containers, with the difference that alignment and positioning of elements now concern only the column in which they appear. However, content elements with an overflow inside a multi-column container will always be clipped at the centre of the column gap, which is unlike other container elements where overflow can be explicitly controlled using the corresponding CSS properties. The specification also foresees basic means to control the column-breaking behaviour, i.e. elements can be forced to remain together as one unit in the same column, which can be useful to avoid the separation of an image and its caption into two different columns. This mechanism, however, is not yet supported by browsers.

4. LARGE-SCREEN ADAPTATION

In order to assess the potential of the aforementioned features to form the basis of an adaptive layout solution, we now take a requirements perspective and analyse the technical and design challenges of adapting web page designs depending on the screen context, with a particular view towards large-display viewing situations.

One of the main differences with large screens is that they usually come with a significantly higher DPI value compared to smaller screens, i.e. the physical size of a pixel becomes smaller on such screens. Now the problem is that many web sites use fixed pixel values to specify the font size [17]. For example, a 15" screen running on a 1024x768 resolution has a DPI value of 85. Text displayed at a font size of 12 pixels has therefore a physical height of 3.59mm on such a screen. In contrast, a 30" screen running on a 2560x1600 resolution has a DPI value of 101 with the result that text using the same font size has a physical height of 3.02mm, which is considerably smaller. Ideally, font sizes should be specified using physical units such as millimetres or inches to guarantee the same font size independently of the screen context. However, as mentioned in the previous section, absolute units in CSS often do not match their physical counterparts as the browsers usually do not work with the real DPI value of the screen. So, to counteract this discrepancy, the font in the large screen setting would need to be increased from 12 to 14 pixels. Many web sites offer a function to manually change the font size, but this requires user intervention. Other sites rely on the browser's default font size and specify relative values using percentage units. Still, this approach is based on the assumption that the default size is appropriately configured for the user's screen context, which may not always be the case since many users are not even aware of this browser setting. A potentially better solution that we investigate in this work, is to use the new media features of CSS3 to adapt the font size relative to the increase in screen width.

As shown in our previous study [17], the majority of existing web sites use static layouts very often with fixed column widths. This is partly understandable if one considers the alternative of fluid or liquid layouts, which while allowing content to flow freely to fill the available screen space also

---

[3]http://www.w3.org/TR/2006/WD-css3-values-20060919/
[4]http://www.w3.org/TR/2010/CR-css3-mediaqueries-20100727/

[5]http://www.w3.org/TR/2009/CR-css3-multicol-20091217/

 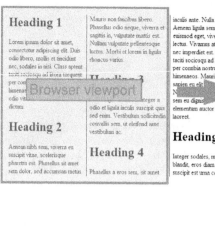 

Figure 1: Comparison of different scroll models in combination with multi-column layout: vertical (left), horizontal (center) and paged, vertical scroll model (right).

lead to excessively long lines of text on large widescreens. We have therefore explored different ways of tackling this problem using the new Multi-column Layout module, since multiple columns divide the screen into smaller portions that can be filled with paragraphs of appropriate width. However, there are two important remaining problems due to the current limitations of the CSS3 module. First, the number of columns and their widths need to be constrained properly so that the problem of long text lines does not occur. Second, while multiple columns work fine for content that fits the browser viewport, problems arise when the height of the columns extends beyond the fold, i.e. exceeds the height of the window, which is typically the case for long, text-intensive content such as news articles or blog entries.

With respect to the first problem, suggested line lengths average around 70 characters [16]; however, the CSS3 specification does not foresee a way of specifying the column width in terms of characters per line. We have explored a possible way that uses approximated em values for specifying the column width. The idea behind this approach is that the width of a character is related to the font height, which em refers to. Having measured the width of an example text (which is basically a lower-case alphabet but taking into account the frequency of letters) in different font sizes and from this calculated the average width of a single character to compare the average character width with the given font size, we found that the ratio between font size and character width is often close to 2 (Table 1). Therefore, by setting the

Table 1: Relation of the font size to the average width of a character for the Arial font.

Font size	Avg. character width	Font size/width ratio
10px	5.01	2.00
12px	5.89	2.04
14px	6.72	2.08
16px	7.35	2.18
18px	8.48	2.12
20px	9.22	2.17
30px	14.08	2.13

column width to 35em, the line length will average around 70 characters per line. We note that this ratio is relatively stable for many font types, e.g. Arial or Times New Roman, but, for example, a monospace font like Courier New has a slightly lower ratio of approximately 1.67, which may require additional balancing.

The second problem that needs to be tackled is when content laid out in multiple columns does not fit the screen so that users have to scroll down and up again to go from the end of one column to the beginning of the next one. This problem can be attributed to the fact that web documents are treated as continuous media and typically grow in the vertical direction (Figure 1, left). Some recent studies [2, 9] suggest adopting a horizontal scroll model for multiple-column designs where the height of the web site is constrained to the height of the browser window and additional content is instead added in the horizontal direction, i.e. usually from left to right (Figure 1, center). Scrolling is then only required for the horizontal axis of the viewport; however, as their study also indicates, we believe that this horizontal scrolling model is very uncommon and unfamiliar to most web users as they have become used to the dominant vertical layout used in the vast majority of web sites. This is further compounded by the fact that mouse wheels, often used for scrolling, only exist in vertical configurations, so a complete change to horizontal scrolling would be rather radical and require many users to change their browsing habits [20].

As an alternative, we have explored an approach based on a paged, vertical scrolling model (Figure 1, right) where content is laid out in fixed-sized virtual pages that fit the viewport so that multiple columns can be used in combination with vertical scrolling. We have decided to position these pages one on top of the other within the same document rather than use a dynamic scroll-less paging model, where content is laid out in a single frame and replaced upon activating associated links for the next or previous page. This is to avoid additional navigation levels and to maintain a continuous reading experience by favouring scrolling over paging. Still, the main principles of our layout adaptation guidelines could also be applied to a dynamic paging model since this would only change how pages are embedded in the document, but not how they are constructed.

Finally, larger viewing sizes may not only require adjusting the font settings and text layout, but also scaling embedded media accordingly. However, media resources such as images and videos are often only available in one size and typically embedded using fixed dimensions [17]. As a result, they often do not adapt well to widescreen viewing contexts. Particularly in the case of ads, which make extensive use of animated GIFs or Flash animations to draw the user's attention, a failure to scale well with the page's remaining content may be to the disadvantage of the advertiser. Modern browsers provide basic support for media scaling which is used when a media resource is embedded in a web page at a smaller or larger size than the original one; however, this can lead to a loss of quality unless the content is available in a vector-based graphics format, e.g. SVG. For raster graphics, we have explored different approaches, again based on the new features of media queries, to replace pictures and videos with high-definition versions as soon as the automatic scaling factor exceeds particular threshold values.

5. ADAPTIVE LAYOUT TEMPLATE

The reported experiments with the aforementioned features and limitations of current web standards have driven us to consider the following features for adaptive web templates to accommodate a range of display settings:

- Automatic font scaling to maintain the physical size of text in different screen contexts.

- Controlled switching between single and multi-column layouts to produce text layouts and line lengths in proportion to the viewing conditions.

- Automatic pagination of content to support multi-column designs in combination with the vertical scrolling model.

- Automatic media scaling and substitution to adjust the size and detail of pictures and videos if available in different resolutions.

The benefit of working with templates for web designers and developers is that, like a framework or a programming library, they can provide general instructions and additional functions, in this case concerning web page layout, which can be tailored to particular application contexts. Technically, the template consists of a number of CSS3-based style enhancements and optimisations, predefined layout classes as well as JavaScript functions to leverage the more advanced features. Target web sites will need to include it in the HTML header definition as well as activate some of the optional functionality directly on selected content elements, as discussed below. The complete source is available for download from our website[6]. Before we move on to show how existing web sites can benefit from building on our template, we first discuss some of the implementation details and the role of the new HTML5 and CSS3 features for supporting the required web site adaptations.

---

[6]http://dev.globis.ethz.ch/adaptiveguardian

5.1 Adaptation of Text and Media

A primary feature of the template is to automatically scale the font size of text elements, i.e. headings, paragraph elements, etc., according to the screen width. The underlying method is based on CSS3 media queries and the assumption that a higher screen resolution usually also means a higher screen DPI value. The implemented countermeasure is therefore to increase the number of pixels used for the font in intervals at larger screen widths. The template automatically applies the new font size to the body of the document. This also sets the base value for all values specified in em (as explained earlier) and therefore automatically applies to headings and paragraph elements unless these define other sizes explicitly.

As for media, it is more difficult to provide fully automatic means using CSS3 media queries only. We have implemented a number of controlled techniques based on using HTML placeholder elements, e.g. for low and high quality, and then using CSS to toggle their visibility according to the screen context. However, this approach is not optimal since all versions of the element would still be loaded by the client browser, including high-definition versions of media resources which may be rather large. Another possible approach is to add the images as background images to a container element. Again, CSS and media queries could be used to switch between the different versions; however, this solution has the drawback that designers would need to mix content and presentation when they define the rules in CSS. We have therefore implemented a technique based on JavaScript that uses only one HTML placeholder element with default content and a customisable threshold to indicate when content of higher quality should be used instead. Our solution automatically checks for alternative versions of pictures and videos based on naming conventions, namely suffixes '-small' and '-large', and replaces the source attribute of the respective DOM element accordingly. The relevant script is called at page-loading time and each time the size of the browser window changes.

5.2 Multi-column Layout

Unlike many of the transformational techniques discussed earlier, our approach is not concerned with content analysis to automatically determine suitable web page elements that could benefit from multi-column layout. Rather, we provide a collection of layout classes that designers can choose from and apply in their web sites in a controlled way. In CSS3, multi-column layout can be specified using a combination of values for the column-width and column-count properties. If only the number of columns is specified, then these are distributed evenly and the column width is computed from the viewport width; if the width is specified, then additional columns will be inserted as long as there is enough space available and the maximum column count (if provided) is not exceeded. Our template defines several multi-column layout classes based on popular grid layouts. For example, we provide 2col-layout and 3col-layout classes that, based on the rules discussed earlier, use an approximate column width of 30em and a maximum of two and three columns, respectively. The number of columns actually used thus varies depending on the size of the browser viewport. This means that if the available width is less than 60em (plus the column gap), the content will be laid out in a single column, otherwise the browser will add, depending on the layout class, one or two additional columns of the same width so long as they can fit. The resulting line lengths for elements of this class therefore range roughly between 60 and 120 characters per column. Some of the classes provided by our template also make use of media queries to provide optimisations with respect to the screen size. This can be especially helpful for main content and side bar elements for which we give two examples in the next section. Finally, our template also provides a special paged-layout class that can be combined with other multi-column classes provided in order to invoke the pagination of content that is described below.

5.3 Pagination of Content

For multi-column layouts to be effective with the vertical scrolling model, we implemented a pagination algorithm that splits content into multiple smaller chunks that each fit into the viewport. We refer to these viewport-sized chunks as pages. The process of splitting content into multiple such pages goes beyond the capabilities of CSS3, where the default behaviour is to grow vertically until a maximum height is reached and then add new columns in the horizontal direction. In particular, there is currently no feature allowing a new line of columns to be added in the vertical direction and overflowing content to be automatically placed in the next row. We have therefore implemented a solution using client-side scripting to segment content into pages, while making sure that the height of each page is smaller than the height of the browser viewport and that new pages are appended in the vertical direction. Special measures can be taken for the first page that often starts at a certain offset below the web site's header elements such as the navigation bar, menus, top banners etc. By setting the height of the first page to the remaining viewport height, we can make sure that all pages have the appropriate dimensions. Additionally, developers can specify a minimum page height, which can be useful to prevent an oversegmentation of content at smaller viewing sizes. The underlying algorithm takes a DOM element's container associated with multi-column layout and performs the following steps in order to paginate its content.

1. If the height of the container is smaller than the viewport height or the container only has one column, nothing is done.

2. The width of one column is determined using the CSS3 column-width property, unless set explicitly. This is done by temporarily appending an element of 100% width to the container and then measuring its width in pixels.

3. Copies of the content elements of the original container are added to an invisible, temporary container whose width is set to the column width determined in the preceding step. The use of clones of the content elements is necessary in order to avoid browser repaints after moving each element to the temporary container.

4. The height of each copied element is measured. As the width of the temporary container is equal to the width of a single column, the height can be determined accurately. In particular, the problem of content elements spanning multiple columns and therefore having an incorrect height does not arise since we do not manipulate the original multi-column container.

5. New pages are created. Each page is filled up with the copied content elements as long as there is enough space available, i.e. the accumulated height of the elements added so far is still smaller than the maximum page height.

6. Only for the first page: the maximum height of the current page is reduced by the offset of the original container element. If the resulting maximum height is smaller than the minimum height, the maximum height is set to the minimum height.

7. The original container is replaced with its new paged version.

After executing these steps, the original container including all of its content elements are still available as objects in the DOM tree. This is required so that the above algorithm can run multiple times, which is necessary to be able to rebuild the pages from the original content when the user changes the size of the browser window. Our script also appends additional container elements between subsequent pages that primarily function as links to scroll to the next page, but can also be styled using CSS, e.g. to draw a line between pages. Such visual enhancements may be appreciated by users to guide their reading flow.

5.4 Page-by-page Scrolling

To take the proposed paged layout even further, our template provides an optional feature that allows users to scroll directly from one page to another with a single action. This page-by-page scrolling mechanism was inspired by a similar function available in common PDF readers and various other document viewing software. Using this kind of scrolling model, the vertical offset of the browser viewport moves directly to the beginning of the next or the previous page created from the pagination algorithm described before. To provide visual feedback to users, scrolling is done with a smooth animation rather than a sudden jump. We implemented three methods to activate page-by-page scrolling in our templates: when clicking a "next page" link inserted between pages, when pressing the Page Up or Page Down key and by using the mouse wheel.

6. APPLICATION

To evaluate the proposed features, we tested our template on a number of existing web pages, in particular The Guardian's news web site[7], which we used for our user study, as we felt it was fairly representative of text-centric web sites with statically laid out columns for navigation and content [17]. Figure 3 (top) shows a screenshot of the web site's original design viewed on a 30" screen at a resolution of 2560x1600 pixels. The content comprises the typical web page elements, such as the header containing the main navigation bar and a slot for advertisements at the top, followed by the main content and the footer at the very bottom of the page. The main part is presented in three columns. The leftmost column is the largest and contains the news article content which typically consists of a header, a picture and the article text itself. A smaller column in the middle of the layout provides various functions to the user, e.g. to

---

[7]http://www.guardian.co.uk/

manually adjust the font size, as well as links to related information. The far right column contains a combination of advertisements, related pages and various services.

As is evident from the screenshot, the web site uses a fixed layout which does not adapt well to the example viewing situation. The main problem is that the overall design is contained in a fixed-width wrapper of 940 pixels which is placed in the middle of the browser viewport, leaving a considerable amount of unused space on both sides when viewed on screens with higher resolutions. Also critical is the default font size of the web site, which is set to 14 pixels. While a function to manually change the font size is available, due to the static layout the column widths do not adjust if the font size is increased, which leads to gradually shorter line lengths. This problem is also observed in many other online news web sites that provide such a function.

The first step in providing a more flexible design was therefore to remove the fixed-size constraints of the web page's layout. We used our template in combination with a fluid layout shown in the middle of Figure 3, where the width of the wrapper was set to 95%so that the document fills most of the available viewport width while still leaving some space at both sides where peripheral vision would set in. As can be seen, the font size automatically adapts to the new viewing context thanks to our template and in order to keep the line lengths proportional to the font size, the width of the three main columns was changed to em values, which also allows those columns to grow if the font size is manually increased. To ensure that the page does not get too narrow, a minimum width was additionally used. We enabled also CSS3 multi-column layout for both the main content area and the sidebar in the third column using features of our template, but we restricted the column numbers to two at this stage in order not to produce a too fragmented web site appearance. To support multi-column layout in combination with a vertical flow of the page, pagination was enabled for the main article container and pages were visually separated by a horizontal line as well as a "Read on" link at the bottom of each page to to scroll directly to the next page.

The second step then focused on adapting the media. As can be seen best when comparing with the bottom of Figure 3, the adaptations concerned both the article image, which now spans the full width of the very first column, as well as the advertisements in order to make sure that the proportions of text and media, and in particular the visibility of ads, are preserved given the relative increase of the spatial coverage of content.

7. EVALUATION

So far, the paper has focused on which adaptations are actually possible based on the new web standards and how they can be complemented technically with the help of a template. Our implementation is based on a lightweight and efficient scripting solution that is carried out only on the client and hence the fluidity of the web page rendering in the web browser is not impacted. Considering performance was not an issue (it might however become one for large web pages viewed on slow machines), we focused our efforts on assessing the effectiveness of our layouts in terms of reading comfort and efficiency for the viewer as well as the perception of the overall look-and-feel of the adaptive web site in comparison with the original version. We therefore conducted a user study with the goal of finding out which of the adaptations contributed to improving the users' overall reading experience on the web.

7.1 Method

Twelve people were asked to read three different news articles taken from The Guardian's web site using different layouts and comment on various aspects of the design. Most participants were at the age of 20 to 29 (two at the age of 30 to 39) with normal or corrected vision and proficient readers of English. Eight of the participants reported reading online news 5-10 hours per week, three declared reading more and one less. When asked which media and devices they regularly used for reading news, ten participants reported using printed newspapers as well as desktop screens and five participants said they also used smartphones.

The three articles that were chosen for the study were *Wimbledon 2010: Ruthless Andy Murray gives Sam Querrey no quarter*[8] (905 words), *Chilean miners: A typical day in the life of a subterranean miner*[9] (1835 words) and *High-speed rail link gets £800m more in state funding*[10] (939 words).

The articles were presented on a 30" desktop screen with a 2560x1600 resolution using three different layouts—one without adaptations and two with different levels of adaptations. The order of the articles and the layouts was rotated for each participant in order to ensure that all articles and layouts were used the same number of times in the whole study. The participants were told that they could freely resize the browser window (initially set at an intermediate size of about 1300x1100 pixels) and change the font size to suit their needs. After reading, the participants were asked simple comprehension questions (e.g. "In which round did Andy Murray win against Sam Querrey?" or "Who will be his next opponent?") for which they were allowed to refer back to the article, if needed, to find the answers. Furthermore, participants were asked to fill in a post-task questionnaire after each article, in which they had to rate reading comfort, positioning of the elements on the web page, image alignment, font size and scrolling behaviour on a five-point Likert scale. At the end of the sessions, participants were asked to compare the three articles in terms of the layout.

The three layouts used for the study were the following:

- **Original.** The original layout of the article with standard scrolling behaviour. Text is presented in a single column at a default font size of 14 pixels (Fig. 3, top).

- **Adaptive 1.** This layout included most of the adaptations described above, i.e. multiple columns and pages, page-by-page scrolling using keyboard or "read on" links and automatic font size adaptation (which evaluated to 16 pixels in this case). Adaptations not available in this layout were page-by-page scrolling using the mouse wheel and the alignment and replacement of images (Fig. 3, middle).

- **Adaptive 2.** Adaptive 1 layout with the addition of page-by-page mouse-wheel scrolling and automatic image scaling and substitution (Fig. 3, bottom).

---

[8]http://www.guardian.co.uk/sport/2010/jun/28/wimbledon-2010-andy-murray
[9]http://www.guardian.co.uk/world/2010/sep/09/chilean-miners-typical-day
[10]http://www.guardian.co.uk/uk/2010/oct/03/high-speed-rail-network-transport

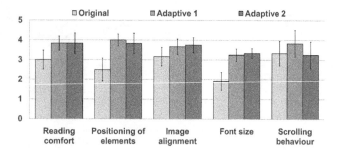

Figure 2: User ratings of various aspects of the layouts on a five-point Likert scale.

The reason for creating two versions of the adaptive layout was that we wanted to test various degrees of adaptations. As we did not want the sessions to last too long, we decided to test only three versions of each article. In all three layouts, a function to change the font size was available at the top navigation as well as in the sidebar. This function had the same behaviour as on the original Guardian web site.

7.2 Results

Since we selected articles of moderate length to keep the study participants interested while attending to the task, we expected to see greater differences in terms of perception and experience than actual reading speed. However, while users took roughly the same amount of time to finish reading the articles regardless of the layout, they were in general faster answering the comprehension questions for articles with adapted layouts, because they could find the relevant passages more easily, thanks to larger fonts and a clearer organisation of the text. Overall, reading comfort was rated higher for the adapted layouts than for the original ones, although a one-way ANOVA statistical analysis reveals that only the difference between the first adapted layout and the original is significant (P=0.031 vs. P=0.193 for the second adapted layout). Indeed, more than half of the participants (seven people) preferred the Adaptive 2 layout, four favoured the Adaptive 1 layout and only one participant liked the original layout best.

Examining the results in more detail, we observe that the greatest rating differences between the original and the adapted layouts are those concerning font-size and positioning of the elements (statistical significance for those factors was confirmed by the ANOVA and pairwise comparisons performed on the data). Participants appreciated the fact that more content was visible on the screen, despite the larger fonts used for the text. For the font size, text on the original layout was perceived as too small by almost all participants (ten), while only two participants perceived it as optimal, or too large. As a consequence, more than half of the users increased the font size when reading with the original layout. With the adaptive layouts, the initial font size was changed less often. Only one person decreased the font size to 14 pixels and two persons increased it to 18 pixels when reading with the Adaptive 2 layout. The initial font size of the adaptive layouts was rated as optimal by half of the participants. Two persons perceived it as too large when reading with the Adaptive 1 layout and four persons stated the same when reading with the Adaptive 2 layout.

Multi-column and multi-page breaking was also rated highly by a majority of testers, who felt that it increased the visi-

bility and clarity of the page. However, there were also a few users who preferred the single-column layout because text could be read in a continuous fashion. The fact that some people were confused about whether to jump to the next column on the same page or continue on the same column but on the next page below, shows that it is important to make an even stronger visual separation between the pages.

Regarding scrolling behaviour, no statistical significance was found in the rating results. When asked about their preferences, three out of four participants chose the standard scrolling behaviour and the rest favoured page-by-page scrolling with the mouse wheel. The other scrolling mechanisms were rarely used or, as in the case of the "Read on" link, never. Two participants did not like the page-by-page scrolling and used the arrow keys for scrolling instead. Another participant used the arrow keys throughout all layouts. Only one person used the adapted Page Up/Page Down keys for scrolling. While users who preferred the page-by-page scrolling appreciated the speed of this scrolling mechanism, those who did not remarked that the jump was too big and sudden and that they lost track of the context after moving to another page. One participant said that scrolling should give feedback if it was successful (e.g. by showing part of the previous page), another person suggested to scroll only by half a page in order to maintain reading context. We tried to counteract this problem when designing the page-by-page scrolling, by smoothing the scrolling with an animation on the one hand and by not making the pages take up the whole viewport height on the other hand. However, the result indicates that further measures are necessary. Another factor for preferring the standard scrolling behaviour is attributable to the reading behaviour of some participants. We observed that these people preferred to adopt a fixed gaze when reading and use gradual scrolling so that the text to be read remained roughly around a fixed area on the screen. The same behaviour was incidentally observed by Braganza et al. [2]. Page-by-page scrolling is inappropriate for this reading behaviour. The same problem applied to multiple columns, where the participants had to relocate their gaze to the right in order to read the second column. However, participants stated that moving their gaze in the horizontal direction is less tedious than moving it in the vertical direction. Another interesting aspect that was not investigated in our study performed in a mouse and keyboard-based desktop context is if and how the reading and scrolling behaviours change when content is viewed on a touch-based device, such as a tablet or an E-book reader. Many document reading applications on those devices are based on page-by-page scrolling models (particularly on E-book readers) and thus our web page-splitting algorithm with tappable "read-on" button would perhaps prove useful for the viewing of web content in such settings.

Concerning image alignment and scaling, the statistical difference between the means is also not significant to draw clear-cut conclusions, but it seems that the adaptive layouts have a slight edge. More importantly perhaps, is the fact that there was no perceived difference between the Adaptive 1 and Adaptive 2 layouts, where for the latter the image was larger and aligned with the column (see Fig. 3). A possible explanation for this result is the fact that images were not a central part of the study and did not contain important information that was necessary for the comprehension of the article. Nevertheless, we believe that having the article image (or other illustrative figures such as charts or fact boxes) remaining in view while reading the article laid out in neighbouring columns is better than seeing it immediately disappear after scrolling down to get to the text.

8. CONCLUSION

In this paper we presented ways of providing more flexible web page layouts which is necessary given the proliferation of small and large-screen devices used to access web content and the fact that many existing web sites still fail to adapt to the viewing context. Unlike prior work in this area, we focused on web standards, especially new features of HTML5 and CSS3, in order to demonstrate how web developers and designers can produce adaptive layouts using technologies they are familiar with.

To accommodate also large-screen contexts, we proposed balanced multi-column solutions with adapted media elements and addressed challenges of readability through automatic adjustment of the font size and line lengths. We also explored pagination of long content, which is necessary if multi-column layouts are to be used in conjunction with the vertical scrolling model employed by the majority of web sites today.

The main outcome of this work is an adaptive layout template that can be used by web developers as well as designers and applied in existing web projects. By means of formative evaluation, we demonstrated that especially text-centric web sites, such as online newspapers, but also wikis, blogs and forums, can directly benefit from the features offered by our template and improve the overall user experience. In future work we plan to extend the proposed template by considering also other types of web sites, such as web mail interfaces, online calendars or task managers, and investigate how we could make more effective use of larger screen real estates in order to support working with web-based applications. At the same time, we hope that the proposed CSS3 standard and in particular the CSS3 multi-column module will continue to evolve and provide more features in the future so that solving issues related to the vertical scroll model and pagination, as demonstrated in this paper, can be solved without the need for client-side scripting.

9. REFERENCES

[1] G. Badros, J. Tirtowidjojo, K. Marriott, B. Meyer, W. Portnoy, and A. Borning. A Constraint Extension to Scalable Vector Graphics. In *Proc. WWW*, 2001.

[2] C. Braganza, K. Marriott, P. Moulder, M. Wybrow, and T. Dwyer. Scrolling Behaviour with Single- and Multi-column Layout. In *Proc. WWW*, 2009.

[3] G. Calvary, J. Coutaz, D. Thevenin, Q. Limbourg, L. Bouillon, and J. Vanderdonckt. A Unifying Reference Framework for Multi- Target User Interfaces. *IWC*, 15, 2003.

[4] S. Ceri, F. Daniel, M. Matera, and F. M. Facca. Model-driven Development of Context-Aware Web Applications. *TOIT*, 7(1), 2007.

[5] Y. Chen, W. Ma, and H. Zhang. Detecting Web Page Structure for Adaptive Viewing on Small Form Factor Devices. In *Proc. WWW*, 2003.

[6] M. Czerwinski, G. Smith, T. Regan, B. Meyers, G. Robertson, and G. Starkweather. Toward Characterizing the Productivity Benefits of Very Large Displays. In *Proc. INTERACT*, 2003.

[7] Z. Fiala, F. Frasincar, M. Hinz, G.-J. Houben, P. Barna, and K. Meißner. Engineering the Presentation Layer of Adaptable Web Information Systems. In *Proc. ICWE*, 2004.

[8] F. Frasincar, G.-J. Houben, and P. Barna. Hypermedia presentation generation in Hera. *IS*, 35(1), 2010.

[9] J. H. Goldberg, J. I. Helfman, and L. Martin. Information Distance and Orientation in Liquid Layout. In *Proc. CHI*, 2008.

[10] G. Hattori, K. Hoashi, K. Matsumoto, and F. Sugaya. Robust Web Page Segmentation for Mobile Terminal Using Content-Distances and Page Layout Information. In *Proc. WWW*, 2007.

[11] J. He, T. Gao, W. Hao, I.-L. Yen, and F. B. Bastani. A Flexible Content Adaptation System Using a Rule-Based Approach. *KDE*, 19(1), 2007.

[12] N. Hurst, W. Li, and K. Marriott. Review of Automatic Document Formatting. In *Proc. DocEng*, 2009.

[13] C. Jacobs, W. Li, E. Schrier, D. Bargeron, and D. Salesin. Adaptive Grid-Based Document Layout. *TOG*, 22(3), 2003.

[14] S. Kopf, B. Guthier, H. Lemelson, and W. Effelsberg. Adaptation of Web Pages and Images for Mobile Applications. In *Proc. IS&T/SPIE*, 2009.

[15] R. Mohan, J. R. Smith, and C.-S. Li. Adapting Multimedia Internet Content for Universal Access. *TMM*, 1(1), 1999.

[16] A. Nanavati and R. Bias. Optimal Line Length in Reading-A Literature Review. *Visible Language*, 39(2), 2005.

[17] M. Nebeling, F. Matulic, and M. C. Norrie. Metrics for the Evaluation of News Site Content Layout in Large-Screen Contexts. In *Proc. CHI*, 2011.

[18] G. Robertson, M. Czerwinski, P. Baudisch, B. Meyers, D. Robbins, G. Smith, and D. Tan. The Large-Display User Experience. *CGA*, 25(4), 2005.

[19] E. Schrier, M. Dontcheva, C. Jacobs, G. Wade, and D. Salesin. Adaptive Layout for Dynamically Aggregated Documents. In *Proc. IUI*, 2008.

[20] H. Weinreich, H. Obendorf, E. Herder, and M. Mayer. Not Quite the Average: An Empirical Study of Web Use. *TWEB*, 2(1), 2008.

Figure 3: Different layouts compared in the study: original layout (top), adaptive layout using a multi-column design and paged, vertical scroll model (middle) and additionally using adaptive media for the article image and advertisements (bottom). All the above screenshots were captured at a resolution of 2560x1600 pixels and scaled down with a common factor for this figure.

Detecting and Resolving Conflicts between Adaptation Aspects in Multi-staged XML Transformations

Sven Karol
Uwe Aßmann
Software Technology Group
Technische Universität
Dresden
Dresden, Germany
sven.karol@tu-dresden.de,
uwe.assmann@tu-dresden.de

Matthias Niederhausen
Klaus Meißner
Multimedia Technology Group
Technische Universität
Dresden
Dresden, Germany
matthias.niederhausen@tu-dresden.de,
klaus.meissner@tu-dresden.de

Daniel Kadner
Professorship of
Geoinformation Systems
Technische Universität
Dresden
Dresden, Germany
daniel.kadner@tu-dresden.de

ABSTRACT

Separation of Concerns (SoC) is a common principle to reduce the complexity of large software and hypermedia systems. Amongst a variety of approaches, *adaptation aspects* are a well-known solution to significantly improve SoC in adaptive hypermedia applications. To model adaptation aspects in XML-based hypermedia applications, we developed *PX-Weave*, a tool which allows to specify and weave such aspects in multi-staged XML transformation environments. However, while aspects increase modularity and thus decrease complexity of software, they do also introduce some complex problems. The most prominent one, aspect interaction, has received a lot of attention from researchers during the last decade. In this paper we investigate the problem of aspect interaction for adaptation aspects. We present a combined approach for static and dynamic detection of aspect interactions in multi-staged XML-based hypermedia applications, which we implemented as an add-on to *PX-Weave*.

Categories and Subject Descriptors

I.7.2 [**Document And Text Processing**]: Document Preparation—*XML, Format and notation, Hypertext/hypermedia, Languages and Systems*; D.1.0 [**Programming Techniques**]: General; H.5.4 [**Information Interfaces and Presentation**]: Hypertext/Hypermedia—*Architectures*

General Terms

Languages, Algorithms

Keywords

Aspect-oriented Programming, Separation of Concerns, aspect interactions, XML, XHTML

DocEng'11, September 19–22, 2011, Mountain View, California, USA.
Copyright 2011 ACM 978-1-4503-0863-2/11/09 ...$10.00.

1. INTRODUCTION

Aspect-oriented Programming (AOP) [15] is a well-known paradigm for increasing modularity and reuse in software systems. Its basic idea is to encapsulate logically associated parts of a program— so-called *cross-cutting concerns*— within independent modules called *aspects*. This way, AOP allows programmers to leave their core algorithms and data model intact, thus keeping them free of code that does not actually contribute to the problem solution (e.g., code belonging to a cross-cutting logging concern). In practice, a widespread implementation of AOP is AspectJ [14], which allows for aspect modules in Java programs.

In web engineering, paradigms are shifted from standard programming languages like Java to a higher level of abstraction and to document-centric approaches, which are often based on the eXtensible Markup Language (XML) or similar languages. The usage of such languages ranges from normal, static websites (XHTML), to visualisation languages (SVG), to staged, XML-based templates languages, to data exchange formats (SOAP, XMI) to simple configuration files. Like traditional programs, such document-centric approaches do equally benefit from AOP and SoC in general. Several groups have carried out research in the past to separate the core application from the adaptation facet [22, 3, 8]. In [11] we presented a generic, XML-based language to develop adaptation aspects for arbitrary XML documents and presented our corresponding tool implementation *PX-Weave*. To produce a final representation or visualization, XML documents in web applications often undergo a series of transformations during the application's life cycle. These transformations and their interconnections are parts of the document's architecture [2]. Besides improving SoC through information hiding and encapsulation of transformations, an explicit architecture of a document also provides several advantages for realizing adaptation aspects. Transformations as distinguished parts of the architecture are cleanly separated from each other and wired only by their incoming and outgoing ports. Tools like PX-Weave can exploit this extra information to apply adaptations at a certain point in the document architecture without reaching the complexity of full-fledged AOP implementations like AspectJ.

While adaptation aspects are an adequate solution to model adaptivity in document-centric web applications, they do also share the disadvantages of AOP. One of the major problems of AOP is aspect interaction. Since the advent of AOP, the problem has been analyzed and investigated in quite a number of research works (e.g., [1, 16, 7, 20]). Aspect interaction occurs whenever two or more aspects refer to overlapping parts of a program and thus influence each other. Since this is the common case in systems developed using AOP, aspect interaction is an expected benavior and accepted by developers. However, problems do arise if the resulting behavior of the AOP program differs from the developer's expectations. Still, an automatic derivation of the correct behaviour in complex interaction scenarios is hard because aspect developers may have widely differing intentions. Therefore, a commonly accepted solution for problematic aspect interactions is to a) automatically detect them and b) provide tooling options for authors to influence aspect ordering [7, 16].

For adaptation aspects, the problem of aspect interactions has not yet been picked up by the research community. In [19] and [11], we sketched the basic problem. Now we want to deepen the discussion on the interaction of adaptation aspects in this paper. We further present an approach for the automatic detection and categorization of aspect interactions, which is based on a specialization of ideas of [7] and [16] to adaptive hypermedia documents. As a third contribution, we present a prototype implementation of our approach, based on our adaptation tool PX-Weave.

This paper is structured as follows. In Sect. 2 we introduce our case study application—a simple portal for music charts—and show how PX-Weave's adaptation aspects can be used to implement adaptivity in this scenario. Sect. 3 gives an introduction to aspect interactions in general and afterwards deepens this problem for adaptation aspects. Section 4 presents our solution for detecting aspect interactions while Sect. 5 provides an overview of our implementation in PX-Weave. Finally, Sect. 6 discusses related work and Sect. 7 concludes the paper, giving an outlook on our ideas for future work.

2. ASPECT-ORIENTED ADAPTATION WITH PX-WEAVE

In this section we present our tool and language PX-Weave for adaptation modelling based on AOP. We designed PX-Weave to blend with existing staged document architectures. To show how PX-Weave works, we first give a brief overview of our example, the pipeline-based hypermedia portal SoundNexus and afterwards explain how PX-Weave extends the underlying architecture and can be used to model adaptation.

2.1 SoundNexus in a Nutshell

SoundNexus is an adaptive, XML-based hypermedia application featuring a music database aggregated from multiple sources, such as traditional SQL data sources or web services. It is based on Cocoon's (http://cocoon.apache.org/) multi-staged transformations to produce an XHTML page, starting from an abstract page description format. This format lets web application authors specify web page components (i.e., fragments) and data sources.

Every request is handled by one of the transformation pipelines defined in a configuration file. For brevity reasons, we do only present the architecture of SoundNexus'

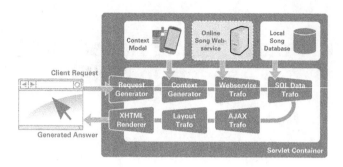

Figure 1: SoundNexus portal architecture.

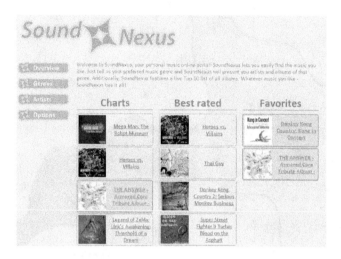

Figure 2: Example SoundNexus output.

main pipeline here. Fig. 1 gives an overview of the different pipeline stages and data sources.

The first stage, titled RequestGenerator, deals with selecting the initial document based on the incoming HTTP request. In the second stage, the ContextGenerator initialises SoundNexus' context model. To this end, it first evaluates standard HTTP request parameters (e.g., language or browser). Second, it integrates data collected by so-called context sensors (for instance, GeoIP services or client-side JavaScript sensors). If a page requires additional data from a web service, e.g., to request the current top 10 charts, the stage WebServiceTrafo translates queries embedded within the document into standardized SOAP requests and sends them to the specified Online Song Database. The result is then added to the document by using a template specified by the author. Similar steps are taken by the SQLDatabase-Trafo which evaluates SQL queries by delegating them to the Local Song Database. After the dynamic content has been injected, the AjaxTrafo component derives a view from the transformed document. This way, only those parts of the page requested by the client will be delivered (local update in the browser). Finally, LayoutTrafo and XHTMLRenderer compile the page into a standard XHTML document that can be rendered by the user's browser. Fig. 2 shows an exemplary render output of the SoundNexus pipeline.

2.2 Enter PX-Weave

In order to add adaptive behaviour to a web application like SoundNexus, PX-Weave employs so-called *Adaptation*

```
<?xml version="1.0" encoding="UTF-8" ?>
<aspect name="ButtonReplacement">
    <adviceGroup>
        <depends>
            <boolExpr>adaptationclass[@id]="Device_Smartphone"</boolExpr>
        </depends>
        <scope>
            <xpath>//aco:AmaSetComponent[@id=navbar]/aco:AmaImageComponent</xpath>
            <joinpoint name="AjaxTrafo" locator="before"/>
        </scope>
        <advices>
            <collapseElement>
                <pointcut>.</pointcut>
                <varDecl>
                    <var>
                        <name>altText</name>
                        <xpath>//amet:alt[1]</xpath>
                    </var>
                </varDecl>
                <textTemplate>$altText</textTemplate>
            </collapseElement>
        </advices>
    </adviceGroup>
</aspect>
```

Figure 3: Adaptation aspect for replacing navigation buttons with plain text.

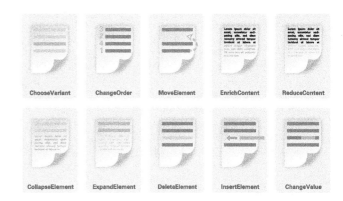

Figure 4: Supported advice primitives in PX-Weave.

Aspects to define transformations on an XML document. Fig. 3 shows an example aspect that SoundNexus uses to replace its navigation buttons with plain text links for smartphone users. One can immediately determine three basic elements of such an aspect: `depends`, `scope` and `advices`.

As in traditional aspect orientation, an aspect must have a pointcut (`scope` element) and one or more advice actions (`advices` element):

A **pointcut** can either address the pipeline state—as a programmatic pointcut—or the content that flows through the pipeline—as a content-based pointcut. By combining both types of pointcuts, an aspect can select a point in time and space of an XML document passing through the pipeline. In PX-Weave's syntax, the content-based pointcut is given by the `xpath` child element, while the programmatic pointcut is defined by the `joinpoint` element. Notice that the programmatic pointcut directly refers to a named stage within the transformation pipeline, as shown in Figure 1.

An **advice** defines the actual transformation of the selected XML content. For this, we provide a set of primitive advice actions, which we discuss in the next section. Our example aspect uses the `CollapseElement` primitive, which reduces an XML element of arbitrary complexity to a mere text node. In this case, an image component is simply replaced by its alternative text (contained in the `amet:alt` child element).

For PX-Weave, we introduce another mandatory part of an aspect: the advice condition (`depends` element), which marks a major difference to traditional aspects. The **advice condition** is evaluated at runtime to determine whether the aspect is actually executed or not. This is a direct consequence of the nature of adaptivity: changes are not performed in any case, but rather depending on the individual user and his context. The example shows a reference to the web application's context model, checking whether the user's device was classified as a smartphone.

2.3 Adaptation Advice Primitives

Adaptation aspects can either be realized as arbitrary transformations or as a sequence of well-defined primitive operations. While the former offers a maximum in flexibility, it lacks an important property: ease of analysis. With two arbitrary adaptation aspects, their combined result is

hard to predict. Therefore, we decided to employ the latter approach, providing a limited set of primitives. These primitives were conceived as to fulfill two opposed goals: On the one hand, they must be generic, independent of the application domain, so they can be utilized in any kind of XML application. On the other hand, they should form more than just some sort of "assembler language" for XML transformations. That is, they had to be powerful enough to easily implement the more complex adaptation patterns widely recognized in the domain of web engineering [6]. The compromise we found, offering moderately powerful operations, but still being generic to all XML domains, are the following ten advice classes. Though extensible, the limited operation set allows a significantly easier conflict analysis.

Figure 4 gives a graphical impression of the different advice primitives explained subsequently.

ChooseVariant selects a variant from a set of exchangeable instances of an element node, according to the current context. The document author must define every variant individually in advance. A pointcut expression can then match the most appropriate variant at runtime. *Example:* A common use case for this is language adaptation. Consider a multilingual document which provides different language versions of each paragraph. Language independence can be achieved by specifying an adaptation aspect that selects the correct variant according to the reader's language—ideally as early as possible in the transformation pipeline.

ChangeOrder re-arranges child nodes of an element, according to a given permutation. *Example:* The order of elements is important when it comes to layout adaptations for different types of devices and their displays. Furthermore, elements may be ordered to reflect user preferences, e.g., after a transformation stage that aggregated data from multiple, inhomogenous sources.

EnrichContent adds or inserts additional text fragments in textual parts of a document, based on a text selection expression. This kind of advice allows the encapsulation of related text parts in adaptation aspects at a very fine granularity. *Example:* An adaptive website may provide different kinds of access levels, e.g., for registered users and unregistered users.

ReduceContent removes text fragments in plain textual parts of a document according to a text selection expression. This advice is the inverse of *EnrichContent*. *Example:* For advanced users, prerequisite steps of an installation tutorial can be omitted to increase readability.

CollapseElement can be used to replace a complete subtree of a document with plain text. *Example:* To provide devices with small screens or little processing power with an appropriate representation of complex content elements (like Flash movies or JavaScript-intense containers), the author may replace these elements with a short, textual abstract.

ExpandElement substitutes a text fragment in the document with a more complex element or subtree. This advice is the inverse of *CollapseElement*. *Example:* A translation service can be extended to provide paying customers not only with the textual translation, but annotate the single words or phrases with the possibility to click them in order to find out how they are pronounced.

DeleteElement is a basic primitive that removes a subtree from a given XML document at a specific joinpoint. It can be compared to the *around* advice in AOP. *Example:* Admin links on a blog can be removed for unregistered users.

InsertElement is a basic primitive that adds an element before or after a specific joinpoint. This kind of advice can be considered the closest relative of the *before* and *after* advice in AOP. *Example:* Additional information can be embedded in popup layers when users request help.

MoveElement removes an element at a joinpoint and inserts it at a different position in the document tree. Although this kind of advice is quite simple, it is very useful for adaptations that perform restructuring. *Example:* Different display formats might require different document layouts, e.g., a generated XHTML page may use a table-based, vertical layout by default which is changed into a horizontally arranged table if a widescreen monitor is used.

ChangeValue can change attribute values at a certain joinpoint. *Example:* According to the user's interests, an adaptation concern may require to highlight different topics within the document by changing text colors or adding boxes around affected text paragraphs.

2.4 Integrating PX-Weave and SoundNexus

Up to this point, we have described how authors can define adaptation aspects in PX-Weave and presented a set of primitive transformations that can be used to achieve this. To give an impression how our conflict resolution works, we will first shortly explain how PX-Weave extends the SoundNexus architecture to implement adaptive behavior into the web application. A more detailed description can be found in our previous work [11].

Upon the web application's deployment, PX-Weave reads the pipeline configuration file and thus determines the available programmatic joinpoints. Using AspectJ, hook methods are then inserted at every such joinpoint. Whenever a request is spawned by an end user and the pipeline gets triggered, PX-Weave evaluates which of the defined aspects will become active during this single request and compiles a list. Advice actions whose advice condition does not match the current context are discarded at this point. As the XML document passes through the pipeline stages, PX-Weave executes the active advice actions in sequence at every joinpoint, thereby achieving context-dependent adaptation of the document.

3. ADAPTATION ASPECT INTERACTION

In this section we discuss the problem of aspect interaction for adaptation aspects. Since a generally accepted termino-

logy does not yet exist, we first introduce some basic terms which we use in this paper. Afterwards, we specify when and under which circumstances adaptation aspects may interact. Finally, we analyze pairs of adaptation primitives with respect to their conflict potential.

3.1 Basic Terminology

Since the advent of AOP, many articles on interaction between aspects have been published, using self-introduced terms to describe their work instead of building upon a common notion. In this paper, we utilize some of the concepts and ideas previously established by [16] in order to avoid that pitfall.

In general, an interaction between two aspects is nothing to fret about but rather a common case, since aspects are rarely ever applied to completely independent parts of a program. Contrariwise, an interaction is often necessary to achieve the intended behavior, as the following example emphasizes (c.f. [16]):

EXAMPLE 1. *Consider a logging aspect A_1 printing each call to a method of a certain class C to a log file. Further, consider an additional aspect A_2 adding an additional call to some method of C. Obviously, A_1 now has to consider this call leading to an interaction between A_1 and A_2. As long as the aspect weaver ensures the intended behavior, no conflict between A_1 and A_2 appears. Now consider an additional aspect A_3 applying some encryption algorithms to some of the parameters of some methods in C. In this case, it is not possible for the aspect system to determine the correct behavior, because a developer may want A_1 to log either plain or encrypted method arguments. Hence, in terms of [16], an interference (or conflict) between A_1 and A_3 occurs, if the resulting program does not show the intended behavior.*

We can further distinguish two special kinds of aspect interactions [16]. An aspect B is *triggered* by an aspect A if some code woven by the advice actions of A is matched by the pointcut expressions of B. In Ex. 1, A_2 triggers A_1 by adding additional method calls to the pointcuts of A_1 such that they are also logged in a log file. In contrast, an aspect B is *inhibited* by an aspect A if A removes some of the joinpoints matched by pointcuts of B. For example, consider a modified version of A_2 in Ex. 1 which replaces[1] a complete method that previously contained calls to methods in class C. Since these calls can no longer be reached during program execution, they do not need to be considered by A_1 and can thus be omitted from A_1's pointcuts.

3.2 A Terminology for Interacting Adaptation Aspects

The terms we introduced above are intended for general AOP and thus do not precisely fit the domain of adaptation aspects. Hence, we now provide some extended definitions for these terms more adequate to the domain of adaptive, document-centric hypermedia applications.

The following definitions are given w.r.t advice actions. Definitions for adaptation aspects can easily be derived and are thus omitted for the sake of brevity. We use *doc* for the document which should be adapted by advice actions. Since we consider tree-structured documents such as XML docu-

---

[1]Replacements can be implemented using the *around* advice in AspectJ.

ments, the set of nodes in *doc* is denoted by N_{doc}. Furthermore, $ppc(a)$ and $cpc(a, doc)$ are used for the programmatic and content-based pointcut of an advice a w.r.t a document *doc*. To represent the application of an adaptation advice, we use $a \bullet doc$.

DEFINITION 1. *Let L_{doc} be the set of valid input documents and Advice be a set of advice actions to be applied. The affects/2 predicate is specified as follows.*

$$\forall a, b \in Advice \, \forall doc \in L_{doc} :$$
$$ppc(a) \cap ppc(b) \neq \emptyset \wedge cpc(a, doc) \neq \emptyset \wedge cpc(b, doc)$$
$$\neq cpc(b, a \bullet doc) \Rightarrow affects(a, b)$$

Given each two advice actions a and b, a *affects* b if an application of a changes the content-based pointcut of b, which is applied later in the document transformation chain.

DEFINITION 2. *Let Advice be a set of advice actions to be applied. The interacts/2 predicate is defined as follows.*

$$\forall a, b \in Advice :$$
$$interacts(a, b) \Leftrightarrow affects(a, b) \vee affects(b, a)$$

Definition 2 states that a *interacts* with b, if a *affects* b or b *affects* a.

DEFINITION 3. *Let L_{doc} be the set of valid input documents and Advice a set of advice actions to be applied. The triggers/2 predicate can be derived as follows.*

$$\forall a, b \in Advice \, \forall doc \in L_{doc} \, \forall n \in N_{doc} :$$
$$affects(a, b) \wedge n \notin cpc(b, doc) \wedge n \in cpc(b, a \bullet doc)$$
$$\Rightarrow triggers(a, b)$$

Given each two advice actions a and b, $triggers(a, b)$ holds if a *affects* b and if b's content-based pointcut is extended by a with some node, e.g., by adding or moving some node.

DEFINITION 4. *Let L_{doc} be the set of valid input documents and Advice a set of advice actions to be applied. The inhibits/2 predicate can be derived as follows.*

$$\forall a, b \in Advice \, \forall doc \in L_{doc} \, \forall n \in N_{doc} :$$
$$affects(a, b) \wedge n \in cpc(b, doc) \wedge n \notin cpc(b, a \bullet doc)$$
$$\Rightarrow inhibits(a, b)$$

Given each two advice actions a and b, $inhibits(a, b)$ holds if a *affects* b and a removes some nodes from b's content-based pointcut, e.g., by moving or deleting some node.

3.3 Interaction Potential of Adaptation Advice Primitives

In the section before, we defined different kinds of interactions in a general way. However, it is important to note that not all advice actions based on the primitives we defined in Sect. 2.3 carry the same potential for triggering or inhibiting other advice actions. As an example, consider *DeleteElement* and *MoveElement* advice actions. While deleting an element from the document obviously removes this element from each pointcut of another advice that matched the element before, moving an element does not necessarily cause

Table 1: **Potential for interactions of adaptation advice primitives in PX-Weave. We use ■ for *true* and □ for *false*. *trg* is short for *triggers*, *inh* stands for *inhibits*, *loc* stands for *location* − *safe* and *val* is short for *value* − *safe*.**

	trg_{loc}	inh_{loc}	trg_{val}	inh_{val}
ChangeOrder	□	□	■	■
ChangeValue	■	■	□	□
ChooseVariant	□	■	□	■
CollapseElement	■	■	■	■
DeleteElement	□	■	□	■
EnrichContent	■	□	□	□
ExpandElement	■	□	■	■
InsertElement	■	□	■	■
MoveElement	□	□	■	■
ReduceContent	■	■	□	□

such effects, since *MoveElement* only changes a document's structure. As a consequence, problems can be avoided by using pointcut expressions which are considered as *location-safe*. Def. 5 explains this term.

DEFINITION 5. *A pointcut expression is called **location-safe** w.r.t a certain element if it matches the same element under every possible context in a document (i.e., in every possible subtree).*

In the case of PX-Weave, we consider XPath expressions independent of an element's context (e.g., by using a unique identifier attribute) as *location-safe* pointcuts.

Another group of advice actions may change primitive values only. As an example, consider *ChangeValue* advice actions, which allows the replacement of arbitrary attribute values by new values. In this case, problems can be avoided by using *value-safe* pointcut expressions, which are explained in Def. 6.

DEFINITION 6. *A pointcut expression is called **value-safe** w.r.t a certain element if it still matches the same element after some attribute value or text value has been changed in any possible way.*

In PX-Weave, XPath expressions using an element's location to reference it are *value-safe*.

Based on the assumptions for *location-safe* and *value-safe* pointcut expressions, we analyzed the interaction potential for each kind of primitive we introduced in Sect. 2.3. Table 1 shows the results of this analysis.

The first pair of columns denotes the affection potential under the assumption of *location-safe* pointcut expressions. It can be read as follows: For each kind of advice a, can there be another advice X that may be triggered or inhibited by a such that $triggers(a, X)$ or $inhibits(a, X)$ hold? The second pair of columns denotes the interaction potential for each kind of advice under the assumption of *value-safe* pointcut expressions.

Note that we omitted the results for arbitrary pointcut expressions, since in these cases each advice has the potential for inhibition or triggering. For *EnrichContent* and *InsertElement* advice actions, this may seem strange at first sight, but consider the following example: An element is inserted into an ordered list of nodes such that the indexes of the list entries change and thus, the nodes may no longer be matched by XPath-based pointcuts using child indexes. Similarly, an *EnrichContent* may only extend plain text content in a document, but still this influences pointcut expressions referring to a certain value of text.

From Table 1 we can draw several conclusions, as advice groups with an equivalent potential for interactions become apparent. There is a first group of advice actions which always inhibit, including *DeleteElement* and *ChooseVariant* advice actions. For *DeleteElement* this is obvious, because a deleted document element can no longer be referenced. In the case of *ChooseVariant*, exactly one element is selected from a list of variants. As a result, subsequent advice actions refering to the excluded variants are *inhibited*. A second group of *value-safe* advice actions using *ChangeValue*, *EnrichContent* and *ReduceContent* are safe w.r.t interactions with other advice actions since they only change attribute or text values. A third group contains *location-safe* advice actions—*MoveElement* and *ChangeOrder* only change element positions within a document. In a fourth group, *ExpandElement* and *CollapseElement* always pose problematic. On the one hand, they add new content while on the other, they remove and replace complete subtree structures, which may both trigger and inhibit a subsequent advice. Finally, An *InsertElement* may trigger other advice actions by creating new elements and values. However, it may also inhibit another advice using non-*location-safe* pointcut expressions, because of possible changes in element sequences.

In the following section, we present our approach for detecting interactions between adaptation aspects and advice actions.

4. INTERACTION DETECTION FOR ADAPTATION ASPECTS

As discussed in Sec. 3, the main cause for XML aspect interactions are aspects working on the same part of a document. As a web application usually has several adaptation concerns, such overlaps cannot be avoided in general. Accordingly, as a main means to detect aspect interactions, one can try to identify these overlapping parts. Going down to XPath as our language for content-based pointcuts, this problem can be treated as finding overlaps between the different aspects' XPath pointcuts.

One basic attempt to detect whether aspects *trigger* (see Definition 3) or *inhibit* (see Definition 4) other ones is to simply apply the XPath expressions onto a given document and compare the resulting node sets. A major drawback of this approach is that it requires an instance XML document to work upon. In the context of an XML pipeline, a document is transformed while traversing the pipeline. Therefore, the initial XML document existing at development time does not resemble the document onto which the aspects are later applied. Consequently, techniques are needed for detecting shared elements of different XPath expressions without the presence of an instance document (*static analysis*). This supports the developer in creating or editing aspects, because he can be made attentive to conflict sources at development time already. However, while giving valuable support during editing, static analysis is restricted by means of static computations over XPath expressions and by the user-intended adaptation semantics. Hence, in addition to static analysis, it is also advisable to support the developer by

tracking conflicts manifested at runtime (*runtime analysis*). In the following, we first investigate when aspect interactions turn into conflicts. Afterwards we present our concepts for static and runtime conflict detection.

4.1 Conflicts between Adaptation Aspects

We distinguish between two kinds of possible conflicts caused by unwanted aspect interactions which are defined and explained subsequently.

DEFINITION 7. *A **syntactic conflict** between an aspect a and an aspect b occurs if b is affected by a such that b's assumptions over the structure of the manipulated parts of a document are violated.*

An example for this kind of conflict is the application of first *CollapseElement* and second *ChangeOrder*. ChangeOrder operates on the chosen element's children. However, CollapseElement replaces the complete subtree of the chosen element by a simple text node and thus violates the assumption of ChangeOrder. From a developer's point of view, syntactic conflicts cause errors in the transformation process and thus should be removed from the aspects.

DEFINITION 8. *A **semantic conflict** between an aspect a and an aspect b occurs if interacts(a, b) holds and if the combined effect of a and b is unintended by the aspect developer.*

This kind of conflict can easily caused by the *MoveElement* advice if no location-safe pointcut expression is used. Since the system cannot make general assumptions about the semantics intended by the developers, they are not treated as errors. From our observations, advice actions which remove or move elements are likely to cause unwanted mismatches. Thus, these are treated as problematic and should be reported to users. However, the final decision whether a semantic conflict leads to a wrong result is left to the aspect author.

4.2 Static Conflict Detection

The detection of conflicts can occur in distinct phases of a web application's life cycle. For static analysis, the identification of these conflicts happens before weaving and uses the XPath expressions used for defining content-based pointcuts. In complement, runtime analysis takes place during the multi-staged transformation.

During the development of a web application, various aspects and advice actions are created. To support aspect developers, a static analysis component can check for conflict potentials resulting from problematic interactions between any two of the deployed advice actions. Fig. 5 illustrates the process of finding interferences between aspects while creating or editing aspects. A developer can always create a new advice or edit an existing one. When editing an advice, it is essential to define the programmatic and content-based pointcuts onto which the advice shall be applied. Using this specification, it becomes possible to check on interferences with already defined advice actions. It is then important to supply the developer with feedback on the detected conflicts. In the detection process, all defined advice actions are pairwisely analyzed. We call such advice pairs *advice combinations*.

In order to reduce the analysis effort, it is advisable to restrict the combinations which can be analyzed:

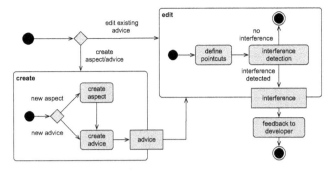

Figure 5: Interference detection for static analysis.

- to advice combinations causing syntactic conflicts or problematic semantic conflicts
- to advice actions sharing the same programmatic pointcut
- to advice actions with the same context condition

First of all, only those combinations need to be examined that are considered as problematic w.r.t Sect. 4.1. The second possible restriction is to inspect only combinations with the "same" programmatic pointcut (c.f. Sec. 2). The argument behind this practice is that developers explicitly express an order by spreading aspects over different stages of the web application's pipeline. Regarding programmatic pointcuts, the joinpoint after a designated pipeline stage and the one before the directly following stage can be considered as equivalent. The last restriction relates to context conditions. As an example, consider an advice which is applied only for desktop PCs and another one which is activated for smartphones only. Since the aspects' context conditions are excluding each other, they will never be applied at the same time.

After creating and restricting advice combinations, the next step of the static analysis is an actual conflict detection. It can be broken down to the *subset problem* for XPath expressions. This means that we test whether, for two XPath expressions p und q, the set of selected nodes of p is always in the set of selected nodes of q, or whether a common intersection exists. To this end, we use the following algorithms: For detecting containment (and from it deriving equivalence) between two XPath expressions, Miklau and Suciu presented an approach leveraging canonical models [17, 18]. The algorithm is restricted to the regular fragment $XP(/,//,[],*)$ of XPath[2]. To check whether there is a non-empty result set for the intersection of two XPath expressions, Hammerschmidt et al. [9] presented an approach which is also restricted to $XP(/,//,[],*)$ of XPath. Given two regular XPath expressions, it constructs the corresponding Deterministic Finite Automatons (DFAs) and computes their product automaton. For an overview of the different fragments of XPath, including time and space complexities for different algorithmic problems, we kindly refer to [5] and [23].

4.3 Detecting Conflicts at Runtime

As already shown, there are differences between detection of conflicts at development time and runtime detection. For

---

[2]The regular XPath fragment consists of nodes, the child- (/) and descendant-axis (//), predicates ([]) as well as wildcards (*).

analysis at runtime, it becomes possible to involve the current context and state of the pipeline. While traversing this pipeline, the final, outgoing document is formed. Therefore, we can check for manifested conflicts while weaving. Another advantage is that during runtime, we can analyze expressions from the full XPath language. Accordingly, the strategy for finding conflicts at runtime is as follows: if two or more advice actions exist, they will be applied to the given document. The resulting set of elements matched by the advice is saved for further analysis. To check whether an interaction between two advice actions exists, one can then check if the same elements exist in both actions' resulting sets. Based on that and according to definitions 7 and 8, we can easily check for conflicts.

4.4 Visualizing Conflicts to Developers

For both variants of conflict detection, it is necessary to present the analysis results to aspect developers. Because static analysis takes place at development time, our solution is integrated in the development process. We propose to mark identified conflicts directly within the aspect editor. Depending on Def. 7 and Def. 8, those marks should distinguish between warnings and errors.

The conflicts found at runtime (i.e., when the web application is running) need also be displayed for further review. For this, we propose using a logger which saves all interaction events that occur. Thereby, we achieve traceability of aspect interactions that occurred during the actual use of a web application.

4.5 Adaptation Conflict Resolution

After detecting conflicts, be it at development time or runtime, it is essential to resolve those conflicts. To this end, two different possibilities exist. The easiest way for resolving conflicts is letting the developer do so manually. To this end, it is necessary to create meaningful clues or solution proposals during the preceeding conflict detection. For that reason, assertions generated by the conflict detection need to be as precisely as possible. Furthermore, we propose to provide guidelines for creating adaptation aspects, also including hints on how to produce safe pointcut expressions depending on the adaptation objectives.

Second, the system can try to automatically resolve the conflicts it has identified. As stated before, it is hard to resolve arbitrary conflicts when the developer's intention with a certain aspect is unknown to the system. Still, some conflicts can be resolved automatically by the system. Further, for unintended semantic conflicts, developers may announce these to the system and instruct the system to also resolve these conflicts automatically. For this to work, the system needs the right to change the aspect application order. A simple reversion of weaving order can then resolve the unintended effect.

The system may use an operator $pre(a, b)$ to determine that an aspect a shall always be woven before another aspect b. With the aid of this operator, the system is able to change the weaving order. However, it is possible that such measures create new interaction effects and, in the worst case, cyclic dependencies can occur. To detect these incidents, we use the directed interaction graph $G(V, E)$ which is induced as follows:

```
<?xml version="1.0" encoding="UTF-8"?>
<PatternConflicts>
    <ChangeOrder>
        <ChangeOrder>
            <status>warning</status>
            <reason>first ChangeOrder is overwritten</reason>
            <solution>no solution yet</solution>
        </ChangeOrder>
        ...
    </ChangeOrder>
</PatternConflicts>
```

Figure 6: XML-based conflict lookup table

- $V := \{a \mid a \in Advice_{PPC}\}$
- $E := \{(a,b) \mid a,b \in V \wedge (triggers(a,b) \vee inhibits(a,b))\}$

The nodes (V) of the interaction graphs are the advice actions which shall be woven into the document at a specific programmatic pointcut (PPC). Between two nodes a and b, there is an edge (E) iff $affects(a,b)$ holds. Each change of the weaving order by using *pre* causes an update to the interaction graph. Automatic resolution by aspect ordering is only possible if the subgraph induced over all conflicting interactions is not cyclic. Despite cyclic dependencies, it remains possible to resolve interactions using corrective measures, e.g., using a $redo(a)$ operator. By using $redo$, the system can re-apply an already woven advice to the document so that a cycle can be broken up.

5. PX-WEAVE AND THE DETECTION OF ASPECT INTERACTIONS

This section describes the integration of both static and runtime interaction detection in PX-Weave. First of all, we provide a glimpse on our data model for specifying problematic conflicts, which we use to retrieve the result of advice combinations and for further handling of found conflicts. Afterwards, we describe the implementation of the concepts presented in section 4.

5.1 Data Models for Conflict Representation

The definition of the conflict "lookup table" is given in an XML representation. An excerpt of this XML file is shown in Figure 6. If two adaptation advice actions have a conflict potential, this is represented with a parent-child-relationship in the XML file. Fig. 6 shows this for two advice actions of the type ChangeOrder. At runtime, we use XML-binding techniques to obtain these definitions for PX-Weave. Hence, we are able to easily add additional advice actions and conflict combinations just by introducing them into the XML structure.

5.2 Integration of Detection

For *static detection* we leverage containment and intersection algorithms to check for conflicts between adaptation aspects. To check whether an advice's XPath is contained in or equivalent to another advice's, we employ the approach of Miklau and Suciu [17, 18]. The algorithm described in their work is based on *TreePatterns*. A corresponding parser we have implemented can generate such TreePatterns out of XPath pointcut expressions. Additionally, we extended Miklau and Suciu's original design with namespace and attribute handling. For identifying intersections between XPath expressions, we build automatons for each given regular XPath expression. In order to check whether the intersection of two

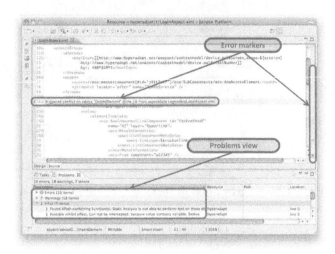

Figure 7: Visualization of conflict potentials

expressions creates a non-empty set of elements, we build a product automaton from the involved automatons. If a transitive closure between the initial and the final state pairs exists, then both expressions do overlap (c.f. [9]). Besides detecting direct conflicts, we have also implemented a module for identifying the effects resulting by each adaptation advice pattern. For example, the advice pattern *MoveElement* has an attribute "to", defining the targeted element's new location in the document. To detect triggering effects, we construct the element's new XPath expression after moving and then check for direct conflicts between this new expression and all other pointcuts.

We integrated static detection into the authoring process with PX-Weave by creating a plug-in for the Eclipse IDE (http://www.eclipse.org) and extending its existing, powerful XML tooling. Our plugin uses an *InteractionBuilder* to start the analysis for conflict potentials whenever a file relevant to adaptation aspects (e.g., an aspect definition file or a document augmented by aspects) has changed, providing the author with immediate feedback. If conflict potentials (of direct or affected nature) have been found, they are displayed in the main editor window as well as in the problems view. Fig. 7 illustrates visualization for an adaptation aspect deployed in the SoundNexus example, which was introduced in Sect. 2.1.

As already discussed, the runtime analysis takes place while traversing the pipeline in PX-Weave. For this, information on the current programmatic pointcut and active advice actions are provided by PX-Weave. Fig. 8 illustrates the integration of the runtime analysis into PX-Weave with a class diagramm. The classes `AspectInterpreter` and `AdviceInterpeter` yield the required information on the current programmatic pointcut and the enlisted advice actions. The module for `DynamicAnalysis` recognizes when PX-Weave is weaving an advice into a document. Having detected such a weaving, the module starts analyzing the advice for conflicts with other actions.

6. RELATED WORK

In the field of adaptive web applications, we are not the first to propose aspect-orientation as a means to factor out adaptivity from the core transformation. The Generic Adap-

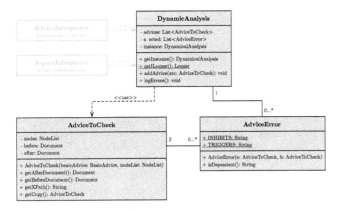

Figure 8: Class diagram for runtime detection module

tation Component (GAC) is an approach to rule-based adaptation of XML documents developed by [8]. Similarily to PX-Weave, GAC uses pre-defined transformation patterns to apply adaptation rules to XML documents, with the execution of rules bound to context conditions. However, GAC does not include support for selecting programmatic joinpoints, hence all adaptation patterns are applied at the same point of execution. Furthermore, GAC does not offer any support for ensuring validity of the transformation result. Especially if there are multiple GAC rules applied, their combined effect can not be predicted. Our approach provides support for static analyses of aspect interactions while GAC only supports ordering of adaptation rules and analyses at run-time.

[3] propose an extension of the (UML-based Web Engineering) UWE framework, treating adaptivity in web applications as a crosscutting concern. They introduce three transformation templates which an author can apply to modify his web application's navigation structure. However, their approach is based on fragments of a specific model (UWE), in contrast to our generic XML weaver. Finally, the authors do not propose any kind of resolution strategies for conflicting rules.

Similar to UWE, aspectWebML [22] is an effort to add aspect-orientation to the development of web applications, namely to the prominent language WebML. To this end, aspectWebML allows authors to extend or replace parts of an existing WebML model using three transformation primitives: enhancement, replacement and deletion. The main idea is that authors define an additional layer of functionality on top of explicitly selected model elements. Therefore, aspectWebML may help to distinguish adaptivity from core application, but does not offer a pointcut language. Given this focus, aspectWebML further does not supply means to detect or resolve interactions— they simply are not anticipated.

Doxpects are an AOP approach for XML-based web services [24], which resembles our approach by using XPath expressions to define content-based pointcuts. A doxpect contains one or more *request* or *response* advice actions to transform a web service message before sending or receiving it. Transformations are based on XML Beans, i.e., statically typed objects representing XML fragments. Weaving is realised by compiling doxpects into web service handlers. In contrast to our approach, doxpects are not applied to

multi-staged XML transformations, but only to web service messages. Furthermore, XML Beans seem less adequate for multi-staged transformation environments. Finally, in contrast to our approach there is no support for detecting interactions or resolving conflicts between doxpects.

In [13], Kellens et al. propose a pointcut language for robust pointcut expressions. The main idea is lifting pointcut expressions to a more abstract conceptual model as a kind of taxonomy or ontology. The conceptual model defines the abstract concepts which may occur in a programm (e.g., accessor methods) and relates them to specific constructs in the actual base program. Fragility of pointcuts with respect to a base program is considered problematic, since changes to the base program may also change a pointcut's semantics. In contrast to our notion of *safe* pointcut expressions, dependencies are moved to the conceptual model, whereas our approach requires aspect authors to design sufficiently robust pointcuts themselves.

Sanen et al. propose a schema for documenting production aspects interactions [21]. For each interaction, the schema requires to name the aspects involved as well as a set of user-defined predicates to formally describe the effects of that interaction. Furthermore, the authors define four kinds of interactions. *Mutual exclusion* allows exactly one of the interacting aspects to be deployed at a time, a *dependency* indicates that an aspect requires another one to be deployed in the system. *Reinforcement* influences another aspect in a positive way, a *conflict* indicates unintended semantic interferences between aspects. In PX-Weave, we currently indicate problematic aspect interactions via error and problem markers. Using an aspect classification and documentation schema, we could improve the user feedback of the PX-Weave analysis components.

In [12], Katz discusses classes of aspects which influence the underlying system and other aspects in different ways and with different potentials for complexity. For example, spectative aspects only read exisiting variables in a system and do not change their values. As a result, *spectative* aspects do not change the underlying system's basic behaviour. *Regulative* aspects can change the basic system's control flow by restricting or avoiding some computations or values. In contrast, *invasive* aspects can influence the behavior and computations of the whole system. It is therefore hard to predict their behavior. While the classification of [12] cannot be mapped directly to our approach, there are some obvious analogies. Since we do not support invasive access to transformations of the underlying document architecture, our adaptation aspects can be considered as non-invasive aspects. However, our approach allows for spectative and regulative aspects, since additional nodes can be inserted and computed results of a transformation can be removed.

7. CONCLUSIONS

In this paper, we discussed the problem of interacting aspects for the domain of adaptive, XML-based documents. We have unfolded the circumstances under which adaptation aspects interact in multi-staged document architectures and how undesired interactions can be avoided by using *location-safe* and *value-safe* pointcuts. Further, we have documented strategies for detecting and displaying conflicts between adaptations, founding on combined static and dynamic analyses. Our static analysis approach is based on a well-known DFA intersection algorithm for a regular fragment of XPath, which we leverage for content-based point-cut expressions. The dynamic analysis works on documents passing through the web application's pipeline to evaluate whether two pointcuts overlap. Finally, by presenting our tool PX-Weave, we have shown the practical feasibility of the presented approach.

However, there still is room for improvements. It remains unclear whether the regular subset of XPath is an acceptable pointcut restriction to aspect authors who wish to utilize static conflict detection facilities. To this end, an evaluation of our approach in a large-scale scenario would be advisable. We also intend to apply our approach to more complex adaptive document or web application architectures, e.g., Java Server Faces (JSF). Furthermore, our approach for static interaction detection could be enhanced by using better algorithms over XPath expressions. For example, XPath expressions can be checked for satisfiability w.r.t a given schema [10, 4]. Hence, a non-empty intersection of two content-based pointcuts may still never match a concrete document, if the document format is restricted by a schema.

8. ACKNOWLEDGMENT

We want to thank our anonymous reviewers for their insightful and valuable comments. This research was funded by the *Deutsche Forschungsgemeinschaft* (DFG, German Research Foundation) within the project *HyperAdapt*.

9. REFERENCES

[1] M. Aksit, A. Rensink, and T. Staijen. A graph-transformation-based simulation approach for analysing aspect interference on shared join points. In *Proceedings of AOSD*, pages 39–50, Charlottesville, Virginia, USA, 2009. ACM.

[2] U. Aßmann. Architectural styles for active documents. *Science of Computer Programming*, 56(1-2):79–98, 2005. New Software Composition Concepts.

[3] H. Baumeister, A. Knapp, N. Koch, and G. Zhang. Modelling adaptivity with aspects. In *Proceedings of ICWE*, number 3579 in LNCS, pages 406–416. Springer, 2005.

[4] M. Benedikt, W. Fan, and F. Geerts. XPath satisfiability in the presence of DTDs. *J. ACM*, 55(2):1–79, 2008.

[5] M. Benedikt, W. Fan, and G. Kuper. Structural properties of XPath fragments. *Theoretical Computer Science*, 336(1):3–31, May 2005.

[6] P. Brusilovsky. Adaptive hypermedia. *User Modeling and User Adapted Interaction*, 11(1-2):87–110, 2001.

[7] R. Douence, P. Fradet, and M. Südholt. A framework for the detection and resolution of aspect interactions. In *Proceedings of GPCE*, volume 2487 of *LNCS*, pages 173–188, London, UK, 2002. Springer-Verlag.

[8] Z. Fiala and G.-J. Houben. A generic transcoding tool for making web applications adaptive. In *Proceedings of CAiSE*, pages 15–20. FEUP, June 2005.

[9] B. C. Hammerschmidt, M. Kempa, and V. Linnemann. On the intersection of xpath expressions. In *Proceedings of IDEAS*, pages 49–57. IEEE, 2005.

[10] J. Hidders. Satisfiability of xpath expressions. In *Database Programming Languages*, volume 2921 of *LNCS*, pages 125–126. Springer Berlin / Heidelberg, 2004.

[11] S. Karol, M. Niederhausen, U. Aßmann, K. Meißner, and M. Steinfeldt. Towards generic weaving of adaptation aspects for XML. In *Proceedings of ICOMP*, volume 1, Las Vegas, USA, July 2010. CSREA Press.

[12] S. Katz. Aspect categories and classes of temporal properties. In *Transactions on Aspect-Oriented Software Development I*, volume 3880 of *LNCS*, pages 106–134. Springer Berlin / Heidelberg, 2006.

[13] A. Kellens, K. Gybels, J. Brichau, and K. Mens. A model-driven pointcut language for more robust pointcuts. In *Software Engineering Properties of Languages for Aspect Technology (SPLAT) workshop collocated with AOSD*, 2006.

[14] G. Kiczales, E. Hilsdale, J. Hugunin, M. Kersten, J. Palm, and W. G. Griswold. An overview of AspectJ. In *Proceedings of ECOOP*, pages 327–353, London, UK, 2001. Springer-Verlag.

[15] G. Kiczales, A. Mendhekar, J. Lamping, C. Maeda, C. V. Lopes, J. Loingtier, and J. Irwin. Aspect-Oriented programming. In *Proceedings of ECOOP*, volume 1241 of *LNCS*, Finland, June 1997. Springer-Verlag.

[16] G. Kniesel. Detection and resolution of weaving interactions. In *Transactions on Aspect-Oriented Software Development V*, volume 5490 of *LNCS*, pages 135–186. Springer Berlin / Heidelberg, 2009.

[17] G. Miklau and D. Suciu. Containment and equivalence for an xpath fragment. In *Proceedings of PODS*, pages 65–76, New York, NY, USA, 2002. ACM.

[18] G. Miklau and D. Suciu. Containment and equivalence for a fragment of xpath. *J. ACM*, 51:2–45, January 2004.

[19] M. Niederhausen, S. Karol, U. Aßmann, and K. Meißner. Hyperadapt: Enabling aspects for xml. In *Proceedings of ICWE*, pages 461–464, Berlin, Heidelberg, 2009. Springer-Verlag.

[20] A. Restivo and A. Aguiar. Towards detecting and solving aspect conflicts and interferences using unit tests. In *Proceedings of SPLAT*, New York, NY, USA, 2007. ACM.

[21] F. Sanen, E. Truyen, W. Joosen, A. Jackson, A. Nedos, S. Clarke, N. Loughran, and A. Rashid. Classifying and documenting aspect interactions. In *Proceedings of the Fifth AOSD Workshop on Aspects, Components, and Patterns for Infrastructure Software*, pages 23–26, Bonn, Germany, 2006.

[22] A. Schauerhuber, M. Wimmer, W. Schwinger, E. Kapsammer, and W. Retschitzegger. Aspect-oriented modeling of ubiquitous web applications: The aspectwebml approach. In *Proceddings of ECBS*, pages 569–576. IEEE Computer Society, 2007.

[23] B. ten Cate and C. Lutz. The complexity of query containment in expressive fragments of XPath 2.0. *J. ACM*, 56(6):1–48, 2009.

[24] E. Wohlstadter and K. D. Volder. Doxpects: aspects supporting XML transformation interfaces. In *Proceedings of AOSD*, pages 99–108. ACM, 2006.

Publicly Posted Composite Documents with Identity Based Encryption

Helen Balinsky
Hewlett-Packard Laboratories
Long Down Avenue
Bristol BS34 8QZ, UK
Helen.Balinsky@hp.com

Liqun Chen
Hewlett-Packard Laboratories
Long Down Avenue
Bristol BS34 8QZ, UK
Liqun.Chen@hp.com

Steven J. Simske
Hewlett-Packard Laboratories
3404 E. Harmony Rd. MS 36
Fort Collins, CO USA 80528
Steven.Simske@hp.com

ABSTRACT

Recently-introduced Publicly Posted Composite Documents (PPCDs) enable composite documents with different formats and differential access control to participate in cross-organizational workflows distributed over potentially non-secure channels. The original PPCD design was based on a Public Key Infrastructure, requiring each workflow participant to own a pair of public and private keys. This solution also required the document master to know the corresponding valid public keys (certificates) of all participants prior to commencement of the workflow. Using Identity Based Encryption (IBE), a recently described cryptographic technique, we eliminate the requirement for the prior knowledge and distribution of the workflow participants' keys. The required public keys for each workflow participant are calculated based on user identities and other relevant factors at workflow onset. The generation of corresponding private keys can be delayed up until the workflow step, when the corresponding workflow participants require access to the document. The solution presented provides automatic workflow order enforcement and the ability to impose multiple document release dates real-time.

Categories and Subject Descriptors

I.7.1 [**Document and Text Editing**]; I.7.4 [**Electronic Publishing**]

General Terms

Algorithms, Design, Security

Keywords

publicly posted composite document; identity based encryption; embedded access control; document security

1. INTRODUCTION

Over the last decade, traditional single-format documents are slowly being replaced by composite ensembles ([10], [7]): dynamically created from multiple differently formatted sources - individual files or fragments - to produce enriched, more appealing results. Disparate parts are combined together into a coherent document presentation using different visual cues: consistent styling, cross-document navigation bars and parts linking. Whilst physically bundled into a document serialization, individual parts retain their original formatting, which is beneficial for their subsequent reuse, re-purposing, remix or mash-up as well as for retaining the ability to access and manipulate individual parts using their native advanced tools. This has only become possible as a result of advanced layout techniques in combination with dramatically increased computing power: document presentation can now be compiled just-in-time and delivered to an awaiting user based on run-time parameters of a displaying device and the user's granted access, whilst taking into account various personal preferences.

In many cases, the problem of providing fine-grained access control to such composite documents can be addressed through on-line or domain services, where document parts are provided for authenticated users subject to the access granted. However, such service, in particularly its administration is fully trusted with both: user authentication and the actual document contents. In today's globalized world economy, when collaborative document workflows more and more often cross the boundaries of secure domains: businesses, academic institutions, government organizations, countries and continents – such fully trusted by all participants side often impossible or impractical to establish.

Not enclosed into a single security domain and not owned by any single entity, these complicated and collaborative workflows are often associated with highly sensitive data: business intelligence, intellectual properties, competitive analysis, and confidential and/or private information. Without a trusted third party to manage and maintain such workflows, documents are simply shipped amongst dispersed participants by low security traditional e-mails, or mailed on CD/DVDs with the fine-grained access control required by the nature of the document often effected through manual data redaction. Subsequent document versions may then be assembled from the disparate individual replies. Sometimes individual document parts are shipped over secure channels (e.g. https) or even encrypted for protection in transition; however, this does not provide for the fine-grained access.

The problem was first raised and addressed in [13] by an introduction of Publicly Posted Composite Documents

(PPCDs): composite documents with an embedded access control and enforced workflow order. The access to a PPCD for workflow participants was provided through Public Key Infrastructure (PKI)[6], and as such the original design inherited existing PKI problems. Prior to commencing a workflow, a pair of public and private keys must be issued for each workflow participant. Then, a document master must gain access to the corresponding trusted public key certificates for each workflow participant. Only then can the document workflow be generated and the document shipped along its workflow. With PKI being the most prevalent key generation system today, even obtaining keys certified by a Certificate Authority (CA), for example by VeriSign [3], is a relatively straightforward task. Despite its straightforward nature, however, the task needs to be accomplished for each workflow participant before a workflow can be commenced. Lack of knowledge of the public key of one or more of the participants results in the inability to create a complete PPCD workflow and initiate (start executing) it.

Even when the precautions are taken and public keys for all of the workflow participants are obtained *a priori*, a document master can still find himself in the situation of being unable to create a PPCD workflow in the very last moment due to:

- The public key certificate of a participant has recently expired;

- The key of a workflow participant was compromised and revoked, which could be established by automatically downloading the Certificate Revocation List, for example, from VeriSign,[2].

- The participant is out of the office for some time and his role has been filled by a proxy, who does not have access to the participant's private key;

- The participant has been replaced in his role by a new hire, who does not have a pair of keys yet;

and many others.

One of the alternatives is to create a shorter workflow trimmed just before the unknown user, and re-visit the situation when this shorter workflow finishes. The requirements of human intervention and/or the introduction of extra steps are likely to introduce extra delays, errors, and unfinished workflows, which may abrogate a contract, slip a deadline or otherwise cause the workflow to not meet its requirements. A problem also surfaces if a workflow participant has keys compromised and revoked after the document is shipped along its workflow.

Identity Based Encryption (IBE) was first proposed by A. Shamir in 1984 [21]; however, it remained an open problem for nearly 2 decades, until The Boneh-Franklin Pairing-Based Encryption Scheme [14] was proposed in 2001. Today, IBE is gradually gaining in popularity, with the RFC Standard being issued in 2007 [8] and the one of the first commercial products made available by [4].

To summarize, for PKI schemes the user's pair of keys is generated randomly, the public key is certified and then the

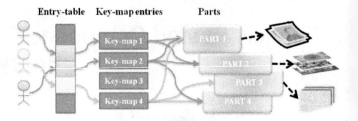

Figure 1: PPCD building blocks, including entry-table, key-map entries and content-parts

certificate is distributed. When using IBE, the keys are calculated, thus providing a unique ability to delay the generation of private keys for a participant until a document reaches the participant. The other major advantage of IBE is that any user's uniquely identifiable information, including publicly known (e.g. user's e-mail), can be used as this workflow participant's public key.

In the current paper, we provide an extension of the original PPCD design to operate over IBE and potentially over a mixed IBE-PKI field.

The paper is organized as follows. In Section 2, we review the structure of Publicly Posted Composite Documents and their mechanism for granting differential access. Challenges posed by the existing PPCD over PKI design are presented in Section 3. Our solution, a special IBE-based scheme designed to address the needs of PPCD is presented in Section 4. For PPCDs to cross the boundaries of different organizations, we provide the generalization of the proposed scheme for multiple Key Generation Centers and even mixed IBE-PPCD scheme (Section 5). In Section 7 we then provide the implementation details of the current prototype, followed by our Conclusions.

2. PPCD TECHNOLOGY

Recall that the Publicly Posted Composite Document (PPCD) technology [13] was developed to support the secure creation and management of composite documents, with variable sensitivity data in multi-user document workflows that are not contained within one secure environment. In its first realization, differential access control and user identity were based on a Public Key Infrastructure (PKI) [6], which is the most prevalent and familiar mechanism for sensitive data exchange – e.g. https, SSL/TLS protocols, and encrypted or signed emails are all make use of PKI today.

PPCD Structure

The PPCD, illustrated in Figure 1, is a serialization consisting of:

1. the entry-table,

2. key-map entries and

3. content-part entries.

Entry-table

The entry-table is a single entry point to a PPC document and its secure gate. The entry-table comprises a set of individual records, with at least one record provided for each workflow participant. Individual access records are assembled into the entry-table in a completely random order to eliminate any correlation with the access order required by the document workflow. A few fake records could be added to conceal the actual number of participants in sensitive workflows. Without a corresponding record in the document entry-table a user cannot access the document.

The workflow participant's private decryption key is required to identify the corresponding record and extract the information from this record. Each individual record includes four separate fields:

1. the ciphertext of the user's access symmetric key K, which is encrypted using the user's public encryption key;

2. a plaintext "magic word" (e.g. any string of characters);

3. the ciphertext of the "magic word" encrypted using symmetric key K;

4. the ciphertext of the encrypted key-map entry name or Id in the document serialization, the key-map entry name or Id is encrypted using key K.

To identify the user's entry, it will be sufficient to decrypt the symmetric key K, then using K to decrypt the encrypted "magic word" and compare it with the plaintext "magic word". The recovered and plaintext "magic words" should match. Once the participant's record in the entry-table is identified, the symmetric key K is used to recover the participant's key-map entry name from the last field in the record. Then, the actual contents of the participant's key-map entry, also encrypted by key K, is loaded from the document serialization. The subset of keys for the document parts, corresponding to the granted access, is recovered by decrypting the participant's key-map entry.

The additional symmetric key K (pre-generated during the PPCD creation) for each participant is deployed to reduce the computational overhead of the potentially slow asymmetric cryptography as well as the potential attack of identifying individual records using known public keys of workflow participants.

Labeling each key-map entry by user-related data or personal identifiable information (PII) could provide an alternative mechanism for identifying the user's key-map entry. However, it is hard to make such identifiers automatically recognizable for each workflow participant without revealing some personal information about participants. This could lead to privacy infringement: revealing the identities of workflow participants or exposing their association with particular document workflows. Protection of PII is a major concern today [18].

In yet another alternative, workflow participants may attempt to decrypt each key-map entry in turn until the correct one – "decryptable" by a particular participant – is found. Whilst robust and privacy friendly, this approach is computationally expensive for large workflows with many different sensitivity parts: such workflows contain relatively large individual key-map entries.

As introduced elsewhere [13], differential access for each content-part in a composite document is enabled by 4 keys: encryption, decryption, signature and verification. Each content-part is first encrypted and then signed. Within the context of a composite document workflow a participant can be granted: Read Write (**RW**), Read Only (**RO**) or Verifiable Authenticity (**VA**) accesses. According to the PPCD protocol, the authenticity of each part is mandatorily verified on receiving a document. The corresponding part's verification key is provided for each part and each workflow participant at each workflow step. **RO** access is further provided by supplying the part's decryption key. To enable part modification **RW**, the remaining two keys (encryption and signature) are required. An item is decrypted, modified as needed, then encrypted and signed. The authenticity of such a modification is immediately validated by the subsequent workflow participant using the corresponding verification key, which is given for any access **RW**, **RO** or **VA**.

The subsets of keys for each participant per a workflow step or for an entire workflow are combined into individual key-map entries. Each of these is individually encrypted and signed and thus made accessible for a corresponding participant only and at a particular workflow step.

3. PPCD CHALLENGES

Built on top of the existing Public Key Infrastructure, [6], publicly-posted composite documents [13] have several potential limitations. The workflow generation mechanism requires a prior knowledge of two tokens per a workflow participant: a valid contact address (e-mail or postal) and a valid public key certificate. This implies that prior commencing a document workflow:

1. each workflow participant must be issued a pair of public and private keys and

2. the public key certificate must be communicated the workflow owner (Figure 2).

Whilst a known public key certificate for a workflow participant is a long term key that can be used for multiple workflows involving the participant, the certificate can expire or be revoked, meaning a new, valid certificate must be obtained.

Due a wide variety of reasons, a workflow participant could be replaced by another user at the very last moment, thus causing a situation wherein the corresponding public key certificate is either not known for a workflow master. Worse, the participant may be required acquire such keys. Any such very real situations may delay or even jeopardize workflow commencement, or require some involvement of a document master at the corresponding workflow steps.

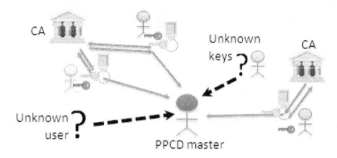

Figure 2: Each workflow participant must own a pair of public and private keys, and the public key certificate (signed by a Certificate Authority) must be communicated to the workflow owner

4. OUR SOLUTION

To alleviate the logistics problem of key distribution, key unavailability (through expiration or revocation) and other problems introduced when using a Public Key Infrastructure (PKI) with a composite document lifecycle, we extend the original PPCD design to accommodate the modern cryptographic technique of Identity Based Encryption (IBE). Using IBE, the user's e-mail address can be used as his public key and the generation of his private keys can be delayed all the way until the participant's access.

The current section is organized as follows. Taking into consideration that IBE is a relatively new cryptographic technique, its main principles are mainly unfamiliar outside of the cryptographic community, in Subsection 4.1 we provide a basic description of generic IBE technique in comparison with well-known PKI scheme. Then, in Subsections 4.2 and 4.3, we provide the amendments required to the current PPCD processes. The details of extending the provided solution for handling special workflow cases, that were hard or often impossible to address by the PKI-based PPCD design, are discussed in Subsection 4.4.

4.1 Identity Based Encryption: Basics

Identity-based encryption (IBE) is a type of public cryptography, where public and private keys are calculated rather than being generated randomly. There are different types of identity-based encryption schemes: based on bilinear pairings on elliptic curves (e.g. [14]), and the quadratic residuosity problem [16] with the latter being computationally inefficient. For PPCD design we used the Boneh-Franklin scheme [8] to generate keys shared between a document master and each workflow participant for each workflow step, for which the corresponding participant is granted one of the accesses: **RW**, **RO** or **VA**.

Some unique user identifier, including publicly known information (IBE public parameters that will be discussed below), are used to calculate the user's public key. In the classic example, the user's e-mail address can be used as his IBE identifier.

The corresponding user's private key is calculated from the user's public key by a trusted third party, known as a Key Generation Center (KGC). A KGC corresponds to a Certificate Authority (CA) in the PKI scheme; however, the functionality is dramatically different. Similar to a CA, a KGC has its own pair of (strong) public and private keys, which are generated once. Then the public key is made publicly known, whilst the private key is kept as a major secret.

However, unlike a CA, each KGC has a third cryptographic object: IBE system (public) parameters, which are also made public. The KGC's IBE public parameters can, for example, specify different IBE algorithms or different key strengths. The KGC's public parameters are required to generate the user's public key from the user's identity. As a result, different pairs of public and private keys will be generated for the same user identity by different KGCs.

Similar to a well-known PKI-based VeriSign [3], a KGC can be run by a trusted third party; alternatively, a relatively large organization or enterprise can run their own KGC. A KGC needs to provide two services (notation according to [8]):

1. A Public Parameter Server (PPS) that provides a well-known location for secure distribution of IBE public parameters.

2. A Private-key Generator (PKG) that

   (a) accepts an IBE user's private key request;
   (b) authenticates the user;
   (c) after successful authentication computes the user's IBE private key using user's identity;
   (d) returns the user's IBE private key.

In both the PKI- and IBE-based PPCD designs, a workflow master is required to know the public keys for each participant in order to generate a PPCD workflow. To generate the user's public key, the KGC's public parameters and its public key are required, as illustrated in Figure 3. The first KGC's service is used by a PPCD master/creator to obtain the KGC's public parameters. A PPCD master is capable of

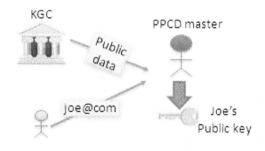

Figure 3: A PPCD master generates participant's public key out of KGC public data (public parameters and public key) and the participant's email address

calculating the public keys for workflow participants after he obtaining the KGC's public parameters. The same or different KGCs can be used for participants of the same workflow:

the public cryptographic material for all used KGCs needs to be known by the document master. In the context of the PPCD, public keys are computed by the PPCD Authoring Tool.

At the same time, the workflow participants in the IBE-PPCD do not need to have their private keys issued by the corresponding KGCs. Using the second KGC's service in time-insensitive and unordered workflows, the participants can retrieve their private keys at any time prior their workflow step, as illustrated in Figure 4. By using an extended

Figure 4: In time-insensitive and unordered workflows, a workflow participant can retrieve his private key at any time

user identifier that, for example, includes the earliest and latest private key release dates, it is possible to enforce different key release and expiry dates, thus providing for time-sensitive workflows. This extension will be described later in Subsection 4.4.

4.2 PPCD: IBE Algorithm and Deployment Scheme

Using the IBE scheme, we establish a shared secret between a document master and each workflow participant at the workflow level or at individual steps -- depending on the workflow type and requirements. In multi-step workflows, where a participant might be required to access a document more than once with different accesses, a new shared secret could be established for every or only some workflow steps, we will discuss this issue in the Subsection 4.4.

In the original design, [13], such shared secrets were also deployed (Section 2, key K); however, they were randomly pre-generated by the workflow master (automatically by the PPCD Authoring tool) at the PPCD creation stage. Each secret was encrypted using the corresponding participant's public encryption key and delivered to the participant within his record in the document's entry-table (Section 2). Each such secret was only recoverable by the corresponding workflow participant using his private decryption key. The secret is subsequently used as a key to symmetric cryptography, e.g. Advanced Encryption Standard (AES [5]), to identify and decrypt the participant's key-map entry.

In the new design, such a unique shared secret is computed on both sides using IBE cryptography:

- at the onset of a workflow by the document master using participants' IBE-public keys (participant's identifier or extended identifier)

- at the point of access by each workflow participant using his IBE-private key.

Using the same key derivation function (KDF), e.g. a Secure Hash Algorithm [9], the shared key for symmetric cryptography is derived from this shared secret. Once the corresponding symmetric encryption key for each participant for each workflow step is computed, it is used to encrypt the participant's key-map entry by the document master and decrypt by the corresponding participant. In the next section, we are going to describe in greater detail how such a shared secret is computed.

4.3 PPCD: Encrypting and Decrypting Key-Map Entries

4.3.1 Authoring: Encrypting a Key-Map Entry

Once the public key for each workflow participant is computed, the Authoring Tool is able to proceed with generating a PPCD document. Similar to the original design, we could generate a symmetric key K (AES key in the original design, see Section 2), which is then encrypted using the computed participant's public key. However, owing to the special properties of IBE, this original scenario can be simplified.

Our solution makes use of a pairing function $e : G_1 \times G_1 \to G_2$, where G_1 and G_2 are finite field groups of a large prime order q, and G_1 has a generator g. The pairing function e takes two elements from G_1 as input and outputs an element of G_2. Our solution also makes use of a symmetric encryption algorithm SE (for example AES), a hash function $H : \{0,1\}^* \to G_1$ and a key derivation function $KDF : G_3 \to \{0,1\}^\ell$, where ℓ is a bit length of a symmetric key used in SE.

In the setup of our system, a trusted third party, namely the Key Generation Center (KGC), generates a private key $x \in Z_q$ and the corresponding public key $(g, y = g^x) \in G_1^2$. In order to create a symmetric key which will be shared between a document owner and a document workflow participant, say SSE, the document owner (acting as an encryptor) first takes the document participant's identity, say ID, creates an ephemeral secret r, which is called the pseudo private key, and then computes $SSE = KDF(e(H(ID)^r, y))$. The document owner also generates $z = g^r$, which is called the pseudo public key and will be made available to the document participant.

The IBE pairing of the user's pseudo private key with the KGC's master public key produces the Shared Secret in Encryption(SSE). The same secret, the Shared Secret in Decryption(SSD) will be computed by the corresponding workflow participant during his workflow step. Using a key derivation function KDF, a random key for AES encryption is further computed out of the SSE, Figure 5.

As such, it is no longer required to pre-generate the symmetric key K (Section 2) at the workflow onset and distribute it to each workflow participant within the corresponding entry in the document entry-table. This results in the modified entry-table (the original entry-table is described in Section 2):

Figure 5: A PPCD creation diagram by the document master

1. a plaintext "magic word" (e.g. any string of characters);

2. the ciphertext of the "magic word" encrypted using symmetric key K, this key is now derived from the run-time computed SSE;

3. the ciphertext of the encrypted key-map entry name in the document serialization, the key-map entry name is encrypted using the same as above key K;

4. the participant's pseudo public key, $z = g^r$.

4.3.2 Workflow: Decrypting User's Key-Map Entry

As it was described in 4.1, in order to access a document at some point in time, each workflow participant needs to contact the corresponding KGC and retrieve his private key. The KGC authenticates the user and generates his private key, Figure 4.

To retrieve the shared symmetric decryption key SSD, given the pseudo public key z, the document workflow participant has to obtain an asymmetric private key, which is based on his identity ID and created by KGC, $k_{ID} = \mathsf{H}(\mathsf{ID})^x \in G_1$, Figure 6 (left). The value SSD is computed as $SSD = \mathsf{KDF}(e(k_{ID}, z))$. Based on the bilinear property of the pairing function, the equation

$$e\left(\mathsf{H}(\mathsf{ID})^r, \mathsf{y}\right) = e\left(k_{ID}, z\right)$$

holds; so $SSD = SSE$.

Then, using the IBE-pairing of the user's private key and the pseudo public key, the user computes the Shared Secret in Decryption (SSD). Applying the same KDF to SSD exactly as described in Subsection 4.3.1, the user derives the shared symmetric (AES) key. Using the simplified entry-table design and the newly derived symmetric key, Figure 6 (right):

1. The user attempts to decrypt the encrypted "magic word" entry from every record in the entry-table and compares the result with the clear text "magic word" from the same record.

2. Once the decrypted and clear text values match, the user (the automatic software on his behalf) identifies his record.

3. The user decrypts the cipher-text of the encrypted key-map entry name, retrieves and decrypts this key-map entry – following the original PPCD access.

4.4 Extending User Identifier

Whilst any unique user information can be used as the participant's IBE identifier, the participant's e-mail address is often the preferred choice as it is always unique and in majority of cases an exclusive accessibility by the address owner can be safely assumed. In a PPCD workflow, using the participant's e-mail address as the identifier provides an additional benefit over the PPCD-PKI implementation: the number of required tokens per participant is reduced to one, as the same e-mail address is used for encrypting and e-mailing a document.

Moreover, a PPCD master can encode some additional information into the user's identifier, for example:

Time-limiting One or two time identifiers limiting the private key release timeframe: the user's private keys may not be generated and given to the user earlier and/or later than the corresponding time identifier. For some sensitive workflows these time identifiers can also be provided in some encrypted form.

Workflow-enforcing A random identifier, nonce, that can be used to enforce the workflow order.

When a workflow participants contacts the corresponding KGC to retrieve his private key for his extended identifier, the KGC can evaluate the identifier and decline the extraction if the time limits are not met. Generally, embedding data in the user's identifier provides an additional control mechanism for a PPCD workflow with authenticity guaranteed through the dependency of the private key on the workflow participant. We are going to discuss the time-limiting and workflow enforcing extensions in greater detail.

4.4.1 Time-Limiting Identifiers

On many occasions, a PPCD workflow can be subject to various time constraints. For example, voting or exam results may not be accessed before a certain date and time. A grant application workflow is another example of a time-sensitive PPCD. With a few instances of PPCDs running along their workflows in parallel, it could be imperative to ensure that

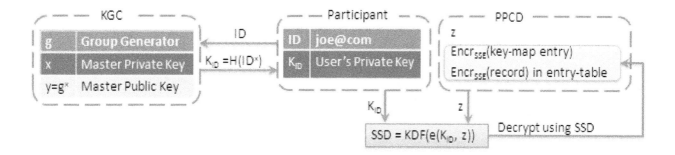

Figure 6: A PPCD access diagram

all applicants receive their results simultaneously. However, some applications can be reviewed and decided upon faster than others, thus causing the final documents to reach some applicants earlier than others. To accommodate this requirement, the IBE identities of the applicants can be extended to include the document release date.

To enforce real time deadlines, release or expiration for a time-sensitive workflow the participant's private key can be made available only within the certain time interval: not earlier than t_1 and nor later than t_2 ($t_1 < t_2$):

$$t_1 < t < t_2$$

by using user identity ID: $ID||t_1||t_2$ as participant's extended identifier. The lower limit t_1 will ensure that the key are not released earlier that time t_1, whilst the upper limit means the key expiration point. If the key expiration point is enforced for the strictly ordered workflow (with only one participant access at each workflow step), the mechanism should be provided to prevent workflow termination once some participant has failed to access the document.

t_1 and t_2 can vary for different workflow participants, and should be ordered according to the workflow logic: keys for early workflow participants should be released earlier. Also a one-sided limitation can be enforced: only release or only expiry dates:

- key release date $ID||t_1||-$;
- key expiry date $ID||-||t_2$.

Apart from providing for time sensitive workflow needs, the delayed key release can provide extra safety, for private keys not being compromised(e.g. exposed or lost), if issued very early. In a similar manner, the key expiry date can prevent any subsequent misuse of the keys. The expiry date can be set tightly to impose a deadline for the workflow participant to provide his contribution; or loosely, after the workflow has finished.

Using the workflow participant's key with different timestamps provides a convenient mechanism to give different access rights for the same workflow participant at the different workflow steps in conjunction with Key Reset Mechanism described in [13].

4.4.2 Sensitivity to Clear Text Time Constraints

Some sensitive workflows might require the real time to not be released within a PPCD in clear: just to prevent an attacker from knowing the document timeline or deliverables. If the document master decides to hide the real time, he can use double encryption processes, for example:

- The real time is encrypted under an IBE key from ID or from ID||nonce, and the document (or the symmetric key for encrypting the document) is encrypted under an IBE key from ID||nonce. The participant is enforced to contact the KGC twice.

- The real time is encrypted under another key belonging to the KGC and the cipher text is sent to the KGC along with the key generation request. In this case the participant only needs to contact the KGC once.

- There is a time releasing KGC, which might not be the same as the document workflow related KGC and which regularly releases a time key, such as every morning, every Monday, every first day of month/year etc. This key is an IBE key with the identifier = time. The document master makes use of two keys (the user's key and time key) to encrypt the document. The time key is shared by everybody in the system and is not a secret, but only available by a certain time. This is a well-known IBE application in the literature [20].

4.4.3 Enforcing Workflow Order

An extended user's identifier can also be used for enforcing the workflow order of access: to ensure the access of all mandatory participants of a previous workflow step prior to participants of the subsequent step are able to access. This can be achieved by adding a random identifier, nonce N, to user's IBE identifier, ID: ID||N. Whilst required by workflow participants in order to retrieve their corresponding private keys, these nonces are not provided within the document openly, but instead hidden within each of the corresponding key-map entries of the previous step participants. Remind, that each key-map entry in a PPCD is encrypted and can only be decrypted by the corresponding workflow participant.

Each workflow participant will receive the nonce, that needed to complete his IBE identifier only when the document is ready for him to access. For instance, the nonce N_i for the

i^{th} participant is released by the $(i-1)^{th}$ participant, and the $(i-1)^{th}$ participant does not have to know who is the next participant, but simply put the nonce into the PPCD before releasing, shipping or e-mailing the document. The IBE key for the ith participant is created by using his identifier and the nonce released by $(i-1)^{th}$ participant $ID_i||N_i$. This will prevent the out of turn access in an ordered workflows.

The user's IBE identifier can be further extended by combining time and nonce. This allows to ensure that the key will only be released when the document is released by the previous participant(s) and when the time is right - when both conditions have to be met.

5. MULTIPLE KGCS AND MIXED KGC-PKI

In the IBE scheme used in a PPCD workflow system, the relationship between participants and their KGCs (key generation centers) is flexible. That means each participant can have his individual KGC, and multiple participants can have either a single KGC (e.g. if they belong to the same organization) or multiple KGCs. The system can be further extended to accommodate access for some participants using public and private keys from PKI. The PPCD entry-table may contain 2 separate parts: one each for IBE and PKI participants. Or, it can be mixed with fake data added to the extra fields in otherwise shorter IBE records.

6. SECURITY VULNERABILITIES: KEY ESCROW PROBLEM

In an ordinary IBE system we cannot escape the property of key escrow - it is obvious that the KGC knows the user's private key. A potential problem from this property is that the KGC may impersonate a user in the system because the KGC is always able to do so. In an ordinary PKI system, however, we have the same problem. A Certificate Authority (CA) can generate a key pair, and (falsely) certify that the public key belongs to a specific user. The CA can then impersonate the user to any other users. In both IBE and PKI we therefore always assume that the trusted authority (KGC or CA) will not impersonate users. However, in PKI, the problem can be solved if we add an extra process - which is actually recommended by many applications - that possession of the user's private key is verified every time when the user uses the key to communicate with others. Unfortunately, we cannot offer the same solution for IBE because key escrow is inherent in IBE. One possible countermeasure is to reduce a single KGC's power. When the key escrow issue is concerned in the document workflow applications, we can make use of multiple KGCs in IBE, i.e., a user has a multiple identities, each of which is associated with a particular KGC. A document is encrypted under multiple IBE keys and the decryptor is required to get these keys from multiple KGCs. Neither of these KGCs is able to obtain the whole set of the keys, except they all collude together.

7. IMPLEMENTATION DETAILS

The current implementation extends the functionality described in [13] to provide an IBE based document workflow. Being a new modern technology, as of today there is no standard implementation of IBE in cryptographic extensions of Java [11] from Oracle, Cryptography Next Genera-

tion (CNG) from Microsoft [12] or BouncyCastle from the Legion of the Bouncy Castle [1]. Our current implementation of PPCD-IBE is based on "The Pairing-Based Cryptography Library" from Stanford [17] – one of the best known implementations of pairing-based cryptography in the public domain today. We used in part "The Java Pairing Based Cryptography Library (jPBC)" [15] for our development.

Remind that the original PKI based implementation of PPCD technology consisted of two parts:

1. The PPCD Authoring Tool – running in a secure environment, this tool allows the author to create the master copy of a composite document and export the secure distribution version;

2. The PPCD Access Tool for accessing a PPCD distribution version out of a secure environment.

For the original PKI-based prototype we used the Java Cryptography Architecture (JCA) [11] libraries for creating, accessing and managing keys and certificates outside of the main application (two parts described above). The created public keys were certified by HP internal Certificate Authority (CA). As it was previously mentioned, IBE is different in essence from a PKI:

- The user's public key needs to be run-time derived by "The PPCD Authoring Tool" out the user's identifier (the long term key) or extended identifier (the short term key) as specified by the document master and/or required by the nature of a workflow.

- The user's private key needs to be run-time computed by the corresponding KGC.

At the same time, the IBE technology is relatively young with no infrastructure developed yet; thus, for the PPCD prototype we needed to develop a KGC prototype implementation:

KGC Initialization A module for generating the public and private keys for the KGC (as described in Subsection 4.3.1).The module runs once for each KGC. The private key of the KGC is kept a secret, whilst the public key and the key generation material are published on-line.

KGC Service 1 *User authentication and private key generation:* A module for authenticating a user and comuting the user's private key out of the user's (extended) identity, the KGC's private key and publicly known key generation material. Following the user's request for his private key and the user's successful authentication, the user's private key is generated using formulas provided in Subsection 4.3.2 and released over a secure SSL-encrypted channel.

KGC Service 2 *KGC's key generation material and KGC's public key:* A module for providing on-line access to the KGC's key generation material and public key. This service is created for an automatic access by the PPCD Authoring Tool.

The PPCD Authoring software was extended to derive the secret encryption key (AES) for each workflow participant according to the formulas provided in Subsection 4.3.1.

7.1 Authoring Tool Prototype: Amendments

The following steps are performed by the new amended Authoring Tool prototype:

Step* 1. All parts and data fragments of the composite document are selected and added to the document master copy (MC).

Step* 2M. Workflow participants are selected with their roles and contacts (e.g. loaded from LDAP) and added to the MC.

Step* 3N. The decision is made whether extended users' identifiers are required by the workflow and if required the type and structure are specified: time-sensitivity, workflow enforcement, role-based, etc.

Step 4N. KGCs for workflow participants are identified and the corresponding public key generation data and public keys are obtained – this is long term data, so the local trusted copy can be used.

Step 5M. Random nonces are generated, assigned to workflow steps, stored in the MC.

Step 6N. The shared secret for each workflow participant is computed using the specified extended ID as described in Subsection 4.3.1.

Step* 7. The required access control is assigned (discretionary in the current prototype).

Step 8. The same access content-items are grouped together to form parts of the distribution version (DV) of the composite document.

Step 9. The access keys for each part are generated, parts are encrypted, signed and placed into the DV; the keys are stored in the MC.

Step 10. Individual key-map entries for each workflow participant for each step are generated.

Step 11. Generated at **Step 5** nonces distributed into the corresponding key-map entries.

Step 12M. The key-map entries are encrypted using the corresponding keys from **Step 6**, then signed and placed into the DV.

Step 13M. The modified entry-table (Subsection 4.3.1) is generated, encrypted by records using the corresponding keys from **Step 6**, signed and placed into the DV.

The DV is ready to be shipped to the first workflow participant, whilst the MC is securely retained for the duration of the workflow and may subsequently be archived or discarded. Steps marked by * require the document master input in the current prototype, but could readily be automated, if required. The steps marked by **N/M** are new or modified in the current prototype. The scematic diagram of PPCD-IBE creation is shown in Figure 7.

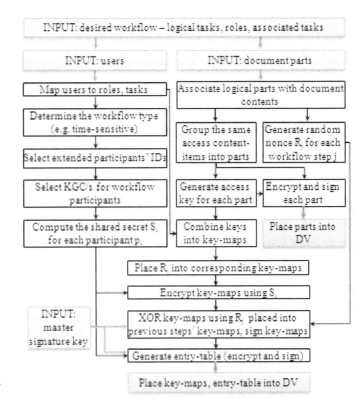

Figure 7: The scematic diagram of PPCD-IBE creation process

7.2 Access Tool Prototype: Amendments

The Access Tool was also amended to derive the AES decryption key from the participant's private key and his pseudo private key, following the formulas from Subsection 4.3.2. SHA256 [9] was used as a key derivation function, which ensures the randomisity of the derived key. However, the output of SHA256 is 32 bytes, whilst only 16 bytes are required for AES key length 128: only the first 16 bytes of SHA256 output were used by both the Authoring and Access Tools.

8. CONCLUSIONS

Using the user's e-mail address as IBE-PPCD identifiers reduces the number of tokens that must be known to a document master per a workflow participant: the same user's e-mail can simultaneously be used as the document delivery address and the user's identifier (to compute the user's public key).

The calculation of both the public and private keys in an IBE-based system can occur as needed, resulting in just-in-time key generation. In contrast, in PKI-design, keys are generated randomly and must be distributed prior to a PPCD creation.

In the IBE scheme these keys are locally derived from publicly known participants' identities or even their roles [19], whilst in the PKI-PPCD the corresponding valid and trusted certificates must obtained either from workflow participants directly or from a trusted directory/service. To provide such certificate for for the document master, each workflow participant in the PKI-PPCD has to be issued a pair of public and private keys.

In the current work we provided extension and generalization of the PPCD technology [13] to IBE. The major benefits of the proposed IBE-based extension are:

- A document can be sent along its workflow to participants who were not issued their public and private keys.

- A document master is not required to know the valid and non-revoked public keys for each workflow participant prior to commencing a workflow. Thus, for example, an employee who has not been hired yet might be included into a particular workflow once some unique identifier is assigned (e.g. his future e-mail address or his role).

- The workflow participant's private keys can be refreshed proactively by the KGC every short period of time.

- A document master can automatically require a workflow participant to use a new access key for a new workflow or even at a workflow step.

- A document master can enforce that a document may only be accessed by a workflow participant at some specified time in the future.

ACKNOWLEDGMENT

The authors gratefully thank Shivaun Albright, Cesare Gritti, Ron Heiney and other colleagues for many helpful discussions.

9. REFERENCES

[1] The Legion of the Bouncy Castle. http://www.bouncycastle.org/index.html.

[2] VeriSign Certificate Revocation List. http://crl.verisign.net/.

[3] VeriSign, Inc. http://www.verisigninc.com/.

[4] Voltage Security, Inc. http://www.voltage.com/products/index.htm.

[5] FIPS-197, Announcing the Advanced Encryption Standard (AES), November 2001. http://csrc.nist.gov/publications/fips/fips197/fips-197.pdf.

[6] ISO/IEC 9594-8:2005, Information technology Open Systems Interconnection The Directory: Public-key and attribute certificate frameworks. International Organization for Standardization, Geneva, Switzerland. 2005.

[7] OEBPS Container Format (OCF) 1.0, Recommended Specification, September, 11 2006. http://www.idpf.org/ocf/ocf1.0/download/ocf10.htm.

[8] Identity-Based Cryptography Standard (IBCS) #1, RFC 5091, December 2007. http://www.rfc-editor.org/rfc/rfc5091.txt.

[9] FIPS-180-3, Secure Hash Standard (SHS). National Institute of Standards and Technology (NIST), October 2008. http://csrc.nist.gov/publications/fips/fips180-3/fips180-3_final.pdf.

[10] HP Dialogue Live Software, Dialog Live Format (dlf), 2010. http://welcome.hp.com/country/uk/en/prodserv/software/eda/products/dialogue-live.html.

[11] Java™ Cryptography Architecture (JCA) Reference Guide, 2010. http://download.oracle.com/javase/6/docs/technotes/guides/security/crypto/CryptoSpec.html.

[12] Cryptography API: Next Generation. Microsoft, March 2011. http://msdn.microsoft.com/en-us/library/aa376210(v=VS.85).aspx.

[13] H. Balinsky and S. Simske. Differential Access for Publicly-Posted Composite Documents with Multiple Workflow Participants. In *Proc. of the 10th ACM Symposium on Document Engineering*, pages 115–124, September, 21-24 2010.

[14] D. Boneh and M. Franklin. Identity-based encryption from the Weil pairing. In *Advances in Cryptology-CRYPTO(Springer)*, volume 2139, pages 213–229, 2001.

[15] A. D. Caro. The Java Pairing Based Cryptography Library (jPBC), 2010. http://www.dia.unisa.it/dottorandi/decaro/projects.html#jpbc.

[16] C. Cocks. An Identity Based Encryption Scheme Based on Quadratic Residues. In *Proceedings of the 8th IMA International Conference on Cryptography and Coding*, 2001.

[17] B. Lynn. The Pairing-Based Cryptography Library (PBC), 2010. http://crypto.stanford.edu/pbc/manual.pdf.

[18] E. McCallister, T. Grance, and K. Scarfone. Guide to Protecting the Confidentiality of Personally Identifiable Information (PII), April 2010. http://csrc.nist.gov/publications/nistpubs/800-122/sp800-122.pdf.

[19] M. C. Mont, P. Bramhall, C. R. Dalton, and K. Harrison. A Flexible Role-based Secure Messaging Service: Exploiting IBE Technology in a Health Care Trial, 2003. http://www.hpl.hp.com/techreports/2003/HPL-2003-21.pdf.

[20] M. C. Mont, K. Harrison, and M. Sadler. The HP Time Vault Service: Innovating the Way Confidential Information is Disclosed, at the Right Time, 2002. http://www.hpl.hp.com/techreports/2002/HPL-2002-243.pdf.

[21] A. Shamir. Identity-Based Cryptosystems and Signature Schemes. In *Advances in Cryptology: Proceedings of CRYPTO 84, Lecture Notes in Computer Science*, volume 7, pages 47–53, 1984.

Citation Pattern Matching Algorithms for Citation-based Plagiarism Detection: Greedy Citation Tiling, Citation Chunking and Longest Common Citation Sequence

Bela Gipp
OvGU, Germany & UC Berkeley, USA
gipp@berkeley.edu

Norman Meuschke
OvGU, Germany & UC Berkeley, USA
meuschke@berkeley.edu

ABSTRACT

Plagiarism Detection Systems have been developed to locate instances of plagiarism e.g. within scientific papers. Studies have shown that the existing approaches deliver reasonable results in identifying copy&paste plagiarism, but fail to detect more sophisticated forms such as paraphrased, translated or idea plagiarism. The authors of this paper demonstrated in recent studies [4, 15] that the detection rate can be significantly improved by not only relying on text analysis, but by additionally analyzing the citations of a document. Citations are valuable language independent markers that are similar to a fingerprint. In fact, our examinations of real world cases have shown that the order of citations in a document often remains similar even if the text has been strongly paraphrased or translated in order to disguise plagiarism.

This paper introduces three algorithms and discusses their suitability for the purpose of Citation-based Plagiarism Detection. Due to the numerous ways in which plagiarism can occur, these algorithms need to be versatile. They must be capable of detecting transpositions, scaling and combinations in a local and global form. The algorithms are coined Greedy Citation Tiling, Citation Chunking and Longest Common Citation Sequence. The evaluation showed that common forms of plagiarism can be detected reliably if these algorithms are combined.

Categories and Subject Descriptors

H.3.3 **[Clustering]**: INFORMATION STORAGE AND RETRIEVAL – *Information Search and Retrieval.*

General Terms

Algorithms, Experimentation, Measurement, Languages

Keywords

Plagiarism Detection Systems, Citation-based, Citation Order Analysis, Citation Pattern Analysis

1. INTRODUCTION

Plagiarism describes the appropriation of other persons' ideas, intellectual or creative work and passing them of as one's own [7]. For including the act of self-plagiarism (see 2.1) we broaden the scope of the term and define academic plagiarism as *using words and/or ideas from other sources without due acknowledgement imposed by academic principles.*

It is a particularly common problem among college students worldwide, but also notably present among established researchers. In a self-report study among ~82,000 students about 40% of undergraduates and ~25% of graduates engaged in plagiarism within 12 months prior to the study [29]. Results of other studies range as high as ~90% of the subjects self-reporting acts of plagiarism [27].

In academia numerous cases of plagiarism have become publicly known. An automated plagiarism check of ~285,000 scientific texts of arXiv.org yielded more than 500 documents very likely to have been plagiarized. In addition, 30.000 documents (~20% of the collection) were found to be very likely duplicates or containing: "[…] excessive self-plagiarism […]" [43, p. 12].

The existing approaches for plagiarism detection have their weaknesses. Using the words of Weber-Wulff, the organizer of regular comparisons for productive Plagiarism Detection Systems (PDS), the current state of available systems can be summarized as follows: "[…] PDS find copies, not plagiarism." [50, p. 6].

The paper is structured as follows. After giving an overview of different forms of plagiarism, the detection approaches currently used and a discussion of their strength and weaknesses, the Citation-based Plagiarism Detection approach is briefly presented. Subsequently, the newly developed algorithms for Citation-based Plagiarism Detection are introduced, evaluated and their suitability for detecting different forms of plagiarism is discussed. Finally, the suitability of the presented approaches is demonstrated using real cases of plagiarism.

2. RELATED WORK

2.1 Forms of Plagiarism

Observations of plagiarism behavior in practice reveal a number of commonly found methods for illegitimate text usage, which are characterized below.

Copy&Paste (c&p) plagiarism specifies the act of taking over text verbatim from another author [49].

Disguised plagiarism subsumes practices intended to mask copied segments [26]. Four different masking techniques have been identified. These are:

· *Shake&Paste (s&p)* plagiarism is characterized by copying and merging sentences or paragraphs from different sources with slight adjustments necessary for forming a coherent text [49];

· *Expansive plagiarism* refers to the insertion of additional text into or in addition to copied segments [26];

· *Contractive plagiarism* describes the summary or trimming of copied material [26];

· *Mosaic plagiarism* encompasses the merge of text segments from different sources and obfuscating the plagiarism by changing word order, substituting words with synonyms or entering/deleting filling words [26, 49];

Technical disguise summarizes techniques for hiding plagiarized content from being automatically detected by exploiting weaknesses off current text-based analysis methods e.g. by substituting characters with graphically identical symbols from foreign alphabets or inserting letters in white font color [20].

Undue paraphrasing defines the intentional rewriting of foreign thoughts in the vocabulary and style of the plagiarist without giving due credit for concealing the original source [26].

Translated plagiarism is defined as the manual or automated conversion of content from one language to another intended to cover its origin [49].

Idea plagiarism encompasses the usage of a broader foreign concept without due source acknowledgement [28]. Examples are the appropriation of research approaches and methods, experimental setups, argumentative structures, background sources etc. [13].

Self-plagiarism characterizes the partial or complete reuse of one's own previous writings not being justified by scientific goals, e.g. for presenting updates or providing access to a larger community, but primarily serving the author, e.g. for artificially increasing citation counts [5, 11].

2.2 Existing Plagiarism Detection Approaches

Plagiarism Detection (PD) is a hypernym for computer-based procedures supporting the identification of plagiarism incidences. Existing PD methods can be categorized into external and intrinsic approaches [26, 45].

External PD methods compare a suspicious document to a collection of genuine works. Different comparison strategies have been proposed in this context.

String matching procedures [2, 32, 52] aim to identify longest pairs of identical text strings. These strings are treated as indicators for potential plagiarism if the share they represent with regard to the overall text exceeds a chosen threshold. Suffix document models, such as suffix trees or suffix arrays, have mostly been used for that purpose in the context of PD.

The strength of substring matching methods is their perfect detection accuracy with regard to literal text overlaps. Their major drawbacks are the relative difficulty of detecting disguised plagiarisms as well as the required computational effort. The former fact is intuitive when recalling the exact matching approach of the detection procedure. The later barrier results from the use of suffix data structures. The most space-efficient suffix tree [25], suffix array [24] and suffix vector [33] implementations allow searching in linear time and require on average ~$8n$ of storage space, with n being the number of symbols in the original document. String B-Trees allow searching in $O(\log n)$, but also require multiple times the storage space of the original document [25]. This renders them impracticable for most large document collections.

Employing *vector space retrieval* based on different term units has been proposed e.g. by [9, 40, 22]. Vector space models (VSM)

are a standard, highly performance tuned Information Retrieval (IR) concept that can overcome the effort-related limitations of elaborate string matching. VSM consider a set of terms, which commonly has been extracted from the whole document or larger parts of the text, for similarity computation. Therefore, vector space retrieval methods just like string matching is classified as global similarity assessments [47].

The well-known cosine measure is a widely used similarity function in PD settings as it is for other IR tasks. More complex similarity functions tend to incorporate semantic information e.g. by considering word synonyms [21] or pre-computing semantic relations [48] between terms. The aforementioned papers show that such considerations can increase detection performance, at the cost of significantly increasing the computational effort required. In the experiments reported in [3] considering synonyms improved the F-measure of the respective detection procedures by 2-3 times. However, the runtime required for doing so increased by more than 27 times.

The detection performance of VSM based PDS is dependent on the individual plagiarism incidence to be analyzed and the parameter configuration, e.g. term unit and term selection strategy, of the specific detection method [18, p. 155]. However, the global similarity assessment of VSMs tends to be detrimental to detection accuracy in PD settings. Verbatim plagiarism is more commonly related to smaller, confined segments of a document, which favors local similarity analysis [47].

Fingerprinting methods, being the most widely used PD approach, perform a local similarity assessment. They aim to form a representative digest of a document by selecting a set of multiple substrings from it. The set represents the fingerprint; its elements are called minutiae [19]. Mathematical, hash-like functions can be applied on minutiae for transforming them into more space efficient byte strings [12].

A suspicious document is checked for plagiarism by computing its fingerprint and querying each minutia with a pre-computed index of fingerprints for all documents of a reference collection. Minutiae found matching with those of other documents indicate shared text segments and suggest potential plagiarism upon exceeding a certain similarity threshold [6].

The inherent challenge of fingerprinting is finding a document representation that reduces computational effort to a suitable dimension, while limiting the information loss incurred to achieve acceptable detection accuracy [31]. A number of parameters, e.g. the chunking strategy, chunk size (granularity of the fingerprint) or number of minutiae (resolution of the fingerprint), reflect that challenge. There is no definite answer to the question of which parameter combination is the best, since this choice is strongly dependent on the nature and size of the collection as well as the amount and form of plagiarism.

Conventional fingerprinting methods implicitly encode the term order of a document in proportion to the length of the chosen text chunk. STEIN proposes an approach, termed fuzzy-fingerprinting, which disregards term order by using a VSM of document terms instead of substrings for minutia computation [44]. Fuzzy-Fingerprints are primarily targeted at reducing computational effort. Compared to fingerprinting using word-3-grams and a MD5 hash function they can be computed >5 times faster, but have been shown to be inferior in detection accuracy [47].

Intrinsic PD methods, opposed to the approaches presented so far, do not depend on the existence of a reference corpus. They statistically examine linguistic features of the suspicious text, a process known as stylometry, without performing comparisons to external documents. They aim at recognizing changes in writing style to indicate potential plagiarism [31].

The linguistic features to be analyzed form a style model. Approximately more than 1.000 individual style markers [38] have been proposed for stylometry, most can be classified as falling into one of the following categories [46]:

- lexical features on character level, e.g. -gram frequency, or word level, e.g. average word lengths or syllables per word;

- syntactic features, e.g. word or part-of-speech frequencies;

- structural features, e.g. average paragraph length or frequency of punctuation.

Style models of intrinsic PDS are generally comprised of an individual combination of multiple linguistic features [31].

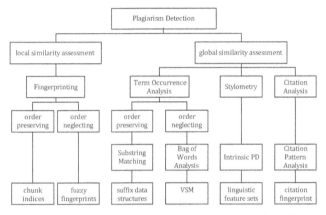

Figure 1: Classification of PD methods (inspired by: [47])

In previous papers [4, 15] we initially proposed employing citation analysis for PD and presented results of initial studies. *Citation-based Plagiarism Detection* (CbPD) is a fundamentally different approach compared to the text-based similarity evaluations described above. It is especially intended for being applied to scientific publications. Being substantially different, it is believed to be capable of overcoming some of the weaknesses of existing techniques. An overview classification of the PD approaches presented above is given in Figure 1.

2.3 Strength and Weaknesses of PDS

As described in [15] objective comparisons of the detection performance achieved by individual PDS are difficult. Authors proposing research prototypes tend to use different collections and evaluation methods. Initiated in 2009 the annual PAN International Competition on Plagiarism Detection (PAN-PC) addresses this lack of comparability. It attempts to benchmark PDS using a standardized collection and a controlled evaluation environment [36]. Results from the latest PAN-PC, held in June 2010, are presented for pointing out the capabilities of state-of-the-art PD prototypes.

Figure 2 displays the plagiarism detection (*plagdet*) scores of the top 5 performing external PDS and the 2 intrinsic PDS of MUHR ET AL. and SUÁREZ ET AL. participating in PAN-PC'10. The *plagdet* score is a measure developed for evaluating PDS in the PAN competitions. It considers the   measure as well as the granularity (*gran*) of a detection method. The granularity reflects

whether a plagiarized section is detected as a whole or in multiple parts: $plagdet = F_1 / log_2(1 + gran)$ [36].

In Figure 2 the scores are plotted dependent on the obfuscation techniques applied to plagiarized text segments. The overall *plagdet* score achieved in all categories is stated in brackets within each legend entry. Note that in the legend "- I" is attached to distinguish the system of MUHR ET AL. participating in the intrinsic from the one in the external task.

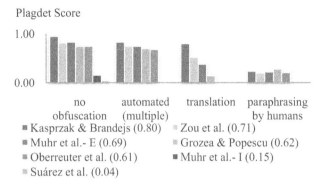

Figure 2: Results of top performing PDS in PAN-PC'10 [35]

The results indicate that unchanged copies of text segments can be detected with high accuracy by state-of-the-art PDS using fingerprinting or bag of words analyses. Detection rates for segments that were plagiarized and disguised by humans are substantially lower for all systems. On average, 76% of those realistically plagiarized segments could not be identified by the top 5 systems. The detection scores for automatically obfuscated plagiarism are 2.5 to 3.7 times higher than those for manually plagiarized sections.

The organizers of the competition judge the results achieved in detecting cross-lingual plagiarism to be misleading. The well-performing systems used automated services for translating foreign-language documents in the reference corpus to English. The employed services, e.g. Google Translate, are similar or identical to those used for constructing the translated, plagiarized sections. It is hypothesized that human-made translations obfuscating real-world plagiarism are more complex and versatile, and hence less detectable by the tested PDS [35].

Intrinsic PDS performed significantly worse than systems using an external approach. The results are in line with those from the prior PAN competition in 2009 [36]. Intrinsic analyses seem to require larger volumes of text for working reliably [46].

3. CITATION-BASED PD

In the academic environment, citations and references of scholarly publications have long been recognized for containing valuable information about the content of a document and its relation to other works [14]. A large volume of semantic information is contained in citation patterns because complete scientific concepts and argumentative structures are compressed into sequences of small text strings. To our knowledge the identification of plagiarism by analyzing the citations[1] and references[2] of

---

[1] Citations are short alphanumeric strings in the body of scientific texts representing sources contained in the bibliography
[2] References denote entries in the bibliography

documents has been first described and successfully applied to PD in [4, 15]. In this context, we proposed this definition:

Citation-based Plagiarism Detection (CbPD) subsumes methods that use citations and references for determining similarities between documents in order to identify plagiarism.

Citations and citation patterns offer unique features that facilitate a PD analysis. They are a comparatively easy to acquire, language independent marker, since more or less well-defined standards for using them are established in the international scientific community. This information can be exploited to detect forms of plagiarism that cannot be detected with text-based approaches.

3.1 Factors for Citation-based Text Similarity

In the following section, factors that influence a similarity assessment for documents based on citations and references are outlined for deriving a suitable document model for CbPD.

3.1.1 *Shared References*

Absolute Number

Having references in common is a well-known similarity criterion for scientific texts called bibliographic coupling [23]. The absolute number of shared references represents the coupling strength, which is used to measure the degree of relatedness.

Relative Number

The fraction that shared references represent with regard to the total number of individual references is another similarity indicator. Two texts, A and B, are more likely to be similar if they share a larger percentage of their references. This assumption is supported by results of text-based PD studies [10].

Both the amount and fraction of shared references depend on a number of factors, most importantly document length and specific document parts to be analyzed. Comprehensive documents contain on average more references than short documents, or certain document parts, e.g. related work sections in academic texts contain more citations per page than e.g. summary parts.

Considering the above factors when using reference counts for CbPD might improve their predictive value.

3.1.2 *Probability of Shared References*

The likelihood that two texts have references in common is not statistically independent. Reference co-occurrences that have a lower probability are more predictive for document similarity. The importance of shared references with regard to document similarity is dependent on a number of factors explained below.

Existing citation counts have been shown to influence future citation counts significantly. If a document is highly cited already, its likelihood of gathering additional citations from other documents increases. The phenomenon has been termed the Matthew effect in science[3]. Imagine a document C that has been widely referenced, e.g. by 500 other documents. Another document D, on the other hand, has been referenced much less frequently, e.g. by 5 other documents. In turn, document D has a smaller probability of being a shared reference of two texts A and B, which are to be analyzed. However, if document D represents a

---

[3] The term refers to a line in the Gospel of Matthew

reference shared by A and B this fact is a comparably stronger indicator for similarity than in the case in which C represents a shared reference of A and B.

Time influences the likelihood of references. As citation counts tend to increase with time [34, 39], so does the probability of a document becoming a shared reference. If texts A and B have been published at different points in time, this fact should be compensated, e.g. by comparing expected citations per unit of time.

The topic of research that two documents A and B deal with also influences the likelihood that A and B share common references. They are more likely to do so if the documents address the same or very similar topics. This assumption can be derived from empirical evaluations using Co-Citation analysis to identify clusters in academic domains [16, 41]. If strong Co-Citation relations exist within a certain academic field, as has been shown, this in turn implies that a higher number of documents share common references within this domain. This is intuitive, since references are often used to illustrate prior research or origins of the ideas presented.

Proximity of authors within a social network increases the probability of respective papers to be referenced. Research showed that a text A is more likely to be referenced by a text B if the author(s) of B is/are personally more closely connected to the author(s) of A. For example, documents are referenced more frequently within texts written by former co-authors or researchers that know each other in person. This effect is sometimes referred to as cronyism [30]. The analysis of co-authorship networks might therefore increase the predictive value of reference co-occurrence assessments.

3.1.3 *Citation Pattern Similarity*

Finding similar patterns in the citations used within two scientific texts is a strong indicator for semantic text similarity and the core idea of CbPD. Patterns are subsequences in the citation tuples C_A and C_B of two texts A and B that (partially) consist of shared references and are therefore similar to each other.

The degree of similarity between patterns depends on the number of citations included in the pattern, and the extent to which their order and/or the range they cover is alike. Thus, literally matching subsequences of citations in two documents are a strong indicator for semantic similarity.

The same is true for texts containing patterns that span over similar ranges, even if the order of citations in the pattern does not necessarily correspond towards each other. The width of the covered range can be expressed with regard to sequential positions of citations in the pattern, textual ranges or combinations of both. Measuring range width in units reflecting some semantics, e.g. paragraphs or sentences, is assumed beneficial compared to considering purely syntactic character or citation counts. For example, documents containing several matching citations, one of them within a single section, the other distributed over several chapters are less likely to be similar. However, if both share identical citations e.g. within a paragraph, then their potential similarity is respectively higher. Alternatively, e.g. the document tree may be used to identify semantic clusters in the form of chapters etc.

A CbPD similarity assessment consists of two subtasks. The first is to identify matching citations and citation patterns. The second is to rate patterns with regard to their likelihood of having resulted from undue practices.

The scope of this paper is limited on presenting algorithms that tackle the first subtask of detecting citation patterns. Results of experiments with regard to the second subtask of ranking identified patterns will be presented in an upcoming paper.

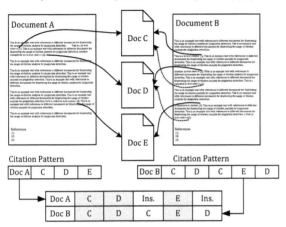

Figure 3: Identifying citation patterns for CbPD

3.1.4 *Challenge of Identifying Citation Patterns*

Detecting citation patterns is a non-trivial task due to the diverse forms of plagiarism. Copy&paste plagiarism results in different citation patterns than e.g. shake&paste plagiarism. Therefore, different algorithms are required to address the specific forms. The following challenges need to be considered.

Unknown pattern constituents – Unlike e.g. in string pattern matching the subsequences of citations to be extracted from a suspicious text and searched for within an original are initially unknown. Citations that are shared by the two documents are easily identified. However, it is unlikely that all of those shared citations represent plagiarized text passages. For instance, two documents might share 8 citations, of which 3 are contained within a plagiarized text section and 4 are distributed over the length of the text and used along with other non-shared citations without representing any form of plagiarism. The citation sequences of the two documents might therefore look like the following:

```
Original:    1 2 3 x x x 4 x x 5 x 6 x 7 8
Plagiarism:  x x 5 x x x 4 x 3 x 1 x 2 x x 7 x 8
```

Numbers 1-8 represent shared citations, the letter x non-shared citations. The shared citations 1-3 are supposed to represent a plagiarized passage.

Transpositions - the order of citations within text segments might be transposed compared to the corresponding original section. Possible causes can be different citation styles or sort orders of the reference list, e.g. alphabetically opposed to sorting it by publication date. Assume an original text segment contains a sentence in the form:

```
Studies show that <finding1>, <finding2> [3,1,2].
```

The semantically identical content might be expressed in the form:

```
Studies show that <finding1>, <finding2> [1-3].
```

Scaling - occurrences of shared citations can be used more than once, which is referred to as being scaled. Assume an original text segment in the form:

```
Study   X   showed   <finding1>,   <finding2>   and
<finding3> [1]. Study Y objected <finding1> [2].
Assessment Z proofed <finding3> [3].
```

This segment might be plagiarized as following:

```
Study X showed <finding1> [1], which was objected
by study Y [2]. Study X also found <finding2> [1].
Assessment Z was able to proof <finding3> [3],
which had already been indicated by study X [1].
```

Missing alignment - potentially plagiarized sections and their corresponding originals need not to be aligned, but can reside in very different parts of the text. For instance, the first paragraph in the first section of an original document *A* might be plagiarized in document *B*, however it may become the fifth paragraph in the third section of *B*. The division of corresponding text segments into paragraphs, sections or chapters might also differ significantly. For instance, a plagiarized text segment might be artificially expanded or reduced to result in a different paragraph split-up in order to conceal the plagiarism.

3.2 Citation-based Similarity Functions

Given the limited empirical knowledge base that exists for CbPD, it is intended to evaluate a balanced mixture of possible similarity functions. The goal is to include global and local similarity assessments as well as functions that focus on the order of citations opposed to functions that ignore citation order, but can handle transpositions and scaling. Besides designing new similarity functions based on the factors outlined above, testing well-proven similarity measures for their applicability to CbPD is a further objective.

The fact that citation sequences of documents can be characterized as strings has been taken as a starting point for identifying existing similarity functions. In this context, string refers to any collection of uniquely identifiable elements that are linked in a way such that each, except for exactly one leftmost and exactly one rightmost element, has one unique predecessor and one unique successor [42]. This definition is broader than the most prominent connotation of the term referring to literal character sequences in the domain of computer science. String processing is a classical and comprehensively researched domain. Thus, multiplicities of possible similarity assessments can be derived from this area see e.g. [17].

	Global Similarity Assessment	Local Similarity Assessment
Order preserving	Longest Common Citation Sequence	Greedy Citation Tiling
Order neglecting	Bibliographic Coupling	Citation Chunking

Figure 4: Categorization of evaluated similarity assessments

According to the objectives outlined above, similarity approaches for each category distinguishable in regard to the scope of the assessment (global vs. local) and consideration of citation order have been defined. In Figure 4, the chosen similarity assessments are outlined.

3.2.1 *Bibliographic Coupling Strength*

Bibliographic coupling is one of the first and best-known citation-based similarity assessments for academic texts.

Similarity is quantified in terms of the absolute number of shared references. Order or positions of citations within the text are ignored. It can be interpreted as a raw measure of global document similarity. Solely considering bibliographic coupling strength is not a sufficient indicator for potential plagiarism and does not allow pinpointing potentially plagiarized text segments.

3.2.2 Longest Common Citation Sequence

The Longest Common Subsequence (LCS) of elements in a string is a traditional similarity measure. The LCS approach has been adapted to citations and comprises the maximal number of citations that can be taken from a citation sequence without changing their order, but allowing for skips over non-matching citations. For instance the sequence (3,4,5) is a subsequence of (2,3,1,4,6,8,5,9) [8, p. 4].

Intuitively, considering the LCS of two citation sequences yields high similarity scores if longer parts of the corresponding text have been adopted without altering the contained citations. Examining the Longest Common Citation Sequence has been chosen because the measure features a clear focus on order relation, opposed to bibliographic coupling. At the same time it offers flexibility for coping with slight transpositions or arbitrary sized gaps of non-matching citations.

It is capable of indicating potential cases of plagiarism in which parts of the text have been copied with no changes, or only slight alterations in the order of citations. This can be the case for copy&paste plagiarism that might have been concealed by basic rewording e.g. through synonym replacements. If significant reordering within plagiarized text segments has taken place (shake&paste plagiarism) or a different citation style has been applied that permutes the sequence of citations, the LCS approach is bound to fail.

3.2.3 Greedy Citation Tiling

Greedy Citation Tiling (GCT) is an adaption of a text string similarity function proposed by WISE [51]. The original procedure called Greedy String Tiling (GST) has explicitly been designed for usage in PD. It has been widely adopted and successfully applied, foremost in PDS for software source code [1, 37].

Greedy String/Citation Tiling aims to identify all matching substrings with individually longest possible size in two sequences. Individual longest matches refer to substrings that are shared by both sequences and cannot be extended to the left or right without encountering an element that is not shared by the two sequences. Corresponding individually longest matches in both sequences are permanently linked with each other and stored as a so called tile.

A tile represents a tuple $t = (s_1, s_2, l)$ consisting of the starting position of a longest match in the first sequence (s_1), the starting position in the second sequence (s_2) and the length of the match (l). The tiling approach is illustrated in Figure 5. Arabic numbers represent equal elements in the sequences to be analyzed, letter x extraneous elements. Individually longest matches are indicated by boxes around elements. Roman numbers above and below the boxes identify the tiles to which the matches have been assigned. As shown in the figure, the tiling approach is able to cope with arbitrary transpositions in the order of individual substrings. A minimum size of matching substrings can be freely chosen.

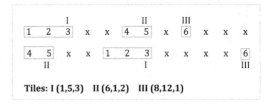

Figure 5: Citation Tiles

The principle of the tiling algorithm is illustrated in Figure 6 assuming a minimum match length of 2. The procedure strictly identifies longer tiles before shorter ones. Auxiliary arrays are used for keeping track of longest tiles and prevent elements from becoming part of multiple tiles. Elements are inserted into the auxiliary arrays at the moment they are assigned to a tile, thus they are "marked" as no longer available for matching and are ignored in future iterations.

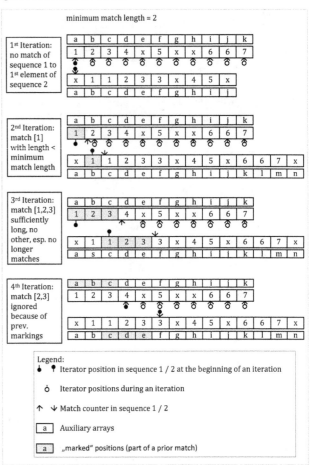

Figure 6: Example flow of the Greedy Citation Tiling algorithm.

The algorithm performs full iterations of both sequences, meaning that sequence 2 is iterated for every element of sequence 1, as long as matches longer than or equal to the specified global minimum length are found in the respective iteration. This indicates that the worst case complexity of the algorithm is $O(n^3)$.

In each iteration only maximal matches are considered for being transformed into tiles. All individual longest matches identified during the same iteration need to be equal to or longer than the maximal match found in the same iteration. If sequence 2 has

254

been traversed for one element of sequence 1, all identified maximal matches are marked in the auxiliary arrays.

For the next iteration the current maximal match length is again set to equal the global minimum match length. This way, the "next-shorter" matches to those marked during the prior iteration are identified. One can see that this repetition continues until no more matches longer than the global minimum match length can be found, which results in the termination of the algorithm. If the minimum match length is set to 1 the GST algorithm is proven to produce the optimal coverage of matching elements with tiles [51].

The GST algorithm has been primarily designed for identifying shake&paste plagiarism. It is able to identify individually longest substrings despite potential rearrangements. Greedy Citation Tiling might serve the same purpose, but opposed to the text-based approach also identifies paraphrased shake&paste plagiarism.

The GCT approach focuses on exact equality with regard to citation order. Finding such patterns provides a strong indication for text similarity. GCT is able to deal with transpositions in the citation sequence that result from rearranging text segments, which is typical for shake&paste plagiarism. However, the approach is not capable of detecting citation scaling or transpositions resulting e.g. from the usage of different citation styles. For covering such cases, another class of detection algorithms has been designed, which is explained in the following section.

3.2.4 *Citation Chunking*

A set of heuristic procedures that aim to identify local citation patterns regardless of potential transpositions and/or scaling have been developed for this study. The approach has been termed Citation Chunking because it is inspired by the feature selection strategies of text-based fingerprinting algorithms. A citation chunk is a variably sized substring of a document's citation sequence.

The main idea of citation chunking is to consider shared citations as textual anchors at which local citation patterns can potentially exist. Starting from an anchor, citation chunks are constructed by dynamically increasing the considered substring of citations based on the characteristics of the chunk under construction as well as the succeeding citations.

Chunking Strategies

Strategies for forming chunks have been derived by imagining potential behaviors of a plagiarist and modeling the resulting citation patterns.

Determining the starting and ending point for a chunk is not a trivial task. There probably does not exist a best solution that fits all plagiarism scenarios. Larger chunks are believed to be better suitable for detecting overall similarities and compensate for transpositions and scaling. Smaller chunks, on the other hand, are more suitable for pinpointing specific areas of highest similarity. In order to experiment with both tendencies, the following procedures for constructing citation chunks have been defined.

1. Only consecutive shared citations form a chunk:

```
Doc A: x,[1,2,3] x,[4,5,3],x,x
Doc B: x,x,[3,2,1] x,[5,3,4],x
```

This is the most restrictive chunking strategy. Its intention is to highlight confined text segments that have a very high citation-based similarity. It is ideal for detecting potential cases in which copy&paste plagiarism might have been concealed by rewording or translation.

2. Chunks are formed dependent on the preceding citation. A citation is included in a chunk if $n \leq 1$ or $1 > n \leq s$ non-shared citations separate it from the last preceding shared citation, with s being the number of citations in the chunk currently under construction:

```
Doc A: x,[1,2,3,x,x,4,5],x,x,x,x,x,x,[6,7]
Doc B: [3,2,x,1,x,x,4],x,x,x,x,x,[5,6,7],x
```

Chunking strategy 2 aims to uncover potential cases in which text segments or logical structures have been taken over from or influenced by another text. It allows for sporadic non-shared citations that may have been inserted to make the resulting text more "genuine". It can also detect potential cases of concealed shake&paste plagiarism by allowing an increasing number of non-shared citations within a chunk, given that a certain number of shared citations have already been included. This process aims to reflect the behavior that text segments (including citations) from different sources are interwoven.

3. Citations exhibiting a textual distance below a certain threshold form a chunk.

Chunking strategy 3 aims to define a textual range inside which possible plagiarism is deemed likely. Studies have shown that plagiarism more frequently affects confined text segments, such as one or two paragraphs, rather than extended text passages or the document as a whole. Building upon this knowledge, the respective chunking strategy only considers citations within a specified range for forming chunks.

Since the split up of a plagiarized text into textual units, such as sentences or paragraphs, might be altered artificially, textual proximity might be analyzed in terms of multiple units. One possibility tested in the study has been to count the characters, words, sentences and paragraphs that separate individual citations. The respective counts have been compared to average numbers expected for a certain textual range. For instance, one paragraph might on average comprise 120 words consisting of 720 characters. If one shared citation is separated from another by 2 paragraphs, but less than 120 words, it will be included in a chunk to be formed. In this manner, even artificially created paragraph split-ups can be dealt with.

Finding a suitable maximal distance for proximity of citations in the text is highly dependent on the individual corpus analyzed. If e.g. the average length of documents is rather short, and individual documents contain smaller number of sections and paragraphs, it is believed to be harder for a plagiarist to artificially alter the textual structure. Consequently, a comparably lower maximal distance should be chosen in this scenario. In contrast, it is believed to be easier to change e.g. the paragraph split-up in longer academic texts.

The complete process of forming chunks according to the outlined chunking strategies is graphically summarized as a flow chart in figure 7. In order to experiment with larger chunk sizes, an optional merging step is tested (dashed box in figure 7).

It is intended to combine supposedly suspicious citation patterns in order to outline longer sections of similar text e.g. as part of an idea plagiarism. Chunks are merged if they are separated by n non-shared citations, $n <= m$ with m being the number of shared citations in the preceding chunk

```
Iteration 1: XXX,x,XX,x,x,XXX,x,x,x,x,x,x,XX
Iteration 2: XXXXX,x,x,XXX,x,x,x,x,x,x,XX
Iteration 3: XXXXXXXX,x,x,x,x,x,x,XX
```

Chunk XX is not merged since its distance to preceding chunks is too large.

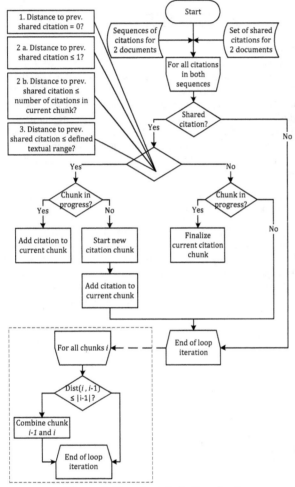

Figure 7: Forming of citation chunks

3.2.5 *Chunk Comparison*

Once chunks have been formed, they are considered in their entirety for comparison. That is, the order of citations within a chunk is disregarded during comparisons in order to account for potential transpositions and/or scaling. The number of shared citations within the units to be compared represents the measure of similarity.

In the following two main strategies for comparing documents based on citation chunks are described. The first is to form chunks for both documents and compare each chunk of the first document with each chunk of the second. The chunk pairs having the highest citation overlap are permanently related to each other and considered a match. If multiple chunks in the documents have an equal overlap, all combinations with maximal overlap are stored.

In the second scenario, chunks are constructed for one document only. Subsequently, each of the chunks is compared to the unaltered citation sequence of the second document by "moving" it as a sliding window over the sequence and assigning it to the position with the maximal citation overlap.

3.2.6 *Strength and Weaknesses of the Algorithms*

In the following, the suitability of the presented algorithms is classified according to their ability to detect different forms of plagiarism.

	Plagiarism type	LCCS	GCT	CitChunk
Local	Identical (c&p segments, translations)	-	++	+
	Transpositions (shake&paste)	-	-	+
	Scaling	-	-	+
	Transpositions & scaling (paraphrases)	-	-	+
Global	Identical	++	++	+
	Transposition	+	-	+
	Scaling	+	-	++
	Transpositions & scaling	+	-	++

Figure 8: Overview of detection capabilities

These classifications are a generalization and should be considered with care. If, for instance, a text is translated word by word then the order of citations will not change much. This case would be classified as "identical" according to the table. In cases of free translations or the existence of several citations within one sentence varying sentence structures resulting from different languages might lead to different citation orders. In such cases a translated plagiarism would be classified as "transposition", "scaling" or even a combination of both.

Moreover, in the table it is distinguished between local and global forms of plagiarism. Local plagiarism can be observed on sentence level, whereas global forms describe document wide plagiarism.

| 242 - 244 | CRS92_Art.V Guttenberg06 | |
| 242 - 244 | CRS92_Art.V Guttenberg06 | |

Figure 9: Example pattern identified in Guttenberg's thesis [53 plagiarism] by applying Citation Chunking

In initial experiments, the described detection procedures have been applied to prominent real world plagiarism cases in doctoral dissertations of German politicians [15]. As the table shows, Citation Chunking yielded best detection results in most cases in our tests. However, for text segments in which large portions of the contained citations were adopted unaltered, e.g. in not-freely translated plagiarisms, Greedy Citation Tiling provided clearer indications for potential plagiarism. Figure 9 shows an example of a citation pattern identified by Citation Chunking.

3.3 Prototype Citation-based PDS

For evaluating the different analysis procedures we developed an Open Source software system in Java coined *CitePlag*. The developed prototype CbPDS consists of three main components.

The first is a Relational Database System (RDBS) termed CbPD database storing data to be acquired from documents as well as detection results. The second is the detection software called

CbPD Detector that retrieves data from the CbPD Database, runs the different analysis algorithms to be evaluated and feeds the resulting output back to the CbPD Database. The third component, the CbPD Report Generator, creates summarized reports of detection results for individual document pairs based on adjustable filter criteria. The three-tier-architecture is illustrated in Figure 10.

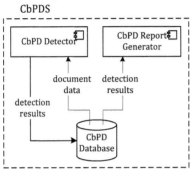

Figure 10: CbPDS system architecture

4. CONCLUSION

Previous studies have shown that CbPD is suitable for detecting forms of plagiarism that remain undetectable for the currently used text-based approaches. This paper has presented three algorithms for identifying citation patterns that have been observed in real-world plagiarism cases. The algorithms are coined Greedy Citation Tiling, Citation Chunking and Longest Common Citation Sequence.

These algorithms are able to detect citation transpositions, citation scaling and their combinations in cases of local and global plagiarism. The algorithms have been evaluated using several plagiarized documents such as the doctoral thesis of Guttenberg and by applying them to the PubMed Central Open Access Subset (PMC OAS). In [15] it was shown that the proposed algorithms could identify 13 out of the 16 sections containing translated plagiarism in the Guttenberg thesis. The tested text-based PDS were unable to detect any of them.

5. REFERENCES

[1]   AHTIAINEN, A., SURAKKA, S., AND RAHIKAINEN, M. Plaggie: GNU-licensed source code plagiarism detection engine for Java exercises. In *Proceedings of the 6th Baltic Sea conference on Computing education research: Koli Calling 2006* (New York, NY, USA, 2006), Baltic Sea '06, ACM, pp. 141–142.

[2]   BAKER, B. S. A Program for Identifying Duplicated Code. In *Computing Science and Statistics: Proceedings of the 24th Symposium on the Interface* (1992).

[3]   BAO, J. P., LYON, C., LANE, P. C. R., AND JI, WEI, M. J. A. Comparing Different Text Similarity Methods. Tech. rep., Science and Technology Research Institute, University of Hertfordshire, May 2007.

[4]   BELA GIPP, AND JOERAN BEEL. Citation Based Plagiarism Detection - A New Approach to Identify Plagiarized Work Language Independently. In *Proceedings of the 21st ACM Conference on Hyptertext and Hypermedia (HT'10)* (New York, NY, USA, June 2010), ACM, pp. 273–274.

[5]   BRETAG, T., AND MAHMUD, S. Self-Plagiarism or Appropriate Textual Re-use? *Journal of Academic Ethics 7* (2009), 193–205.

[6]   BRIN, S., DAVIS, J., AND GARCIA MOLINA, H. Copy Detection Mechanisms for Digital Documents. In *Proceedings of the 1995 ACM SIGMOD international conference on Management of data* (New York, NY, USA, May 1995), M. Carey and D. Schneider, Eds., ACM, pp. 398–409.

[7]   COCEL. *Concise Oxford Companion to the English Language [electronic resource].* Oxford Reference Online. Oxford University Press, 1998.

[8]   CROCHEMORE, M., AND RYTTER, W. *Jewels of Stringology.* World Scientific Publishing, 2002.

[9]   DREHER, H. Automatic Conceptual Analysis for Plagiarism Detection. *Information and Beyond: The Journal of Issues in Informing Science and Information Technology 4* (2007).

[10]   ERRAMI, M., HICKS, J. M., FISHER, W., TRUSTY, D., WREN, J. D., LONG, T. C., AND GARNER, H. R. Déjà vu—A study of duplicate citations in Medline. *Bioinformatics 24*, 2 (2008).

[11]   ERRAMI, M., SUN, Z., LONG, T. C., GEORGE, A. C., AND GARNER, H. R. Déjà vu: a database of highly similar citations in the scientific literature. *Nucleic Acids Research 37*, suppl 1 (2009), D921–D924.

[12]   FINKEL, R. A., ZASLAVSKY, A. B., MONOSTORI, K., AND SCHMIDT, H. W. Signature extraction for overlap detection in documents. In *Computer Science 2002, Twenty-Fifth Australasian Computer Science Conference (ACSC2002), Monash University, Melbourne, Victoria, January/February 2002* (Darlinghurst, Australia, 2002), M. J. Oudshoorn, Ed., vol. 4 of *Conferences in Research and Practice in Information Technology*, Australian Computer Society Inc., pp. 59–64.

[13]   FRÖHLICH, G. Plagiate und unethische Autorenschaften. *Information - Wissenschaft & Praxis 57*, 2 (2006), 81–89.

[14]   GARFIELD, E. Citation Indexes for Science: A New Dimension in Documentation through Association of Ideas. *Science 122*, 3159 (July 1955), 108–111.

[15]   GIPP, B., MEUSCHKE, N., AND BEEL, J. Comparative Evaluation of Text- and Citation-based Plagiarism Detection Approaches using GuttenPlag. In *Proceedings of 11th ACM/IEEE-CS Joint Conference on Digital Libraries (JCDL'11)* (Ottawa, Canada, June 2011).

[16]   GRIFFITH, B. C., SMALL, H. G., STONEHILL, J. A., AND DEY, S. The Structure of Scientific Literatures II: Toward a Macro- and Microstructure for Science. *Science Studies 4*, 4 (1974), p 339.

[17]   GUSFIELD, D. *Algorithms on Strings, Trees, and Sequences: Computer Science and Computational Biology.* Cambridge University Press, New York, NY, USA, 1997.

[18]   GUTBROD, M. A. *Nachhaltiges E-Learning durch sekundäre Dienste.* Dissertation, Technischen Universität Braunschweig Institut für Betriebssysteme und Rechnerverbund, Jan. 2007.

[19]   HOAD, T. C., AND ZOBEL, J. Methods for Identifying Versioned and Plagiarised Documents. *Journal of the American Society for Information Science and Technology 54*, 3 (2003).

[20]   KAKKONEN, T., AND MOZGOVOY, M. Hermetic and Web Plagiarism Detection Systems for Student Essays—An Evaluation of the State-of-the-Art. *Journal of Educational Computing Research 42*, 2 (2010), 135–159.

[21]   KANG, N., GELBUKH, A., AND HAN, S. PPChecker: Plagiarism Pattern Checker in Document Copy Detection. In *Text, Speech and Dialogue*, P. Sojka, I. Kopecek, and K. Pala, Eds., vol. 4188 of *Lecture Notes in Computer Science*. Springer Berlin / Heidelberg, 2006, pp. 661–667.

[22]   KASPRZAK, J., AND BRANDEJS, M. Improving the Reliability of the Plagiarism Detection System - Lab Report for PAN at CLEF 2010. In *Notebook Papers of CLEF 2010 LABs and Workshops, 22-23 September, Padua, Italy* (2010), M. Braschler, D. Harman, and E. Pianta, Eds.

[23] KESSLER, M. M. Concerning some problems of intrascience communication. Lincoln laboratory group report, Massachusetts Institute of Technology. Lincoln Laboratory, 1958. Cited according to: B.H. Weinberg. BIBLIOGRAPHIC COUPLING: A REVIEW. Information Storage Retrieval, 10: 189-196, 1974.

[24] KO, P., AND ALURU, S. Space Efficient Linear Time Construction of Suffix Arrays. In *Combinatorial Pattern Matching* (2003), R. Baeza-Yates, E. Chávez, and M. Crochemore, Eds., vol. 2676 of *Lecture Notes in Computer Science*, Springer Berlin / Heidelberg, pp. 200–210.

[25] KURTZ, S. Reducing the Space Requirement of Suffix Trees. *Software-Practice and Experience 29*, 13 (1999), 1149–1171.

[26] LANCASTER, T. *Effective and Efficient Plagiarism Detection*. PhD thesis, School of Computing, Information Systems and Mathematics South Bank University, 2003.

[27] LIM, V. K. G., AND SEE, S. K. B. Attitudes Toward, and Intentions to Report, Academic Cheating Among Students in Singapore. *Ethics & Behavior 11*, 3 (2001), 261–274.

[28] MAURER, H., KAPPE, F., AND ZAKA, B. Plagiarism - A Survey. *Journal of Universal Computer Science 12*, 8 (Aug. 2006), 1050–1084.

[29] MCCABE, D. L. Cheating among college and university students: A North American perspective. *International Journal for Academic Integrity 1*, 1 (2005), 1–11.

[30] MEHO, L., AND YANG, K. Impact of data sources on citation counts and rankings of LIS faculty: Web of Science vs. Scopus and Google Scholar. *Journal of the American Society for Information Science and Technology 58*, 13 (2007), 2105–25.

[31] MEYER ZU EISSEN, S., AND STEIN, B. Intrinsic Plagiarism Detection. In *Advances in Information Retrieval 28th European Conference on IR Research, ECIR 2006, London, UK, April 10-12, 2006. Proceedings* (2006), M. Lalmas, A. MacFarlane, S. M. Rüger, A. Tombros, T. Tsikrika, and A. Yavlinsky, Eds., vol. 3936 of *Lecture Notes in Computer Science*, Springer Berlin Heidelberg, pp. 565–569.

[32] MONOSTORI, K., ZASLAVSKY, A., AND SCHMIDT, H. Document Overlap Detection System for Distributed Digital Libraries. In *DL '00: Proceedings of the fifth ACM conference on Digital libraries* (New York, NY, USA, 2000), ACM, p. 226.

[33] MONOSTORI, K., ZASLAVSKY, A., AND SCHMIDT, H. Suffix vector: space- and time-efficient alternative to suffix trees. *Aust. Comput. Sci. Commun. 24*, 1 (2002), 157–165.

[34] PHELAN, T. A compendium of issues for citation analysis. *Scientometrics 45* (1999), 117–136. 10.1007/BF02458472.

[35] POTTHAST, M., BARRÓN CEDEÑO, A., EISELT, A., STEIN, B., AND ROSSO, P. Overview of the 2nd International Competition on Plagiarism Detection. In *Notebook Papers of CLEF 2010 LABs and Workshops, 22-23 September, Padua, Italy* (Sept. 2010), M. Braschler, D. Harman, and E. Pianta, Eds.

[36] POTTHAST, M., STEIN, B., EISELT, A., BARRÓN CEDEÑO, A., AND ROSSO, P. Overview of the 1st International Competition on Plagiarism Detection. In *PAN09 - 3rd Workshop on Uncovering Plagiarism, Authorship and Social Software Misuse and 1st International Competition on Plagiarism Detection* (2009), B. Stein, P. Rosso, E. Stamatatos, M. Koppel, and A. Eneko, Eds.

[37] PRECHELT, L., PHILIPPSEN, M., AND MALPOHL, G. JPlag: Finding plagiarisms among a set of programs. Technical Report 2000-1, Universität Karlsruhe, 2000.

[38] RUDMAN, J. The State of Authorship Attribution Studies: Some Problems and Solutions. *Computers and the Humanities 31* (1997), 351–365.

[39] SEGLEN, P. O. Why the impact factor of journals should not be used for evaluating research. *BMJ 314*, 7079 (1997), 497.

[40] SI, A., LEONG, H. V., AND LAU, R. W. H. CHECK: A Document Plagiarism Detection System. In *SAC '97: Proceedings of the 1997 ACM symposium on Applied computing* (New York, NY, USA, 1997), B. Bryant, J. Carroll, J. Hightower, and K. M. George, Eds., ACM, pp. 70–77.

[41] SMALL, H., AND GRIFFITH, B. C. The Structure of Scientific Literatures I: Identifying and Graphing Specialties. *Science Studies 4*, 1 (1974), pp. 17–40.

[42] SMYTH, B. *Computing Patterns in Strings*. Pearson Addison-Wesley, Harlow, England; New York, 2003.

[43] SOROKINA, D., GEHRKE, J., WARNER, S., AND GINSPARG, P. Plagiarism Detection in arXiv. Technical report computer science, Cornell University TR2006-2046, Dec. 2006.

[44] STEIN, B. Fuzzy-Fingerprints for Text-Based Information Retrieval. In *Proceedings of the I-KNOW '05, 5th International Conference on Knowledge Management, Graz, Austria* (July 2005), K. Tochtermann and H. Maurer, Eds., vol. Special Issue of *Journal of Universal Computer Science*, Springer-Verlag, Know-Center, pp. 572–579.

[45] STEIN, B., KOPPEL, M., AND STAMATATOS, E., Eds. *Proceedings of the SIGIR 2007 International Workshop on Plagiarism Analysis, Authorship Identification, and NearDuplicate Detection, PAN 2007, Amsterdam, Netherlands, July 27, 2007* (2007), vol. 276 of *CEUR Workshop Proceedings*, CEUR-WS.org.

[46] STEIN, B., LIPKA, N., AND PRETTENHOFER, P. Intrinsic Plagiarism Analysis. *Language Resources and Evaluation [Online Resource]* (2010), 1–20.

[47] STEIN, B., AND MEYER ZU EISSEN, S. Near Similarity Search and Plagiarism Analysis. In *From Data and Information Analysis to Knowledge Engineering Proceedings of the 29th Annual Conference of the Gesellschaft für Klassifikation e.V. University of Magdeburg, March 9–11, 2005* (2006), M. Spiliopoulou, R. Kruse, C. Borgelt, A. Nürnberger, and W. Gaul, Eds., Springer Berlin Heidelberg, pp. 430–437.

[48] TSATSARONIS, G., VARLAMIS, I., GIANNAKOULOPOULOS, A., AND KANELLOPOULOS, N. Identifying free text plagiarism based on semantic similarity. In *Proceedings of the 4th International Plagiarism Conference* (2010).

[49] WEBER WULFF, D. Copy, Shake, and Paste - A blog about plagiarism from a German professor, written in English. Online Source, Nov. 2010. Retrieved Nov. 28, 2010 from: http://copy-shake-paste.blogspot.com.

[50] WEBER WULFF, D. Test cases for plagiarism detection software. In *Proceedings of the 4th International Plagiarism Conference* (Newcastle Upon Tyne, 2010).

[51] WISE, M. J. String Similarity via Greedy String Tiling and Running Karp-Rabin Matching. Online Preprint, Dec. 1993. Retrieved from: http://vernix.org/marcel/share/RKR_GST.ps.

[52] ZHAN SU, BYUNG-RYUL AHN, KI-YOL EOM, MIN-KOO KANG, JIN-PYUNG KIM, AND MOON-KYUN KIM. Plagiarism Detection Using the Levenshtein Distance and Smith-Waterman Algorithm. In *Innovative Computing Information and Control, 2008. ICICIC '08. 3rd International Conference on* (June 2008), pp. 569 –569.

[53 plagiarism] GUTTENBERG, K.-T. F. *Verfassung und Verfassungsvertrag : Konstitutionelle Entwicklungsstufen in den USA und der EU*. Dissertation (**Retracted as plagiarism**), Universität Bayreuth, Berlin, 2009.

Contributions to the Study of SMS Spam Filtering: New Collection and Results

Tiago A. Almeida
School of Electrical and
Computer Engineering
University of Campinas
Campinas, Sao Paulo, Brazil
tiago@dt.fee.unicamp.br

José María Gómez
Hidalgo
R&D Department
Optenet
Las Rozas, Madrid, Spain
jgomez@optenet.com

Akebo Yamakami
School of Electrical and
Computer Engineering
University of Campinas
Campinas, Sao Paulo, Brazil
akebo@dt.fee.unicamp.br

ABSTRACT

The growth of mobile phone users has lead to a dramatic increasing of SMS spam messages. In practice, fighting mobile phone spam is difficult by several factors, including the lower rate of SMS that has allowed many users and service providers to ignore the issue, and the limited availability of mobile phone spam-filtering software. On the other hand, in academic settings, a major handicap is the scarcity of public SMS spam datasets, that are sorely needed for validation and comparison of different classifiers. Moreover, as SMS messages are fairly short, content-based spam filters may have their performance degraded. In this paper, we offer a new real, public and non-encoded SMS spam collection that is the largest one as far as we know. Moreover, we compare the performance achieved by several established machine learning methods. The results indicate that Support Vector Machine outperforms other evaluated classifiers and, hence, it can be used as a good baseline for further comparison.

Categories and Subject Descriptors

H.3.3 [**Information Storage and Retrieval**]: Information Search and retrieval—*information filtering*; H.3.4 [**Information Storage and Retrieval**]: Systems and Software—*performance evaluation (efficiency and effectiveness)*

General Terms

Performance, Experimentation, Security, Standardization

Keywords

Spam filtering, mobile spam, classification

1. INTRODUCTION

Short Message Service (SMS) is the text communication service component of phone, web or mobile communication

systems, using standardized communications protocols that allow the exchange of short text messages between fixed line or mobile phone devices. According to the International Telecommunication Union (ITU)[1], SMS has become a massive commercial industry, worth over 81 billion dollars globally as of 2006.

The downside is that cell phones are becoming the latest target of electronic junk mail, with a growing number of marketers using text messages to target subscribers. SMS spam (sometimes also called mobile phone spam) is any junk message delivered to a mobile phone as text messaging. Although this practice is rare in North America, it has been very common in some parts of Asia.

Apparently, SMS spam is not as cost-prohibitive to spammers as it used to be, as the popularity of SMS has led to messaging charges dropping below US$ 0.001 in markets like China, and even free of charge in others. According to Cloudmark stats[2], the amount of mobile phone spam varies widely from region to region. For instance, in North America, much less than 1% of SMS messages were spam in 2010, while in parts of Asia up to 30% of messages were represented by spam.

The main problem with SMS spam is that it is not only annoying, but it can also be expensive since some people pay to receive text messages. Moreover, there is a limited availability of mobile phone spam-filtering software. Other concern is that important legitimate messages as of emergency nature could be blocked. Nonetheless, many providers offer their subscribers means for mitigating unsolicited SMS messages.

In the same way that carriers are facing many problems in dealing with SMS spam, academic researchers in this field are also experiencing difficulties. For instance, the lack of real and public databases can compromise the evaluation of different approaches. So, although there has been significant effort to generate public benchmark datasets for anti-spam filtering, unlike email spam, which has available a large variety of datasets, the mobile spam filtering still has very few corpora usually of small size. Other concern is that established email spam filters may have their performance seriously degraded when directly employed to dealing with mobile spam, since the standard SMS messaging is limited

---

[1]See http://www.itu.int/dms_pub/itu-s/opb/pol/S-POL-IR.DL-2-2006-R1-SUM-PDF-E.pdf.
[2]See http://www.cloudmark.com/en/article/mobile-operators-brace-for-global-surge-in-...mobile-messaging-abuse

to 140 bytes, which translates to 160 characters of the English alphabet. Moreover, their text is rife with idioms and abbreviations.

To fill this important gap, in this paper, we make available a new real, public and non-encoded SMS spam corpus that is the largest one as far as we know. Moreover, we compare the performance achieved by several established machine learning methods in order to provide good baseline results for further comparison.

The remainder of this paper is organized as follows. Section 2 offers details about the newly-created SMS Spam Collection. In Section 3, we present a comprehensive performance evaluation for comparing several established machine learning approaches. Finally, Section 4 presents the main conclusions and outlines for future works.

2. THE NEW SMS SPAM COLLECTION

Reliable data are essential in any scientific research. The absence of representative data can seriously impact the processes of evaluation and comparison of methods. In this way, areas of more recent studies are generally affected by the lack of public available data.

Regarding studies of mobile spam filtering, although there are few databases of legitimate SMS messages available in the Internet, it is very hard to find real samples of mobile phone spam. Thus, to create the corpus for the purposes of this work, we use data derived from several sources.

A collection of 425 SMS spam messages was manually extracted from the Grumbletext Web site. This is a UK forum in which cell phone users make public claims about SMS spam messages, most of them without reporting the very spam message received. The identification of the text of spam messages in the claims is a very hard and time-consuming task, and it involved carefully scanning hundreds of web pages. The Grumbletext Web site is: `http://www.grumbletext.co.uk/`.

We have also included in our corpus a subset of 3,375 SMS randomly chosen ham messages of the NUS SMS Corpus, which is a dataset of about 10,000 legitimate messages collected for research at the Department of Computer Science at the National University of Singapore. These messages were collected from volunteers who were made aware that their contributions were going to be made publicly available. The NUS SMS Corpus is available at: `http://www.comp.nus.edu.sg/~rpnlpir/downloads/corpora/smsCorpus/`.

We have added legitimate samples by inserting 450 SMS messages collected from Caroline Tag's PhD Thesis available at `http://etheses.bham.ac.uk/253/1/Tagg09PhD.pdf`.

Finally, we have incorporated the SMS Spam Corpus v.0.1 Big. It has 1,002 SMS ham messages and 322 spam messages and it is public available at: `http://www.esp.uem.es/jmgomez/smsspamcorpus/`. This corpus has been used in the following academic research efforts: [6], [7], and [14]. The sources used in this corpus are also the Grumbletext Web site and the NUS SMS Corpus.

The created corpus is composed by just one text file, where each line has the correct class followed by the raw message.

The SMS Spam Collection is public available at `http://www.dt.fee.unicamp.br/~tiago/smsspamcollection`.

The new collection is composed by 4,827 legitimate messages and 747 mobile spam messages, a total of 5,574 short messages. To the best of our knowledge, it is the largest available SMS spam corpus that currently exists. Table 1 shows the basic statistics of the created database.

Table 1: Basic statistics

Msg	Amount	%
Hams	4,827	86.60
Spams	747	13.40
Total	5,574	100.00

Table 2 presents the statistics related to the tokens extracted from the corpus. Note that, the proposed dataset has a total of 81,175 tokens and mobile phone spam has in average ten tokens more than legitimate messages.

Table 2: Token statistics

Hams	63,632
Spams	17,543
Total	81,175
Avg per Msg	14.56
Avg in Hams	13.18
Avg in Spams	23.48

2.1 Message Duplicates Analysis

As the newly collected messages in the SMS Spam Collection have been augmented with a previously existing database built using roughly the same sources (GrumbleText, NUS SMS Corpus), it is sensible to check if there are some duplicates coming from both databases.

To address this issue, we have performed a duplicates analysis based on plagiarism detection techniques [9]. In [9], a number of techniques for plagiarism are presented, and those based on "String-of-Text", and implemented by the tool WCopyfind[3], can be considered as reasonable baseline for the purpose of detecting near-duplicated messages in our collection. The "String-of-Text" methods involve scanning suspect texts for approximately matching character sequences. Texts are compared searching for N-grams for relatively big sizes (e.g. $N = 6$), with additional parameters (length of match in number of characters, etc.). We have simplified this method to N-gram matches after text normalization involving replacing all token separators by white spaces, lowercasing all characters, and replacing digits by the character 'N' (to preserve phone numbers structure).

We searched for near-duplicates within three subcollections: the previously existing SMS Spam Corpus v.0.1 Big (**INIT**), the additional messages from Grumbletext, the NUS SMS Corpus, and Tag's PhD Thesis (**ADD**), and the actual new SMS Spam Collection (**FINAL**). In order to assess the overlap between both collections, we have compared each pair of messages within each subcollection and in common between both subcollections, stored all N size matches (N-grams with $N = 5$, 6, and 10), and sorted the N-grams according to their frequencies.

We have found 5-grams already presented in the **INIT** and the **ADD** collections do not collapse to greatly increase their frequencies, and they typically correspond to templates often presented in cell phones, and used in legitimate messages (e.g. "sorry i ll call later"). The 5-grams that co-occur in **INIT** and **ADD**, so they get their frequencies increased

---

[3]See: `http://plagiarism.phys.virginia.edu`

in **FINAL**, are new instances of spam probably sent by the same organization. In 6-grams results, we have found that there are not significant near-duplicates except for those already presented in each subcollection. Moreover, the results achieved with 10-grams are very similar to the 5- and 6-grams ones.

In consequence, we believe it is safe to say that merging the subcollections, although they have roughly the same sources, does not lead to near-duplicates that may ease the task of detecting SMS spam.

3. EXPERIMENTS

We have tested several well-known machine learning methods in the task of automatic spam filtering using the created SMS Spam Collection. The main goal of this performance evaluation is to provide good baseline results for further comparison, since established email spam filters may have their performance seriously impacted when employed to classify short messages. In addition, mobile phone messages often have a lot of abbreviations and idioms that may also affect the filters accuracy.

We consider in this work two different tokenizers:

1. tok1: tokens start with a printable character, followed by any number of alphanumeric characters, excluding dots, commas and colons from the middle of the pattern. With this pattern, domain names and mail addresses will be split at dots, so the classifier can recognize a domain even if subdomains vary [16].

2. tok2: any sequence of characters separated by blanks, tabs, returns, dots, commas, colons and dashes are considered as tokens. This simple tokenizer intends to preserve other symbols that may help to separate spam and legitimate messages.

In addition, we did not perform language-specific preprocessing techniques such as stop word removal or word stemming, since other researchers found that such techniques tend to hurt spam-filtering accuracy [5, 17].

The list of all evaluated classifiers are presented in Table 3[4].

Given that the corpus is biased to the ham class, an obvious baseline is the trivial rejector (TR) for the spam class.

As the most of the tokens with the highest Information Gain score often occur in the spam class, it is sensible to expect that messages may get automatically segregated into two classes on the basis of those tokens. In consequence, we provide an additional baseline in the form of the results of the Expectation-Maximization (EM) clustering algorithm [8], over a vector representation based on the tokenizer tok2. EM is an iterative soft clusterer that estimates cluster densities. Basically, cluster membership is a hidden latent variable that the maximum likelihood EM method estimates.

In our experiments, we have set up a maximum of 20 iterations and used the rest of the default values for EM parameters in WEKA.

---

[4]Some of the implementations of the described classifiers are provided by the Machine Learning library WEKA, available at http://www.cs.waikato.ac.nz/ml/weka/. The algorithms have been used with their default parameters except when otherwise is specified.

Table 3: Evaluated classifiers

Classifiers
Basic Naïve Bayes (NB) – Basic NB [2]
Multinomial term frequency NB – MN TF NB [2]
Multinomial Boolean NB – MN Bool NB [2]
Multivariate Bernoulli NB – Bern NB [2]
Boolean NB – Bool NB [2]
Multivariate Gauss NB – Gauss NB [2]
Flexible Bayes – Flex NB [2]
Boosted NB [12]
Linear Support Vector Machine – SVM [10, 13]
Minimum Description Length – MDL [4]
K-Nearest Neighbors – KNN [1, 14] (K = 1, 3 or 5)
C4.5 [15, 14]
Boosted C4.5 [14]
PART [11, 14]

3.1 Results

We carried out this study using the following experiment protocol. We divided the corpus in two parts: the first 30% of the messages were separated for training (1,674 messages) and the remainder ones for testing (3,900 messages). As all the messages are fairly short, we did not use any kind of method to reduce the dimensionality of the training space, *e.g.*, terms selection techniques.

To compare the results we employed the following well-known performance measures: Spam Caught ($SC\%$), Blocked Hams ($BH\%$), Accuracy ($Acc\%$), and Matthews Correlation Coefficient (MCC) [3].

Table 4 presents the best fifteen results achieved by each evaluated classifier and tokenizer. Note that the results are sorted in descending order of MCC.

Table 4: The fifteen best results achieved by combinations of classifiers + tokenizers and the baselines Expectation-Maximization (EM) and trivial rejection (TR)

Classifier	$SC\%$	$BH\%$	$Acc\%$	MCC
SVM + tok1	83.10	0.18	97.64	0.893
Boosted NB + tok2	84.48	0.53	97.50	0.887
Boosted C4.5 + tok2	82.91	0.29	97.50	0.887
PART + tok2	82.91	0.29	97.50	0.887
MDL + tok1	75.44	0.35	96.26	0.826
C4.5 + tok2	75.25	2.03	95.00	0.770
Bern NB + tok1	54.03	0.00	94.00	0.711
MN TF NB + tok1	52.06	0.00	93.74	0.697
MN Bool NB + tok1	51.87	0.00	93.72	0.695
1NN + tok2	43.81	0.00	92.70	0.636
Basic NB + tok1	48.53	1.42	92.05	0.600
Gauss NB + tok1	47.54	1.39	91.95	0.594
Flex NB + tok1	47.35	2.77	90.72	0.536
Boolean NB + tok1	98.04	26.01	77.13	0.507
3NN + tok2	23.77	0.00	90.10	0.462
EM + tok2	17.09	4.18	85.54	0.185
TR	0.00	0.00	86.95	–

It is notable that the linear SVM achieved the best results and outperformed the other evaluated methods. It caught 83.10% of all spams with the cost of blocking only 0.18%

of legitimate messages, acquiring an accuracy rate higher than 97.5%. It is a remarkable performance considering the EM and TR baselines and the high difficulty of classifying mobile phone messages. However, the results also indicate that the best four algorithms achieved similar performance with no statistical difference. All of them accomplished an accuracy rate superior than 97%, that can be considered as a very good baseline in a such context.

It is important to point out that MDL and C4.5 techniques also achieved good results since they found a good balance between false and true positive rates. On the other hand, the remainder evaluated approaches had an unsatisfying performance. Note that, although the most of them have obtained accuracy rate superior than 90%, they have correctly filtered about only 50% of spams or even less.

Therefore, based on the achieved results, we can certainly conclude that the linear SVM offers the best baseline performance for further comparison.

4. CONCLUSIONS

The task of automatic filtering SMS spam still is a real challenge nowadays. There are three main problems hindering the development of algorithms in this specific field of research: the lack of public and real datasets, the low number of features that can be extracted per message, and the fact that the text is rife with idioms and abbreviations.

To fill some of those gaps, in this paper we presented a new mobile phone spam collection that is the largest one as far as we know. Besides being large, it is also publicly available and composed by only non-encoded and real messages.

Moreover, we offered statistics relating to the proposed corpus, as tokens frequencies and since the corpus is composed by subsets of messages extracted from the same sources, we also presented a study regarding the message duplicates and the found results indicate that the proposed collection is reliable because there are no more duplicates than those ones already presented within the used subsets.

Finally, we compared the performance achieved by several established machine learning methods and the results indicate that SVM outperforms other classifiers and, hence, it can be used as a good baseline for further comparison.

Future work should consider to use different strategies to increase the dimensionality of the feature space. Well-known techniques, such as orthogonal sparse bigrams (OSB), 2-grams, 3-grams, among many others could be employed with the standard tokenizers to produce a larger number of tokens and patterns which can assist the classifier to separate ham messages from spams.

5. ACKNOWLEDGMENTS

The authors would like to thank the financial support of Brazilian agencies FAPESP and CAPES/PRODOC.

6. REFERENCES

[1] D. Aha and D. Kibler. Instance-based learning algorithms. *Machine Learning*, 6:37–66, 1991.

[2] T. A. Almeida, J. Almeida, and A. Yamakami. Spam Filtering: How the Dimensionality Reduction Affects the Accuracy of Naive Bayes Classifiers . *JISA*, 1(3):183–200, 2011.

[3] T. A. Almeida, A. Yamakami, and J. Almeida. Evaluation of Approaches for Dimensionality Reduction Applied with Naive Bayes Anti-Spam Filters. In *Proc. of the 8th IEEE ICMLA*, pages 517–522, Miami, FL, USA, 2009.

[4] T. A. Almeida, A. Yamakami, and J. Almeida. Filtering Spams using the Minimum Description Length Principle. In *Proc. of the 25th ACM SAC*, pages 1856–1860, Sierre, Switzerland, 2010.

[5] G. Cormack. Email Spam Filtering: A Systematic Review. *Foundations and Trends in Information Retrieval*, 1(4):335–455, 2008.

[6] G. V. Cormack, J. M. Gómez Hidalgo, and E. Puertas Sanz. Feature Engineering for Mobile (SMS) Spam Filtering. In *Proc. of the 30th ACM SIGIR*, pages 871–872, New York, NY, USA, 2007.

[7] G. V. Cormack, J. M. Gómez Hidalgo, and E. Puertas Sanz. Spam Filtering for Short Messages. In *Proc. of the 16th ACM CIKM*, pages 313–320, Lisbon, Portugal, 2007.

[8] A. P. Dempster, N. M. Laird, and D. B. Rubin. Maximum Likelihood from Incomplete Data via the EM Algorithm. *Journal of the Royal Statistical Society.*, 39(1):1–38, 1977.

[9] H. Dreher. Automatic Conceptual Analysis for Plagiarism Detection. *Issues in Informing Science and Information Technology*, 4:601–628, 2007.

[10] G. Forman, M. Scholz, and S. Rajaram. Feature Shaping for Linear SVM Classifiers. In *Proc. of the 15th ACM SIGKDD*, pages 299–308, 2009.

[11] E. Frank and I. H. Witten. Generating Accurate Rule Sets Without Global Optimization. In *Proc. of the 15th ICML*, pages 144–151, Madison, WI, USA, 1998.

[12] Y. Freund and R. E. Schapire. Experiments with a new boosting algorithm. In *Proc. of the 13th ICML*, pages 148–156, San Francisco, 1996. Morgan Kaufmann.

[13] J. M. Gómez Hidalgo. Evaluating Cost-Sensitive Unsolicited Bulk Email Categorization. In *Proc. of the 17th ACM SAC*, pages 615–620, Madrid, Spain, 2002.

[14] J. M. Gómez Hidalgo, G. Cajigas Bringas, E. Puertas Sanz, and F. Carrero García. Content Based SMS Spam Filtering. In *Proc. of the 2006 ACM DocEng*, pages 107–114, Amsterdam, The Netherlands, 2006.

[15] J. R. Quinlan. *C4.5: Programs for Machine Learning.* Morgan Kaufmann Publishers Inc., 1993.

[16] C. Siefkes, F. Assis, S. Chhabra, and W. Yerazunis. Combining Winnow and Orthogonal Sparse Bigrams for Incremental Spam Filtering. In *Proc. of the 8th ECML PKDD*, pages 410–421, Pisa, Italy, 2004.

[17] L. Zhang, J. Zhu, and T. Yao. An Evaluation of Statistical Spam Filtering Techniques. *ACM TALIP*, 3(4):243–269, 2004.

A Study of the Interaction of Paper Substrates on Printed Forensic Imaging

Guy Adams
Hewlett-Packard Labs
Longdown Ave. Stoke Gifford
Bristol UK BS34 8QZ
guy.adams@hp.com

Stephen Pollard
Hewlett-Packard Labs
Longdown Ave. Stoke Gifford
Bristol UK BS34 8QZ
stephen.pollard@hp.com

Steven Simske
Hewlett-Packard Labs
3404 E. Harmony Rd.
Fort Collins CO 80528 USA
steven.simske@hp.com

ABSTRACT

At the microscopic level, printing on a substrate exhibits imperfections that can be used as a unique identifier for labels, documents and other printed items. In previous work, we have demonstrated using these minute imperfections around a simple forensic mark such as a single printed character for robust authentication of the character with a low cost (and mobile) system. This approach allows for product authentication even when there is only minimal printing (e.g. on a small label or medallion), supporting a variety of secure document workflows. In this paper, we present an investigation on the influence that the substrate type has on the imperfections of the printing process that are used to derive the character 'signature'. We also make a comparison between two printing processes, dry electro photographic process (laser) and (thermal) inkjet. Understanding the sensitivity of our methods to these factors is important so that we know the limitations of the approach for document forensics.

Categories and Subject Descriptors

I.4.1 [**Image Processing and Computer Vision**]: Digitization and Image Capture—*Scanning*. I.4.10 [**Image Processing and Computer Vision**]: Image Representation—*Statistical*. K.6.5 [**Management of Computing and Information Systems**]: Security and Protection—*Authentication*.

General Terms

Algorithms, Security

Keywords

Security, Forensics, Model Matching, Document Fraud, High-resolution Imaging.

1. INTRODUCTION

Forensic analysis of printed material including documents, packaging and labels, can be classified into two broad categories: 1) device forensics/ballistics [1]-[3] where a document (or set of documents) is analyzed to see if it was printed on a specific device or class of devices; 2) print forensics [4]-[7] wherein individual printed artifacts are uniquely identified. This second class, which is of interest here, allows the differentiation of individual instances of the same or highly similar documents - including high quality copies. For the majority of printing technologies unique properties result from the unrepeatable statistical aspects of the print process itself and the interaction with the underlying structural properties of the substrate material on which it is printed. Thus a forensic mark can be any form of glyph, character or printed shape of sufficient size to carry information to determine if the forensic mark under investigation is the exact same unique forensic mark that was previously printed [8]. In this way, print is used as a security mechanism preventing the counterfeit and copy of documents and product packaging [9].

We have previously [10] demonstrated a low-cost USB-powered mini-appliance capable of resolving 3.5 microns (7257 dpi) with 1:1 magnification. This is accomplished using a Dyson relay lens which comprises a single refractive surface in series with a mirror and a low cost 3M pixel CMOS image sensor. With a self-contained (white LED) illumination source, this Dyson relay CMOS imaging device (Dr CID) affords the capture of individual typed characters with printing "parasitics"—such as the absorbance of ink into the fibers of the substrate (e.g. paper, cardstock, etc.) along with the droplet "tails" that exhibit micro-random aberrations. The laser process has different characteristics such as toner particle size variation, splatter, charge bleed etc.

In [11]-[13] we adopted a model-based approach that separates the truly random part of the outline of the individual printed character, which we termed a model based signature profile (MBSP), from the shared shape-conveying component. We showed that the MBSP allows forensic (highly statistically significant) levels of authentication to be achieved (even in the case where multiple imaging modalities are employed). Furthermore for many forensic marks the MBSP is extracted in an order fixed by the frame of reference defined by the model. This in turn allowed the introduction (see [12], [13]) of a simple fixed order shape warp descriptor (which is typically less than a few hundred bits long) which is extracted from the MBSP. This is more compact than the MBSP and easier to manipulate and yet retains most of the forensic authentication power of the former.

In this paper, we present an investigation on the influence that the substrate type has on the imperfections of the printing process that are used to derive the character 'signature'. We also make a comparison between two printing processes, dry electro photographic process (laser) and (thermal) inkjet. Understanding the sensitivity of our methods to these factors is important so that we know the limitations of the approach for document forensics. It is also important so that we can determine how many printed characters are required to forensically identify a document. We also go beyond previous work and show that it is possible to base authentication entirely on an analysis of the substrate material in the vicinity of the printed character.

2. METHOD

2.1 Model Based Signature Profiles (MBSP)

We define our models simply as a set of N uniformly spaced points (x, y coordinates) defining the outer edge of a character glyph and associated unit normal vectors. Figure 1a shows an illustration of a model of the outer contour of a Times Roman lowercase 'a'. The process of MBSP extraction is illustrated in the remainder of figure 1 (see also [13] for details). Importantly during the extraction of the profile image 1e it is preferable to low-pass filter the underlying image 1b using a standard Gaussian convolution kernel. This has the effect of removing imaging noise and more importantly avoiding sampling artifacts.

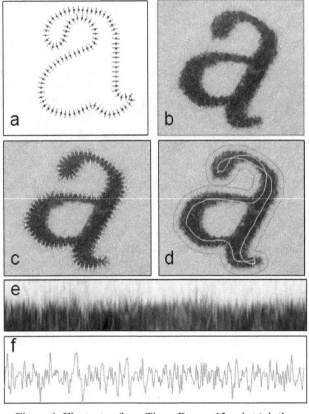

Figure 1. Illustrates, for a Times Roman 12 point 'a', the extraction of a MBSP. Where: (a) is a simplified model for the outline of the character composed of 100 feature points shown with associated normals (note that in practice to avoid sampling artifacts models are an order of magnitude more dense than shown in this figure; 2000 model points are used in the experiments presented in this paper); (b) is a 900x800 (wide x tall) image of a 12 point character captured by Dr CID; (c) shows superimposed transformed model; (d) shows the loci of sampled regions for the extracted normal profile images in (e). Each column of (e) corresponds to sampling on a vector between the loci along the normal vector for each individual (x, y) contour point of the model; Finally (f) shows the MBSP extracted from (e).

Many methods can be used to recover the signature profile from the profile image, including simple thresholding or maximum edge detection. We have found the following grayscale edge metric that combines all the data in the profile image to work well. For each column in the profile image the signature profile is defined as:

$$p_i = \sum_j jw_j e_{ij} \Big/ \sum_j w_j |e_{ij}|$$

where e_{ij} is an edge strength corresponding to the digital derivative of the profile image along the column i and w_j is a windowing function (in our case a Gaussian with standard deviation ¼ the column height centered on the mid-point of the column). Dividing by a normalizing sum of windowed absolute edge strength results in a measure that achieves robustness to both scene content and illumination variation.

2.2 Shape Warp Coding (SWC)

Introduced in [12] for micro-color-tile inspection and generalized in [13] for any printed forensic mark (provided a suitably irregular shape for which the matching process recovers a unique model location) the SWC is derived from the MBSP as follows. First we divide the signature profile into N equal length segments. Then for each, compute a sum squared error (SSE) of the residual (which is akin to a local variance):

$$SSE_j = \sum_{p_i \in segment(j)} (p_i - \mu_j)^2$$

where p_i is the signature profile in the segment j and μ_j is the mean value of that profile over the whole of the segment. We then compute the mean value (or alternatively the median) of this error metric, SSE_{mean}, over all N segments. This provides us with an atomic unit of encoding (a "digit") and it is possible to use it to form an N-position string which is the SWC:

$$SWC(j) = \left\| SSE_j \Big/ SSE_{mean} \right\|$$

where $\|.\|$ is a rounding function. The SDED, for comparing the SWCs of any two forensic marks, is thus defined as:

$$SDED = \sum_j \min(| SWC_1(j) - SWC_2(j) |, T_{max})$$

where T_{max} is an optional threshold to improve robustness. The SDED can be considered a form of modified Hamming Distance where the expected value of SWC(*) is 1 at each digit due to the normalization process described. For example, a pair of SWCs (N=50) extracted from Dr CID data for the same printed 'a' and their absolute difference are:

```
SWC1 = 11011111201101111211211112111111011212112111110101111
SWC2 = 11111111121010010121121112111211101111111121111011210
DIFF = 00100000011001010000000000001000001010000000001101
```

for which the SDED is 11 (or 0.22 when normalized by N).

3. EXPERIMENTS

In this section, we report the results of 2 experiments. The first applies the previously presented methodology (outlined in section 2) to a number of substrate types and includes both laser and thermal inkjet prints where appropriate. The second set of experiments involves substrate-only comparisons using a modified form of the SWC where the SSE is replaced by the variance of the substrate in the top quartile of an extended profile image (chosen to ensure that the region over which the variance is measured is not close to the intentionally inked part of the print). Thus, for the latter experiment, the sole purpose of the model is to provide a unique frame of reference to allow consistent measurement of the properties of the substrate.

After considering a wide range of paper types we chose to test 6 distinct types; 5 for laser (HP 80g office, HP 160g Matte, HP 200g Photo Matte, HP 120g Soft Gloss and Handmade Lokta by Wild Paper) and 3 for inkjet (HP 80g office, HP Premium Photo Glossy and Handmade Lokta) with two in common. These represent a broad cross-section of the paper types available. The laser printer used was an HP CP6015 and the inkjet was an HP K5400 and the printers were configured for the specific paper type being used.

For each print/paper-type combination we printed 40 Times Roman 12 point letter 'a's. Each is scanned twice using different Dr CID devices resulting in 640 individual images. Examples of each are shown in figure 2.

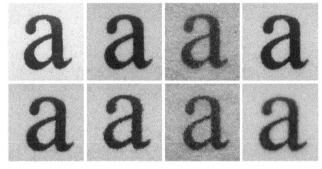

Figure2. Top left to bottom right are Laser HP 6015 on Soft Gloss, Photo Matte, Handmade, Matte and Office, followed by inkjet HP K5400 on Office, Handmade and Premium Photo.

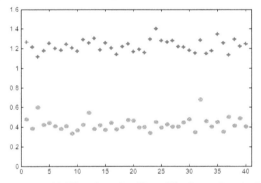

Figure 3. Plots SDED values for 40 veridical matches (red) and 40 random false matches (blue) for CP6015 laser print on 80g office paper. The sigma of Gaussian smoothing was 5.0.

3.1 Substrate Comparison

The first set of experiments use the standard methodology outlined in section 2. In each experiment the 40 SWC's derived from printed letters captured by one Dr CID device are each compared with an SWC derived from the same printed character captured with the other Dr CID device (veridical match) and with a random incorrectly matching SWC (false match) also captured with the second device (for the same print/substrate combination). The SDED values that result from one such experiment (laser on 80g plain paper) are shown in figure 3. It is clear from this figure that the distribution of SDED values of the veridical matches is well separated from that of the false matches.

Summary statistics (means with standard deviation error bars) of a series of such experiments are shown in figure 4 for our 8 print and paper combinations. Of these only the Handmade Lokta (for both laser and inkjet) shows significantly different distribution statistics to those illustrated in figure 3.

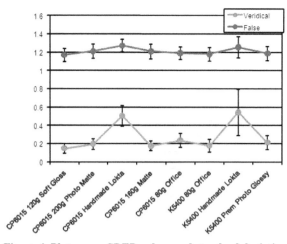

Figure 4. Plots mean SDED values and standard deviation error bars for veridical and false matches for each paper/print combination. The sigma of Gaussian smoothing was 5.0.

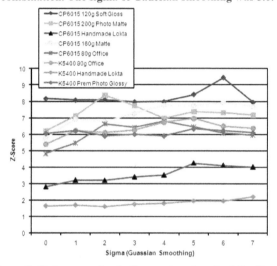

Figure 5. Plots approximate Z-score trend against the Sigma used in low-pass filter for each paper/print combination.

If we assume that the distributions of veridical and false matches are Gaussian (for which we have presented compelling evidence in [13]) then we can use an approximate a Z-score (approximate as these are small sample, rather than population, statistics) to measure the separation of the two populations:

$$Z = \left| \overline{S}_V - \overline{S}_F \right| / \left(\sigma_V + \sigma_F \right)$$

that is the absolute difference of the mean SDED scores for veridical and false matches divided by the sum of their standard deviations. The relationship between Z-score and the probability of false-positive/negative is highly non-linear, while a Z-score of 3 corresponds to a probability of 0.001 a Z-score of 6 relates to a probability of 10^{-9}. Figure 5 shows how approximate Z-scores vary with the degree of low-pass filtering applied to image during the construction of the profile image – illustrating the need to overcome sampling artifacts for effective matching. Note that a sigma of 5.0 results in Z-scores above 6.0 for all but the handmade paper types.

3.2 Substrate Authentication

As discussed in the introduction to this section the second set of experiments investigates the use of substrate variance alone to achieve forensic authentication. Figure 6 shows summary statistics for each of the 8 paper types. Remarkably the results are very similar to those of figure 4 which included both ink/toner and substrate in the region under analysis. Only the Premium Photo Glossy inkjet paper is significantly impaired due to its highly uniform and specular surface properties.

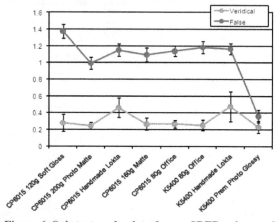

Figure 6. Substrate only plot of mean SDED values and standard deviation error bars for veridical and false matches for each paper/print combination (Sigma 5.0).

Figure 7 shows for the substrate only experiments the Z-score trend with the degree of low-pass filtering. In this case there are more significant relationships that are paper specific. Most surprising is the large difference between the two 80g office papers – given that this experiment uses only the substrate for SDED comparisons we would expect the identical paper types to give identical results. However for all the laser printed documents the substrate will inevitably include flecks of stray toner that add to the microscopic texture that is measured in the analysis of the Dr CID images.

4. DISCUSSION

Our results show that for the majority of print and substrate combinations forensic levels of authentication can be achieved with the analysis of a single Dr CID image of a single printed character whether or not the ink/toner mass of the character is included in the analysis. This is a significant and interesting result that adds to our understanding of print/substrate as a forensic signature of an individual document. In those cases where the statistical significance is reduced it is necessary to use a number of printed characters to achieve forensic-level identification. In general, if the probability of a false positive identification for a given character is p, and the desired forensic-level certainty is F, then n characters are required to achieve forensic-level certainty governed by the equation:

$$p^n = F$$

So, if p=0.022 (as is the case for a Z-score of 2) and F=10^{-9}, then we need 6 characters (that is, n=5.4) to achieve forensic-level validation of inkjet print on handmade paper - which while being richly textured is also highly specular and hence a small difference in the capture conditions tends to dominate the image statistics.

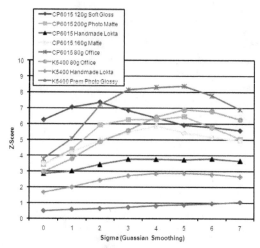

Figure 7. Substrate only plot of Z-score against the Sigma used in low-pass filter for each paper/print combination.

5. REFERENCES

[1] E. Kee and H. Faird, "Printer profiling for forensics and ballistics", *ACM MM&Sec*, 2008.

[2] S.J. Simske, M. Sturgill, P. Everest, and G. Guillory, "A system for forensic analysis of large image sets," *Proc. IEEE WIFS*, 2009.

[3] P-J. Chiang, N. Khanna, A. K. Mikkilineni, M. V. Ortiz Segovia, S. Suh, J. P. Allebach, G. T-C. Chiu & E. J. Delp, "Printer and scanner forensics: examining the security mechanisms for a unique interface", *IEEE Signal Processing Magazine* **72**, 2009.

[4] J. J. Plimmer, "Choosing correct forensic marker(s) in currency, document and product protection", *SPIE-IS&T* 6075, 2008.

[5] B. Zhu, J. WU, M. S. Kankanhalli, "Print Signatures for Document authentication", *CCS ACM*, 2003.

[6] A. Idrissa, T. Fournel & Alain Aubert, "Secure embedded verification of print signatures", *J. Phys.: Conf. Ser.* **206** 012036, 2010.

[7] S.J. Simske and G. Adams, "High-Resolution Glyph-Inspection Based Security System", *IEEE ICASSP*, 2010.

[8] L. Hindus, "Image-Based "Fingerprinting", *Advanced Imaging*, 1998.

[9] D.E. Bicknell and G.M. Laporte, "Forged and Counterfeit Documents", in Wiley Encyclopedia of Forensic Science, 3104, June 2009.

[10] G. Adams, "Hand held Dyson relay lens for anti-counterfeiting", *Proc. IEEE IST*, 2010.

[11] S.B. Pollard, G. Adams and S.J. Simske, "Resolving distortion between linear and area sensors for forensic print inspection", *IEEE ICIP*, 2010.

[12] S.J. Simske, S.B. Pollard & G. Adams, "An imaging system for simultaneous inspection, authentication and forensics", *IEEE IST*, 2010.

[13] S.B. Pollard, S.J. Simske, G.B. Adams, "Model Based Print Signature Profile Extraction For Forensic Analysis of Individual Text," Proc. IEEE WIFS 2010, pp. 1-6, 2010

Version Control Workshop

Neil Fraser

Google

1600 Amphitheatre Parkway

Mountain View, CA, 94043

fraser@google.com

ABSTRACT

This three hour workshop takes participants on a tour of popular Version Control systems, particularly Subversion and Git. By the end of the workshop each participant will be proficient in using both of these systems. The focus is on solving real-world problems, such as resolving conflicting changes or rolling back a change. This workshop is not about the theory or academic underpinnings of such systems. Participants are required to bring a Macintosh, Linux or Windows laptop.

Categories and Subject Descriptors

D.2.7 [**Software Engineering**]: Distribution, Maintenance, and Enhancement – *Version control*

I.7.1 [**Document and Text Processing**]: Document and Text Editing – *Version control*

General Terms: Management.

Keywords: Subversion, Git.

1. INTRODUCTION

Version Control systems have become an indispensable tool for organizations that maintain digital resources. Often used for maintaining source code, Version Control is also used for text documents, engineering diagrams, content management, and wikis. Version Control allows multiple people to work on a project simultaneously without fear of overwriting each other's work. Version Control also allows one to trace the origin of any change back to the original author and the context in which it was made.

Despite its utility, Version Control remains underused. Due to their steep learning curves, owners of smaller projects sometimes shy away from the overhead of setting up Version Control. Likewise, many users of Version Control do not exploit the more advanced features such as personal branching due to unfamiliarity.

GOALS

This workshop is intended to teach how to use Version Control in small to medium-scale environments. Unlike most documentation, the workshop is not structured around particular features of the software. Instead a series of real-world scenarios are posited and one sees how they might be resolved with Version Control.

Subversion and Git have been chosen for this workshop since they represent opposite designs within the world of Version Control. Subversion follows a client-server architecture with rigidly defined workflows. Subversion's methodology is similar to CVS and Perforce. Git follows a distributed architecture with more ad-hoc workflows. Git's methodology is similar to BitKeeper and Mercurial. Gaining hands-on experience with both Subversion and Git will give participants a solid understanding of the options and trade-offs present in Version Control systems.

The workshop starts by examining single-person (non-collaborative) scenarios using Subversion:

- Single-edit workflow. Check out, edit a file, commit. This is the most common use case.

- Multiple-edit workflow. Managing a change list across several files including additions and deletions.

- Parallel edits. Handling two edits by the same user at the same time (such as a large edit that is interrupted by a high-priority small edit).

Next the workshop examines multiple-person (collaborative) scenarios where the participants share a single repository.

- Conflicting changes. Dealing with collisions and three-way merges.

- History. Identifying who was responsible for a particular change and how to dig through the repository history to answer questions.

- Rollback. Undoing another contributor's changes.

Each of these scenarios uses one or more plain-text documents, with each step building upon the previous step.

After learning how to use an existing repository, the workshop then examines how to setup a new repository (up to now the participants would have been using a private repository running on a local server).

The workshop then repeats all of the above scenarios with Git instead of Subversion.

Finally the workshop looks at a few of the graphical interfaces for Subversion and Git, as well as a brief overview of some of the other popular Version Control systems that are available.

DocEng'11, September 19–22, 2011, Mountain View, California, USA.
ACM 978-1-4503-0863-2/11/09.

Participants leave with a solid understanding of how to use both Subversion and Git in the real world.

2. TECHNICALITIES

Each participant is required to bring a Windows, Macintosh or Linux laptop upon which they have rights to install software. Chrome OS is not suitable. Participants will install clients for Subversion and Git, both of which are free and open source. Laptops must have wireless capabilities (either WiFi or cellular). Free WiFi will be available.

The workshop lasts for three hours, with a short break at the midpoint. Punctuality in starting is encouraged as catching up is problematic if one doesn't have the required software installed. However, joining at the midpoint when we switch from Subversion to Git is possible, should one already be familiar with Subversion.

3. PRESENTER

Neil Fraser is a software engineer at Google. In addition to using Version Control systems professionally, Mr Fraser is also a contributor to Google Docs, MobWrite, the Diff Match Patch code library, and the Differential Synchronization algorithm – all of which have elements of Version Control at their core.

4. PREPARATION

To speed up getting started with the workshop, it would be appreciated if participants could follow a few preparatory steps on their laptop prior to the workshop. This should take about ten minutes. Email fraser@google.com if there are any problems.

1. Check for Subversion. Open a command prompt and verify that you get a help screen when you type: `svn help`

   If not, then install it:

   - Linux: `apt-get install subversion`
   - Macintosh: http://www.open.collab.net/downloads/community/
   - Windows: http://sourceforge.net/projects/win32svn/

2. Check for Git. Open a command prompt and verify that you get a help screen when you type: `git help`

   If not, then install it:

   - Linux: `apt-get install git`
   - Macintosh: http://code.google.com/p/git-osx-installer/downloads/list?can=3
   - Windows: "Full installer for official Git 1.7.x"

     http://code.google.com/p/msysgit/downloads/list?can=3

3. Fill out the form at http://goo.gl/ikQQL

5. REFERENCES

[1] Collins-Sussman, B., Fitzpatrick, B. W., and Pilato, C. M. 2011. *Version Control with Subversion.* O'Reilly Media. http://svnbook.red-bean.com/

[2] Chacon, S. 2011. *Git Documentation.* http://git-scm.com/documentation

Secure Document Engineering

Helen Balinsky
Hewlett-Packard Labs
Longdown Ave. Stoke Gifford
Bristol UK BS34 8QZ
helen.balinsky@hp.com

Steven Simske
Hewlett-Packard Labs
3404 E. Harmony Rd.
Fort Collins CO 80528 USA
steven.simske@hp.com

ABSTRACT

With the boom in interactive and composite documents and the increased coupling between the on-line and physical worlds, the need for secure document engineering is greater than ever. Four important factors contribute to the need to re-engineer document lifecycles and the associated workflows. The first is the rapid increase in mobile access to documents. The second is the movement of documents from private directories, shared directories and intranets to the cloud. The third is the increased generation of – and expectation for – document content, context and use analytics. Finally, the proliferation of social website applications and services over the past half-decade have created for many a constant state of log-in. Each of these trends creates a significantly increased "attack surface" for individuals, organizations and governments interested in breaching the privacy and security of web users. Combined, these transformations create as big a change to content security as that of the browser in the 1990s. In this workshop, we consider the impact of these ongoing transformations in document creation and interaction, and consider what the best approaches will be to provide privacy and security in light of these transformations.

Categories and Subject Descriptors

I.7.1 [**Document and Text Editing**]; I.7.4 [**Electronic Publishing**]; K.6.5 [**Management of Computing and Information Systems**]: Security and Protection—*Authentication.*

General Terms

Algorithms, Design, Security

Keywords

Security, Publicly Posted Composite Document, Access Control, Identity Based Encryption

1. INTRODUCTION

Document engineering continues to gain importance in the Internet era as a consequence of at least four important factors. First, mobile access to document workflows—including document creation, routing, sharing and printing—has made access to documents formerly "stranded" on hard drives, intranets, and

shared directories "fair game" for ubiquitous access, modification, reading and feedback. In effect, the social internet has driven the "socialization" of documents—the more timely and easy the access, the better.

Secondly, the movement of consumer and enterprise applications and services to the "cloud" has resulted in an emphasis on access security and simplicity. Content is more and more expected to be available—and, if one is authorized, nothing should be inaccessible.

Third, data is no longer allowed to lie idly. It is, instead, expected to be converted into information. The broad term "analytics" covers this upgrading of data into information, and is of huge importance as the amount of data created continues to increase exponentially [1].

A fourth important factor to consider is the transformation in Internet behavior caused by the explosion of social websites over the past 5 years. For the next generation, posting real-time information on social web sites is second nature, and they have grown up expecting any and all content they create, participate in, or otherwise wish to consumer (print, mash-up), etc.

Looking for commonality among these sea changes in document engineering, one that immediately comes to mind is security. Any time there are changes in a system, there are new security risks. As an example of heightened security risks, attacks on Sony, the financial industry, and other targets have been prominent in the news the entire first half of 2011 [2]. These attacks are often done for social/"ethical" rationales.

One "logical" target for attack by terrorist organizations are the businesses supporting the military of their victims. In [3], for example, Lockheed Martin's network was attacked since it is known to contain sensitive data about contracts and defense technology currently under development. The Lockheed Martin network also holds sensitive information about cutting edge technology that is deployed in Iraq and Afghanistan — possibly similar to the stealth helicopter used in a raid that lead to the death of Osama bin Laden [4].

The attack surface available to these groups is now much larger – instead of looking for weaknesses in firewalls, the attackers can use for targets:

(1) The always-streaming back and forth data.

(2) The less secure – often unencrypted or even not password protected – mobile devices.

(3) The more distributed – often outsourced – nature of enterprise data management.

2. DOCUMENT SECURITY NEEDS

2.1 Document Lifecycle

Documents are no longer single-version, static content. Instead, they are "live"—dynamic, multi-user, multi-media and thus multi-version. Document security must address, simultaneously:

(1) Document creation. Here is where access rights must be established for subsequent participants in the document workflow.

(2) Document workflow. Who are the intended users? Can they be aggregated by role? Will they change—will actors/roles be added, deleted or modified?

(3) Will access be given by document type? Fixed document properties? Changing document property? Or by document parts?

(4) Can the workflows and access throughout the workflows be audited? How will access, versioning and other content, context and use information be logged? How will logging be authenticated?

(5) What will be the nature of storage? Will all versions be stored? Will versions be stored separate from logging/auditing information? Will data and its access be tokenized? If not, how will access rights be maintained?

(6) When will documents be deleted? How will this be guaranteed?

2.2 Access Control

Access control within the document workflow can be garnered in multiple ways. It can be fixed by policy, by document type, or by fixed rules for the specific document. These are "traditional" means for assigning access rights, and they are pervasive because of their simplicity and – presumed – security. However, access can also be defined by the participant's identity or role in a workflow [5] and so defined such that it reflects the current state of a workflow, in which case the access changes as a document propagates along its workflow. Access rights can also be driven through the interaction of document content with the directing policies applicable to both document access and sharing. The policy-driven compliance [6] approach has value in that it can be performed real-time without significant impact on performance and with no policy training requirements for the participants. For example, system call "eavesdropping" can be used to target the specific system calls correlated with inadvertent transgression of policy [7].

2.3 Cloud

As mentioned above, cloud access to documents create a shifting paradigm for document access, sharing, management and storage. Virtualization increases the attack surface, because there are potentially more locations, interfaces and protocols to access the same data.

In the workshop, we will discuss the potential security risks of GoogleDocs and Apple iCloud. Apple iCloud, for example, avers "Apps make it possible to create amazing presentations, write reports, and more right from your iOS device. You don't have to manage your documents in a complicated file system or remember to save your work. Your documents are just there, stored in your apps, and ready whenever you need them. And now, your apps can store that information in iCloud. Which means you can access your documents — with your latest updates — on whichever device you happen to be using at the time. Even better, this all happens automatically, without any effort from you."

We will also consider the IBM SmartCloud and Amazonws. IBM notes that "Industries like healthcare, banking, retail and manufacturing simply cannot be accommodated by one-size-fits-all cloud computing environments. The IBM SmartCloud provides IT communities with unprecedented control, speed, security and cost savings."

Since cloud applications are relatively new and still going through adolescence, it is hard to gauge the legal implications of breakdowns in cloud security. What happens if a cloud is broken and – among other possibilities – data is stolen [8]? As a case in point, what are the privacy laws governing storage on Dropbox [9]? "Dropbox employees are prohibited from viewing the content of files you store in your Dropbox account, and are only permitted to view file metadata (e.g., file names and locations). Like most online services, we have a small number of employees who must be able to access user data for the reasons stated in our privacy policy (e.g., when legally required to do so). But that's the rare exception, not the rule. **We have strict policy and technical access controls that prohibit employee access except in these rare circumstances.** In addition, we employ a number of physical and electronic security measures to protect user information from unauthorized access." What are the rare circumstances? What is the policy? How is it enforced? Are we willing to accept a world in which it is ambiguous – guesswork! – how often and by whom a document has been accessed?

As stated in [10], "There is a risk of data theft from machines in the cloud, by rogue employees of cloud service providers or by data thieves breaking into service providers' machines, or even by other customers of the same service if there is inadequate separation of different customers' data in a machine that they share in the cloud. Governments in the countries where the data is processed or stored may have legal rights to view the data under some circumstances … There is also a risk that the data may be put to unauthorized uses. It is part of the standard business model of cloud computing that the service provider may gain revenue from authorized secondary uses of user's data, most commonly the targeting of advertisements. However, some secondary data uses would be very unwelcome to the data owner (such as, for example, the resale of detailed sales data to their competitors). At present there are no technological barriers to such secondary uses."

Global Secure Systems, in fact, argue that "the more effective the controls at the entrance, the less one has to worry about having distributed controls. Practice has shown that once the number of individuals who may access a system exceeds three, it becomes exponentially difficult to manage the process." [11]. But is access restriction the best route? Or does building security into the data itself [5] provide a more reliable, more scalable, even more convenient approach? This is an important question we will address in the workshop.

In short, we fully agree with [8] as it states: "This paper is about legal questions connected with cloud computing, the business trend in which computation is carried out on behalf of a user on remote machines, using software accessed through the Internet. The user may not know where these machines are; they are "somewhere in the cloud". Some of these legal issues will be resolved by standard agreements between buyers and vendors. I will give some examples from current agreements from prominent cloud service providers. Other issues will probably end up in court. It makes sense to consider these questions now, before they become urgent."

2.4 Mobile

Mobile access provides an additional set of security threats. As [12] notes, in the "post-perimeter world, changes in malware attacks have rendered network firewalls and perimeter-centric security an ineffective defense…the increasingly mobile workforce makes an on-premise approach even more futile. Walls can no longer keep the bad guys out, nor can they keep the good guys in."

Indeed, the security threat surface increases exponentially once mobile devices are allowed to access information, because of the increasing number of interactions that can occur between – not to mention within –devices. In the workshop, we will compare and contrast the security threats of:

(1) Multi-user log on devices

(2) Device access where secure passwords are not required/enforced.

(3) Improper usage of biometric access

(4) User "password reuse"

(5) Machine-machine communication

2.5 Analytics

The term "analytics" is currently the champion buzzword. But one thing is certain—analytics are about data transactions. And data transactions have long been one of the most successful routes to security transgression—from SQL injection to buffer overflow, using data as a means for attacking a system has a long and proud history.

3. METHODS TO ADDRESS NEEDS

In the workshop, we will discuss and defend/attack methods to address the security needs for each of the four major areas identified. For these areas, representative discussion items are listed below.

3.1 Access Control

(1) Encrypt everything? Use PKI? Use Identity-Based Encryption?

(2) Security intimately linked/part of the document?

(3) Partition the data store and provide no separate security?

3.2 Cloud

(1) Should a single cloud architecture be used to minimize the threat surface or should different architectures be used based on the security needs of the data?

(2) What is the best method to reduce the number and severity of threats caused by the cloud? What effect does cloud architecture have on security? What is the impact of the choice of operating system?

3.3 Mobile

(1) What approach to access control minimizes the risk to security imposed by mobile devices?

(2) Can security policy be imposed on mobile users by the nature of the data by refusing data transmission when the device is non-compliant?

3.4 Analytics

(1) What data should be collected to provide the best security? The best response to security?

(2) Does the data itself pose a threat to security?

(3) How can analytics be enabled without creating new security risks?

4. DISCUSSION

In the workshop, we will discuss secure document engineering for fully electronic workflows as emphasized above, and for mixed electronic/paper workflows.

For mixed electronic/the latter, security is concerned with authenticating the physical (printed) item, either through authentication of the print process [13] or forensic identification of the interaction of the ink and the printing substrate [14]. As part of this workshop, we will discuss the nature of security provided by:

(1) Document inspection and quality

(2) Document track and trace

(3) Document authentication

(4) Document forensics

(5) Document inference

Interesting and challenging questions arise from the simultaneous consideration of wholly-electronic and mixed-print/electronic workflows. What is the most secure system? Redundant electronic? Paper and electronic with different access rules? Redundant electronic with paper backup? We believe the latter, but expect an excellent discussion around this topic at the workshop. We look forward to your participation.

5. REFERENCES

[1] "An Exabyte-a-Day: RayOnStorage Blog, Storage, Strategy & Systems, http://silvertonconsulting.com/blog/2009/10/02/an-exabyte-a-day/, last accessed 8 July 2011.

[2] "Sony Investigating Another Hack," http://www.bbc.co.uk/news/business-13636704, last accessed 7 July 2011.

[3] "Lockheed Martin hit by Security Breach," http://online.wsj.com/article/SB10001424052702303654804576350083016866022.html, last accessed 7 July 2011.

[4] "Top Secret Stealth Helicopter Program Revealed in Osama Bin Laden Raid: Experts," http://abcnews.go.com/Blotter/top-secret-stealth-helicopter-program-revealed-osama-bin/story?id=13530693, last accessed 7 July 2011.

[5] H. Balinsky and S. J. Simske, "Differential Access for Publicly-Posted Composite Documents with Multiple Workflow Participants," ACM DocEng 2010, pp. 115-124, 2010.

[6] S. J. Simske and H. Balinsky, "APEX: Automated Policy Enforcement eXchange," ACM DocEng 2010, pp. 139-142, 2010.

[7] H. Balinsky, D. Perez, S. J. Simske, "System Call Interception Framework ", the Fifteenth IEEE International EDOC Conference (EDOC 2011), Helsinki, Finland, August, 2011.

[8] M. Mowbray "The Fog over the Grimpen Mire: Cloud Computing and the Law", 2009, vol. 6, issue 1, pp.132-146. Available at http://www.law.ed.ac.uk/ahrc/script-ed/vol6-1/mowbray.pdf.

[9] Dropbox, http://www.dropbox.com/privacy, last accessed 8 July 2011.

[10] S. Pearson, Y. Shen, M. Mowbray "A Privacy Manager for Cloud Computing", HPL Technical Report http://www.hpl.hp.com/techreports/2009/HPL-2009-156.pdf

[11] Global Secure Systems, "Sensitive document vault," http://www.gss.co.uk/products/index/Identity_and_Security/Cyber-Ark/Sensitive_Document_Vault/, last accessed 8 July 2011.

[12] White Paper, "Beyond the Wall: Security in a Post-Perimeter World," http://www.informationweek.com/whitepaper/Security/Attacks-Breaches/beyond-the-wall-security-in-a-post-perimeter-wor-wp1286372808028, last accessed 8 July 2011.

[13] S.J. Simske, M. Sturgill, P. Everest, and G. Guillory, "A system for forensic analysis of large image sets," *Proc. IEEE WIFS*, pp. 16-20, 2009.

[14] S.J. Simske, S.B. Pollard & G. Adams, "An imaging system for simultaneous inspection, authentication and forensics," *IEEE IST*, pp. 266-269, 2010.

Multimedia Document Processing in an HTML5 World

Dick C. A. Bulterman
CWI*
Kruislaan 413
P.O. Box 94079
1090 GB Amsterdam
The Netherlands
Dick.Bulterman@cwi.nl

*also Vrije Universiteit Amsterdam and W3C

Rodrigo Laiola Guimarães
CWI
Science Park 123
1098 XG Amsterdam
The Netherlands
rlaiola@cwi.nl

Pablo Cesar
CWI
Science Park 123
1098 XG Amsterdam
The Netherlands
P.S.Cesar@cwi.nl

Ethan Munson
Department of Electrical Engineering and
Computer Science
P.O. Box 784
University of Wisconsin - Milwaukee
Milwaukee, WI 53201
munson@uwm.edu

Maria da Graça C. Pimentel
Instituto de Ciências Matemáticas e de
Computação - USP
Av. Trabalhador São-carlense, 400 - Centro
CEP: 13566-590 –
São Carlos – SP Brazil
mgp@icmc.usp.br

Abstract

The evolution in media support within W3C standards has led to the development of HMTL5. HTML5 provides extensive support for audio/video/timed-text within an interoperable browser context.

This workshop examines the impact of HTML5 on research and systems support for multimedia documents. We will consider issues such as extensibility, adaptivity and maintenance, and will discuss the future needs for Multimedia in a Web context.

Categories & Subject Descriptors: H.5.1 [**Information Interfaces and Presentations**]: Multimedia Information Systems—*Audio, Video*. I.7.2 [**Document and Text Processing**]: Document Preparation—*Markup languages, Multi/mixed media, Standards*

General Terms: Design, standardization.

Keywords: Multimedia, Timed-text, Web browser, HTML5, W3C

Making Accessible PDF Documents

Heather Devine
Adobe Systems
345 Park Ave
San Jose CA 95110
hdevine@adobe.com

Andres Gonzalez
Adobe Systems
345 Park Ave
San Jose CA 95110
andgonza@adobe.com

Dr. Matthew Hardy
Adobe Systems
345 Park Ave
San Jose CA 95110
mahardy@adobe.com

ABSTRACT

Accessibility features in the Adobe Portable Document Format (PDF) help facilitate access to electronic information for people with disabilities. This workshop explores how to create accessible PDF documents, from within Adobe Acrobat and other applications; how to use the Adobe Acrobat PDF accessibility checker and repair workflow; best practices for accessibility; and how accessibility has been built into forthcoming ISO standards (PDF/UA, PDF 32000-2).

Categories and Subject Descriptors

I.7.2 Document Preparation, Adobe Acrobat.

General Terms

Documentation, Standardization.

Keywords

Accessibility, Content Reuse, Logical Structure, PDF, PDF/UA.

1. INTRODUCTION

Electronic document accessibility is crucial to allow users with disabilities access to electronic information. Making accessible PDF documents not only enables access through assistive technologies (AT), such as screen readers and magnifiers, but also provides logical structure to a document which can enable better reflow, display of documents and content reuse.

2. CREATING ACCESSIBLE PDF DOCUMENTS

There are two primary workflows for creating accessible PDF documents. The first is to add logical structure and text extraction capabilities to an existing PDF file. This process is often referred to as tagging. Tagging is, in most cases, a semi-automated process where a software inference engine creates a basic structure, and a user makes corrections and additions to that structure using manual tools.

The second is to generate an accessible PDF file from other software application source files, such as a word processor or desktop publishing application. This process is mostly automatic; the semantic information contained in the source file about the document structure is translated into a tagged PDF.

2.1 Existing PDF Document

In order to make an existing PDF document accessible, one must consider the characteristics of the document to determine which changes are necessary to increase accessibility. For example, scanned images of text must undergo optical character recognition (OCR) to add searchable text. Document logical structure must be identified, and its correspondence to document content and layout must be established.

Document logical structure includes an appropriate reading order, and the identification of semantic structural elements such as sections, headings, paragraphs, tables, and images. Documents should contain alternative text descriptions for images and graphics, as well as a specification of the document's language. Documents can also include navigational aids, such as bookmarks, which are particularly helpful for longer documents.

2.2 Converting Documents to PDF

The key to creating accessible PDF documents from other source files, such as files from a word processor or desktop publishing application, is to design the source document with accessibility in mind. The author of a document needs to create the proper logical structure in the original format that is being used, and the document structure will be translated into the corresponding PDF tags.

Similar guidelines apply: sections, headings, and paragraphs should be identified using the software's styles; images, graphics or animated elements should include alternative text descriptions; and navigational aids, such as bookmarks, should be included. Content that is not relevant to the text, such as background images or page numbers, should be tagged as artifacts. Document metadata can be included, such as title, author, keywords, and language. Additionally, some programs allow tab order and reading order to be defined, which can aid in accessibility.

Options for creating PDF files from programs may also impact the accessibility of the generated document. Conversion settings which effect accessibility include tag generation, standards compliance, bookmark conversion, file security, link conversion, and animation settings. If the conversion results in accessibility problems in the generated PDF file, these issues can be addressed using accessibility checker repair tools, or in some cases in the source document itself. Keep in mind that repairs made to the generated file will not be saved if the file is regenerated. Similarly, once the PDF file is generated, its accessibility should be evaluated using an accessibility checker so that any remaining accessibility issues can be repaired.

2.3 Supporting Assistive Technologies

The logical structure and content of a properly tagged PDF can be conveyed to assistive technologies such as screen readers and magnifiers via accessibility APIs. Adobe Acrobat and Reader support the Microsoft Active Accessibility (MSAA) API. In addition, the Adobe IPDDom API is used to render to assistive technologies those aspects of the structure of PDF documents that are not covered by MSAA. For details about conveying document structure and content through accessibility APIs see Gonzalez and Guarino-Reid (2005) [1].

3. ACCESSIBILITY CHECKER

An accessibility checker can help identify parts of a document which do not meet accessibility requirements. An accessibility checker can iterate through each accessibility requirement to determine whether the document meets that requirement. For example, it can examine the document for scanned text which should be made searchable, identify images and graphics which require alternative text descriptions, determine whether the document includes structure information, locate sections headings to add navigation aids, and ensure that document text is available to AT.

4. REPAIR WORKFLOW

The repair workflow can be used in conjunction with an accessibility checker to create an accessible PDF document. After using the checker to determine which parts of the document do not meet accessibility requirements, repairs can be made to improve accessibility. After evaluating the characteristics of the document and its accessibility level, there are several steps which can then be taken to improve accessibility based on the results: (1) perform OCR to convert scanned text to searchable text; (2) add navigation aids, such as bookmarks; (3) ensure that document security does not limit access to document text; (4) add tags and structure to the file; and (5) re-evaluate the document to determine if there are other areas which need improvement.

5. BEST PRACTICES

There are a number of best practices which support the creation of accessible PDF documents. These practices should be kept in mind when making an existing PDF file accessible or generating an accessible PDF file from other software programs:

- Ensure the document text is available to AT
- Identify document structure through the use of tags
- Include tooltips or alternative text for form fields, graphics, and animated elements
- Create navigation aids
- Verify the reading order for document elements
- Verify the document's accessibility

6. ACCESSIBILITY IN PDF STANDARDS

In 2008, PDF became an ISO standard (ISO 32000-1:2008). While PDF already defined mechanisms by which a PDF document could be made accessible, there was no standard for accessible PDF. However, work is now underway on a new ISO PDF subset called PDF Universal Accessibility (ISO 14289-1, PDF/UA). PDF/UA enforces the optional structures provided in the PDF standard to ensure than any PDF generated to the PDF/UA specification is potentially accessible. It also restricts certain types of content which would be difficult to interact with for someone using AT, e.g. scripts that require user input that has timing requirements.

PDF/UA further defines what information an interactive PDF processor (e.g. a PDF reading application) must make available to AT and in some cases specifies how it should interact with AT. An example of this is with multimedia playback, where the controls to a sound clip must be made available to someone who cannot see to use the controls with a mouse.

Finally, PDF/UA describes requirements for AT, to ensure that no matter which combination of interactive PDF processor and AT a user chooses, they have similar levels of access to the content and interactive elements of the document.

PDF/UA is still a work in progress and is currently an ISO draft international standard (DIS). The expectation is that it will be published as a final standard in 2012.

During the creation of PDF/UA, a number of needs were identified in PDF to facilitate accessibility. These additional needs are being addressed in PDF 2.0, ISO 32000-2. The changes primarily consist of extensions to the logical structure tags, which can now represent more types of content. Examples of these new tags include support for mathematics and document titles.

7. ACKNOWLEDGMENTS

Our thanks to Adobe Systems and Kirk Gould for supporting our participation in *DocEng'11*.

8. REFERENCES

[1] Andres Gonzalez and Loretta Guarino Reid. 2005. Platform-independent accessibility API: accessible document object model. In *Proceedings of the 2005 International Cross-Disciplinary Workshop on Web Accessibility (W4A)* (W4A '05). ACM, New York, NY, USA, 63-71. DOI=http://doi.acm.org/10.1145/1061811.1061824

Documenting Social Networks

Maria da Graça Pimentel
Universidade de São Paulo
São Carlos, SP, Brazil
mgp@icmc.usp.br

ABSTRACT

The many social networks available on the Web offer users several facilities involving the sharing of media as a way of allow communication. This workshop will discuss the role of document engineering in social networks, targeting at issues such as: what documents can we create by analyzing the available information; what documents users manipulate/add/refer to in social networks; what are the roles of authors and readers.

Categories and Subject Descriptors

H.5.3 [**Group and Organization Interfaces**]: Collaborative Computing; H.5.1 [**Multimedia Information Systems**]: Evaluation/ Methodology; H.3.m [**Information Storage and Retrieval**]: Miscellaneous; D.2.8 [**Software Engineering**]: Metrics.

General Terms

Algorithms, Design, Documentation, Economics, Experimentation, Human Factors, Languages, Legal Aspects, Management, Measurement, Performance, Reliability, Security, Standardization, Theory, Verification.

Keywords

Social Networks, Document Engineering, Research Issues.

1. INTRODUCTION

Web Supported Social Networks, or simply Social Networks, have been around since the late 90´s, the earlier examples including SixDegrees.com and LiveJournal. Later systems such as LinkedIn (Fig. 1) and Orkut (Fig. 2) were accompanied by media-oriented ones such as SoundCloud for music (Fig. 3), Flicker for photos (Fig. 4) and YouTube for video (Fig. 5), to name a few.

Many organizations are users of Twitter[1] (Fig. 6) and Facebook (Fig. 7), the latter expected to be the number one in display advertisement in 2011[2] after being in use by a large numbers of individual users (Fig. 8), groups of users (e.g.Fig. 9) and events (Fig. 10). While facilitated communication has been crucial for the success of Twitter, the easy integration with web-based applications and sites has been key to Facebook. Recently, Google+ (Fig. 11) has joined this growing market, investing in a cleaner user interface and novel circles of friends.

In their article comprising a definition, the history, and research issues for Social Network Sites, Boyd and Ellison offer the following definition for Social Networks [2]: "web-based services that allow individuals to (1) construct a public or semi-public profile within a bounded system, (2) articulate a list of other users with whom they share a connection, and (3) view and traverse their list of connections and those made by others within the system". Among the research issues involved, the authors point to privacy, friendship performance and maintenance, network structure and performance, as well as to aspects involving identity, self-presentation, and civic engagement.

Several other research groups have been giving attention to Social Networks. As an example, "Social network analysis" [21] is a recent research field that studies social entities (people, actors or users) and their interactions and relationships. Some models for the analysis of social networks and the evaluation of the interactions among users have been proposed based on graph theories [15] [20], on individual network's usage patterns [1] [9], and on semantics [4] [12].

Associated with the small-world principle [13] [19], authors have proposed models to represent the relationships associated with the interactions among users, for instance via the application of data mining techniques involving clustering analysis [10]. In this scenario, it may of interest to investigate which documents can be produced to allow the representation of situations involving users in social interactions has also been valuable [7] [8]. This, in turn, leads to questions related to how to represent these interactions in the forms of documents — possibly for interchange.

Social Networks may impact society, as demonstrated in the recent conflicts in the Middle East[3], even though the

---

[1] http://www.searchenginejournal.com/16-examples-of-huge-brands-using-twitter-for-business/7792/
[2] http://www.businessinsider.com/facebook-display-advertising-2011-6
[3] http://www.un.org/News/Press/docs/2011/pal2145.doc.htm

Figure 1: LinkedIn page or an individual user

Figure 2: Orkut page or an individual user

role Social Networks in this case is not unanimous[4]. Giving the importance of the theme, it is worth to review previous results (Section 2) and discussing the role of Document Engineering in Social Networks, as suggested in Section 3.

2. RELATED RESULTS

From the Document Engineering point of view, an example of document that can created using the information available in Facebook is "The Museum of Me" application offered by Intel[5]: once a user authorizes access to his account, the application rapidly creates a video mapping the user´s entries in Facebook (including photos, videos, places, friends and pages the user "liked") to rooms in a Museum Exhibition associated with the user (Figure 12).

From previous editions of the DocEng Symposium, results include the study of tools to facilitate the semi-automatic annotation of YouTube videos [5] [6] [16], and the proposal of a model for creating and sharing personalized time-based annotations of videos on the web [14].

Associated efforts include investigating the use of social tagging [11] and folksonomies [17], as well as from specialized shared databases [18]. These are very associated with the proposal by Cesat et al., regarding enhancing the social sharing of videos [3].

3. DISCUSSION

This workshop aims at discussing Document Engineering in the context of Social Network scenario, starting with:

- what documents can we create by analyzing the available information?

- what documents users manipulate/add/refer to in social networks?

- what are the roles of authors and readers?

Participants will be invited to provide their insights relative to these issues, along with others derived from applying

Figure 3: Soundcloud page or an individual user

Figure 4: Flicker page or an individual user

---

[4]E.g. http://www.cbsnews.com/stories/2011/02/14/opinion/main20031739.shtml and http://www.nytimes.com/2009/01/25/magazine/25bloggers-t.html
[5]http://www.intel.com/museumofme

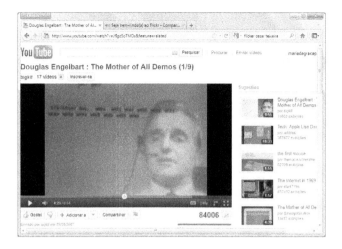

Figure 5: Youtube Engerlart's *Mother of alll demos*

Figure 6: Twitter for an institution

Figure 7: Facebook for an institution

Figure 8: Facebook for an individual user

Figure 9: Facebook for an group of users

Figure 10: Facebook for a symposium

Figure 11: Google+ page or an individual user

Figure 12: MuseumOfMe video for a user

Document Engineering to Social Networks from a general computer science point of view. A tentative model is to investigate issues associated with Algorithms, Design, Documentation, Security and Standardization (and other ACM General Terms).

Acknowledgments. I thank my students Alan K. Gomes, Roberto Fagá Jr. and Diogo S. Martins for their hard work on this theme. I also thank the several organizations which support our research: CAPES, CNPq, FAPESP, FINEP, MCT and RNP.

4. REFERENCES

[1] F. Benevenuto, A. Pereira, T. Rodrigues, V. Almeida, J. Almeida, and M. Gonçalves. Evaluation of users access and navigation profiles on web video sharing environments. In *ACM XV Brazilian Symposium on Multimedia and the Web (WebMedia)*, pages 31:1–31:8, 2009.

[2] d. m. boyd and N. B. Ellison. Social network sites: Definition, history, and scholarship. *Journal of Computer-Mediated Communication*, 13, 2007.

[3] P. Cesar, D. C. Bulterman, D. Geerts, J. Jansen, H. Knoche, and W. Seager. Enhancing social sharing of videos: fragment, annotate, enrich, and share. In *Proceeding of the 16th ACM International Conference on Multimedia*, MM '08, pages 11–20. ACM, 2008.

[4] B. Chakraborty and T. Hashimoto. Topic extraction from messages in social computing services: Determining the number of topic clusters. In *IEEE Int. Conf. on Semantic Computing (ICSC)*, pages 232–235, 2010.

[5] R. Fagá, Jr., B. C. Furtado, F. Maximino, R. G. Cattelan, and M. d. G. C. Pimentel. Context information exchange and sharing in a peer-to-peer community: a video annotation scenario. In *Proceedings of the 27th ACM International Conference on Design of Communication*, SIGDOC '09, pages 265–272. ACM, 2009.

[6] R. Fagá, Jr., V. G. Motti, R. G. Cattelan, C. A. C. Teixeira, and M. d. G. C. Pimentel. A social approach to authoring media annotations. In *DocEng '10: Proceedings of the 10th Symposium on Document engineering*, pages 17–26, 2010.

[7] A. K. Gomes, D. C. Pedrosa, and M. G. C. Pimentel. Evaluating asynchronous sharing of links and annotations as social interaction on internet videos. In *IEEE International Symposium on Applications for Internet (SAINT´2011)*, 2011.

[8] A. K. Gomes, D. C. Pedrosa, and M. G. C. Pimentel. Measuring synchronous and asynchronous sharing of collaborative annotations sessions on ubi-videos as social interactions. In *International Conference on Ubi-media Computing (U-Media 2011)*, pages 122–129, 2011.

[9] L. Gyarmati and T. A. Trinh. Measuring user behavior in online social networks. *IEEE Network*, 24(5):26–31, 2010.

[10] J. Han and M. Kamber. *Data Mining: Concepts and Techniques, 2nd Ed.* Morgan Kaufmann Publishers Inc., 2005.

[11] A. W.-C. Huang and T.-R. Chuang. Social tagging, online communication, and peircean semiotics: a conceptual framework. *J. Inf. Sci.*, 35:340–357, June 2009.

[12] S. Huang. Mixed group discovery: Incorporating group linkage with alternatively consistent social network analysis. In *IEEE International Conference on Semantic Computing (ICSC)*, pages 369–376, 2010.

[13] J. Kleinberg. The small-world phenomenon: an algorithm perspective. In *ACM Symp. on Theory of Computing (STOC)*, pages 163–170, 2000.

[14] R. Laiola Guimarães, P. Cesar, and D. C. Bulterman. Creating and sharing personalized time-based annotations of videos on the web. In *Proceedings of the 10th ACM Symposium on Document Engineering*, DocEng '10, pages 27–36. ACM, 2010.

[15] A. Mislove, M. Marcon, K. P. Gummadi, P. Druschel, and B. Bhattacharjee. Measurement and analysis of online social networks. In *ACM Conference on Internet Measurement (ICM)*, pages 29–42, 2007.

[16] V. G. Motti, R. Fagá, Jr., R. G. Catellan, M. d. G. C. Pimentel, and C. A. Teixeira. Collaborative annotation of videos: watching and commenting (youtube) videos. In *EuroITV '10: 8th European Conference on Interactive TV and Video*, New York, NY, USA, 2010. Demonstration - Conference Program - page 23.

[17] M. G. Noll, C.-m. Au Yeung, N. Gibbins, C. Meinel, and N. Shadbolt. Telling experts from spammers: expertise ranking in folksonomies. In *Proceedings of the 32nd International ACM SIGIR Conference on Research and Development in Information Retrieval*, SIGIR '09, pages 612–619. ACM, 2009.

[18] D. A. Pereira, B. Ribeiro-Neto, N. Ziviani, A. H. Laender, M. A. Gonçalves, and A. A. Ferreira. Using web information for author name disambiguation. In *Proceedings of the 9th ACM/IEEE-CS Joint Conference on Digital libraries*, JCDL '09, pages 49–58. ACM, 2009.

[19] D. J. Watts. *Small worlds: the dynamics of networks between order and randomness*. Princeton U. Press, 1999.

[20] C. Wilson, B. Boe, A. Sala, K. P. Puttaswamy, and B. Y. Zhao. User interactions in social networks and their implications. In *ACM European conference on Computer Systems*, EuroSys '09, pages 205–218, 2009.

[21] G. Xu, Y. Zhang, and L. Li. *Web Mining and Social Networking: Techniques and Applications*. Springer-Verlag, 1st edition, 2010.

Google Mystery Workshop

John Day-Richter
Google, Inc.
1600 Amphitheatre Parkway
Mountain View, CA 94043
johndayrichter@google.com

Abstract

This workshop will explore a new Google technology involving documents and programming (not yet unannounced by the paper submission deadline).

Participants will require a laptop with a Java development environment configured.

Categories & Subject Descriptors: H.4.0 [**Information Systems Applications**]: General

General Terms: Design.

Keywords: Documents, Programming, Java

Author Index

www.ingramcontent.com/pod-product-compliance
Lightning Source LLC
Chambersburg PA
CBHW080356060326
40689CB00019B/4034